D1268442

Fishes of the Middle Savannah River Basin

Fishes of the Middle Savannah River Basin

With Emphasis on the Savannah River Site

BARTON C. MARCY JR.
DEAN E. FLETCHER
F. DOUGLAS MARTIN
MICHAEL H. PALLER
MARCEL J. M. REICHERT

PHOTOGRAPHS BY
DAVID E. SCOTT

The University of Georgia Press
Athens and London

Designed by April Leidig-Higgins
Set in Adobe Caslon by Copperline Book Services, Inc.
Printed and bound by Four Colour Imports, LTD

The paper in this book meets the guidelines for permanence and durability of the Committee on Production Guidelines for Book Longevity of the Council on Library Resources.

Printed in China
09 08 07 06 05 C 5 4 3 2 1

Library of Congress Cataloging-in-Publication Data
Marcy, Barton C.
Fishes of the Middle Savannah River Basin : with emphasis on the Savannah River site / Barton C. Marcy Jr., [et al.] ; photographs by David E. Scott.
p. cm.
Includes index.
ISBN 0-8203-2535-X (hardcover : alk. paper)
1. Fishes—Savannah River Watershed (Ga. and S.C.)
I. Title.
QL628.S28M37 2005
597'.09758'1—dc22 2004007658

British Library Cataloging-in-Publication Data available

Contents

Foreword

Susan Acuff Dyer and John Mark Dean

The Savannah River has values far beyond the obvious industrial, navigational, recreational, educational, and scientific uses typically seen today. This great river was the cultural base for Native Americans of the Southeast and a highway for the settlement of the American colonies. The modern river has been subjected to a multitude of environmental modifications due to changing weather patterns, dam construction and operations, human development, channelization, and recreational pressures. The character and value of the river and its tributary streams are evidenced in their use by boaters, anglers, researchers, and naturalists. The watershed evaluations required for the operation of the Savannah River Site (SRS) allow—no, require—us to communicate on issues associated with the Savannah River with regulatory agencies, natural resource trustee agencies, university researchers, and local fishermen. These interactions involve people with diverse perspectives and social backgrounds that shape their values and interest in the river and the fisheries resources within it.

We have the pleasure of working in a field in which river users such as recreational fishermen can be more knowledgeable in some aspects of the fishery resources than trained researchers. Knowing that others outside the realm of science appreciate and can relate to the technical issues at hand makes the work that much more appealing. The appreciation of fishes is open to the strict academician, scientific researcher, and outdoor enthusiast alike.

The resources necessary to make this publication possible are to a great extent the result of the development of the SRS. The SRS nuclear production facility brought science and technology to an isolated and undeveloped rural area of the southeastern United States at a time of immense global tension. Researchers have been involved ever since in determining the effects of SRS operations on the local and regional environments. As a result, the middle reaches of the Savannah River (including the water and the fishes) have been the subject of long-term scientific study. From its inception, the 777 km^2 SRS required high levels of security and was shrouded in secrecy and suspicion. Gossip and rumors abounded in the local communities, and some older residents still refer to it as "the bomb plant." The SRS remains a necessarily secure area, but its roles and responsibilities are now well known, and there is public participation in determining the management policies for the facility. The guarded gates, fences, and signs that secure the perimeter act as reminders, even to boaters along the Savannah River, that the area is of concern. SRS operations now support active research programs conducted in the interest of protecting the aquatic resources of the middle Savannah River. Many are envious of the ability of SRS researchers to establish stations and know that their instrumentation will not be tampered with, stolen, or destroyed. Since all of the SRS is within the drainage basin of the middle Savannah River, the security has been to the advantage of scientific study.

When the land for the SRS was acquired, most of the surrounding area was farmland and homesteads. Within the boundaries of the SRS, floodplain swamps, bottomland hardwoods, and pine forests dominated uninhabited areas. The SRS looked to scientists from the Academy of Natural Sciences in Philadelphia, the University of Georgia, and the University of South Carolina for assistance in preoperational baseline studies, and those organizations and numerous other institutions are involved in the environmental research that continues today. At the time the SRS began operations, the evaluation of a nuclear facility's impacts on the environment was unique, and it opened the door for environmental study in new fields of science, principally radioecology. Impacts associated with radiological releases, thermal discharges, water intake, conversion of farmland to forest, and the mass relocation of entire communities were a few of the components that influenced those early research efforts. The initial baseline studies, under the control of the Atomic Energy Commission (AEC), began in the summer of 1951 and have evolved to support current missions of the Department of Energy (DOE). Current research efforts include tracking largemouth bass movement using radiotelemetry, determining fish escapement from Par Pond dam, evaluating fish and invertebrate community structures in Coastal Plain stream systems, fish and mammal tissue body burden bio-uptake assessments, wildlife distribution modeling, and human health fish consumption risk evaluations. The research efforts now are collaborative and involve university students and faculty, regulatory agencies, natural resource agencies, and the general public. The diversity of the people involved in decisions associated with the SRS and the connection of the site streams with the Savannah River have initiated a progression of data and information needs that reaches far beyond the SRS boundaries.

A major emphasis of the early impact evaluations and ecological studies dealt with nuclear reactor operations that significantly altered fish habitats. There were major changes in flow velocity, periodicity, and temperatures. Elevated flows inundated floodplains with thermal discharges that killed existing vegetation. This led to stream widening through erosional processes and subsequent stream incision. Sediments carried downstream were deposited at the stream–Savannah River–swamp interface as deltaic deposits. The morphology of the stream systems was modified, and studies of restored versus nonrestored systems were a major research emphasis when reactor operations ceased. Cooling-water reservoirs were created, influencing major waterfowl flyways. The opening of bottomland hardwood and cypress-tupelo swamp canopies, and chemical and radioactively contaminated discharges are examples of the site-related impacts that affected the aquatic systems at SRS in positive and negative ways. Those impacts continue to influence the regional environment, including the Savannah River, today.

The middle Savannah River is an attractive resource for many reasons. Augusta, Georgia, and neighboring communities take full advantage of the river for commerce, industry, recreation, drinking water, and waste disposal. Local residents benefit from the river's aesthetic and recreational aspects, in particular its fisheries. The natural and man-made systems influencing the Savannah River —its tributaries, floodplains, and lock and dam systems—provide a diversity of habitats. In contrast, the Savannah River Site has been protected from the impacts of large-scale development for more than 50 years. Where once farmlands dominated the landscape, managed pine plantations and healthy stream systems now exist. The isolation has produced a sanctuary for wildlife, an environmental oasis of presently undisturbed forests, floodplains, swamps, streams, and Carolina bays. The ecosystems within the SRS provide a glimpse of how resilient and valuable protected lands can be. Walking through the woods at the SRS, one can happen upon stairs that lead nowhere or scattered bricks that are the faintly visible footprint of a long forgotten family. The stream systems offer similar silent reminders of days past. Old pilings from roads long since forgotten and farm ponds now breached and left to succession are visible to even casual observers.

Fishery biologists, as a professional group, generally relish life. They are fun to be around, good traveling companions, and accomplished storytellers; they enjoy good food and drink, and many are excellent cooks. We think that these criteria characterize the authors of this book quite well, perhaps because fishery biologists spend so much time in the field and on the water. They are in contact with recreational anglers and professional fishers who are certain that they know more about fish than any scientist or fishery manager. A good scientist/fishery manager can exchange information with such people without appearing to be arrogant—and running the risk of embarrassment when proven wrong.

Many books on the fish of a particular region speak only to the very narrow interests of taxonomists and systematists or are useful for only the most general audience. This book should prove a valuable addition to many collections because it brings regional data and information to the forefront in an accessible and usable format. Workers who must make quick but considered and informed decisions and management recommendations will find the information they need easily accessible. Researchers involved in fisheries science will find this book useful, but academicians, anglers, students, and those with merely a casual interest in local fisheries will appreciate the wealth of information within its pages as well.

The aquatic resources of the Savannah River influence us all. Few of us can drive over a bridge, walk along the banks of the river, fish along the shore, paddle a canoe, or spend a day (or night) collecting fishes and not develop an enduring affection for the river's physical characteristics and spirit. Upon contact or sight, we indulge in a lasting relationship with the ever-changing body and spirit of the river. And that is one of the great pleasures of being in the profession of fisheries science; work and pleasure are inextricably entwined. We do not deny it but rather revel in it and relish our good fortune.

Acknowledgments

In this book we treat the life histories and ecology of 98 species using information assimilated from many sources. We would like to express our gratitude to those involved in the book's preparation, review, and production. Especially invaluable were the numerous researchers from many institutions who contributed to the body of ichthyological information and the scientists in the field who collected the specimens. The assemblage of fisheries information here spans more than 120 years for the middle Savannah River basin (MSRB), and 50 years for the Savannah River Site (SRS). The SRS has contributed more than 500 publications to fish studies. It is impossible to describe the importance of all those who played a role in completing this book, and we apologize to those not mentioned by name. The assistance and support of all are much appreciated.

Numerous contributions to our knowledge of South Carolina and Georgia fishes came through the support of two biologists who work with state fish resources. Chris Thomason of the South Carolina Department of Natural Resources (SCDNR) was a tremendous help in providing species lists, loaning specimens, offering advice for and assistance with fish collections, and locating pond and stream fish data within the MSRB. David Allen (SCDNR) also provided extensive agency support. Chris and David accepted the tremendous task of reviewing the first draft of the entire manuscript. We thank them for their thorough job; their constructive comments were instrumental in improving this book.

Paul Bertsch (Director), Rebecca Sharitz, and Whit Gibbons of the Savannah River Ecology Laboratory (SREL), and John Gladden of the Savannah River National Laboratory (SRNL) provided significant encouragement and advice. We are indebted to Michelle Standora for her unwavering efforts, including many long days collecting fish; preparing specimens, keys, and plates; editing figures; and assisting with the literature cited.

Our special thanks to Laura Janecek of the SREL, who designed the initial promotional material and greatly assisted us with the layout and graphics throughout the book's preparation and production. David Scott, who spent untold hours and shot more than 5,000 photographs to give us the outstanding photos of live fish and various habitats used here, also deserves special recognition. The dazzling results of his efforts significantly enhance the quality of this book.

We are very grateful to John Mark Dean of the University of South Carolina and Susan A. Dyer of the Westinghouse Savannah River Company (WSRC) for agreeing to write the foreword. John Mark Dean also offered continuous encouragement and support, and Susan Dyer kindly provided essential information on fish and literature.

Thanks to Dennis Ryan of the U.S. Department of Energy for supporting the book. Jack Mayer of the WSRC receives our thanks for his unfailing encouragement and continuous prodding, which more than once pushed us along in our efforts to complete the book. We also thank J Vaun McArthur for his support of our efforts.

All those who were involved in the field collections and contributed to the body of ichthyological information receive our gratitude. A number of people assisted Dean Fletcher with collecting fish for photography, particularly Mike Bailey, Clemson University (CU); Beth Dakin, University of Georgia (UGA); Emily and Nathaniel Fletcher; Tim Grabowski, CU; Jeff Isely, CU; Tucker Jones, CU; David Kling, SREL; Curt Ouzts, University of South Carolina; Brady Porter, UGA; Cameron Tuckfield, SREL; and David Wilkins, South Carolina Aquarium.

Joseph Nelson of the University of Alberta graciously provided advice on numerous taxonomic issues. Jim Bulak of SCDNR searched the South Carolina Fisheries Information Network System (SC-FINS) for species occurrence. Mark Collins of SCDNR assisted in research of saltwater and riverine species. Robbi Bowen, Ed Betross, and Tim Barrett of the Georgia Department of Natural Resources assisted with fisheries information for the Georgia portion of the MSRB. Gerald Dinkins, Dinkins Biological Consulting, provided Georgia Department of Transportation stream fish occurrence and distribution survey information. Jan Hoover, U.S. Army Corps of Engineers, provided fish species occurrence data for the Fort Gordon ponds and local stream systems in the Georgia area of the MSRB. Bruce Saul, Augusta State University, provided species occurrence data for the River Watch Parkway catch basins in Augusta, Georgia. Justin Congdon and Joel Snodgrass of the SREL provided species occurrence and distribution data for reservoirs and many of the small ponds and Carolina bays of the SRS. Lynn Wike of the SRNL, WSRC, provided valuable insight on SRS fish species collections and life history information on the grass carp.

Our thanks go as well to the many people involved in the preparation and production of the book, in particular the University of Georgia Press, Nicole Mitchell, Director; Jane Kobres, Assistant to the Director; Sandra Hudson, Design and Production Manager; Jennifer Reichlin, Managing Editor; and Walton Harris, Production Coordinator and Designer, for their patience and guidance during the planning and production phases. Dan Dieter, William Driggers, Karen Ellet, Daniel Foy, Frank Helies, and Christian Jones, all former students at the University of South Carolina, assisted in the preparation of the glossary, indexes, and literature cited, and also tested an earlier version of the taxonomic key. We thank them for their many efforts. We thank the Geographic Information Processing Laboratory of the Baruch Institute for Marine and Coastal Sciences at the University of South Carolina for providing the baseline of the regional map. Map 5 was adapted from Figure 2-6 in WSRC (1995). We thank Jennifer Keller of HelixGraphics for the excellent original artwork of three fish species (Plate 1). A few photographs were procured from colleagues. For photographs of some species we thank Hank Bart, Tulane University (*Carpiodes velifer*); Noel Burkhead, USFWS National Fish Research Laboratory, Gainesville, Florida (*Acipenser oxyrinchus*); Brooks Burr, Southern Illinois University (*Notropis maculatus*); Mark Collins, SCDNR (*A. brevirostrum* and ventral head photographs of *A. oxyrinchus* and *A. brevirostrum*); Robert Jacobs, Connecticut Department of Environmental Protection (*Petromyzon marinus, P. marinus* ventral head, and *Pomoxis annularis*); Kenneth McLeod, SREL (thermally impacted stream); William Pflieger, Missouri Department of Conservation (*Morone chrysops*); F. Douglas Martin (*Agonostomus monticola*). Dean Fletcher provided 39 key plate photos, 1 habitat photo, and 7 chapter heading photos.

We thank Jim Clark, Melissa Salmon, and Dan Hagley from the Riverbanks Zoo and Garden in Columbia, South Carolina, for their assistance and access to fish for photographs of *Acipenser brevirostrum*, *Ictalurus furcatus*, *Elassoma boehlkei* and *E. evergladei*. We thank David Wilkins and the South Carolina Aquarium, Charleston, for assistance and access to displays for our photography of *Lepisosteus osseus* (family heading photo), *Megalops atlanticus*, *Moxostoma robustum* juvenile, *Ictalurus punctatus*, *Oncorhyncus mykiss*, *Salmo trutta*, *Morone americana*, *Centrarchus macropterus*, and *Lepomis microlophus*. Joe Quattro, University of South Carolina, graciously provided specimens of *E. boehlkei* for the identification key photos.

The publication of this book at an affordable price would have been impossible without the generous support of the Savannah River Ecology Laboratory, University of Georgia; the Electric Power Research Institute (EPRI)(special thanks to Doug Dixon of EPRI, who was instrumental in providing a major source of funding through EPRI with matching electric utility funding from South Carolina Electric and Gas, Inc., and Duke Power); Westinghouse Savannah River Company; and the University of South Carolina. This is also Publication SRO-NERP-27 of the U.S. Department of Energy National Environmental Research Park Program and Contribution 1369 of the Belle W. Baruch Institute for Marine and Coastal Sciences.

Family Key and Species Accounts Reviewers

We appreciate the efforts of the many people willing to review the various family and species accounts. The changes they suggested strengthened the book. We thank all who contributed to make the species accounts as accurate and relevant as possible.

Family and species account reviewers.

Family	Reviewer: Key	Reviewer: Species account
General family key	Clark Hubbs, Univ. of Texas, Austin Thomas M. Buchanan, Westark College, Fort Smith, Ark.	
Acipenseridae—sturgeons		Mark R. Collins, SCDNR,[1] Charleston
Lepisosteidae—gars		Anthony A. Echelle, Oklahoma State Univ., Stillwater
Amiidae—bowfins		Dan Crochet, SCDNR, Florence
Megalopidae—tarpons		John M. Dean, Univ. of South Carolina, Columbia
Anguillidae—freshwater eels		Douglas A. Dixon, Gloucester Point, Va.
Clupeidae—herrings and shads	Thomas A. Monroe, NMFS,[2] Systematics Laboratory, MNH,[3] Washington, D.C.	Douglas A. Dixon, Gloucester Point, Va. Ray R. Drenner, Texas Christian Univ., Fort Worth
Cyprinidae—carps and minnows	Noel M. Burkhead, U.S. Fish and Wildlife Service, Gainesville, Fla. Carter R. Gilbert, Univ. of Florida, Gainesville	Lynn D. Wike, Westinghouse Savannah River Co., SRNL,[4] Aiken, S.C. Carol Johnston, Auburn Univ., Auburn, Ala.
Catostomidae—suckers	Robert E. Jenkins, Roanoke College, Salem, Va.	Robert E. Jenkins, Roanoke College, Salem, Va.
Ictaluridae—bullhead catfishes		Brooks M. Burr, Southern Illinois Univ., Carbondale Fred C. Rohde, NCDMF,[5] Wilmington
Esocidae—pikes and pickerels		Ron Ahle, SCDNR, Columbia E. J. Crossman, Royal Ontario Museum, Toronto
Umbridae—mudminnows		Fred C. Rohde, NCDMF, Wilmington
Salmonidae—salmons and trouts	David Allen, SCDNR, Barnwell	
Amblyopsidae—cavefishes		Brooks M. Burr, Southern Illinois Univ., Carbondale
Belonidae—needlefishes		Dennis Allen, W. Belle Baruch Inst. of Marine Biology and Coastal Research, Univ. of South Carolina, Columbia Douglas Cooke, SCDNR, Columbia
Fundulidae—topminnows		Jamie E. Thomerson, Southern Illinois Univ., Edwardsville
Poeciliidae—livebearers		Gary K. Meffe, Univ. of Florida, Gainesville
Atherinopsidae—New World silversides		Anthony A. Echelle, Oklahoma State Univ., Stillwater
Moronidae—temperate basses	Raymond P. Morgan II, Appalachian Laboratory, Center for Environmental Studies, Frostburg, Md. Edward D. Houde, Univ. of Maryland, Chesapeake Biological Laboratory, Solomons	James S. Bulak, SCDNR, Columbia

Family and species account reviewers (continued).

Family	Reviewer: Key	Reviewer: Species account
Elassomatidae—pygmy sunfishes	Wayne C. Starnes, N.C. State Museum of Natural History, Raleigh	Joseph Quattro, Univ. of South Carolina, Columbia Christopher Anderson, Univ. of South Carolina, Columbia James Clark and colleagues, Riverbank Zoo and Garden, Columbia, S.C.
Centrarchidae—sunfishes	Wayne C. Starnes, N.C. State Museum of Natural History, Raleigh Lawrence M. Page, Illinois Natural History Survey, Champaign	Mark C. Belk, Brigham Young Univ., Provo, Utah Jeff Boxrucker, Oklahoma Fishery Research Laboratory, ODWC,[6] Norman J. Andrew DeWoody, Purdue Univ., West Lafayette, Ind. Roy C. Heidinger, Southern Illinois Univ., Carbondale Casey J. Huckins, Michigan Tech. Univ., Houghton Mark S. Peterson, Univ. of Southern Mississippi, Ocean Springs Wayne Starnes, N.C. State Museum of Natural History, Raleigh
Percidae—perches and darters	Lawrence M. Page, Illinois Natural History Survey, Champaign	Lawrence M. Page, Illinois Natural History Survey, Champaign
Mugilidae—mullets		William F. Loftus, USGS,[7] Biological Resources Division, Homestead, Fla.
Achiridae—New World soles		Carla Curran, Savannah State Univ., Savannah

[1] South Carolina Department of Natural Resources
[2] National Marine Fisheries Service
[3] Museum of Natural History
[4] Savannah River National Laboratory
[5] North Carolina Division of Marine Fisheries
[6] Oklahoma Department of Wildlife Conservation
[7] U.S. Geological Survey

We would like to acknowledge the people listed in the table on pages xiii–xiv for their review and valuable comments on the various keys to identify families and species.

And finally, last but certainly not least, we thank our families for years of patience and support while we completed this book.

Species Distribution Map Sources

We gratefully acknowledge the sources for the data for the fish distribution maps from the 759 collections contained in museums and state agencies that were made in the MSRB from 1879 to 2002. Please see the appendix for specific information.

1785

The University of Georgia

Savannah River Ecology Laboratory

Fishes of the Middle Savannah River Basin

Introduction

Our book is based primarily on fisheries studies from 1879 to 2002 in the middle Savannah River basin (MSRB) and studies since 1950 on the U.S. Department of Energy's Savannah River Site (SRS). This book significantly expands *The Fishes of the Savannah River Plant: National Environmental Research Park* (Bennett and McFarlane 1983), which reported on 79 fish species collected on the SRS. Our book enlarges the geographical focus to include the portion of the Savannah River basin that is hydrologically and physiographically similar to the SRS and is the subject of more studies and the source of more data.

Our book comprises habitat characterizations, family descriptions, species accounts, habitat and species photographs, and a taxonomic identification key. We address 24 fish families that include 98 native or introduced fish species historically collected in the MSRB, and list 86 fish species collected specifically on the SRS since the 1950s. Of the 98 species covered in this book, 84 are native (Warren et al. 2000). Thus, this book covers approximately 70% of the native species found in the entire Savannah River drainage. A brief account of each family is followed by accounts of all species of that family found in the MSRB.

The study area comprises the stream reaches corresponding to the U.S. Geological Survey's hydrological unit codes (HUC) 03060106 (the Middle Savannah River) and 03060108 (Brier Creek). This area includes the Savannah River and its tributaries from the Fall Line just above Kiokee Creek in Columbia County (about 6 km below Clarks Hill Dam), Georgia, to the mouth of Brier Creek (river kilometers [rkms] 156–355) and all of the Brier Creek drainage. This represents all of the Savannah River drainage area on the Upper Coastal Plain plus small areas at the edge of the Lower Coastal Plain. Stevens Creek, in Edgefield and McCormick counties, South Carolina, while part of HUC 03060106, is not included because it is a Piedmont stream with a fish species assemblage more similar to that of the upper Savannah River basin.

Geographical and Historical Perspectives

The Savannah River lies in an area of the United States with high fish species diversity. The Savannah River basin is home to at least 118 native fish species (Warren et al. 2000), equal to the number of native freshwater fish species found in all of Kansas and more than the number found in 14 western states and 8 northeastern states (Warren and Burr 1994). The Edisto River basin, which adjoins the Savannah River basin to the east, has a reported (collected and identified) fish fauna of 87 species (25 families) in its freshwater portion and 120 species (52 families) in its saltwater portion (Marcy and O'Brien-White 1995).

Fifteen of the 98 species that are found in the MSRB and covered here were introduced into the area, mostly for fisheries purposes. The effects of these fish on the 84 native species are largely unknown. Several species should have no effect because they have little chance of becoming established in the MSRB. Examples are the grass carp, which is being introduced as triploid sterile indi-

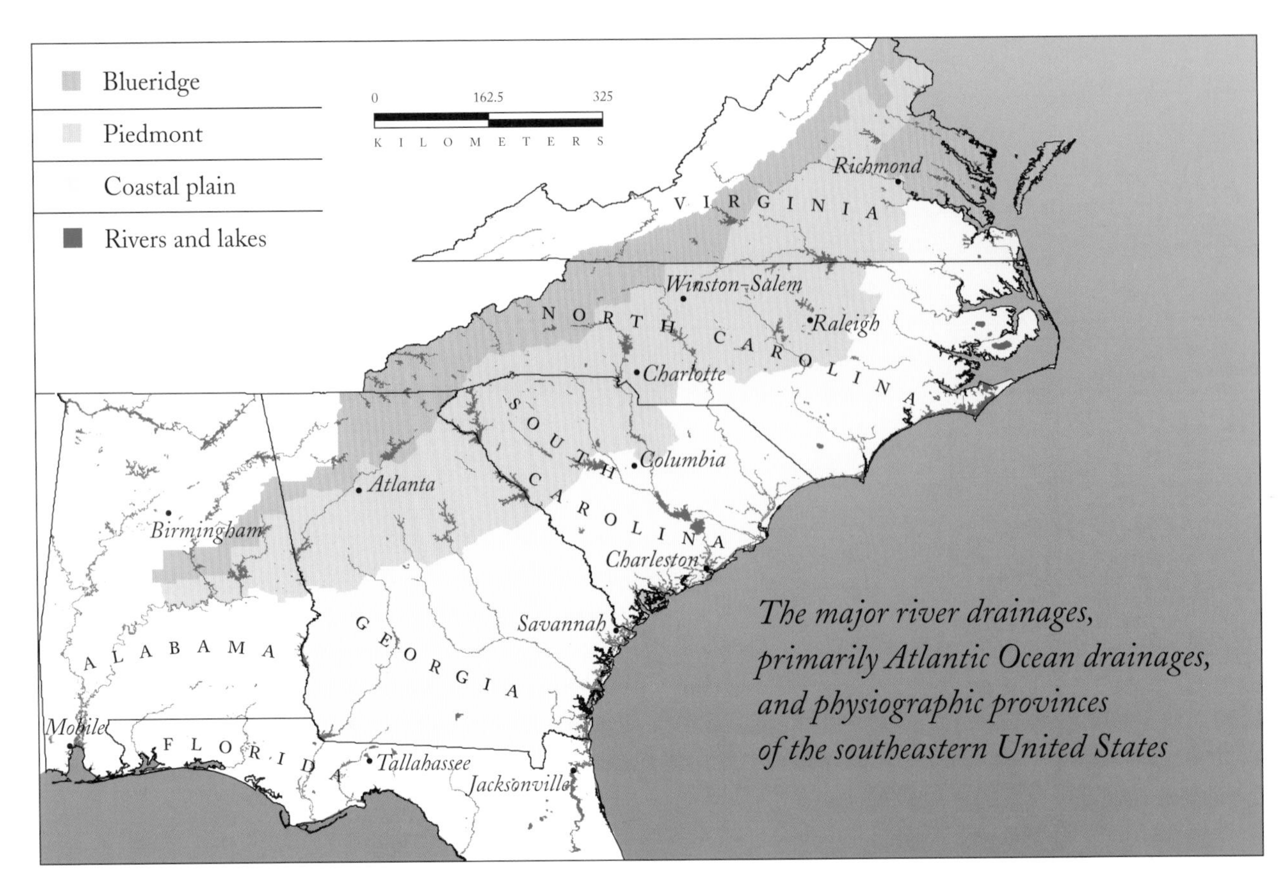

MAP 1

viduals and probably does not have suitable habitat for egg incubation anyway, and the rainbow trout, which is unable to survive normal summer water temperatures in the MSRB. On the other hand, flathead catfish should do well here and may have negative effects on redbreast sunfish and shortnose sturgeon populations similar to those speculated for the Edisto and Combahee rivers (Allen and Thomason 1993; Thomason et al. 1993).

Construction on the Savannah River Plant, as the SRS was initially called, by the U.S. Atomic Energy Commission (AEC) began in 1950 and covered an area of about 777 km^2. By 1953, R-Reactor went critical and began using river water as coolant. Between 1954 and 1971, four other reactors were brought on line, each cooled by river water that was then discharged into streams onsite or into two reservoirs, Par Pond and L Lake. Early photographs of the receiving streams show considerable localized environmental damage in the form of heat-killed vegetation and fauna and highly increased stream flow with accompanying erosion. Unlike the other streams and lakes in the MSRB, the SRS streams and lakes are unique because no public access has been allowed for more than 50 years, thus reducing impacts from nonpoint runoff and fish mortality from fishing pressure. They are also unique in the time period and degree of thermal contamination, and in the subsequent recovery evaluations. Discussions of these streams and reservoirs along with the environmental damage and subsequent remediation and recovery are presented in the Savannah River Tributaries and Reservoirs/Ponds/Isolated Wetlands chapters.

The Savannah River originates in the mountains of North Carolina, South Carolina, and Georgia and flows 505 km to the Atlantic Ocean, for most of its length forming the boundary between South Carolina and Georgia (see Map 1). Over this distance the river and its tributaries drain more than 24,475 km^2. The river originates in the Blue Ridge physiographic province as the Seneca and Tugaloo rivers, which join at Lake Hartwell to form the Savannah River (see Map 2). Approximately 2,580 km^2 of drainage area lies within the Blue Ridge physiographic province, which is characterized by Precambrian

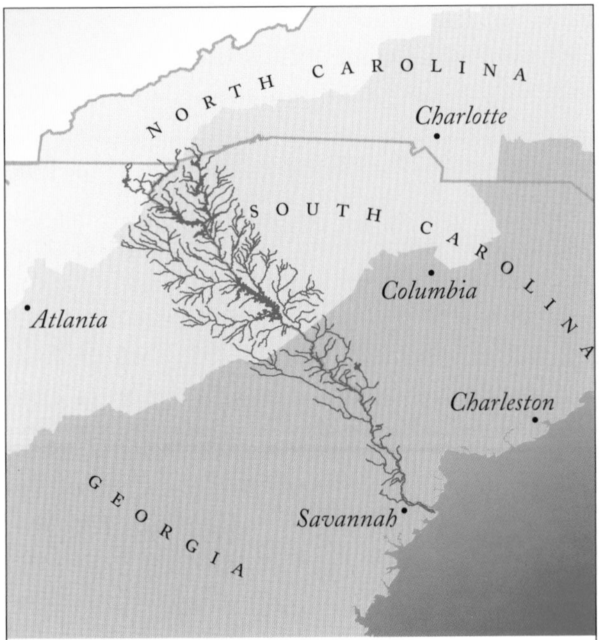

The Savannah River in relationship to the physiographic provinces

Savannah River

Blueridge physiographic province

Piedmont physiographic province

Coastal physiographic province

MAP 2

and Paleozoic metamorphic rocks, irregular topography, and dendritic stream patterns. Slopes are often steep, and soils may be loamy in texture.

The largest drainage area for this river system, roughly 12,480 km^2, lies within the Piedmont physiographic province. On the west, the Piedmont physiographic province is bounded by the Brevard Fault Zone, which separates it from the Blue Ridge physiographic province; to the east, the Fall Line separates it from the Coastal Plain physiographic province. Irregular topography, Precambrian and Paleozoic metamorphic rocks, frequent areas of largely impermeable and highly weathered red clay soil, and dendritic stream patterns characterize the Piedmont physiographic province. Slopes are usually intermediate, and soil depth can vary greatly within short distances.

Just upriver from Augusta, Georgia, lie the Savannah Rapids (also called the Savannah Shoals), where the river crosses the Fall Line and leaves the Piedmont physiographic province to flow onto the Coastal Plain. The rocks near the surface of the Coastal Plain are mostly sedimentary, porous and permeable, and of more recent origin than the deeper "basement rocks," which are often crystalline metamorphic rocks identical with those of the Piedmont. Typical features of the Coastal Plain between Florida and southern Virginia are the oval to elliptical, isolated depressional wetlands called Carolina bays. These usually occur in interstream areas well away from floodplains and have a long axis oriented northwest to southeast with a low, sandy rim that is best developed on the southeastern margin. The bays vary in hydroperiod—the number of days per year when they hold water—ranging from just a few days per year to nearly year-round. If they are near enough to stream floodplains where flooding can populate them, these wetlands may have semipermanent fish populations.

The Coastal Plain in this portion of the Atlantic slope is divided into the Upper Coastal Plain and the Lower Coastal Plain. The Upper Coastal Plain is characterized by rolling topography and, in the portion nearest the Fall Line, the rapidly draining soils of the Carolina Sandhills region of South Carolina, which is continuous with the Fall Line Hills district of Georgia. The Savannah River drains approximately 7,000 km^2 of the Upper Coastal Plain (see Map 3). While the basic pattern of streams in this region is dendritic, frequently with long, relatively narrow drainage networks, considerable variation does exist in the size of the drainage area for streams of a given stream order. Generally, a single channel with variable sinuosity and floodplain width forms. Where there are deltas, and downstream from major erosional sources of sediment, dichotomic or braided stream systems may develop.

The Lower Coastal Plain lies east of the 46 m elevation contour. The geologic features that define the boundary between the Upper and the Lower Coastal Plain are Trail Ridge south of the Altamaha River and the base of the Orangeburg Escarpment north of the Altamaha. The Lower Coastal Plain was formed as sea level changed during the Pleistocene ice ages and appears as a series of barrier island–salt marsh sequences. This portion of the Coastal Plain is thus a series of steplike, relatively flat

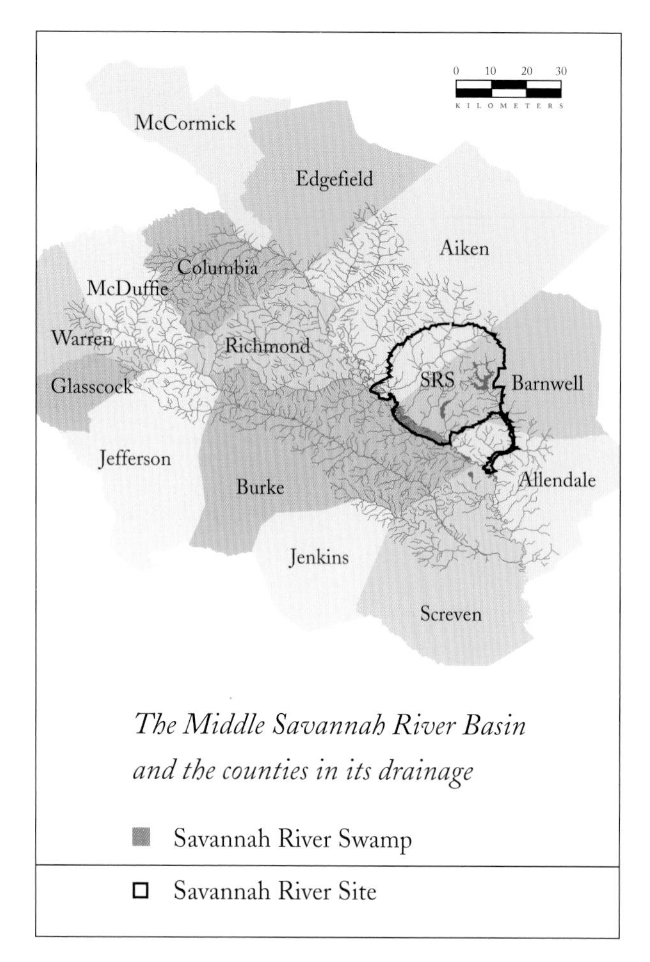

MAP 3

areas decreasing in elevation from west to east. Streams in this area are often braided or dichotomic in form. Soils are usually deep, often with interbedding of sandy or loamy soil with clay. The Savannah River drains roughly 2,420 km^2 of the Lower Coastal Plain.

To place the Savannah River in its zoogeographic context, it is important to compare its fish fauna with the faunas of the four drainage systems that border the Savannah River basin. For the purposes of this discussion, the Santee and Cooper river systems are treated as a single entity, though they were joined only recently by human intervention. Also, while the Ogeechee and Altamaha rivers are not joined, they are treated as a single entity because of their close geographical proximity, and comparable separate faunal data for each river system are not easily available. These drainage systems are, then, to the north, the Santee-Cooper drainage system and the Edisto-Combahee drainage system and, to the south, the Ogeechee-Altamaha drainage system and the Apalachicola drainage system. The Edisto River system is almost completely confined to the Coastal Plain. The Apalachicola and Santee-Cooper drainages drain through all three physiographic provinces, while the Ogeechee-Altamaha system originates in the Piedmont rather than in the Blue Ridge. The Apalachicola system lies on and west of the Brevard Fault Zone and drains south into the Gulf of Mexico; the other systems drain eastward into the Atlantic. Several authors have suggested stream capture between headwaters of the Savannah River and the Apalachicola River system (Gilbert 1964; Swift et al. 1986) and between the headwaters of the Savannah River and the Santee-Cooper and/or Tennessee drainage (Tsai and Rainey 1974; Lee et al. 1980).

Table 1 summarizes the comparisons made among these drainage systems. While the area drained has no significant correlation with the number of native fish species found in any of these river systems, the number of physiographic provinces through which the river system runs may be related to the number of native species. Warren et al. (1997) noted the same phenomenon and attributed the correlation between number of physiographic provinces crossed and number of species to the greater stream type diversity when more physiographic provinces are crossed.

The direction of drainage appears to influence the number of species these drainages have in common in that the lone Gulf of Mexico drainage system, the Apalachicola, has the lowest similarity index value with all other stream systems (Table 2). The pattern of the Apalachicola drainage having lower species similarity values compared with the Savannah, Ogeechee, and Altamaha rivers was also noted by Gilbert (1987), who postulated that it indicates long-term isolation. Based on these two sets of data, it appears that the Savannah River is not unusual in species composition or number of species for the geographic region in which it lies.

The past history of land use has long-term effects on stream biodiversity. Harding et al. (1998) reported that in the southeastern United States, 1950s land use patterns predict the diversity of invertebrates and fish in streams better than 1990s land use patterns. For this reason we offer a thumbnail sketch of historical land use and river obstructions (e.g., dams) within the MSRB.

At the time of European settlement in the MSRB the land was forested except for areas where Native Ameri-

TABLE 1. Comparison of the Savannah River system with adjacent river systems.

	Santee-Cooper	Edisto	Savannah	Ogeechee-Altamaha	Apalachicola
Physiographic Province					
Blue Ridge	X		X		X
Piedmont	X		X	X	X
Coastal Plain	X	X	X	X	X
Drainage Area (km²)	45,090	8,080	24,480	52,000	52,470
Number of Species					
Total*	117	83	118	109	119
Single Drainage**	18	1	3	4	34
Endemic	2	0	0	2	6

*Not including introduced species or probably introduced species.
**Species not endemic, but found in only one of the drainages under consideration.
Source: Species data derived from Warren et al. 2000.

TABLE 2. Matrix of similarity indexes.*

	Santee-Cooper	Edisto	Savannah	Ogeechee-Altamaha
Edisto	0.574			
Savannah	0.715	0.642		
Ogeechee-Altamaha	0.569	0.684	0.707	
Apalachicola	0.372	0.453	0.454	0.530

* Based on the Jaccard Index, which is calculated as:

$$JI = \frac{a}{a + b + c}$$

where a is the number of species occurring at both locations, b is the number of species found only at the first location, and c is the number of species found only at the second location.

cans, mostly villages of the Creek Confederation, had cleared plots for growing corn, pumpkins, beans, tobacco, and squash. Creek villages tended to be located along streams, so lands further from rivers and permanent creeks tended to remain forested.

The first European contact in the MSRB occurred in 1540 when Hernando de Soto crossed the Savannah River near present-day Augusta while searching for rumored gold. There is evidence that diseases of European origin greatly reduced the Native American populations of South Carolina and Georgia so that anthropogenic disturbances were reduced for some time. For the next 200 years Europeans passed through the area only as explorers and traders or settled only along the margins of our area of concern. The city of Augusta was founded in 1736, and as late as the 1770s Augusta was the only major city in Georgia away from the coast. Most of the MSRB was forested, with few areas cleared for field crops. Right along the river, plantations began to be built in this period, and larger tracts of land were cleared for row crops. By the 1820s cotton was grown extensively in Georgia. In order to industrialize Augusta and keep cotton profits in the area, textile mills were built. As early as 1828 the Vaucluse Mill on Horse Creek in Aiken County was in operation. The success of this and other textile ventures led in 1846 to the construction of the Augusta Canal, with its wing dam to direct water into the canal, to take advantage of the more than 15 m drop in elevation from the head of the canal to downtown Augusta. The wing dam was extended and made taller, and by the 1870s it extended completely across the river and was capped with rocks and timber at its current height.

The next engineered change in the Savannah River's flow was the Stevens Creek Dam, which despite its name is on the Savannah River. This dam, completed in 1914 to serve as a source for hydropower, is located approximately 1.5 km above the mouth of the Augusta Canal and re-regulates some of the flow released from upstream dams. Clarks Hill/J. Strom Thurmond Lake re-regulates flow from the dams upstream of it, so that the only re-regulation required by Stevens Creek Dam is of water released through Clarks Hill/J. StromThurmond Lake.

The New Savannah Bluff Lock and Dam, completed in 1937, was constructed to provide water deep enough for commercial boat traffic between Savannah and Augusta. Although commercial traffic ceased in 1979, the locks are opened during the spawning season to allow American shad access to the river up to the Savannah Rapids.

The most recent engineered structure that controls water flow in the Savannah River is the Clarks Hill/J. Strom Thurmond Dam located a little more than 4.6 km above Kiokee Creek. This dam was completed in 1954 and forms a multipurpose reservoir. When hydroelectric power is being generated, flow rates below the dam may be many multiples of the base flow, so downstream structures are necessary to re-regulate river flow.

Two other factors are important in controlling water quality, and perhaps flow, in the Savannah River: urbanization within the watershed and the SRS.

Land use has been shown to have large effects on fish species assemblages in southeastern streams. In North Carolina, for example, most species of fish are either less abundant or missing altogether from urbanized sections of watersheds compared with areas where the land is used primarily for silviculture or agriculture (Lenat and Crawford 1994). Urbanization results in more variability in dissolved oxygen, an especially important factor in the survival of fish eggs and larvae, and a concomitant reduction in the spawning success of fishes, especially anadromous species (Limburg and Schmidt 1990). Even low-intensity urbanization can have both large and subtle effects on fish species assemblages, with fish that feed primarily on terrestrial prey being strongly affected (Weaver and Garman 1994). Interestingly, human disturbances that affect water and/or habitat quality appear to be major factors in the success of introduced or invading species, while biological factors may be of much less importance (Ross et al. 2001).

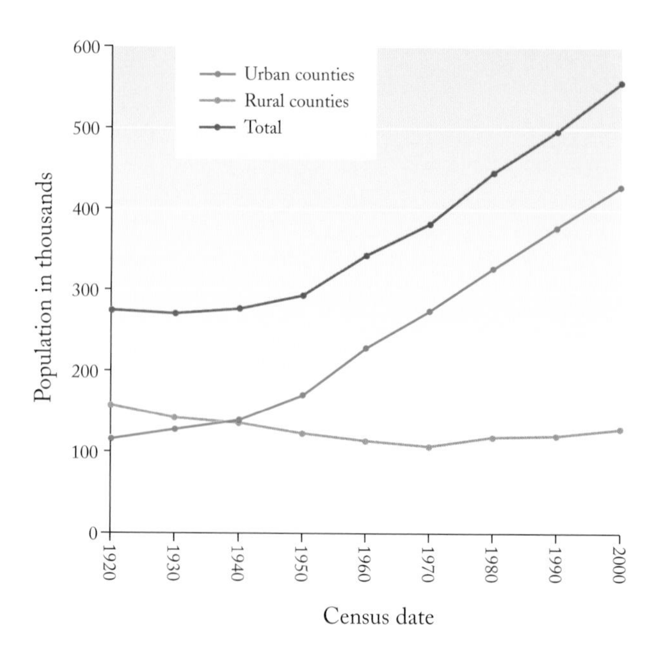

FIGURE 1. Population trends for the MSRB counties (Aiken, Allendale, and Barnwell counties, South Carolina; and Burke, Columbia, Glasscock, Jefferson, Jenkins, McDuffie, Richmond, Screven, and Warren counties, Georgia) based on U.S. Census Bureau data. Data are unavailable for two counties for 1900 and unavailable for one county for 1910, so the analysis is based on censuses with data for all counties.

Augusta was settled as a trading center with the Indians, but the surrounding area along the river and the larger creeks was quickly brought into agriculture, largely cotton, with the concomitant construction of textile mills mentioned above. By the 1830s Augusta was becoming the manufacturing center of the area, with textile mills and various smaller industries in operation. Despite this industrialization, the region's economy was based primarily on agriculture well into the twentieth century. Figure 1 shows population trends since 1920 in the MSRB. Around 1940, the population of the three urbanized counties—Columbia and Richmond in Georgia and Aiken in South Carolina—combined became higher than the population of the combined rural counties. Along with this shift to an urbanized population came increased industrial use of the river, higher sewage and other waste loading in the river, more roofs and pavement to increase surface runoff and decrease soil penetration of precipitation, and more channelization and other modification of streams. High levels of metals and other noxious chemicals characterize the urban runoff into the river system today. The rural counties are characterized by increased

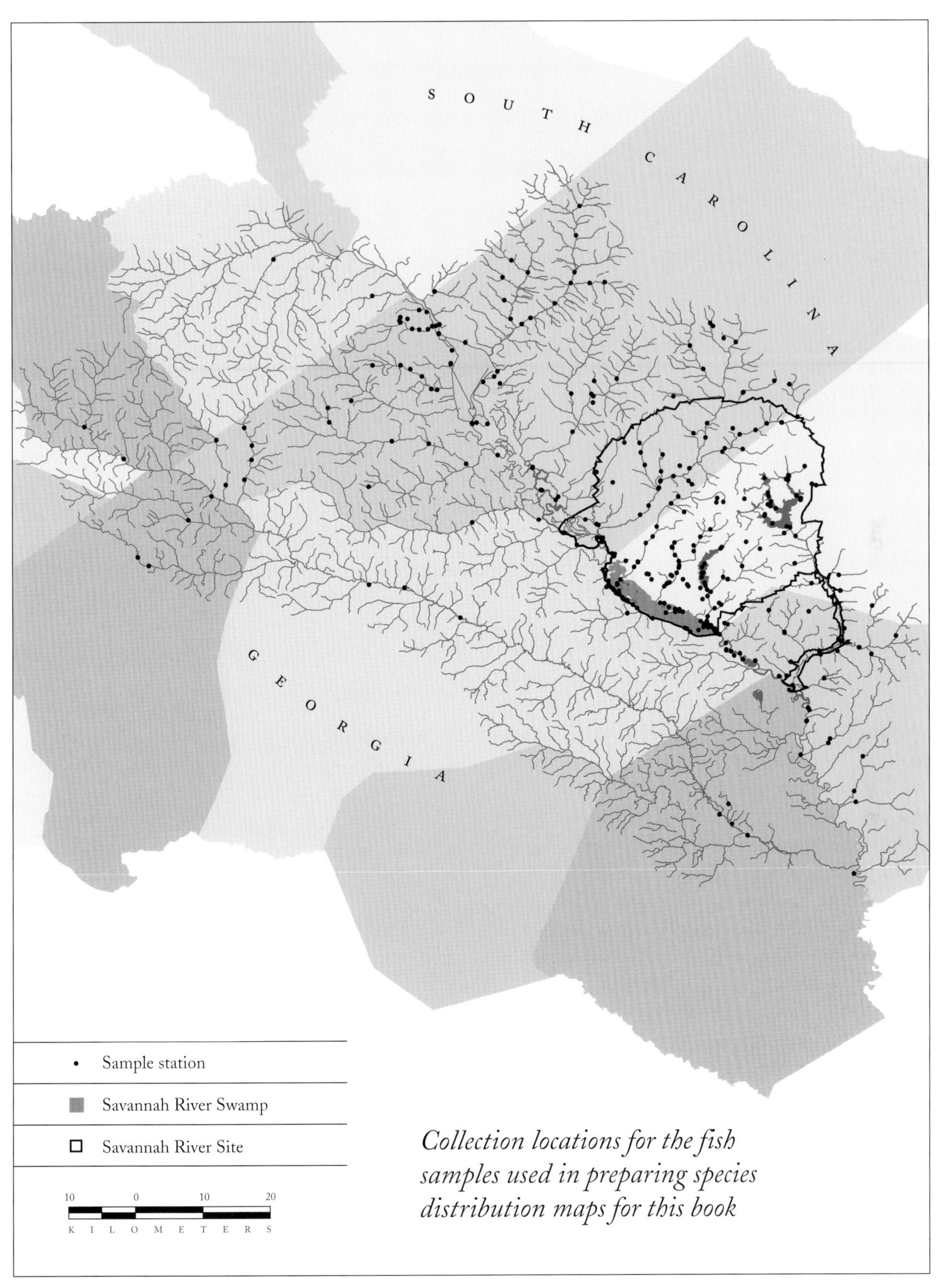

Collection locations for the fish samples used in preparing species distribution maps for this book

MAP 4

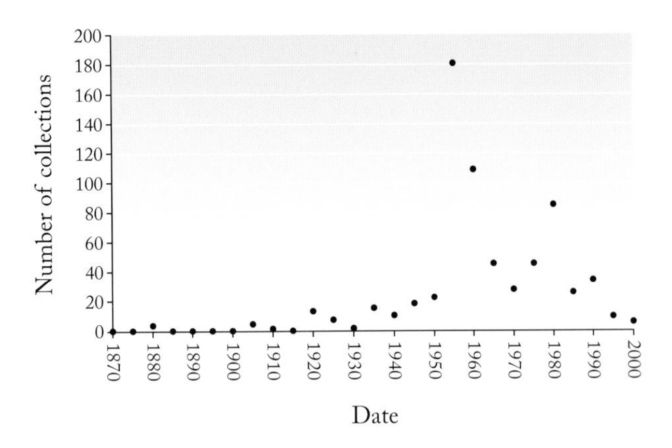

FIGURE 2. Number of individual collections of MSRB fishes held in museums. *Note*: A collection is considered here to be all species of fish taken from one location on one date. Collection numbers are grouped by 5-year intervals.

use of fertilizer and other agricultural chemicals as well as some shifting to silviculture and development of smaller "bedroom communities." Major industries that might influence water quality in the Savannah River in addition to SRS activities are textile mills, polystyrene foam and paper product plants, chemical processing plants, and a commercial nuclear power plant, Plant Vogtle.

The first surveys of the fish fauna in the middle Savannah River began in the late 1800s. Map 4 shows the locations where the historical fish samples (1879–2002) that were used in preparing fish species distribution maps for our book were collected. Our sources of data were museum collections, reports in peer-reviewed journals, and unpublished reports from SRS monitoring programs and South Carolina, Georgia, and Plant Vogtle environmental monitoring reports and databases. The appendix lists museum and agency sources of collection data supporting the species distribution data and maps.

Few of the museum specimens from this geographical area were collected prior to the 1920s, and most, as indicated by museum holdings, were collected post-1945 (see Figure 2). The greatest sampling efforts seem to have been in conjunction with various projects to characterize the fauna of the SRS and the environmental impacts of its operation. These projects began just prior to the start of SRS construction (1950).

In addition to specimens placed in museums, the Savannah River Ecology Laboratory, the Savannah River National Laboratory, and various environmental subcontractors have created published and unpublished reports (about 500). Those that give exact locations for species sampled and were judged to be probably correct in species identifications were used in our analysis of local distributions. In addition, South Carolina and Georgia state agencies have sampled this area, and they made reports and data files available to us for these analyses. The appendix lists the reports, data files, and publications used.

Native Fish Species

The native species of the MSRB can be divided into four categories: (1) resident species that are present for all life stages, (2) diadromous (catadromous or anadromous) species that are present only for certain life stages, (3) marine species that penetrate rivers up to the Fall Line either regularly or as strays, and (4) upland species that sometimes stray below the Fall Line. Table 3 summarizes the native species falling into each category. Of the marine species included in our book, the mountain mullet and the tarpon are rare or infrequently collected and the others are part of the regular summer and fall fauna of the MSRB.

Introduced Fish Species

The introduced species can likewise be divided into categories (Table 4): (1) persistently abundant and established, (2) rare and possibly not established or established only in restricted areas, (3) definitely not established, and (4) too little information to determine status.

General Distribution within the Region

Each of the species considered in this book has its own distribution pattern within the MSRB, but most fit into one of four categories. For the benefit of this discussion we use the following definitions taken from Conner and Suttkus 1986:

"river"—flowing body of water with an average width of 70 m or more at mean low stage;
"stream"—flowing body of water with an average width of 10–70 m at mean low stage;
"creek"—flowing body of water with an average width less than 10 m at mean low stage; and
"swamp"— sluggishly flowing or lentic features asso-

TABLE 3. Native resident, diadromous, marine, and upland fish species of the MSRB (listed in phylogenetic order).

	Scientific name	Common name
Resident species		
Lepisosteidae	*Lepisosteus osseus*	longnose gar
	Lepisosteus platyrhincus	Florida gar
Amiidae	*Amia calva*	bowfin
Clupeidae	*Dorosoma cepedianum*	gizzard shad
Cyprinidae	*Cyprinella leedsi*	bannerfin shiner
	Cyprinella nivea	whitefin shiner
	Hybognathus regius	eastern silvery minnow
	Hybopsis rubrifrons	rosyface chub
	Nocomis leptocephalus	blueheaded chub
	Notemigonus crysoleucas	golden shiner
	Notropis chalybaeus	ironcolor shiner
	Notropis cummingsae	dusky shiner
	Notropis hudsonius	spottail shiner
	Notropis lutipinnis	yellowfin shiner
	Notropis maculatus	taillight shiner
	Notropis petersoni	coastal shiner
	Opsopoeodus emiliae	pugnose shiner
	Pteronotropis stonei	lowland shiner
	Semotilus atromaculatus	creek chub
Catostomidae	*Carpiodes cyprinus*	quillback
	Carpiodes velifer	highfin carpsucker
	Erimyzon oblongus	creek chubsucker
	Erimyzon sucetta	lake chubsucker
	Hypentelium nigricans	northern hogsucker
	Minytrema melanops	spotted sucker
	Moxostoma collapsum	notchlip redhorse
	Moxostoma robustum	robust redhorse
	Scartomyzon sp. cf. *lachneri*	brassy jumprock
Ictaluridae	*Ameiurus brunneus*	snail bullhead
	Ameiurus catus	white catfish
	Ameiurus natalis	yellow bullhead
	Ameiurus nebulosus	brown bullhead
	Ameiurus platycephalus	flat bullhead
	Noturus gyrinus	tadpole madtom
	Noturus insignis	margined madtom
	Noturus leptacanthus	speckled madtom
Esocidae	*Esox americanus*	redfin pickerel
	Esox niger	chain pickerel
Umbridae	*Umbra pygmaea*	eastern mudminnow
Aphredoderidae	*Aphredoderus sayanus*	pirate perch
Amblyopsidae	*Chologaster cornuta*	swampfish
Fundulidae	*Fundulus chrysotus*	golden topminnow
	Fundulus lineolatus	lined topminnow
Poeciliidae	*Gambusia holbrooki*	eastern mosquitofish
Atherinopsidae	*Labidesthes sicculus*	brook silverside
Centrarchidae	*Acantharchus pomotis*	mud sunfish
	Centrarchus macropterus	flier
	Enneacanthus chaetodon	blackbanded sunfish

TABLE 3. (continued)

	Scientific name	Common name
Resident species		
Centrarchidae	*Enneacanthus gloriosus*	bluespotted sunfish
	Enneacanthus obesus	banded sunfish
	Lepomis auritus	redbreast sunfish
	Lepomis gibbosus	pumpkinseed
	Lepomis gulosus	warmouth
	Lepomis macrochirus	bluegill
	Lepomis marginatus	dollar sunfish
	Lepomis microlophus	redear sunfish
	Lepomis punctatus	spotted sunfish
	Micropterus salmoides	largemouth bass
	Pomoxis nigromaculatus	black crappie
Elassomatidae	*Elassoma evergladei*	Everglades pygmy sunfish
	Elassoma okatie	bluebarred pygmy sunfish
	Elassoma zonatum	banded pigmy sunfish
Percidae	*Etheostoma fricksium*	Savannah darter
	Etheostoma fusiforme	swamp darter
	Etheostoma hopkinsi	Christmas darter
	Etheostoma inscriptum	turquoise darter
	Etheostoma olmstedi	tessellated darter
	Etheostoma serrifer	sawcheek darter
	Percina nigrofasciata	blackbanded darter
Diadromous species		
Acipenseridae	*Acipenser brevirostrum*	shortnose sturgeon
	Acipenser oxyrinchus	Atlantic sturgeon
Anguillidae	*Anguilla rostrata*	American eel
Clupeidae	*Alosa aestivalis*	blueback herring
	Alosa mediocris	hickory shad
	Alosa sapidissima	American shad
Moronidae	*Morone saxatilis*	striped bass
Marine species		
Megalopidae	*Megalops atlanticus*	tarpon
Belonidae	*Strongylura marina*	Atlantic needlefish
Mugilidae	*Agonostomus monticola*	mountain mullet
	Mugil cephalus	striped mullet
Achiridae	*Trinectes maculatus*	hogchoker
Upland species		
	Micropterus coosae	redeye bass[1]

[1] The Savannah River is the only area of the redeye bass's range where it occurs below the Fall Line.

TABLE 4. Introduced fish species in the MSRB and their status.

	Scientific name	Common name
Clearly established		
Clupeidae	*Dorosoma petenense*	threadfin shad
Cyprinidae	*Cyprinus carpio*	common carp
Ictaluridae	*Ictalurus punctatus*	channel catfish
Percidae	*Perca flavescens*	yellow perch
Rare and possibly not established		
Cyprinidae	*Carassius auratus*	goldfish
Moronidae	*Morone americana*	white perch
	Morone chrysops	white bass
Centrarchidae	*Lepomis cyanellus*	green sunfish
	Pomoxis annularis	white crappie
Clearly not established		
Cyprinidae	*Ctenopharyngodon idella*	grass carp
Salmonidae	*Oncorhynchus mykiss*	rainbow trout
Too little information		
Ictaluridae	*Ictalurus furcatus*	blue catfish
	Pylodictis olivaris	flathead catfish

ciated with floodplains of lowland streams, includes marshes and sloughs.

The distribution patterns that we have noted are: (1) species living primarily in river mainstream including impoundments and side channels; (2) swamp, marsh, oxbow lake, or slough inhabitants; (3) primarily stream or creek inhabitants; and (4) generalists. There are no large permanent natural lakes in the MSRB, so it is not surprising that there are few species that are primarily lake and pond dwellers. The distribution pattern is noted in the account for each species.

Summary

At this time it appears that the fish fauna of the middle Savannah River basin region is still in relatively good condition. The major concerns for the future are continued urbanization with increased runoff, channelization, siltation, and chemical and nutrient input. Also, if Harding et al. (1998) are correct, the full effects on the fauna of recent conditions may not yet be apparent. Unfortunately, the data set we have had to use in our analyses lacks the information necessary to assess sampling effort and relative species abundance. Anderson et al. (1995), using a data set with consistent sampling effort and species relative abundance estimates, were able to show significant changes in species assemblages for freshwater fishes of Texas and the effects of dams and other local impactors.

The effects of introduced species in the MSRB are not well known at the present time either, but to date they seem to have been subtle, or at least minor. On a more positive note, seriously depleted species such as the American shad, the shortnose sturgeon, and the robust redhorse (Figure 3) are spawning in this subdrainage. Their reproductive success is still under evaluation.

FIGURE 3. The robust redhorse, *Moxostoma robustum*. This endangered species found in the MSRB appears to be making a comeback with help from the Robust Redhorse Recovery Plan.

Savannah River and Swamps

The MSRB is in many ways typical of southeastern river basins. It is home to a diverse fish fauna, and its associated tributaries and wetlands are important spawning and nursery areas. And like other southeastern rivers, the Savannah River watershed is increasingly affected by the region's growing human population.

Fish Habitats of the Middle Savannah River Basin and Associated Swamps

The Savannah River itself has several habitat types that are utilized by the fish populations of the MSRB. The most obvious of these habitats is the main river channel (Figure 4). The main river channel within the MSRB generally has a sandy substrate, but a variety of bottom substrate types may be found: cobbles and rock shelves at and near the Savannah River Rapids, sand or gravel where there is moderate flow, and mud and plant detritus in backwaters.

In addition to the main channel there are a number of "cutoff bends" and "dead rivers" (former river channels that are still connected to the main channel). Many cutoff bends and dead rivers were created when the Augusta Navigational Project modified the original river channel for barge traffic. Dead rivers and cutoff bends have characteristics intermediate between those of flowing lotic systems and still-water lentic systems, and they form oxbow lakes when both ends become disconnected from the main river channel. In this discussion, dead rivers are considered part of the lotic environment, and oxbows are treated as lentic environments. Historically there has been little sampling in dead rivers of the MSRB. Side channels are segments of former river channels that are still connected to the main channel at each end but have reduced flow compared with the main channel. Side channel habitat, which is limited, has likewise been sampled little but is similar to that of cutoff bends.

The major swamp habitats in the MSRB are in two locations: Phinezy Swamp, adjacent to Augusta and the Savannah River, and the Savannah River Swamp, located mostly within the SRS adjacent to the Savannah River (Figures 5 and 6). The Savannah River Swamp receives water from Upper Three Runs (rkm 251), Fourmile Branch (rkm 241), Beaverdam Creek (rkm 243), Pen Branch (which does not directly drain into the Savannah River), Steel Creek (rkm 227), and Lower Three Runs (rkm 206). Across the river from the SRS there is swamp habitat, essentially a continuation of the Savannah River Swamp, associated with Sweetwater Creek (rkm 214). While there has been sporadic sampling of the fish fauna in the Savannah River Swamp, there are virtually no fisheries data for Phinezy Swamp or for the Sweetwater Creek area. There are some data from a swampy area on McBean Creek in Georgia, and some samples from the lower part of Brier Creek and Beaverdam Creek in Screven County, Georgia. These latter samples should probably be considered swamp samples, but the site description information was limited.

There are other swamp or swamplike habitats in the MSRB in addition to those identified above. Each of the streams and creeks has associated floodplains with bottomland hardwood forest. Major rain events normally

FIGURE 4. Savannah River main channel at about the middle of the Savannah River Site boundary.

cause flooding of these bottomland forests with a slow return of the stream to its main channel. Several fish species enter such flooded areas to feed or to spawn. Floodplain ponds and pools are common on these floodplains. These wetlands are connected with the mainstream during flooding events but are isolated during normal flows. Several fish species colonize these ponds during flood events and remain until either the next floodwater incursion or the area dries up. Very little information is available on the fishes of the floodplain ponds.

In the lower reaches of Steel Creek, Pen Branch, and Fourmile Branch on the SRS, heated reactor effluent eliminated the natural vegetation, and sediments from parts of the upper stream basin and upper floodplain were deposited onto the lower floodplain where the stream entered the Savannah River Swamp, forming deltas. After reactor shutdown in 1986, these bottomland forests were partially replaced with grassy marshes through natural regeneration. The marshes have been sampled; the limited fish data for these specific marsh habitats are discussed more fully elsewhere.

The term "slough" has several meanings, but in the MSRB context it denotes linear wetlands that may represent sediment-filled oxbow lakes or other former river or stream channels. This habitat is quite limited, and the fish fauna data for it are proportionally smaller.

Human Influences on the Fish Fauna of the Savannah River and Associated Swamps

There are many sources of indirect influences on local fish fauna. Those whose effects are best documented are discussed here.

URBANIZATION Three urbanized counties—Richmond and Columbia counties in Georgia and Aiken County in South Carolina—are the center of population for the MSRB. Increased urban population brings with it an in-

FIGURE 5. Savannah River main channel and the Savannah River Swamp near the mouth of Steel Creek.

FIGURE 6. Savannah River Swamp.

crease in water-impervious surfaces such as roofs, roads, and paved driveways, and a decrease in natural vegetation. Water cannot infiltrate such surfaces and instead flows directly into streams. This runoff increases the frequency of flash floods and decreases the predictability of flooding. In doing so it interferes with the reproduction of the species that breed in the floodplains, which require predictability in the timing and intensity of flooding for successful reproduction.

FISHERIES The sport, subsistence, and commercial fisheries have a direct influence on fish populations and species assemblages. Sportfishing in the MSRB largely targets largemouth bass, black crappie, sunfishes, American shad, chain pickerel, larger catfishes such as the white catfish or channel catfish, and striped bass and its hybrids. Local newspaper fishing reports indicate that the area just below the New Savannah Bluff Lock and Dam is one of the most popular sportfishing areas within the MSRB.

Significant commercial fisheries are those on American shad, channel catfish, and white catfish. Commercial fisheries for blueback herring formerly existed in South Carolina (Ulrich et al. 1978), but no herring are taken in Georgia because of netting restrictions. Music (1981) reported that the Savannah River American shad catches represented 51% of Georgia landings in 1980, although only 13% of Georgia's commercial shad fishermen operated in the Savannah River. Striped bass are considered sport fish, and no commercial bass fishery is allowed in the Savannah River. Because they are highly regarded as a recreational species, considerable effort has been expended by government agencies to improve striped bass population levels. While Gilbert et al. (1986) suggested that striped bass spawn outside the MSRB, primarily in the tidal portions of the Savannah River, Paller et al. (1984, 1985, 1986b) have documented spawning well up into the river basin.

INDUSTRIAL ACTIVITIES Use of river water for industrial purposes, such as cooling water, has affected MSRB fish populations through entrainment (in which fish eggs and larvae are caught up in the current of a water intake device) and impingement (the removal of juvenile and adult fish from the intake stream by means of a small-mesh [0.95 cm] screen). Entrainment occurs wherever large volumes of water are removed, such as at domestic water treatment plants, or used in industrial processes. Mortality due to entrainment varies according to the species of fish, its life stage, and physical parameters of water flow such as current speed and turbulence. Changes in temperature or other water quality parameters and amelioration devices such as traveling screens that return the entrained animal to the water away from the intake device also play a role in survival. See Schubel and Marcy 1978 for biological assessment of entrainment impacts. Historically, the largest sources of entrainment in the MSRB have been the reactor cooling water intakes for the SRS (9.8% of Savannah River flow) and the Plant Vogtle nuclear power station (4.2% of river flow; Wiltz 1981; DOE 1990).

SAVANNAH RIVER SITE Historically, the SRS has affected populations of commercially and recreationally important fish species in the river primarily through impingement and entrainment losses of fish eggs, larvae, and adults during intake of cooling water (McFarlane et al. 1978). The overall rates of impingement at the SRS intakes were low relative to those of other cooling-water intake facilities in the Southeast (DOE 1988). Cessation of reactor operations and the concomitant lack of need for cooling water withdrawals from the Savannah River reduced entrainment impacts substantially.

FLOOD/WATER CONTROL Dams, levees, and locks all influence fish species assemblages. Dams and their associated locks convert lotic environments into lentic ones. Levees restrict overbank flooding, preventing the filling of riparian wetlands that are important to the life cycles of numerous riverine fish species and altering the frequency of floodplain inundation, which has profound effects on species assemblages utilizing floodplain ponds and sloughs. Overbank flooding also affects the biological community structure of rivers through introducing nutrients in the form of plant-derived materials into the main channel. The Clarks Hill/J. Strom Thurmond Dam, the most downriver of the U.S. Army Corps of Engineers dams, probably has the strongest effect due to its regulation of flooding events. Additionally, the water released below the Clarks Hill/J. Strom Thurmond Dam is from below the thermocline and thus has a profound effect on temperature and dissolved oxygen levels for considerable distances downstream.

The New Savannah River Lock and Dam, the first dam on the Savannah River fish migrating upstream encounter, is a major obstruction for commercial and recreational species such as American shad, blueback herring, striped bass, and Atlantic sturgeon. It also affects the movements of such threatened and endangered species as the robust redhorse and shortnose sturgeon. In the early 1900s, some of these species migrated annually to the headwaters of the Savannah River—all the way to Tallulah Falls, 614 km from the Atlantic Ocean. Construction of reservoirs, dams, and hydropower facilities eliminated essential riverine habitat for these species. Access to historic spawning habitats has been reduced so that migrating fish can go only half as far upriver as formerly, and this is a major cause of the decline in the Savannah River's migratory fish populations (U.S. Fish and Wildlife Service 2001).

CHANNELIZATION Portions of the Savannah River, including 31 meander cutoffs, were channelized by the Corps of Engineers to enhance navigation between Augusta and Savannah, reducing the river length between these cities by 24.1 km (Wike et al. 1994). In addition to eliminating the meander cutoffs, the channelization increased the water velocity, likely decreased habitat heterogeneity in the straightened main channel, and increased backwater habitat in the form of dead rivers and man-made oxbows. These cutoffs are prominent in the lower portion of the MSRB. The work was completed by 1965, but some dredging continued through 1985 (Wike et al. 1994). Additionally, pile dikes increase the water velocity of the main channel by restricting flows. Eddies behind and around these dikes provide refuges from the high water velocities where fish frequently congregate. These areas are frequently favorite spots for anglers.

SEDIMENTATION AND POLLUTION The introduction of commercial agriculture in the eighteenth century increased the potential for erosion and thus increased sedimentation. The recent shift toward silviculture and pastureland is working to ameliorate this process. Countering this trend is the increased urbanization of the MSRB, which tends to increase sedimentation from construction activities and stormwater runoff. Sedimentation eliminates breeding habitat for fishes requiring clean substrates, extirpates invertebrates that may serve as food sources, and leads to wholesale changes in flora and fauna in affected areas (Berkman and Rabeni 1987; Waters 1995; Jones et al. 1999). In southern Appalachian streams, sedimentation resulting from deforestation shifts fish species assemblages away from species that spawn in rock crevices and clean gravel or cobbles to mound-building minnows, their nest associates, and fish that excavate nests in soft sediments (sunfishes and basses) (Sutherland et al. 2002).

Along with the aforementioned changes in runoff and erosion there is the problem of eutrophication. The Public Interest Research Group lists the Savannah River as the eighth most polluted river in the United States (National Science Center's Fort Discovery 2002). The expanding human population in the area has increased the amount of organic waste concentrated in a relatively small area. Even with sewage treatment, nutrients such as nitrogen compounds and phosphates are entering waterways at high enough levels to cause increased plant growth in some bodies of water. Additionally, even though less land area is under cultivation now than in the nineteenth century, agricultural fields are receiving higher levels of fertilizers and other chemicals. Water percolating through the soils of these fields picks up fertilizer, and the soil particles, often accompanied by absorbed inorganic fertilizer, enter waterways through erosion. This effect is most obvious in certain smaller bodies of water because of lack of dilution.

The deteriorating water quality (especially reduced levels of dissolved oxygen) appears to be degrading the southern river nurseries and summer habitats of the shortnose sturgeon and the Atlantic sturgeon; protection of these essential habitats is critical (Collins et al. 2000b). Influences of the deterioration on other species are not as well documented.

The factors that cause increased nutrient loading in the local waterways also introduce toxic substances. Currently, the Environmental Protection Agency (EPA) advises against eating fish from most streams and other water bodies in the MSRB due to mercury concentrations. Although there are many possible sources of this mercury, about 99% of it appears to be entering the river through atmospheric deposition (EPA 2000). Other contaminants include pesticides from agricultural practices, radiocesium (^{137}Cs) and tritium from previous SRS operations, toxic chemicals from lawns, fecal coliform bacteria from domestic animals, and other urban pollutants.

Fish Habitat in Savannah River and Associated Wetlands

Natural flooding regimes are important in organizing "natural" fish communities. For instance, in Missouri, both the stream fish assemblages and the assemblages occurring in the bottomland hardwood wetlands were very different in areas with natural flood regimes and those with flood regimes regulated by dam release (Finger and Stewart 1987). Several of the species that were abundant in the naturally flooded areas were absent from the regulated areas. Unfortunately, we can say little about this with regard to the MSRB as we found no reliable data for the Savannah River collected before completion of the Clarks Hill/J. Strom Thurmond Dam around 1951; probably none exist predating the Stevens Creek Dam (rkm 333) completed in 1914.

Floodplain ponds are important habitats for MSRB fish species such as the pickerels and pirate perch. While these are not, strictly speaking, lotic environments, they are integral parts of swamp and marsh habitats, and when inundated they are part of the river itself. Adult fishes utilize the available habitats to varying degrees. A number of species are either largely confined to the main river channel habitat or use river channels most of the time but use swamp habitat or lower reaches of tributaries as available (Figures 7 and 8; Table 5).

While some of the numbers in Table 5 may be biased by the number of available collections for particular habitats (e.g., *Lepomis marginatus* and *Notropis chalybaeus* are not typical of large river habitat), some species clearly are most prevalent in the river. A smaller number of species are primarily swamp or marsh dwellers that may use the river channel or stream channels periodically. One complicating factor is that neither time of day of capture nor life stage is known for most of the specimens contained in our derived historical database. Ross and Baker (1983) showed that during spring inundations of floodplains in

FIGURE 7. The whitefin shiner, *Cyprinella nivea*, is largely confined to the Savannah River and the lower reaches of tributary streams.

FIGURE 8. The snail bullhead, *Ameiurus brunneus*, is common in the Savannah River.

Mississippi, a number of "flood-exploitative" species were taken frequently on the floodplain in daylight hours but were more common near the channel at night, while "flood-quiescent" species were largely restricted to the channel and were almost never taken on the floodplain.

River and Swamp Spawning Habitat

In general, there are specific guilds of fish that preferentially spawn in channels rather than in backwater habitats such as oxbows, sloughs, "dead rivers," and floodplain pools (Copp 1989). Light trap data from the Tallahatchie River of Mississippi indicated that larval *Lepomis* spp., *Pomoxis* spp., and *Dorosoma* spp., and juvenile *Gambusia affinis* were much more abundant in sloughs and low-current floodplain tributaries, while minnow larvae were more abundant in the higher-current (channel) areas (Turner et al. 1994). The bowfin, *Amia calva*, is one of the most common species caught by subsistence fishermen. In the SRS, larval fish appear to use submerged macrophytes for cover in floodplain braided streams of the Steel Creek Swamp more during daylight hours than during nighttime (Paller 1987).

Seasonal movement into and out of floodplain pools has been documented for a number of species. Flooding of backwater habitats during spawning periods increases year class production of fish that spawn in these habitats, especially largemouth bass (Lambou 1959). In the Kankakee River of Illinois, the following species found in the MSRB moved into and out of backwater habitats during the spring and early summer (the period of most spawning): redfin pickerel, common carp, golden shiner, pugnose minnow, yellow bullhead, tadpole madtom, pirate perch, largemouth bass, green sunfish, bluegill, white crappie, and black crappie (Kwak 1988). Similar seasonal movements have been noted for largemouth bass in the Savannah River (Jones 2001). In the early winter, when water temperatures in the Savannah River Swamp drop below temperatures in the main river channel, largemouth bass and apparently many other species leave portions of the swamp and the entire Steel Creek Inlet and move out into the river. The fish return to the inlet in the spring when this temperature differential disappears; however, extreme drought conditions may reduce water quality in the swamp and inlet, preventing their use.

There are a number of species that either spawn on the inundated floodplain or whose larvae are more frequently captured there when these habitats are sampled. In the bottomlands along the Cache River of Arkansas, the larvae of the following MSRB species were found more frequently in the flooded bottomland forest than in the main channel: gizzard shad, pugnose minnow, tadpole madtom, pirate perch, flier, and black crappie. In addition, spotted sucker and channel catfish larvae were found about as frequently in the flooded bottomland forest as in the main channel (Killgore and Baker 1996).

Seasonal inundations, in addition to furnishing access to feeding areas and spawning sites, also provide highways for fish (Whitehurst 1981), and even species that spend long periods sedentary in swamps may move long

TABLE 5. Fish represented in collections from river and swamp habitats significantly more than would be represented by chance (more than 30% of records from river habitat or more than 15% of records from swamp habitat; species are listed in phylogenetic order).

Scientific name	Common name	Number[1]	River[2]	Swamp[2]	Lake[2]	Stream[2]
Acipenser brevirostrum	shortnose sturgeon	7	100			
Acipenser oxyrinchus	Atlantic sturgeon	7	100			
Lepisosteus osseus	longnose gar	96	68	21		11
Lepisosteus platyrhincus	Florida gar	17	24	76		
Amia calva	bowfin	77	45	27	5	22
Anguilla rostrata	American eel	142	42	10	2	45
Alosa aestivalis	blueback herring	37	92		8	
Alosa mediocris	hickory shad	7	57	43		
Alosa sapidissima	American shad	38	92	8		
Dorosoma cepedianum	gizzard shad	90	74	14	7	4
Dorosoma petenense	threadfin shad	19	68		32	
Cyprinella leedsi	bannerfin shiner	60	88	2	5	5
Cyprinella nivea	whitefin shiner	88	88		2	10
Cyprinus carpio	common carp	10	70	20	10	
Hybognathus regius	eastern silvery minnow	124	94	1	3	2
Hybopsis rubrifrons	rosyface chub	39	87			13
Notemigonus crysoleucas	golden shiner	197	46	10	19	25
Notropis chalybaeus	ironcolor shiner	135	50	13	5	31
Notropis hudsonius	spottail shiner	165	91	1	2	6
Notropis maculatus	taillight shiner	151	77	6	6	11
Notropis petersoni	coastal shiner	210	53	9	3	35
Opsopoeodus emiliae	pugnose minnow	78	82	12	1	5
Carpiodes cyprinus	quillback	11	100			
Carpiodes velifer	highfin carpsucker	3	100			
Erimyzon oblongus	creek chubsucker	111	36	9	3	52
Minytrema melanops	spotted sucker	178	57	11	2	30
Moxostoma collapsum	notchlip redhorse	39	167	9		24
Moxostoma robustum	robust redhorse	2	100			
Scartomyzon sp. cf. *lachneri*	brassy jumprock	2	100			
Ameiurus brunneus	snail bullhead	49	65			35
Ameiurus catus	white catfish	24	67	4	8	21
Ameiurus natalis	yellow bullhead	122	33	13	12	42
Ameiurus nebulosus	brown bullhead	78	54	6	8	32
Ameiurus platycephalus	flat bullhead	120	48	7	18	28
Ictalurus punctatus	channel catfish	40	58	15	2	25
Noturus gyrinus	tadpole madtom	106	40	8	8	44
Pylodictis olivaris	flathead catfish	1	100			
Esox americanus	redfin pickerel	231	38	7	10	44
Esox niger	chain pickerel	209	42	10	12	35
Umbra pygmaea	eastern mudminnow	38	26	21	18	34
Aphredoderus sayanus	pirate perch	227	38	10	4	48
Chologaster cornuta	swampfish	31	22	35		42
Strongylura marina	Atlantic needlefish	9	100			
Fundulus chrysotus	golden topminnow	18	89	11		
Fundulus lineolatus	lined topminnow	108	56	6	10	29
Gambusia holbrooki	eastern mosquitofish	236	45	8	8	38
Labidesthes sicculus	brook silverside	205	58	10	10	21

TABLE 5. (continued)

Scientific name	Common name	Number[1]	River[2]	Swamp[2]	Lake[2]	Stream[2]
Morone americana	white perch	2	50		50	
Morone chrysops	white bass	3	100			
Morone saxatilis	striped bass	5	100			
Elassoma evergladei	Everglades pygmy sunfish	1	100			
Elassoma okatie	bluebarred pygmy sunfish	5	60		40	
Elassoma zonatum	banded pygmy sunfish	100	62	16	1	21
Centrarchus macropterus	flier	85	58	10	8	24
Enneacanthus gloriosus	bluespotted sunfish	98	67	12	5	15
Lepomis auritus	redbreast sunfish	293	44	5	13	37
Lepomis cyanellus	green sunfish	9	78		11	11
Lepomis gibbosus	pumpkinseed	79	86		4	10
Lepomis gulosus	warmouth	203	48	7	18	27
Lepomis macrochirus	bluegill	271	47	7	15	31
Lepomis marginatus	dollar sunfish	185	42	7	22	30
Lepomis microlophus	redear sunfish	113	64	14	4	18
Lepomis punctatus	spotted sunfish	250	38	10	11	42
Micropterus coosae	redeye bass	4	100			
Micropterus salmoides	largemouth bass	250	42	11	15	32
Pomoxis annularis	white crappie	39	79	5	8	8
Pomoxis nigromaculatus	black crappie	135	70	4	16	9
Etheostoma fusiforme	swamp darter	139	66	6	10	17
Etheostoma olmstedi	tesselated darter	205	51	2	2	44
Perca flavescens	yellow perch	116	66	7	15	12
Agonostomus monticola	mountain mullet	2	100			
Mugil cephalus	striped mullet	21	52	43		5
Trinectes maculatus	hogchoker	44	100			

[1] The number of collections in our historical data set in which a given species occurred.
[2] The number is the percentage of records from the given habitat type, rounded to the nearest whole number.

distances (measured in kilometers) during spring floods. For example, tagged white catfish moved 23 km downstream and 61 km upstream during spring and early summer in the Connecticut River (Marcy 1976).

Because sloughs, oxbow lakes, and floodplain ponds are populated with subsets of the adjacent river's fauna, these habitats may serve an important function as refuges in the event of catastrophic stream mortality (Halyk and Balon 1983). On the other hand, floodplain sloughs and ponds may serve as population sinks during droughts, drying up much as large areas of the Everglades do each year, with concomitant heavy mortality of fishes (Carlson and Duever 1977).

Savannah River Tributaries

Because of its unique history as a U.S. Department of Energy reservation, the Savannah River Site (SRS) in Aiken, South Carolina, has been the subject of extensive ecological research and monitoring for more than 50 years. This research has included a number of surveys of SRS streams to monitor the possible impacts of SRS operations on fish assemblages and to bolster our knowledge of the ecology of southeastern Coastal Plain streams. Because of the wealth of fisheries information from SRS streams, because SRS streams are diverse in habitat and ecology, and because the SRS occupies a relatively large area centrally located in the MSRB, most of the following discussion concerning stream fish assemblages is based on data collected on and near the SRS. However, it also includes information from Fort Gordon streams in Georgia, which were the subject of an extensive fisheries survey in 1995 and 1996 (Hoover and Killgore 1999). Fort Gordon is located just below the Fall Line that marks the transition from the Piedmont to the Upper Coastal Plain and thus represents a geographically different part of the MSRB. Another source of information that will soon be available for MSRB stream fish assemblages is the Fisheries Information Network System (FINS), a database being developed by the South Carolina Department of Natural Resources that will include recent and historical fish survey information from the entire state.

Physiography

Much of the SRS lies on the Aiken Plateau, an upland area that reaches a maximum elevation of more than 100 m and is bordered to the southwest by the Savannah River Valley. Tributary streams on the SRS and in much of the MSRB are often parallel, with narrow, elongate drainage basins, although their headwaters are sometimes dendritic. Along the portion of the Savannah River that borders the SRS, tributary streams entering from the east typically have larger drainage basins than those entering from the west (Geomatrix 1993). Tributary stream valleys, especially those of the uplands, are often asymmetrical, with a steep bank on one side (usually the east side) and a gently sloping bank on the other. Examples on the SRS include Upper Three Runs, Lower Three Runs, Tinker Creek, and Fourmile Branch. Sand hill areas of the uplands prevalent in the eastern SRS are less dissected by stream drainages than other areas on the site, possibly because the coarse soils are so permeable that runoff is insufficient to form streams.

Much of the southern boundary of the SRS is defined by the Savannah River, which has cut a wide valley into the Aiken Plateau. The river has gradually migrated southwesterly in this area, often cutting steep, high banks on the Georgia side. As it migrated away from the uplands of the Aiken Plateau and progressively cut more deeply into the valley floor, the river formed a series of steplike

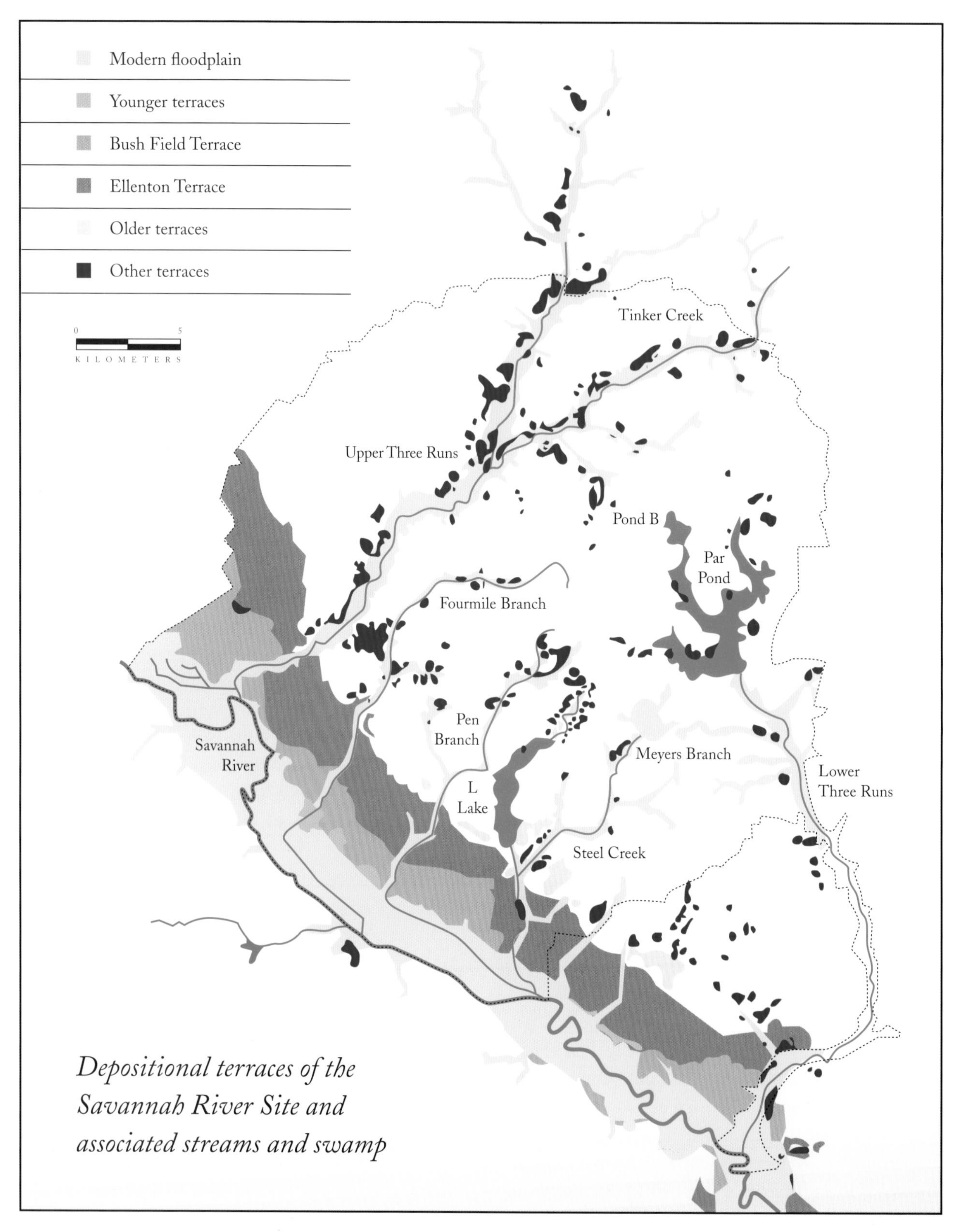

Depositional terraces of the Savannah River Site and associated streams and swamp

MAP 5

fluvial terraces on the South Carolina side (Map 5). Consequently, the oldest terraces occur at higher elevations adjacent to the Aiken Plateau to the northwest, and the youngest occur at lower elevations nearest the current river channel to the southeast. Upstream and downstream of the SRS, paired terraces often exist on east and west sides of the river, suggesting less consistent lateral migration in these areas.

Two terraces are particularly prominent on the SRS, although each is often composed of multiple terrace surfaces differing slightly in elevation (Geomatrix 1993). The upper one, Ellenton Terrace, ranges from about 17 to 25 m above the modern river level and borders the Aiken Plateau to the northeast. All five tributary streams flow across this terrace, which is about as wide as the modern Savannah River floodplain. The Bush Field Terrace, named for Bush Field Airport in Augusta, is lower; elevations there range from about 8 to 13 m above the Savannah River. All of the major SRS tributary streams but Steel Creek cross this terrace. Stream channels are typically highly braided on the Bush Field Terrace and only moderately braided or unbraided on the Ellenton Terrace. Remnants of older terraces have been identified at higher elevations to the west of the Ellenton Terrace, and remnants of younger terraces occur between the Bush Field Terrace and the modern floodplain and on the modern floodplain itself. However, these are difficult to discern and do not have major influences on the SRS tributaries.

Below and to the southwest of the fluvial terraces lies the modern Savannah River floodplain, which is about 2–3 m above the modern river channel and generally shows little topographical relief. The floodplain is generally 2–4 km wide along the SRS and includes much of the Savannah River Swamp, which is contiguous with the river to the east and is periodically inundated when the river rises. Lower portions of the Bush Field Terrace and the remnants of more recent terraces are also sometimes inundated. All five SRS tributaries cross the modern floodplain before joining the Savannah River. Some streams from the Georgia side empty directly into the Savannah River without crossing an extensive floodplain. Near the mouths of these streams, fishes more typical of small streams (e.g., bluehead chubs) are sometimes caught in the mainstream of the Savannah River.

Major Savannah River Site Tributaries

Upper Three Runs, the largest SRS tributary, originates well offsite near Aiken. It has a total length of approximately 40 km, drains about 545 km^2, and has an average discharge near its confluence with the Savannah River of about 6.8 cubic meters per second (cms) (Halverson et al. 1997) (Map 5). The lower 28 km of Upper Three Runs drains the northwestern portion of the SRS. Although it began receiving effluents from the F/H Area Effluent Treatment Facility in 1988 (Wike et al. 1994), the portion of Upper Three Runs on the SRS is considered relatively undisturbed and is known for its exceptional aquatic insect diversity (Morse et al. 1980, 1983). Upper Three Runs has two important tributaries: Tinker Creek, a comparatively large and mostly undisturbed stream that drains the north-central SRS, and Tims Branch, a smaller stream that was contaminated during earlier SRS operations (Halverson et al. 1997). Tinker Creek in turn has several smaller tributaries, including Mill Creek, Reedy Branch, Crouch Branch, and McQueens Branch. The former two are largely undisturbed while the latter two have been affected to varying degrees by industrial operations in their headwaters. Portions of Upper Three Runs and Tinker Creek are among the more deeply incised streams on the SRS as a result of long-term geological uplift of the surrounding area (Geomatrix 1993).

Fourmile Branch is located to the east of Upper Three Runs and originates near the center of the SRS (Map 5). It is about 24 km long, has a watershed of approximately 57 km^2, and has an average discharge in its lower reaches of about 1 cms. Fourmile Branch received heated reactor cooling water from 1955 to 1985 and still receives low concentrations of various contaminants from seepage basins located in its headwaters. The heated water, discharged into Fourmile Branch from C-Reactor via Castor Creek, killed riparian vegetation and most of the stream fauna, and reshaped the stream channel by eroding sediments from the upper and middle reaches and depositing them further downstream, especially where Fourmile Branch enters the Savannah River floodplain. Fourmile Branch is now undergoing secondary succession as riparian vegetation becomes reestablished in the stream corridor.

Pen Branch, to the east of Fourmile Branch, flows southwesterly for about 24 km and drains an area of approximately 55 km^2 (Halverson et al. 1997) (Map 5). Its average discharge between 1990 and 2000 was about 1.7 cms. Unlike other SRS streams, Pen Branch does not discharge directly into the Savannah River but instead empties into the Savannah River Swamp, where it flows through various channels parallel to the Savannah River before joining the lower reaches of Steel Creek. Like Fourmile Branch, Pen Branch received large volumes of heated nuclear reactor cooling water from 1954 to 1988 that scoured its stream channel and killed riparian vegetation. This water entered Indian Grave Branch before discharging into the mid-reaches of Pen Branch, leaving the headwaters of Pen Branch largely undisturbed. Portions of Pen Branch and the Savannah River floodplain swamp that received heated water began to recover when K-Reactor was shut down in 1988. Like Fourmile Branch, they are currently undergoing secondary succession and are characterized by a more open canopy and a denser growth of aquatic macrophytes than undisturbed streams. An experimental stream restoration has been conducted on the lower portion of Pen Branch (Nelson et al. 2000).

Steel Creek, near the eastern side of the SRS, was impounded in its upper reaches to make L Lake, a 418 ha reservoir formerly used to cool L-Reactor. Steel Creek flows about 3 km before entering L Lake (Halverson et al. 1997). From the L Lake dam, it flows another 5 km before entering the Savannah River floodplain swamp and an additional 3 km before entering the Savannah River. At base flows, waters from the swamp and Steel Creek coalesce into a single, large 1.8 km channel that joins the river. A 10-km-long tributary, Meyers Branch, joins Steel Creek below the L Lake dam (Figure 9). The total watershed of Steel Creek and Meyers Branch is about 91 km^2, and the average 1990–2000 discharge from Steel Creek into the Savannah River was about 2.2 cms. Steel Creek received heated cooling water from L-Reactor from 1954 to 1968, resulting in the same type of damage that occurred in Fourmile Branch and Pen Branch. Recovery of the stream began in 1968 following the cessation of L-Reactor operations. L Lake was constructed in 1985 to protect Steel Creek from further thermal impacts when L-Reactor was restarted in 1986. L-Reactor again ceased operation in 1988. Steel Creek is currently recovering like the other reactor cooling streams, although it is in a more advanced state of secondary succession because its recovery period has been longer. Water temperatures of Steel Creek are strongly affected by releases from the L Lake dam, although downstream sections are buffered to some degree by inflow from Meyers Branch. Apart from selective logging in the riparian zone during the 1940s (Giese et al. 2000), Meyers Branch has remained largely undisturbed.

FIGURE 9. An upper reach of Meyers Branch, an undisturbed headwater stream.

The easternmost and second largest stream on the SRS is Lower Three Runs. This stream, like Steel Creek, was impounded in its headwaters to produce Par Pond, a reactor cooling reservoir. Lower Three Runs travels about 39 km between the Par Pond dam and its confluence with the Savannah River and drains about 460 km^2 (Halverson et al. 1997). From 1990 to 2000 its average discharge was about 3.9 cms. Unlike Steel Creek, Lower Three Runs never directly received heated reactor cooling water because Par Pond was completed before reactor operations started. Also, unlike other SRS streams, most of Lower Three Runs is located outside the main body of the SRS within a narrow corridor of land owned by the U.S. Department of Energy.

Several streams near the SRS have also been sampled as part of SRS-related fisheries programs. These include three offsite tributaries of Lower Three Runs: Furse Mill, Gantts Mill, and Miller Creek; Cedar Creek, a tributary of Upper Three Runs; two tributaries of the Salkehatchie River: Rosemary Creek and Buck Creek; and Hollow Creek, a direct tributary of the Savannah River. The most significant of these is Hollow Creek, a fairly large blackwater stream located to the west of Upper Three Runs. The fish assemblages and habitats supported by these streams are generally similar to those in SRS streams.

Four streams on Fort Gordon in Georgia have been surveyed in considerable detail (Hoover and Killgore 1999): Spirit Creek, Sandy Run, Boggy Gut, and Brier Creek. Spirit Creek is a 27-km-long tributary of the Savannah River that has been affected by erosion in its headwaters, the discharge of treated sewage, and impoundment to form several ponds. Sandy Run (17 km long) and Boggy Gut (20 km long) are tributaries of Brier Creek (115 km long), the largest Savannah River tributary in the MSRB. Sandy Run is impounded in several places, but the areas surveyed for fishes in the two other streams were largely unimpacted (Hoover and Killgore 1999).

Horse Creek, in Aiken County, South Carolina, north of the SRS, represents another major drainage in the MSRB. During the past century it was heavily polluted by textile mill wastes. Recently, however, this stream has been recovering as a result of improved waste treatment.

Habitat

Our descriptions of MSRB stream habitats will focus on the SRS streams from which the most habitat data are available. SRS streams vary in width from less than 2 m in small first- and second-order streams to more than 15 m in the lower portion of Upper Three Runs. Average depths vary from several centimeters to more than 1 m. Elevation gradients are relatively low (0.6–3.0 m/km), and currents are slow to moderate (from several to 30 or 40 cm/s). Most streams consist of alternating pools and runs or glides. Pools are mostly lateral scour pools created by deflection of water by trees or stumps along the bank, but debris pools occur as well. Plunge pools created by root dams growing across the channel may be found in the smallest headwaters. Although rare on the SRS, a few shallow riffles, generally with a cobble substrate, exist in several streams. Streams typically have sandy substrates, sometimes with gravel or pebbles in areas of more rapid flow. Substantial amounts of silt may be found in slow-flow areas such as wider, deeper pools and backwaters.

Because most streams in the MSRB traverse bottomland hardwood forests, they are littered with sunken logs, stumps, branches, twigs, and leaves. This material provides both habitat for the stream-dwelling invertebrates eaten by fish (Benke et al. 1991; Braccia and Batzer 2001) and shelter for the fishes themselves (Benke and Wallace 1990; Lobb and Orth 1991). Also, wood debris, shoreline trees, and stumps are important in shaping stream channels. Smaller undisturbed streams on the SRS are often 80–90% shaded by trees that line their banks (Table 6). In contrast, the trees that lined the banks of streams that formerly received reactor cooling water were killed by high temperatures. The banks of these streams were colonized by willows and shrubs following reactor shutdown (by 1986 or earlier), but they have not yet produced the shading characteristic of undisturbed streams (Fletcher et al. 2000). The sunlight that reaches the surface of these "post-thermal" streams has produced considerable emergent and submerged aquatic vegetation (Table 6) that has a substantial effect on ecological functioning because it alters the source of primary productivity from allocthonous (coming from outside the stream as imported organic debris), as is typical of most undisturbed streams, to autochthonous (produced within the stream) (Lakly and McArthur 2000). It also produces differences in physical habitat structure that influence the species composition of fishes and invertebrates.

The water in most SRS streams is well oxygenated, mildly acidic, and of moderate conductivity (a measure of ionic strength; see Table 6). Upper Three Runs has the lowest pHs and conductivities of any SRS stream. It is a "blackwater" stream with relatively clear but tea-colored water that is strongly influenced by the leaching of organic acids from decomposing plant material in the watershed. Its low conductivity indicates the presence of low concentrations of most dissolved ions. Hollow Creek, located just to the west of Upper Three Runs, has similar water quality. Other SRS streams share these characteristics to varying degrees, although several (e.g., Lower

TABLE 6. Chemical and physical measurements for typical Savannah River Site streams, listed from first order through fourth order.

Stream	Stream order	Type of disturbance[1]	Average width (m)	Average depth (cm)	Average current (m/s)	Percentage macrophytes[2]	Percentage tree canopy[2]	Dissolved oxygen (mg/l)	pH	Conductivity (µmhos/cm)
Crouch Branch	1	O	1.7	6	0.4	0	90	10	6.8	70
McQueens Branch	1	U	2.0	12	0.2	1	89	10	6.5	50
Fourmile Branch (upper)	2	O	3.8	35	0.1	12	73	9	5.2	30
Steel Creek (above L Lake)	2	O	3.3	16	0.3	1	72	10	6.5	40
Reedy Branch	2	U	2.0	15	0.3	7	82	—	—	—
Tims Branch	2	O	2.0	14	0.2	1	88	12	6.8	60
Pen Branch (upper) site 1	2	U	1.7	13	0.0	0	98	11	6.4	40
Pen Branch (upper) site 2	2	U	4.0	11	0.1	0	86	10	6.8	60
Meyers Branch (upper)	2	U	3.2	16	0.3	0	90	9	6.4	50
Mill Creek	2	U	3.7	22	0.2	1	87	11	6.8	60
Fourmile Branch (middle)	3	O	5.3	40	0.3	14	68	11	6.5	70
Fourmile Branch (lower)	3	R	5.5	56	0.3	8	36	10	6.6	70
Pen Branch (lower)	3	R	9.4	37	0.1	59	38	10	6.7	60
Meyers Branch (lower)	3	U	4.9	24	0.2	0	81	—	—	—
Steel Creek (lower)	3	R	9.5	47	0.3	25	34	—	—	—
Lower Three Runs (near Par Pond)	3	O	7.7	62	0.4	1	62	—	—	—
Upper Three Runs (middle)	4	U	11.9	46	0.4	24	49	11	5.7	20
Upper Three Runs (middle)	4	U	13.9	44	0.2	19	20	—	—	—
Upper Three Runs (lower)	4	U	15.6	107	0.4	0	73	12	5.9	30

[1] U = undisturbed stream, R = a recovering stream that formerly received heated reactor water, O = other type of disturbance.

[2] Percentage coverage of the stream bottom area (with macrophytes) or stream surface area (by tree canopy).

Three Runs, Crouch Branch) are also influenced by the introduction of Savannah River water added to meet legally mandated minimum flow requirements or by discharges from various industrial areas. These streams typically have somewhat higher conductivities and pHs than undisturbed streams. Within Fort Gordon, Georgia, Sandy Run and Boggy Gut are blackwater streams with relatively clear, acidic (pH 5.1–5.5) water of low conductivity (19.6–20.3 μSiemen/cm). Spirit Creek and Brier Creek are more turbid, slightly acidic (pH 6.4), and have higher conductivities (100–107 μSiemen/cm).

Relationship between Habitat and Fish Assemblages

SRS streams are inhabited by a varied assemblage of fishes totaling more than 60 species; 44 species have been collected from the four streams on Fort Gordon (Table 7). Generally, species abundance in SRS and Fort Gordon streams follows a log-normal pattern of distribution (Hoover and Killgore 1999). In such distributions, which are generally typical of undisturbed habitats, most species are intermediate in abundance and relatively few are very numerous or very rare. Deviations from the log-normal pattern occasionally occur at sites dominated by large schools of minnows or at highly disturbed sites dominated by one or two taxa.

The species found in MSRB streams are partitioned into distinguishable but overlapping assemblages on the basis of habitat, which can be considered on three physical scales: watershed, mesohabitat, and microhabitat.

On a watershed scale, fish assemblages typically exhibit longitudinal zonation as streams progress from small, shallow headwaters to wider, deeper reaches downstream (Rahel and Hubert 1991). The number of fish species typically increases with stream size, although it may plateau as stream order increases above fourth, fifth, or sixth (Horwitz 1978). Larger streams typically support more large piscivorous fish such as largemouth bass and chain pickerel and more large benthic insectivorous fish such as spotted suckers than do smaller streams, which typically support more surface or midwater insectivores and general insectivores such as most minnows, shiners, and small sunfishes (Horwitz 1978; Schlosser 1987). These changes in fish assemblage composition occur with progression downstream as species are added and replaced. Research on SRS streams ranging from 1-m-wide McQueens Branch to 15-m-wide Upper Three Runs has shown that this pattern holds true in SRS streams as well (Paller 1994). However, SRS headwater streams support more species than headwater streams in many other parts of the country, possibly because of the region's comparatively mild climate and lack of steep elevation gradients. Because runoff into streams is buffered by the generally sandy soils, discharges also tend to be more stable than those in many other geographic areas.

Fish assemblage diversity and composition are also affected by habitat changes that occur on the mesohabitat scale—that is, the scale of individual pools, runs, and riffles. The composition of stream fish assemblages on the SRS is strongly influenced by gradients of current velocity, stream width, and stream depth (Meffe and Sheldon 1988). Pools provide deeper, slower-moving water and often have a fine sand or silty bottom littered with submerged logs, branches, leaves, and other woody debris. Runs provide shallower, swifter water and coarser sand or, less commonly, gravel bottoms frequently littered with larger wood debris such as sunken logs and branches. Although many species may opportunistically use more than one type of mesohabitat, there are fairly distinct assemblages associated with particular mesohabitats. For example, adult spotted sunfish, redbreast sunfish, and dusky shiners typically occur in slower-flowing pools with deeper water and fine sand or silty bottoms. Pools in larger streams may also support largemouth bass, spotted suckers, and longnose gar. More rapidly flowing and shallower runs are often occupied by fewer species, although they are the preferred habitat of some, such as the northern hogsucker, margined madtom, and Savannah darter. Preference for mesohabitat type may change with age; small redbreast sunfish often occur in shallow areas while adults prefer relatively deep pools (Meffe and Sheldon 1988). Similarly, dusky shiners spawn in relatively still waters, and their young remain there until they are large enough to join the adults in eddies of swifter water (Fletcher 1993).

The smallest habitat scale is the microhabitat, the "fish eye" view of local differences in aquatic habitat. Some microhabitat features that can affect fish distribution include depth, position in the water column (bottom, middepth, near the surface), current velocity, substrate size and composition, distance from the bank, amount and

TABLE 7. Fish species that have been collected from streams on the Savannah River Site (SRS), South Carolina, and Fort Gordon (FG), Georgia (listed in phylogenetic order).

Scientific name	Common name	SRS	FG[1]
Lepisosteus osseus	longnose gar	X	
Lepisosteus platyrhincus	Florida gar	X	
Amia calva	bowfin	X	
Anguilla rostrata	American eel	X	
Alosa aestivalis	blueback herring	X	
Alosa sapidissima	American shad	X	
Ctenopharyngodon idella	grass carp		X
Cyprinella nivea	whitefin shiner	X	
Hybognathus regius	eastern silvery minnow	X	
Hybopsis rubrifrons	rosyface chub	X	
Nocomis leptocephalus	bluehead chub	X	X
Notemigonus crysoleucas	golden shiner	X	X
Notropis chalybaeus	ironcolor shiner	X	X
Notropis cummingsae	dusky shiner	X	X
Notropis hudsonius	spottail shiner	X	
Notropis lutipinnis	yellowfin shiner	X	X
Notropis maculatus	taillight shiner	X	
Notropis petersoni	coastal shiner	X	X
Opsopoeodus emiliae	pugnose minnow		X
Pteronotropis stonei	lowland shiner	X	X
Semotilus atromaculatus	creek chub	X	
Erimyzon oblongus	creek chubsucker	X	X
Erimyzon sucetta	lake chubsucker	X	X
Hypentelium nigricans	northern hogsucker	X	
Minytrema melanops	spotted sucker	X	X
Ameiurus brunneus	snail bullhead	X	X
Ameiurus catus	white catfish	X	
Ameiurus natalis	yellow bullhead	X	X
Ameiurus nebulosus	brown bullhead	X	X
Ameiurus platycephalus	flat bullhead	X	X
Ictalurus punctatus	channel catfish	X	
Noturus gyrinus	tadpole madtom	X	X
Noturus insignis	margined madtom	X	
Noturus leptacanthus	speckled madtom	X	X
Esox americanus	redfin pickerel	X	X
Esox niger	chain pickerel	X	X
Umbra pygmaea	eastern mudminnow	X	
Aphredoderus sayanus	pirate perch	X	X
Chologaster cornuta	swampfish		X
Fundulus chrysotus	golden topminnow	X	
Fundulus lineolatus	lined topminnow	X	X
Gambusia holbrooki	eastern mosquitofish	X	X
Labidesthes sicculus	brook silverside	X	X
Morone saxatilis	striped bass	X	
Elassoma evergladei	Everglades pygmy sunfish		X
Elassoma okatie	bluebarred pygmy sunfish		X
Elassoma zonatum	banded pygmy sunfish	X	X
Acantharchus pomotis	mud sunfish	X	X

TABLE 7. (continued)

Scientific name	Common name	SRS	FG[1]
Centrarchus macropterus	flier	X	
Enneacanthus chaetodon	blackbanded sunfish	X	
Enneacanthus gloriosus	bluespotted sunfish	X	X
Enneacanthus obesus	banded sunfish	X	
Lepomis auritus	redbreast sunfish	X	X
Lepomis gibbosus	pumpkinseed	X	
Lepomis gulosus	warmouth	X	X
Lepomis macrochirus	bluegill	X	X
Lepomis marginatus	dollar sunfish	X	X
Lepomis microlophus	redear sunfish	X	X
Lepomis punctatus	spotted sunfish	X	X
Micropterus coosae	redeye bass		X
Micropterus salmoides	largemouth bass	X	X
Pomoxis annularis	white crappie	X	
Pomoxis nigromaculatus	black crappie	X	
Etheostoma fricksium	Savannah darter	X	X
Etheostoma fusiforme	swamp darter		X
Etheostoma hopkinsi	Christmas darter	X	
Etheostoma inscriptum	turquoise darter	X	X
Etheostoma olmstedi	tesselated darter	X	X
Etheostoma serrifer	sawcheek darter	X	X
Perca flavescens	yellow perch	X	
Percina nigrofasciata	blackbanded darter	X	X

[1] Fort Gordon data are from Hoover and Killgore 1999.

type of in-stream cover, and overhanging bank vegetation (Lobb and Orth 1991; Grossman and Ratajczak 1998). Important microhabitat differences can occur over a distance of centimeters, especially for small fishes. Fish are flexible in their use of microhabitats, and microhabitat preference changes with age (Lobb and Orth 1991; Grossman and Ratajczak 1998). The microhabitat preferences of fish have not been extensively studied on the SRS or in other MSRB streams. However, the complex channel morphometry of these streams combined with the abundant and complex in-stream structure provided by woody debris indicate the presence of a variety of potential microhabitats that can support many species.

In any habitat with a number of interacting species there is the possibility of symbiotic species associations, and several such relationships have been observed in SRS streams. It is not unusual to observe schools of spawning yellowfin shiners hovering over the pebble mounds constructed by male bluehead chubs as nests for their own eggs and larvae (Figure 10). Eggs deposited by shiners in the bluehead chub nests benefit from both the parental care provided by the chub and the shelter provided by the nest (Wallin 1992). The relationship is obligatory on the part of the shiners, which apparently fail to reproduce in the absence of bluehead chubs. It may also be mutualistic in the sense that the early life stages of both species may benefit from reduced predation due to a dilution effect. A less benign relationship exists between the dusky shiner and the redbreast sunfish. Dusky shiner adults spawn in redbreast sunfish nests and feed voraciously on redbreast sunfish eggs and larvae (Fletcher 1993). Other minnow species of the MSRB known to spawn on nests of other species include taillight shiners, golden shiners, and rosyface chubs (see species accounts for details).

Typical Species Assemblages

Stream size is a major determinant of fish assemblage structure in the MSRB. However, other factors that influence assemblage composition, such as gradient, po-

FIGURE 10. Bluehead chub, *Nocomis leptocephalus*, with an aggregation of yellowfin shiners, *Notropis lutipinnis*.

sition in watershed, and anthropogenic disturbance, are considered as well. Although it is influenced by basin shape, stream size is generally strongly correlated with stream order, the relative rank of a stream channel segment in a drainage network.

First- and second-order streams are considered headwaters. These relatively small, shallow streams can be efficiently sampled by electrofishing, making it possible to obtain a relatively accurate and complete list of the species present and their abundance (Paller 1995). SRS studies indicate that such streams are populated primarily by small fishes including minnows (Cyprinidae), bullheads and madtoms (Ictaluridae), sunfishes (Centrarchidae), and darters (Percidae) (Meffe and Sheldon 1988; Paller 1994) (see Table 8). Among the most common species in a typical relatively undisturbed headwater stream with moderate flow and sandy bottom are yellowfin shiner, bluehead chub, and pirate perch. Others include creek chub, dollar sunfish, spotted sunfish, redbreast sunfish, dusky shiner, tesselated darter, yellow bullhead, speckled madtom, margined madtom, creek chubsucker, and redfin pickerel.

Very-low-gradient headwater streams with slow currents and silty bottoms may depart somewhat from this pattern by having fewer yellowfin shiners and bluehead chubs and more pirate perch, creek chubsuckers, and redfin pickerels. They may also support mud sunfish and pygmy sunfishes. Sampling a typical 100–200 m stream segment in undisturbed first- and second-order SRS streams may yield anything from about 5 species in very shallow, narrow headwaters to 15–20 species in larger second-order habitats (Paller 1995). Examples of undisturbed first- and second-order streams on the SRS include the upper reaches of Pen Branch, the upper reaches of Meyers Branch, Reedy Branch, and Mill Creek. These streams are valuable resources because they support comparatively intact assemblages of headwater species that differ from those in larger streams.

Several headwater streams on the SRS have been affected by industrial discharges or runoff (Specht and

TABLE 8. Changes in the abundance of some common species of fish with changes in stream order on the Savannah River Site (figures are percentages).

Scientific name	Common name	Orders 1 & 2	Order 3	Order 4
Surface water insectivores				
Notropis lutipinnis	yellowfin shiner	28.0	17.8	7.7
Notropis cummingsae	dusky shiner	5.4	6.1	15.0
Notropis petersoni	coastal shiner	1.4	3.3	4.1
Pteronotropis stonei	lowland shiner	1.3	0.6	5.7
Generalized insectivores				
Aphredoderus sayanus	pirate perch	23.1	9.4	3.1
Nocomis leptocephalus	bluehead chub	8.7	7.1	0.4
Lepomis auritus	redbreast sunfish	3.8	7.3	7.1
Lepomis punctatus	spotted sunfish	2.1	7.5	4.4
Semotilus atromaculatus	creek chub	1.9	0.2	0.0
Lepomis marginatus	dollar sunfish	1.7	1.2	0.4
Lepomis macrochirus	bluegill	0.8	1.1	2.9
Benthic insectivores				
Noturus sp.	madtoms	3.2	2.1	0.9
Erimyzon oblongus	creek chubsucker	2.7	1.9	2.5
Ameiurus natalis	yellow bullhead	2.2	1.7	0.3
Etheostoma olmstedi	tesselated darter	1.7	2.2	0.2
Percina nigrofasciata	blackbanded darter	0.6	1.8	1.9
Minytrema melanops	spotted sucker	0.0	0.5	12.4
Insectivores-piscivores				
Esox americanus	redfin pickerel	3.0	3.0	1.1
Anguilla rostrata	American eel	0.6	2.5	6.3
Lepomis gulosus	warmouth	0.5	0.7	0.8
Micropterus salmoides	largemouth bass	0.4	1.2	7.1
Esox niger	chain pickerel	<0.1	1.2	2.5

Source: Data are from Paller 1994.

Paller 2001); examples include Crouch Branch and possibly the upper reaches of McQueens Branch. Other streams, such as portions of Tims Branch, may be affected by the low dissolved oxygen in water discharged from relatively stagnant beaver impoundments in the summer. The fish assemblages in these streams exhibit symptoms of degradation that may include reduced species richness, loss of sensitive species such as darters and madtoms, increased relative abundance of tolerant species such as bullheads and mosquitofish, changes in the relative proportions of different trophic groups, and reductions in overall fish abundance (Paller et al. 1996). Comparisons of variables such as these between disturbed and undisturbed streams can serve as the basis for multimetric indexes such as the Index of Biotic Integrity that are useful tools for evaluating stream health (Karr et al. 1986).

Humans are not the only organisms that have altered the habitat in SRS headwater streams. The SRS supports a substantial population of beavers that have created numerous impoundments in headwater streams as well as in the side channels of higher-order streams. The more lentic environment of impounded reaches supports a different assemblage of fishes than is characteristic of flowing waters (Snodgrass and Meffe 1998) (Figure 11). Large beaver ponds, in particular, favor a shift from lotic to lentic species and replacement of small-bodied minnows by larger predators. Beaver ponds may also serve as breed-

FIGURE 11. Blackbanded sunfish, *Enneacanthus chaetodon*, are frequently encountered in upland beaver ponds.

ing and nursery areas for stream fishes (Snodgrass and Meffe 1999). As beaver ponds age and habitats change through secondary succession, the species composition shifts. The overall effect of beaver ponds of various ages is to increase species richness of the stream reach.

The middle and lower reaches of SRS streams support somewhat different species assemblages than those found in the headwaters. The greater depth and habitat space provided by third-, fourth-, and fifth-order streams permits occupation by larger fish including largemouth bass, spotted sucker, redbreast sunfish, chain pickerel, and large American eel (Paller 1994) (Table 8). Other species common in these streams include pirate perch, spotted sunfish, blackbanded darter, flat bullhead, and a variety of minnows including dusky shiner, coastal shiner, ironcolor shiner, yellowfin shiner, eastern silvery minnow, and spottail shiner. Lowland shiners and speckled madtoms may be locally abundant in areas with higher current velocities. Lower-gradient streams may support warmouth, bluespotted sunfish, and lined topminnow. Bluegills are more common in larger streams, especially downstream from impoundments or other lentic habitats that serve as a source of immigrants. Sampling a typical 100–200 m reach in a larger SRS stream will often produce 20–30 species (Paller 1995). Examples of undisturbed third- and fourth-order streams on the SRS include Tinker Creek and Upper Three Runs (Figure 13).

Deeper pools in the lower reaches of SRS streams sometimes hold surprisingly large fish. Longnose gar approaching 1 m have been collected from pools in Pen Branch and Fourmile Branch on the Bush Field Terrace and in Steel Creek below L Lake. Large bowfins, channel catfish, and largemouth bass have been collected from Upper Three Runs and the lower reaches of Steel Creek. These fish may move between the creeks and the Savannah River in search of food, suitable temperatures, and spawning sites (Jones 2001). Perhaps most surprising is the periodic occurrence of striped bass exceeding 5–10 kg in the lower portion of Upper Three Runs. During the summer, this species enters Upper Three Runs to take advantage of the relatively cool temperatures produced by inflowing groundwater. Areas such as this provide critical thermal refuges for adult striped bass, which return to the Savannah River when river temperatures cool later in the year. Other species typical of the Savannah River also commonly invade lower portions of SRS streams, especially reaches on the modern Savannah River floodplain and the Bush Field Terrace.

The lower reaches of larger Savannah River tributaries also serve as spawning sites for anadromous fishes. Blueback herring and American shad eggs and larvae have been collected at the mouths of some of the larger MSRB Savannah River tributaries (Paller et al. 1986b). The early life stages of other taxa such as *Notropis* spp., *Lepomis* spp., *Erimyzon* spp., *Etheostoma* spp., *Pomoxis* spp., and *Dorosoma* spp. have also been collected from these habitats. These larvae may be found in high densities in macrophyte beds and other protected areas in the lower stream reaches (Paller 1987). Some are washed into the Savannah River, where they contribute to the ichthyoplankton assemblage in the river itself.

FIGURE 12. Middle reach of Meyers Branch.

FIGURE 13. Lower reach of Upper Three Runs, an undisturbed stream, on a fluvial terrace of the Savannah River.

Thermal Effluent Impacts

The middle and lower reaches of several SRS streams were strongly affected by the discharge of hot cooling water from nuclear reactors up until 1988. The affected streams included Fourmile Branch, Steel Creek, Indian Grave Branch, and the middle and lower reaches of Pen Branch. Discharge temperatures and volumes were exceedingly high (up to 70 °C and 11.3 m^3/s), resulting in the elimination of fish and aquatic invertebrates from the main channels, the death of riparian vegetation, and extensive scouring and erosion of the streambed (Figure 14). The only fish found in these streams during reactor operations were mosquitofish and a few sunfishes, which were confined to side channels and backwaters where the water had cooled enough to permit survival (Paller and Saul 1986). During periods of reactor shutdown, fish from the Savannah River would enter reactor cooling streams, sometimes in large numbers, setting the stage for fish kills when operations were resumed (Aho et al. 1986a). Direct discharge of high-temperature reactor cooling water began in 1954–1955 and ceased in 1968 to Steel Creek, in 1985 to Fourmile Branch, and in 1988 to Pen Branch.

FIGURE 14. Thermal stream on the SRS during plant operation.

FIGURE 15. Lower reach of Fourmile Branch on a fluvial terrace of the Savannah River. This stream formerly received heated effluent and was highly disturbed.

Fish rapidly recolonized the cooling-water-receiving streams after the reactors were shut down (Aho et al. 1986a). The fish assemblages in these "postthermal" streams initially consisted mostly of larger species such as spotted sucker, redbreast sunfish, largemouth bass, and longnose gar, possibly because their greater mobility facilitated recolonization and because they preferred the relatively deep pools and runs scoured by the reactor flows.

With time, secondary succession has brought numerous changes to the postthermal streams, including the development of riparian vegetation (primarily herbaceous plants, willows, and shrubs), the growth of emergent and submerged aquatic plants, changes in channel width and depth, and accumulation of smaller woody debris (Figure 15) (Fletcher et al. 2000). Because the canopy over postthermal streams is relatively open, aquatic macrophytes have proliferated, resulting in an increase in autochthonous production. The switch to shrub/willow vegetation, the erosion of forest soils, and the deposition of alluvial sands have also altered the quantity and quality of forest floor organic matter in the riparian zone, with subsequent effects on allochthonous inputs (Wigginton et al. 2000; Kolka et al. 2002). Evapotranspiration rates and water table levels also differ from those of undisturbed systems (Kolka et al. 2000, 2002).

As a result of secondary succession, stream reaches in postthermal areas on the Bush Field Terrace—and to a lesser degree, the Ellenton Terrace—have become physically complex with a high level of habitat heterogeneity. They often consist of one or two rather wide, deep, and rapidly flowing main channels as well as small, shallow, often slowly flowing or still side channels. Numerous fish species occupy these disturbed habitats, and the postthermal streams often have greater species richness now than the undisturbed streams have (Paller et al. 2000). Electrofishing samples collected from undisturbed sites in lower Meyers Branch and Upper Three Runs and postthermal sites in lower Pen Branch and Fourmile Branch indicated greater average numbers of fish species and fish abundance in the postthermal streams (Paller et al. 2000) (Table 9). Species that were particularly abundant at the postthermal sites included dusky shiner, spotted sunfish, redbreast sunfish, spotted sucker, and creek chubsucker. Smaller side channels in the postthermal sites supported large numbers of eastern mosquitofish, dollar sunfish, redfin pickerel, and pirate perch and served as seasonal nursery areas for the young of other species. This abundance of fishes is presumably related to the high productivity and wide variety of micro- and mesohabitats in the postthermal streams, although the functional ecology of such streams has not been studied in detail. And while these habitats currently support species-rich fish assemblages, their recovery will not be complete until an intact canopy of mature bottomland trees has been restored.

Contaminants

Steel Creek, Lower Three Runs, and Fourmile Branch were contaminated with radiocesium (^{137}Cs) as a result of accidental releases that occurred during the 1960s and

TABLE 9. Relative density of fish (no./100 m^2) in streams recovering from thermal impacts (Pen Branch and Fourmile Branch) and in undisturbed streams (Meyers Branch and Upper Three Runs) on the SRS (species are listed in phylogenetic order).

		Recovering streams		Undisturbed streams	
Scientific name	Common name	Main channel	Side channel	Main channel	Side channel
Lepisosteus platyrhincus	Florida gar	0.0	0.0	0.0	0.0
Amia calva	bowfin	0.0	0.0	0.0	0.0
Anguilla rostrata	American eel	0.3	1.2	1.3	0.8
Cyprinella nivea	whitefin shiner	0.4	0.6	0.0	0.0
Hybognathus regius	eastern silvery minnow	0.5	0.0	0.0	0.0
Hybopsis rubrifrons	rosyface chub	0.2	0.0	0.0	0.0
Nocomis leptocephalus	bluehead chub	0.0	0.3	0.1	0.1
Notemigonus crysoleucas	golden shiner	0.2	0.6	0.0	0.0
Notropis chalybaeus	ironcolor shiner	0.0	1.9	0.2	0.0
Notropis cummingsae	dusky shiner	35.7	77.4	2.2	4.9
Notropis lutipinnis	yellowfin shiner	7.8	21.7	7.7	6.0
Notropis petersoni	coastal shiner	2.7	6.8	0.7	5.0
Pteronotropis stonei	lowland shiner	0.0	0.0	0.4	2.5
Semotilus atromaculatus	creek chub	0.0	0.0	0.0	0.2
Erimyzon oblongus	creek chubsucker	1.6	6.8	0.2	0.3
Erimyzon sucetta	lake chubsucker	0.5	1.6	0.0	0.0
Hypentelium nigricans	northern hogsucker	0.0	0.0	0.2	0.0
Minytrema melanops	spotted sucker	2.7	1.6	0.0	0.0
Ameiurus natalis	yellow bullhead	0.3	0.7	0.0	0.5
Ameiurus platycephalus	flat bullhead	0.2	0.0	0.0	0.0
Noturus gyrinus	tadpole madtom	0.1	0.7	0.0	0.4
Noturus insignis	margined madtom	0.0	0.0	0.1	0.2
Noturus leptacanthus	speckled madtom	1.1	6.1	1.2	3.2
Esox americanus	redfin pickerel	0.1	1.0	0.1	1.8
Esox niger	chain pickerel	0.0	0.0	0.1	0.0
Umbra pygmaea	eastern mudminnow	0.0	0.1	0.0	0.0
Aphredoderus sayanus	pirate perch	1.9	4.4	1.4	9.7
Fundulus lineolatus	lined topminnow	0.0	0.1	0.0	0.0
Gambusia holbrooki	eastern mosquitofish	2.7	6.0	0.1	0.7
Labidesthes sicculus	brook silverside	0.1	0.1	0.0	0.0
Enneacanthus gloriosus	bluespotted sunfish	0.1	0.4	0.0	0.7
Lepomis auritus	redbreast sunfish	1.6	7.1	0.1	0.3
Lepomis gulosus	warmouth	0.1	0.4	0.0	1.1
Lepomis marginatus	dollar sunfish	0.1	2.4	0.0	2.2
Lepomis punctatus	spotted sunfish	5.7	10.4	0.3	0.9
Micropterus salmoides	largemouth bass	0.8	1.9	0.5	0.0
Etheostoma fricksium	Savannah darter	0.0	0.0	0.4	0.1
Etheostoma olmstedi	tesselated darter	0.3	3.0	1.3	5.0
Etheostoma serrifer	sawcheek darter	0.0	0.1	0.0	0.0
Percina nigrofasciata	blackbanded darter	1.2	7.7	1.0	0.1

early 1970s. Extensive surveys demonstrated that this radionuclide bioaccumulated in fish, reaching concentrations significantly higher than those found in the water (Paller et al. 1999). Concentrations were especially high in predators as a result of food chain bioaccumulation. Since the 1970s, radiocesium concentrations in fish have declined strongly, with ecological half-lives ranging from 3.2 to 16.7 years, as a result of physical, chemical, and biological processes that remove ^{137}Cs from aquatic ecosystems (Paller et al. 1999). Other radiological contaminants, notably strontium, were also released into some SRS streams and entered aquatic food chains. Concentrations of radionuclides in SRS streams are currently insufficient to cause discernible effects on fish abundance, species composition, and appearance (Paller and Dyer 1999).

Mercury is another contaminant commonly found in fishes from SRS and other MSRB streams. Food chain bioaccumulation causes mercury concentrations to be higher in predatory fishes such as largemouth bass and longnose gar than in fishes that feed lower on the food chain. Unlike ^{137}Cs, the mercury in SRS fish probably originated offsite and entered SRS streams primarily through aerial deposition in the watershed (EPA 2000). Mercury contamination of fish is a pervasive problem throughout much of the South Carolina Coastal Plain.

Other types of disturbance to MSRB streams both on and off the SRS include the discharge of effluents from industrial and sanitary facilities and erosion and siltation resulting from construction and other land uses that alter watershed cover. Spirit Creek, located on Fort Gordon in Georgia, receives sewage effluent in its middle reaches in addition to suffering erosion in its upper reaches (Hoover and Killgore 1999). The fish assemblage in Spirit Creek differs from that in other Fort Gordon streams in having fewer species and being numerically dominated by the golden shiner, a relatively tolerant minnow.

The Horse Creek–Langley Pond system in Aiken County, South Carolina, has been subjected to large quantities of untreated and partially treated wastes, primarily textile by-products, since the late 1800s. The system nearly became a "biological desert," and sediment data revealed high levels of chromium, mercury, PCBs, and other metals and organics. In 1979, a regional wastewater treatment facility was completed that collected all point-source wastewater. Fish have since repopulated Horse Creek, but a fish consumption advisory is still in effect due to excessive levels of mercury and PCBs in fish tissue (Darr 1987, 1988).

A remaining disturbance to some MSRB streams is artificial impoundment to produce ponds and reservoirs. This type of disturbance has been investigated on the SRS, where the upper reaches of Steel Creek and Lower Three Runs were impounded to produce reactor cooling reservoirs. Such reservoirs have both upstream and downstream impacts. Fish assemblages upstream from the reservoirs are isolated from downstream assemblages by the presence of the reservoir. That factor may be partly responsible for a reduction in the number of lotic fish species in Steel Creek upstream from L Lake (Specht and Paller 2001). Downstream effects include alterations of stream water temperatures and flow regimes, blockage of fish movements, and invasion by lentic fishes such as bluegills. Shortly after the construction of L Lake, large numbers of juvenile bluegills entered Steel Creek and Meyers Branch in spillover from the L Lake dam. They largely disappeared within several years with no lasting effects on the native fish assemblages (Meffe 1991). Substantial numbers of adult largemouth bass and other large species are sometimes found in the plunge basins below the L Lake and Par Pond dams. It is not known whether these fishes originate downstream and are blocked from going further upstream, or if they emigrate from the reservoir through discharge structures in the dams. During the warmer months, cold, deoxygenated, hypolimnetic water is sometimes released from the reservoirs into Steel Creek and Lower Three Runs, reducing stream temperatures and oxygen levels as a result. Summer temperatures below 15 °C have been recorded near the L Lake dam. These conditions cause the emigration of fishes from affected stream reaches and undoubtedly affect growth and reproductive success as well.

Although most have been influenced in many ways by human activities, MSRB streams still support a distinctive and diverse fish fauna that has been studied extensively only on the SRS and, to a lesser extent, Fort Gordon. Several SRS streams are unique in having been relatively free of human influences for 50 years or more. There are few other places on the south Atlantic Upper Coastal Plain where such resources still exist. These undisturbed streams together with the other MSRB streams offer a unique resource for research and observation of fishes and other aquatic life.

Reservoirs/Ponds/Isolated Wetlands

The lakes and reservoirs in the MSRB—all man-made—range in size from over 1,053 ha to about 40.5 ha. The only permanent natural lakes in the MSRB are the Carolina bays. On the Georgia side of the Savannah River of the MSRB, most of the lakes are lightly stocked with catfish and sunfish species for recreational fishing. Most of the lakes and ponds on the South Carolina side of the Savannah River in the MSRB lie within the Savannah River Site or in the Horse Creek watershed. The SRS lakes and ponds are unique compared with the rest of the lakes in the MSRB—or even in the southeastern United States—as no public access has been allowed for more than 50 years and neither urbanization nor row crop agriculture dominates their watersheds. This isolation has significantly reduced the nonpoint runoff from industrial and agricultural sources and fish mortality caused by commercial fishing or sportfishing. The lakes are also unique in terms of the thermal and contaminant input they received from past SRS nuclear reactor operations, from which they are still recovering.

Most of the lakes in the MSRB deeper than about 6 m stratify into three thermal zones during the warmest months of the year, with the warmest water at the surface and the coldest at the bottom. This stratification results in considerable variation in the amount of oxygen available in the various layers. The epilimnion (the shallow-water zone typically from the surface to around 3.6 m deep) has good light penetration allowing algae and other plants to produce oxygen. The metalimnion (the middle layer, or thermocline, usually between 3.6 and 7.6 m) water temperature declines with depth, and there is still enough light penetration for photosynthesis to occur and provide necessary oxygen for fish survival. The deepwater zone, or hypolimnion, contains little or no oxygen during the summer, and fish can survive there only for short periods.

During the fall, wind action and changing water temperatures cause surface waters to mix with deeper water, resulting in a "turnover" that mixes the waters of the three layers. This usually occurs by early November in the MSRB. Fish can generally thrive in all areas of a lake during these times because the deep water has sufficient oxygen.

Most of the lakes in the MSRB are classified as eutrophic because they are older water bodies with relatively high nutrient input and are turbid with high algal growth. Eutrophication has accelerated in many lakes, especially in the suburban areas around Augusta, Georgia, due to increased nutrient input from human activity.

Most of the ponds in the MSRB are man-made farm ponds or recreational fishing ponds. Some, like the Merry Brothers Brick and Tile Company's 27 ponds in south Augusta, were restored from past pit mining operations. The old reservoirs on Horse Creek were constructed for textile plant operations. Many are still polluted with chemicals from mill discharges and for many years were not capable of supporting any aquatic life at all.

Many of the lentic habitats (impounded still waters)

in the MSRB are concentrated on the SRS. The site has many bodies of water, ranging from small depressions with seasonal water supply to large reservoirs (405–1,069 ha) that contain a variety of fish species assemblages. The SRS has two large reservoirs (Par Pond, including connected Ponds B and C, and L Lake) and 299 isolated upland wetland depressions (194 Carolina bays, 28 small ponds, and various other basins). The water bodies vary in degree of disturbance; both "disturbed" (i.e., affected by thermal discharges from operating nuclear production reactors before 1989, contaminants, drawdowns, etc.) and "undisturbed" (i.e., relatively free of man-made disturbance and contamination) conditions have existed in these lentic habitats since the SRS was constructed in 1951.

Various fish species assemblages have been monitored in each of these lentic habitats for several years. Much of the data were collected to assess the impacts of thermal discharges and contamination sources from the site's operating reactors and are therefore sometimes comparative rather than quantitative. See Kilgo and Blake (in press) for a 50-year overview of natural resource studies at SRS.

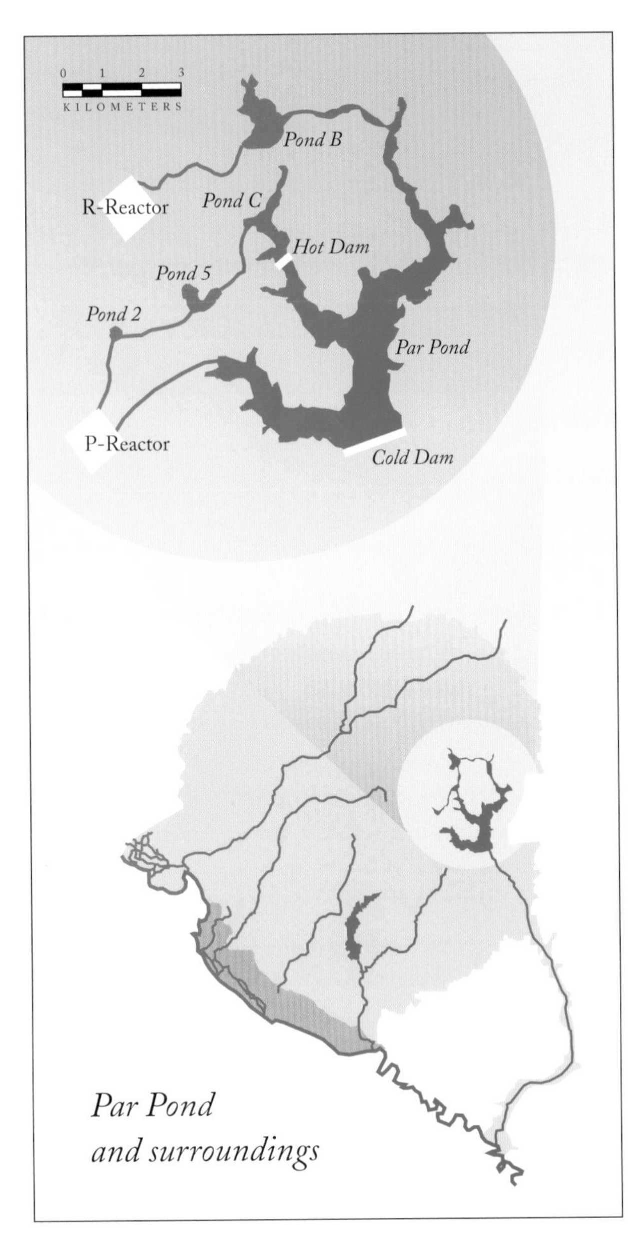

MAP 6

Upland Reservoirs

The largest reservoir within the MSRB is located on the SRS. Par Pond is a 1,069 ha recirculating cooling-water reservoir created in 1958 by the construction of an earthen dam on Lower Three Runs Creek. Releases from R-Reactor contaminated Par Pond with low levels of radioactive materials, primarily ^{137}Cs, when a fuel element failed in 1957. All radioactive isotope releases from R-Reactor except tritium ceased following the shutdown of R-Reactor in 1964. Most of the ^{137}Cs resides in the upper 0.3 m of fine sediments; elevated levels of mercury have also accumulated in sediments through pumping water containing mercury from the Savannah River (Du Pont 1987; Mohler et al. 1997).

Both P-Reactor and R-Reactor (hence the name "Par") used Par Pond from November 1961 to June 1964. During this period, R-Reactor discharged effluent to the north arm through Pond B, and P-Reactor discharged effluent into the middle arm of the reservoir through a series of canals and precooler ponds, including Pond C (Figure 16; Map 6). P-Reactor operation, and therefore thermal effluent to Par Pond, ceased in 1988.

During reactor operations in the summer, the temperatures near the inflow from Pond C in Par Pond typically ranged from 22.2 to 42.2 °C and up to 70 °C. The thermal effluent cooled rapidly as it dispersed, primarily through the southern half of the reservoir. The north and south arms of Par Pond had temperatures at or only slightly above those typical for the region (Wilde 1985, 1987).

For the past three decades, the water level in Par Pond has been maintained at a nominal 61 m above mean sea level by the addition of water from the Savannah River, permitting the development of a rich and diversified ecosystem (Figure 17). Although Par Pond is man-made, it

FIGURE 16. A precooler pond that formerly received heated effluent before allowing it to flow to Par Pond.

has a unique ecology and is known worldwide through more than 250 scientific publications, 25 master's theses and doctoral dissertations, and some 60 papers/reports on fishes (Marcy et al. 1994). Most of these studies emphasized the effects of elevated temperatures on fish behavior, physiology, and ecology. There also have been community-level studies emphasizing the abundance and distribution of Par Pond fishes (SREL 1991). The fish population in Par Pond developed from stream populations present before the dam was built and from Savannah River fish brought into the pond with water pumped from the river as part of the once-through cooling system.

Gibbons et al. (1978) reviewed the body condition factor, or K (weight/length^3), of more than 10,000 largemouth bass from Par Pond from 1967 to 1976; in general, high K values indicate well-fed and healthy fish. Their data demonstrated significantly lower adult largemouth bass K values in the areas of highest temperature compared with other areas of Par Pond, and significantly

FIGURE 17. Black crappie, *Pomoxis nigromaculatus*, a species common in reservoirs in the Southeast.

lower K values in summer compared with winter in all areas of the reservoir. Analysis of temporal changes in gonadal weight during 1983 and 1984 (Paller and Saul 1985) indicated that largemouth bass spawned earlier in the areas with the highest temperatures.

Paller and Saul (1985) reported that thermal effects resulting from the operation of P-Reactor included localized reductions in the number of species in the heated areas and early spawning of some species. Other possible thermal effects included deterioration in the condition of largemouth bass and aggregation of largemouth bass and black crappies in the discharge area.

Based on collections of early life stages and abundance within fish communities, Halverson et al. (1997) concluded that Par Pond's condition was no worse than that of other southeastern reservoirs that had not been exposed to reactor coolant.

Par Pond has never been stocked. General surveys by Clugston (1973), Siler (1975), Hogan (1977), Martin (1980), and Bennett and McFarlane (1983) have identified 30 fish species from Par Pond. All of these species have also been reported from Lower Three Runs, which along with the Savannah River represents the original source of all fish species in the three reservoirs.

Cove rotenone (a sampling technique using an oxygen-reducing chemical to collect fish) studies conducted from 1969 to 1980 by Clugston (1973), Siler (1975), Hogan (1977), and Martin (1980) consistently demonstrated that bluegill, lake chubsucker, largemouth bass, chain pickerel, other species of sunfishes, and yellow perch were the most abundant species in Par Pond (Table 10).

TABLE 10. Percentage composition (ordered by weight) of fishes from Par Pond during 1969–1980.

Scientific name	Common name	1969	1972	1977	1980
Lepomis macrochirus	bluegill	25.3	13.7	25.1	30.5
Micropterus salmoides	largemouth bass	22.8	5.6	8.0	17.4
Erimyzon sucetta	lake chubsucker	15.4	42.3	33.8	19.7
	other sunfishes	13.1	13.7	10.8	15.0
Perca flavescens	yellow perch	5.8	0.9	3.4	1.2
	unidentified fish	4.6	—	0.2	0.4
Esox niger	chain pickerel	4.5	12.9	7.5	5.4
Pomoxis nigromaculatus	black crappie	2.6	1.7	1.1	2.5
Amia calva	bowfin	2.5	1.8	2.7	0.4
Minytrema melanops	spotted sucker	2.2	—	—	—
Ameiurus sp.	bullheads	0.8	2.3	2.4	0.5
Notemigonus crysoleucas	golden shiner	0.4	5.2	5.0	7.3
Alosa aestivalis	blueback herring	0.1	—	—	0.1

Sources: Data were obtained from various cove rotenone studies: 1969 and 1972 (Clugston 1973; Hogan 1977); 1980 (Martin 1980) from Paller 1997.

Paller and Saul (1985) reported at least 23 species from Par Pond based on a sample of 13,166 adult and juvenile fish; 2 other species (eastern mosquitofish and swamp darter) were observed or captured as ichthyoplankton but not collected as adult fishes. The dominant species in Par Pond were brook silverside (38% by number), lake chubsucker (18%), largemouth bass (18%), bluegill (14%), and black crappie (2%).

Largemouth bass and bluegills were more abundant in Par Pond than in most other reservoirs in the Southeast. Estimates of largemouth bass populations in the pond have ranged from 29,000 to about 100,000 (up to about 40/ha) (Gibbons and Bennett 1971; Martin 1980; Gilbert and Hightower 1981). In contrast, Pond C had a largemouth bass population of about 5 individuals/ha, or about 833 altogether (Siler and Clugston 1975). Many of the largemouth bass in Par Pond suffered from red-sore disease (Esch et al. 1976). Outbreaks of red-sore disease, caused by bacteria (*Aeromonas* spp.), occur throughout rivers, lakes, ponds, and reservoirs in the Southeast.

Approximately 165,980 fish larvae and eggs, representing at least 11 taxa, were collected from Par Pond by Paller and Saul (1985). Black crappie was the most abundant taxon (36.0%), followed by various sunfishes (32.7%) and darters (14.9%). The heated areas had lower percentages of darters and higher percentages of largemouth bass (Halverson et al. 1997).

Comparisons between Par Pond and other reservoirs in the United States indicated that the Par Pond fish community was comparable to other reservoir communities in terms of species number, diversity, and standing crop of all species summed together (Paller and Saul 1985). However, Par Pond differed in having more largemouth bass and lake chubsucker and fewer gizzard shad and carp.

When Par Pond was drawn down for dam repair in 1991–1994 from 61 m above mean sea level (MSL) to 55 m MSL, the reservoir's surface area was reduced by 50% and the volume was reduced by 65% (DOE 1993, 1995). Electrofishing data from before, during, and after the drawdown showed that the procedure reduced both the number of species and species abundance, particularly of those species dependent on littoral zone vegetation. The size structure of individual species also was affected. During the first months after the Par Pond drawdown, predation from largemouth bass (some exceeding 6.8 kg), bluegill, and pickerel significantly depleted small baitfish in the exposed, newly created littoral zone with little macrophyte cover (Whicker et al. 1993a, 1993b).

For comparative purposes, 17 species of fish were collected from Par Pond by electrofishing during the predrawdown period (Table 11). The most abundant by number were brook silverside (51%), bluegill (18%), and largemouth bass (16%). Within 9 months of the refill, the

TABLE 11. Percentage composition of fishes (by number) collected by electrofishing from Par Pond, South Carolina, before, during, and after a drawdown event of 3.5 years' duration.[1]

		Pre-Drawdown	Drawdown		Refill	Post-refill		
Scientific name	Common name	1984–1985	1991	1992	1995	Spring 1995	Fall 1995	Fall 1996
Labidesthes sicculus	brook silverside	50.7	3.3	2.8	90.8	9.1	21.7	33
Lepomis macrochirus	bluegill	17.9	46.7	45.8	2.8	21.4	17.9	21.7
Micropterus salmoides	largemouth bass	15.6	16.7	22.2	2.8	20.6	8.3	2.6
Erimyzon sucetta	lake chubsucker	6.1	2.5	1.4	1.6	5.8	4.7	6
Notropis petersoni	coastal shiner	3.4	—	11.1	—	9.7	4.8	6.4
Notemigonus crysoleucas	golden shiner	1.8	3.3	—	0.2	14.8	13.2	8.4
Esox niger	chain pickerel	1.3	5	1.4	1.6	2.9	4.5	7.3
Perca flavescens	yellow perch	0.9	—	—	—	7.6	6	0.4
Lepomis auritus	redbreast sunfish	0.7	1.7	—	—	0.8	0.6	2.3
Pomoxis nigromaculatus	black crappie	0.4	0.8	1.4	0.2	0.4	1.4	0.4
Lepomis gulosus	warmouth	0.4	2.5	—	—	0.2	1.9	2.2
Amia calva	bowfin	0.2	0.8	1.4	—	0.8	0.4	—
Lepomis punctatus	spotted sunfish	0.1	0.8	—	—	1.2	2	0.4
Ameiurus natalis	yellow bullhead	0.1	—	—	—	—	—	—
Lepomis marginatus	dollar sunfish	0.1	—	—	—	4.7	1.2	5.8
Enneacanthus gloriosus	bluespotted sunfish	<0.1	14.2	12.5	—	—	11	2.5
Dorosoma cepedianum	gizzard shad	<0.1	—	—	—	—	0.1	0.5
Gambusia holbrooki	eastern mosquitofish	—	—	—	—	—	0.1	—
Etheostoma fusiforme	swamp darter	—	—	—	—	—	0.1	—
Ameiurus platycephalus	flat bullhead	—	1.7	—	—	—	0.1	—

[1] From Paller 1997.

fish community had recovered in terms of numbers of species and overall fish abundance, and had nearly recovered in terms of species composition (Paller 1997).

In 1996, flow from the Savannah River was terminated, and it was predicted that the reduction in nutrient inputs would result in the development of aquatic communities (i.e., plankton and fish) more like those of other southeastern reservoirs that do not receive substantial nutrient inputs. Paller (1997) noted that Par Pond is indeed now generally similar to other southeastern reservoirs.

L Lake (Figure 18), another large reservoir in the MSRB, was constructed by the U.S. Department of Energy in 1984 by impounding the headwaters of Steel Creek to form a cooling-water reservoir to dissipate the thermal effluent discharged from the operation of L-Reactor (DOE 1984). At the normal pool elevation of 58 m above MSL, the L Lake dam impounds about 419 ha (U.S. Army Corps of Engineers 1987). A flow of at least 0.28 m^3/sec is maintained from the dam to Steel Creek to maintain the stream's biological community. L-Reactor was shut down in 1988. Temperatures in excess of 40 °C entered the lake during periods of reactor operation. In the lake, however, the maximum recorded water temperature was approximately 33.9 °C near the outfall in June 1986, approximately 2 °C higher than the maximum temperatures in southeastern reservoirs without thermal inputs.

High temperatures (>40 °C) precluded the survival of fish near the inflow of heated water in the lake during periods of reactor operation. During the extended summer reactor outages, however, which lasted approximately six months, temperatures decreased to ambient levels, permitting fish to invade the upper portion of L Lake. When L-Reactor was restarted in the fall, large numbers of these fish were killed by the resulting elevated water temperatures (Paller et al. 1988). There was no indication that these fish kills affected community structures in the lower portion of L Lake.

Cooling-water discharges were managed by varying reactor power levels to maintain a balanced biological community in the lake as required by the South Carolina

FIGURE 18. L Lake, a 418 ha reservoir that formerly received heated effluent from L-Reactor (in background).

Department of Health and Environmental Control (SCDHEC) in the SRS's National Discharge Elimination System (NPDES) permit (i.e., about 50% of the lake would not exceed 32.2°C). L Lake is a productive reservoir, and its ecology and water quality were extensively monitored both during the input of thermal effluent from 1985 to 1987 and afterward. Gladden et al. (1985) and Wike et al. (1989) have published the most comprehensive summaries of the data collected on the water chemistry and ecology of L Lake.

In September 1988 and 1989, 10 South Carolina reservoirs were intensively sampled for trophic status, community structure, and biologically balanced community criteria (Bowers 1992). Based on the results of this study, L Lake had nitrogen and phosphorus concentrations characteristic of a eutrophic system.

The fish assemblages in L Lake were sampled from early 1986 until late 1998. Although sampling frequency and intensity changed over the years, the sampling methods were consistent enough to provide an accurate record of changes in fish assemblage structure over a 13-year period. The data from L Lake thus constitute an unusual record of the long-term development of a fish assemblage in a largely unfished southeastern reservoir (Paller, n.d.). The lake's fish community was sampled extensively from January 1986 to December 1989; the sampling began approximately two months after the lake was filled in November 1985.

L Lake was stocked with approximately 40,000 juvenile (19–25 mm) bluegills in the fall of 1985 and approximately 4,000 juvenile largemouth bass in the spring of 1986. Largemouth bass, bluegill, redbreast sunfish, and threadfin shad dominated the L Lake fish community between 1987 and 1992. The most important trends in the fish community in recent years have involved changes in the abundance of these species and interactions between the dominant fish species and lower trophic levels, particularly zooplankton, due to changes in feeding preference of the dominant predators.

The fish community in the lower half of L Lake devel-

TABLE 12. Fish species and changes in abundance in L Lake from 1986 to 1998.[1]

Scientific name	Common name	Number collected			
		1986	1990	1992	1998
Lepomis macrochirus	bluegill	549	318	297	194
Lepomis auritus	redbreast sunfish	133	509	548	7
Notropis petersoni	coastal shiner	53	63	373	11
Lepomis punctatus	spotted sunfish	44	—	—	10
Notemigonus crysoleucas	golden shiner	36	4	—	10
Micropterus salmoides	largemouth bass	24	133	225	25
Lepomis marginatus	dollar sunfish	11	—	—	2
Gambusia holbrooki	mosquitofish	11	—	—	—
Ameiurus natalis	yellow bullhead	10	3	4	—
Semotilus atromaculatus	creek chub	7	2	—	—
Enneacanthus gloriosus	bluespotted sunfish	6	—	—	3
Lepomis gulosus	warmouth	5	7	16	8
Labidesthes sicculus	brook silverside	4	63	241	36
Nocomis leptocephalus	bluehead chub	4	—	—	—
Pomoxis nigromaculatus	black crappie	3	5	1	10
Ameiurus platycephalus	flat bullhead	2	23	7	1
Erimyzon sucetta	lake chubsucker	2	1	—	1
Perca flavescens	yellow perch	1	54	214	49
Esox niger	chain pickerel	1	1	20	8
Lepomis microlophus	redear sunfish	1	—	—	—
Dorosoma petenense	threadfin shad	—	—	290	—
Dorosoma cepedianum	gizzard shad	—	31	20	—
Pomoxis annularis	white crappie	—	—	—	9
Cyprinus carpio	common carp	—	—	2	—
Alosa aestivalis	blueback herring	—	—	—	1

[1] Represents four (1986, 1990, 1992, and 1998) of the nine years (1986–1992, 1995, and 1998) of collections made at L Lake (Paller, n.d.).
Note: Additional species collected in other years included brown bullhead (*Ameiurus nebulosus*) and snail bullhead (*A. brunneus*).

oped as expected during the lake's early years. However, the warmer temperatures near the reactor discharge point precluded normal community development in the upper half. The community has continued to change since L-Reactor ceased operating in 1988 and subsequent nutrient loading as a result of Savannah River input was reduced.

A total of 27 species were collected from 1986 through 1998, representing 30 collections (Table 12). The most prominent patterns in fish communities/assemblages were exhibited by small littoral zone fishes. They were initially abundant as a result of the successful colonization of L Lake by fishes from Steel Creek, but their numbers decreased dramatically as populations of larger predators and/or competitors (largemouth bass and bluegill) increased (Paller et al. 1992). Subsequently their densities increased again, concurrent with the proliferation of aquatic vegetation. Increases in the abundance of small littoral zone species apparently resulted in greater fish assemblage diversity and abundance over time.

It appears that slightly more than 10 years (from 1986 to 1998) were required for a relatively stable fish assemblage structure to become established in L Lake. This period probably represents a minimum for southeastern reservoirs, because artificial introductions likely accelerated the process of assemblage development in L Lake. Currently, most of the species that occupy L Lake are also common in other established South Carolina reservoirs, suggesting that the process of assemblage development in L Lake may be fairly typical.

The L Lake fish community presently includes at least 19 species, with the most abundant being brook silverside, yellow perch, bluegill, redbreast sunfish, coastal shiner, largemouth bass, chain pickerel, and spotted sunfish. These species are generally common in southeastern reservoirs with abundant aquatic vegetation. Most or all appear to have successfully reproducing and self-sustaining populations in L Lake.

Another SRS pond affected by heated effluent is Pond B (Map 6), which was filled in July 1961 after completion of an earthen dam across Joyce Branch of Lower Three Runs Creek. Pond B received inputs of heated effluent from R-Reactor operations during the period 1961–1964. The reservoir's water level has been maintained to act as a shield because its soils and water column are contaminated, principally by ^{137}Cs from a failed fuel element in R-Reactor in 1957 (DOE 1999).

Pond B, which covers 87 ha, has an average depth of 4.3 m and a maximum depth of 12.5 m. It has four small islands, six bays, and 9 km of shoreline (Whicker et al. 1990). Pond B stratifies from April through October (Alberts et al. 1988). Since the cessation of thermal input, it has developed extensive stands of aquatic macrophytes (Kelly 1989).

Fish populations became established in waters that did not directly receive thermal discharges and survived the heated water input by remaining in the cooler-water refuges of the pond (Parker et al. 1973; Block et al. 1984). General surveys by Clugston (1973), Siler (1975), Hogan (1977), Martin (1980), Bennett and McFarlane (1983), and Paller and Saul (1985) established the presence of 15 fish species in Pond B.

Paller and Saul (1985), for example, collected 1,340 adult and juvenile fish representing 15 species. Numerically dominant species were gizzard shad (15.9%), largemouth bass (17.7%), brook silverside (34.1%), yellow bullhead (7.4%), bluegill (4.9%), and flat bullhead (3.0%). Eastern mosquitofish, yellow perch, brown bullhead, warmouth, redbreast sunfish, dollar sunfish, redfin pickerel, swamp darter, and black crappie were also collected. The lake chubsucker, a dominant species in nearby Par Pond, was absent from Pond B. In addition, Paller and Saul (1985) examined approximately 48,300 fish eggs and larvae representing six taxa collected in Pond B between January 1984 and June 1985. The most abundant taxa were various sunfishes (57%), black crappie (19%), and darters (11%). Whicker et al. (1990) estimated the largemouth bass population size in the pond at 5,400 in 1986 using standard mark-and-recapture methods.

Paller and Saul (1985) compared the impacts of high temperatures and entrainment on Par Pond and Pond B by sampling fish populations. Par Pond was similar to Pond B on the basis of most of the fisheries parameters they measured. Number of species, total number of fish collected, and diversity—parameters often depressed in thermally stressed ecosystems—were higher in Par Pond than in Pond B. Largemouth bass, an important sportfish and the principal predator fish in both Pond B and Par Pond, were larger and in better condition in Par Pond. The mean condition factor (K) of largemouth bass from all sample areas in Par Pond ($K = 1.15$) was lower than the average for other U.S. reservoirs ($K = 1.41$). However, the mean condition of largemouth bass from Pond B ($K = 1.05$) was even lower than that found in Par Pond.

Pond C, which was constructed at the same time as Par Pond, was also affected by thermal effluent. Pond C has an area of 67.2 ha, a mean depth of 3.9 m, and a maximum depth of 11 m (Map 6) (Wilde 1985). During reactor operations, water temperatures of approximately 70 °C were released into Pond C through a series of precooler ponds and canals (Wilde and Tilly 1985). This precooling system accounted for approximately 86% of the total cooling in the Par Pond system (Wilde 1985).

During reactor operations, the fish assemblage in Pond C consisted primarily of bluegill, largemouth bass (which together constituted 95% of the fishes in the reservoir), and eastern mosquitofish (Clugston 1973). The fish survived within the reservoir by occupying one of four relatively small thermal refuges cooled by water from springs or streams.

The fish assemblage that has survived in Pond C is not as diverse as assemblages in representative ponds of similar size in the southeastern United States. Research data from the Savannah River Ecology Laboratory (SREL) indicated that from the time of P-Reactor shutdown in 1987 to Par Pond drawdown in 1991, the structural complexity in Pond C habitats increased, the number of individuals within species increased, and the number of species increased from three to seven.

Eastern mosquitofish were reportedly dip-netted from portions of Pond C at temperatures above 40 °C. Fish kills due to changing water temperatures caused by intermittent reactor operation were observed in the pond. Kills occurred following a restart of P-Reactor after the annual two-week outage and involved primarily juvenile bluegills and eastern mosquitofish (Wilde 1985).

Seventeen species have been reported from Pond C: blueback herring, pirate perch, gizzard shad, banded pygmy sunfish, lake chubsucker, redfin pickerel, swamp darter, eastern mosquitofish, yellow bullhead, redbreast sunfish, warmouth, bluegill, dollar sunfish, redear sunfish, largemouth bass, golden shiner, and black crappie (Halverson et al. 1997). During warm-water discharges, only 4 species were known to reproduce there (bluegill, eastern mosquitofish, redbreast sunfish, and largemouth bass) (McCort et al. 1984).

Sampling in Pond C during 1994 and 1995 (J. Congdon unpub. data) produced 5,092 fish of 19 species. The species composition was dominated by largemouth bass, lake chubsucker, bluegill, redbreast sunfish, warmouth, and brook silverside (Table 13).

TABLE 13. Fish species and numbers collected in Pond C during 1994 and 1995 (listed in order of frequency).

Scientific name	Common name	Number collected
Micropterus salmoides	largemouth bass	1,213
Lepomis macrochirus	bluegill	1,147
Erimyzon sucetta	lake chubsucker	884
Lepomis auritus	redbreast sunfish	688
Lepomis gulosus	warmouth	497
Labidesthes sicculus	brook silverside	166
Perca flavescens	yellow perch	157
Pomoxis nigromaculatus	black crappie	122
Dorosoma cepedianum	gizzard shad	105
Minytrema melanops	spotted sucker	37
Notropis petersoni	coastal shiner	32
Ameiurus natalis	yellow bullhead	26
Notemigonus crysoleucas	golden shiner	6
Aphredoderus sayanus	pirate perch	4
Ictalurus punctatus	channel catfish	2
Esox niger	chain pickerel	2
Esox americanus	redfin pickerel	2
Etheostoma fusiforme	swamp darter	1
Gambusia holbrooki	eastern mosquitofish	1

Source: J. Congdon, unpublished data.

Ponds

Shields et al. (1982) identified 28 ponds (bodies of water formed by the blockage or obstruction of a natural drainage pattern of streams) on the SRS. They ranged in size from 0.16 ha to 82 ha and included natural and man-made farm ponds, oxbows, and beaver ponds (Figure 19).

Skinface Pond, a 3.2 ha pond located in the Crackerneck Wildlife Management Area, is currently managed for public fishing under a 1995 agreement between DOE, the South Carolina Department of Natural Resources (SCDNR), and the U.S. Forest Service (USFS). It was opened to public fishing and stocked by USFS from 1977 until 1984, when DOE reclaimed the Crackerneck land area (M. Caudell, SCDNR pers. comm.). The pond has been open on a limited basis since that time; it was drawn down and restocked with bluegill, redear sunfish, and largemouth bass in April 2000, and was reopened to fishing in September 2001 (C. Thomason, SCDNR pers. comm.). Grass carp were stocked there at a density of 8–10 per ha in 1999 as well (L. Wike pers. comm.). Other ponds of about the same size and species composition are Risher, Fire, and Dick's ponds.

Other impounded areas on the SRS are the reactor basins (called "186 basins") created when water was pumped from the Savannah River to the basins and stored to be used as a secondary cooling system in case of loss of water from the river (DOE 1984). Fish were sampled for ^{137}Cs concentrations in four compartments of the R-Reactor 186-R (18,927 m^3) cooling-water basin and in the 183-1R basin in 1994. The sample comprised 166 fish: 76 gizzard shads, 23 catfish, 35 bluegills, 3 American eels, 26 largemouth bass, 2 crappies, and 1 warmouth (Figure 20) (C. Mitchell pers. comm.).

In Aiken County, South Carolina, near Clearwater, the Graniteville Company (now Avondale Mills) owns the 101 ha Langley Pond on Horse Creek in the Savannah River drainage. Table 14 lists fish species occurrence in other publicly maintained ponds (Vaucluse and Clearwater).

In Georgia, the ponds on Fort Gordon (west of Augusta) were sampled in 1996–1998 (J. J. Hoover pers. comm.). Mirror Pond (6.1 ha), in the Spirit Creek drainage, was sampled during a drawdown in 1996. Leitner (11.3 ha), Lower Leitner (10.1 ha), and Union (7.7 ha) ponds (Sandy Run Creek, Savannah River drainage) were sampled using seines and light-traps in 1997–1998 (F. Perry pers. comm.). The four ponds, which are used for recreational fishing, contained almost the same number of species, with a total of 23 species collected in all four in the late 1990s (Table 15).

FIGURE 19. Risher Pond, a typical small farm pond found on the Savannah River Site. Risher Pond drained in June 2003 when the dam washed out.

FIGURE 20. Warmouth, *Lepomis gulosus*, a species frequently found in ponds and reservoirs.

TABLE 14. Occurrence of 18 fish species noted in collections in Langley, Vaucluse, and Clearwater ponds from 1983 to 1997 (listed in phylogenetic order).

		Location		
Scientific name	Common name	Langley[1]	Vaucluse[2]	Clearwater[3]
Anguilla rostrata	American eel	X	—	X
Notemigonus crysoleucas	golden shiner	X	—	X
Erimyzon sp.	chubsucker	X	X	X
Minytrema melanops	spotted sucker	X	—	—
Ameiurus catus	white catfish	X	X	X
Ameiurus natalis	yellow bullhead	X	—	—
Ameiurus nebulosus	brown bullhead	X	—	X
Ameiurus platycephalus	flat bullhead	X	—	—
Ictalurus punctatus	channel catfish	X	—	—
Esox niger	chain pickerel	X	—	X
Gambusia holbrooki	eastern mosquitofish	—	—	X
Labidesthes sicculus	brook silverside	X	—	X
	unidentified sunfish	X	—	—
Lepomis gulosus	warmouth	X	X	X
Lepomis macrochirus	bluegill	X	X	X
Lepomis microlophus	redear sunfish	X	X	X
Micropterus salmoides	largemouth bass	X	X	X
Pomoxis nigromaculatus	black crappie	X	—	—
Perca flavescens	yellow perch	X	X	X

[1] Langley Pond (101 ha) data from Gaymon 1983; Darr 1987, 1988; SCDNR 1992, 1997.
[2] Vaucluse Pond (40.5 ha) data from Shealy Environmental Services, Inc. 1990.
[3] Clearwater Pond (16.2 ha) data from Shealy Environmental Services, Inc. 1990.

TABLE 15. Occurrence of 23 fish species collected in four ponds on Fort Gordon, Georgia, 1996–1998.

Scientific name	Common name	Mirror	Leitner	Lower Leitner	Union
Ctenopharyngodon idella	grass carp	X	—	—	—
Notemigonus crysoleucas	golden shiner	X	—	—	—
Erimyzon sucetta	lake chubsucker	X	X	—	—
Ameiurus natalis	yellow bullhead	X	—	—	—
Ameiurus nebulosus	brown bullhead	X	—	—	—
Esox americanus	redfin pickerel	X	X	—	—
Fundulus lineolatus	lined topminnow	X	X	X	X
Lepomis gulosus	warmouth	X	X	X	X
Lepomis macrochirus	bluegill	X	X	X	—
Lepomis marginatus	dollar sunfish	X	—	—	—
Lepomis microlophus	redear sunfish	X	—	—	—
Micropterus salmoides	largemouth bass	X	X	—	—
Etheostoma fusiforme	swamp darter	X	X	X	X
Erimyzon oblongus	creek chubsucker	—	X	—	X
Esox niger	chain pickerel	—	X	X	X
Gambusia holbrooki	eastern mosquitofish	—	X	X	X
Labidesthes sicculus	brook silverside	—	X	X	X
Elassoma okatie	bluebarred pygmy sunfish	—	X	X	X
Lepomis auritus	redbreast sunfish	—	X	X	X
Erimyzon sp.	chubsuckers	—	—	X	—
Notropis sp.	shiners	—	—	—	X
Enneacanthus gloriosus	bluespotted sunfish	—	—	—	X
Etheostoma serrifer	sawcheek darter	—	—	—	X
Total number of species documented		13	13	10	12

Source: J. J. Hoover, pers. comm.

Lake Olmstead, located in Augusta, Georgia, lies adjacent to and is connected to the Augusta Canal. Approximately 40.5 ha in size, the lake was created in 1870 as part of a project to enlarge the Augusta Canal. Fish were sampled in Raes Creek (Lake Olmstead and outfall) at Lakeshore Road, Richmond County, Georgia, on 23 March 1997 (Gerald Dinkins pers. comm.). The most abundant species were eastern mosquitofish, redbreast sunfish, and bluegill. Catfish are a common sportfish here (Table 16).

The five small Sears lakes are located near the Gordon highway, just southwest of Augusta. Two of the lakes are regularly stocked with catfish. Both channel catfish and blue catfish were caught in one 4.9 ha lake (*Augusta Chronicle*, 13 January 1999). The Merry Brothers Brick and Tile Company ponds in Richmond County, south of Augusta, are a conglomerate of 27 ponds, many more than 100 years old, totaling 810 ha. The principal sport species caught by anglers are bluegill and redear sunfish (shellcrackers). Fifteen species were collected in these ponds in 1922, 1939, and 1941 (Table 16).

Another small lake in north Augusta is Lake Aumond. Nine species were collected in sampling there in 1941. The impoundment was outside Augusta at the time of the collections but is now within the city limits and is completely surrounded by suburban housing. Located along the Savannah River near Augusta are several River Watch Parkway catchment basins. Students from Augusta State University collected fish in these six interconnected 1–2.4 ha ponds from 1994 to 2002 (Bruce Saul pers. comm.). The fish species found in these ponds are listed in Table 16.

According to the *Augusta Chronicle* (23 June 2001), the Georgia Department of Natural Resources stocked the approximately 2 ha county-owned Mayor's Pond in June 2001 with about 12,000 catfish. The catfish share the pond with sunfishes, crappies, and bass. The pond is located

TABLE 16. Fish species collected in the Merry Brothers Brick and Tile Company ponds, Lake Aumond, Lake Olmstead, and the River Watch Parkway catch basins near Augusta, Georgia (listed in phylogenetic order).

		Location			
Scientific name	Common name	MBBP[1]	LA[2]	LO[3]	RPCB[4]
Anguilla rostrata	American eel	—	—	X	—
Dorosoma petenense	threadfin shad	—	—	—	X
Cyprinus carpio	common carp	—	—	—	X
Notemigonus crysoleucas	golden shiner	X	X	—	—
Notropis lutipinnis	yellowfin shiner	—	—	—	X
Notropis maculatus	taillight shiner	X	—	—	—
Erimyzon sucetta	lake chubsucker	—	—	—	X
Ameiurus catus	white catfish	—	X	—	—
Ameiurus melas	black bullhead	—	—	X	—
Ameiurus natalis	yellow bullhead	X	—	—	—
Ameiurus nebulosus	brown bullhead	—	—	—	X
Ictalurus punctatus	channel catfish	—	—	—	X
Esox americanus	redfin pickerel	—	—	—	X
Esox niger	chain pickerel	—	X	—	—
Umbra pygmaea	eastern mudminnow	X	—	—	—
Aphredoderus sayanus	pirate perch	—	X	—	—
Fundulus lineolatus	lined topminnow	—	—	X	—
Gambusia holbrooki	eastern mosquitofish	X	—	X	—
Labidesthes sicculus	brook silverside	X	—	X	—
Morone americana	white perch	—	—	X	—
Enneacanthus gloriosus	bluespotted sunfish	X	—	—	—
Lepomis auritus	redbreast sunfish	X	X	X	—
Lepomis cyanellus	green sunfish	—	—	—	X
Lepomis gibbosus	pumpkinseed	—	—	—	X
Lepomis gulosus	warmouth	—	—	X	—
Lepomis macrochirus	bluegill	X	X	X	—
Lepomis marginatus	dollar sunfish	X	—	X	—
Lepomis microlophus	redear sunfish	—	—	X	—
Micropterus salmoides	largemouth bass	X	—	X	—
Pomoxis annularis	white crappie	X	—	—	—
Pomoxis nigromaculatus	black crappie	—	X	X	—
Etheostoma fusiforme	swamp darter	X	—	X	X
Etheostoma olmstedi	tessellated darter	—	—	—	X
Perca flavescens	yellow perch	—	—	X	—

[1] MBBP = Merry Brothers Brick and Tile Company ponds, 1922, 1939, and 1941.
[2] LA = Lake Aumond, 1941.
[3] LO = Lake Olmstead, 1941 and 1950, 1997.
[4] RPCB = River Watch Parkway catch basins, 1994–2002.

FIGURE 21. Two Carolina bays on the Savannah River Site. Carolina bays are restricted to the Coastal Plain of the Southeast.

near the Augusta airport next to Phinizy Swamp Nature Area south of Augusta.

Isolated Wetlands

Isolated wetlands are a common feature of the MSRB and SRS landscapes. Shields et al. (1982) identified 194 upland wetland depressions (called Carolina bays) at the SRS that support a variety of aquatic and wetland communities. Carolina bays are distributed across the SRS in clusters and broad bands at elevations ranging from 36 to 104 m above MSL (Figure 21). Their surface areas range from less than 0.08 to 2 ha. Carolina bays are typically isolated wetlands fed largely by rainfall or shallow, low-solute groundwater (Schalles et al. 1989; Lide 1991). Thus, they have a nutrient-poor, soft-water, acidic chemistry that restricts primary and secondary productivity and use of these systems to tolerant species. In addition, fluctuations of their hydrology make these bays relatively unpredictable habitats. Most of the bays contain water at least seasonally (Schalles et al. 1989; Kirkman 1992).

Fishes have been observed in several Carolina bays on the SRS (Bennett and McFarlane 1983). Redfin pickerel, mud sunfish, sunfish (*Lepomis* spp.), lake chubsucker, and mosquitofish were observed in four Carolina bays on the SRS during 1978–1983. Fewer than 10% of the Carolina bays on the SRS are known to have permanent fish populations, although overwash from neighboring swamps or streams may reestablish the ichthyofauna of formerly dry basins (Schalles et al. 1989).

Snodgrass et al. (1996) randomly sampled 63 isolated wetlands with baited minnow and hoop nets. Thirteen, or 21%, of the isolated wetland areas sampled contained fish populations. Fishes were limited to wetlands that dried infrequently, were relatively close to intermediate aquatic habitats, and varied little in elevation from the nearest permanent aquatic habitat. Of the 12 species collected, dollar sunfish and lake chubsucker dominated in the upper portions of drainage basins, and mud sunfish and eastern mosquitofish dominated in the downstream portions of drainage systems. The species collected included lake chubsucker (474), dollar sunfish (418), eastern mosquitofish (165), mud sunfish (59), flier (36), warmouth (25), golden shiner (24), eastern mudminnow (19), redfin pickerel (8), yellow bullhead (6), brown bullhead (6), and bluespotted sunfish (1).

In a more complete multiyear study from 1994 to 1997, Snodgrass et al. (1998) collected 6,888 fish representing 13 species by passive trapping in 25 isolated wetland areas. The species collected included lake chubsucker (1,792), mud sunfish (1,285), eastern mosquitofish (1,049), dollar sunfish (753), flier (662), eastern mudminnow (492), redfin pickerel (342), golden shiner (127), warmouth (103), bluespotted sunfish (100), yellow bullhead (83), swamp darter (66), and lined topminnow (34).

Many isolated wetlands were ditched and drained for agricultural use prior to the acquisition of the land for the SRS (Christel-Rose 1994). As few have been disturbed since the early 1950s, most of the altered bays have undergone successional revegetation (Schalles et al. 1989). The SRS recently began a program to restore some bays to their former hydrology.

Storm-based retention basins are typically engineered and fenced ditches located in developed areas to control and channel stormwater runoff. These areas can include both open and shrub-scrub wetland habitat (Mayer and Wike 1997). In the spring of 1996, Mayer and Wike (1997) sampled 10 storm retention basins at the SRS with minnow traps as part of the Urban Wildlife Study. About 100 goldfish approximately 25 mm long were collected in the Central Shops storm retention basin, the first record of this species at the SRS and in the MSRB. We collected additional individuals in 2003, including the one pictured in the species account, so it appears that a relatively stable population has been established.

Introduction to Fish Identification

This chapter introduces the elements of fish identification and explains how to use identification keys. The information it provides should be sufficient to allow even those unfamiliar with fish biology and dichotomous keys to identify any fish caught in the middle Savannah River basin. Many publications provide additional details (e.g., Menhinick 1991; Page and Burr 1991; Etnier and Starnes 1993; Jenkins and Burkhead 1993).

Fish Morphology

The following general description of some important features of fish morphology will be helpful in using the identification keys and interpreting the species accounts. The various body shapes of fishes are generally described as elongate, moderately deep, and deep (Plate 2 a–d), as well as rounded, moderately compressed, and compressed (Plate 2 e–h). Among the more easily recognizable features are the type and shape of the tail and the caudal fin. Most modern fishes have a homocercal tail, in which all the principal rays of the fin are attached to the last, modified vertebra (hypural plate) (Plate 3 c–f). Some primitive families, like the sturgeons and gars, have a vertebral column that turns upward into the dorsal lobe resulting in a heterocercal (Plate 3 a) or abbreviated heterocercal tail (Plate 3 b). The margin of the caudal fin can be rounded, straight, emarginate, or forked (Plate 3 c–f). The caudal, dorsal, and anal fins are called the median or unpaired fins (Plate 1 a, b, c). The members of some families have only one dorsal fin (e.g., Cyprinidae), while others have two (e.g., Percidae). The dorsal fins may be widely separated, as in striped mullet (*Mugil cephalus*), or touching, as in white perch (*Morone americana*), or somewhere in between. Catfishes (Ictaluridae) and some other families have an adipose fin, an additional fleshy median fin that has no rays and is located between the dorsal and caudal fins. The pectoral and pelvic fins are paired fins. The membranes of the fins are supported by soft rays and, in some species, spines. Soft rays develop from paired embryonic structures and are usually segmented, branched, and soft, and are not pointed. Spines develop from unpaired embryonic structures and are unsegmented, unbranched, and usually hard and pointed. Rays can become calcified so that they are indistinguishable from spines except during early development (e.g., the "spine" in the carp's dorsal fin is really a calcified ray). In such cases we treat the calcified ray as if it were a spine. Rays are present in all fins, and spines are usually present in the dorsal, pectoral, and anal fins. Fish identification requires recognizing and counting fin rays and spines, and the numbers of each are given in the "meristics" section in the species accounts. In catfish (Ictaluridae) and pickerels (Esocidae), all rays should be counted, even the smaller rudimentary ray at the beginning of the fin. In the other species, the first (few) rudimentary rays and the first segmented (much longer) ray are counted as one. The last ray is counted as one if it looks as if the branches will join just below the skin surface (e.g., Hubbs and Lagler 1958; Page and Burr 1991).

The position and shape of the mouth are important indicators of the fish's feeding behavior and diet. Fish with superior mouths often live and forage near the surface. Most fish with terminal mouths feed in the water

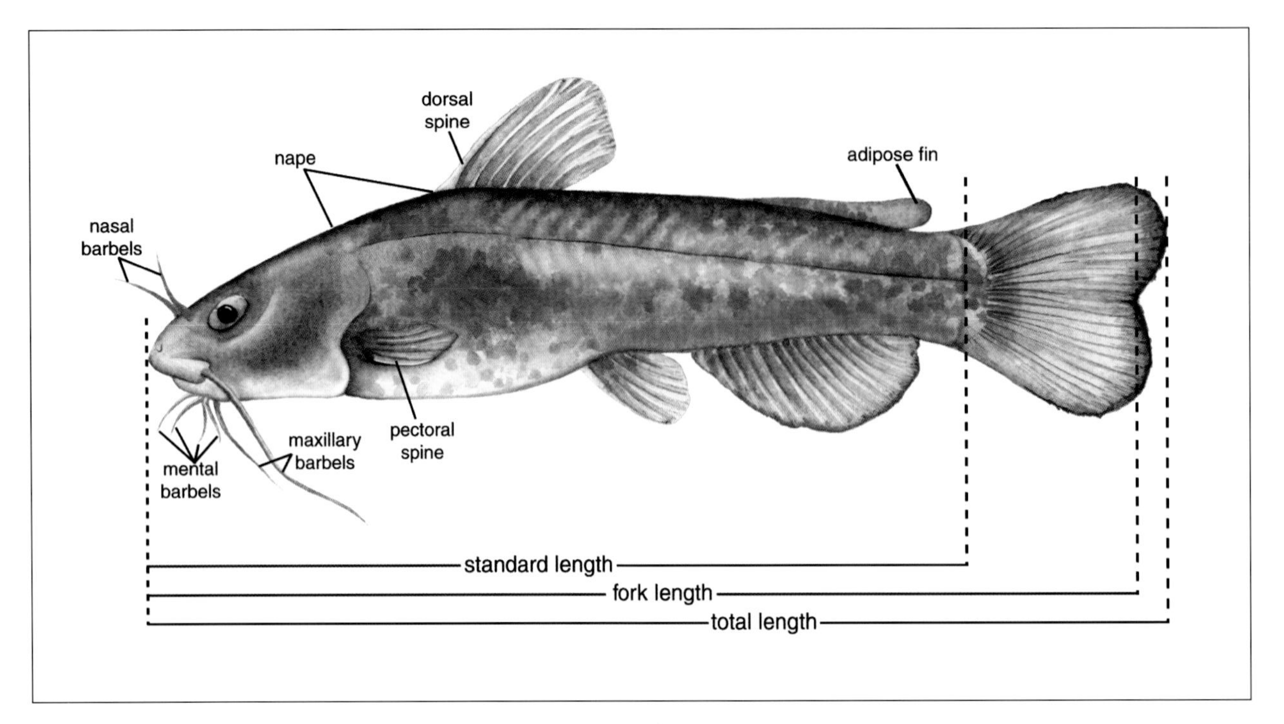

1A

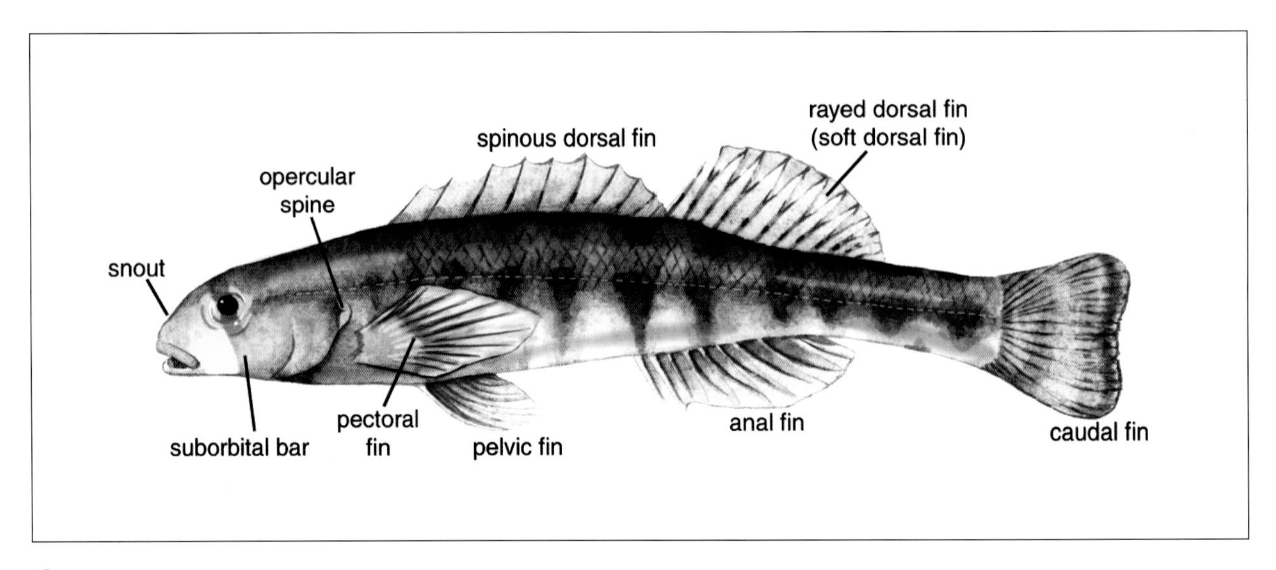

1B

column on other fish or zooplankton, but may also be found near the surface, near the bottom, or near structures (Plate 4). Species with subterminal mouths mostly live near or on the bottom. Many feed on prey that is buried in the sediment or lives on the sediment or substrate, while others scrape algae from the substrate. The shape of the mouth and lips can be an important morphological feature when identifying species, especially the suckers (Catostomidae, Plate 12). Some species have teeth on the tongue or the roof of the mouth (e.g., Plate 18 e). These teeth can be important in identification and are easily detected by using a blunt probe, finger, or pen to explore the tongue—especially the back—and the roof of the mouth. Pharyngeal teeth (bony projections from the fifth gill arch) can be important in the identification of sunfishes, suckers, and in particular minnows. It is necessary to sacrifice the fish and remove the pharyngeal arch to examine the pharyngeal teeth. Menhinick 1991 provides detailed instructions on the removal and cleaning of pharyngeal teeth. Each pharyngeal arch has one or

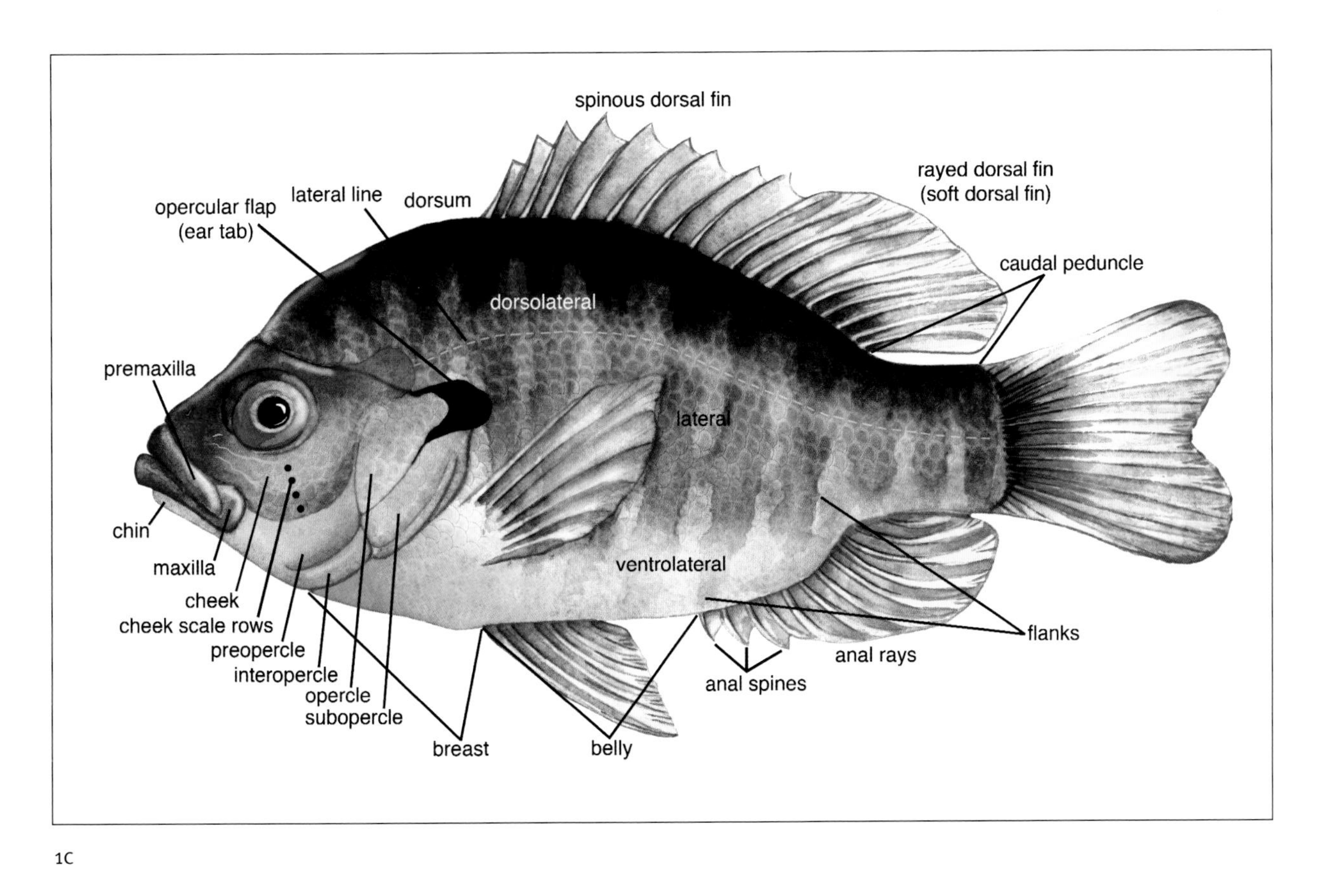

1C

PLATE 1. External anatomical features and measurements used in identifying fish: (a) flat bullhead, *Ameiurus platycephalus*; (b) blackbanded darter, *Percina nigrofasciata*; (c) redbreast sunfish, *Lepomis auritus*.

two rows of teeth, and the pharyngeal teeth formula reflects the number of teeth in each row on each of the two arches. Numbers of teeth on the left and right arches are separated by a hyphen and rows on the same arch by a comma (left outer row, left inner row-right inner row, right outer row). Examples of pharyngeal teeth formulas are 2,4-4,2 (each pharyngeal arch has 2 rows of teeth, one with 2 teeth and one with 4 teeth; Plate 11 h, k, m, o) and 1,4-4,2 (each pharyngeal arch has 2 rows of teeth, one with 1 or 2 teeth and one with 4 teeth).

The body of most fish is covered with scales—flattened, stiff, bony or horny structures on the skin, embedded in scale pockets. The two most abundant scale types are ctenoid scales, thin scales that bear a patch of tiny spinelike prickles (ctenii) on the exposed surface; and cycloid scales, more or less rounded scales that are flat and bear no ctenii. Other scale types include sawtooth scales, sharply pointed scales present on the belly of herrings/shads; and scutes, modified scales in the form of horny or bony plates that are often spiny or keeled. The number of scales along various parts of the body is often an important feature for identification. Scale counts along the lateral line, around the caudal peduncle (Plate 4 i), and on the cheek (e.g., in sunfishes; Plate 1 c) are most commonly used. The number of scales listed in the various meristics sections refer to the scales present in adult fish. However, because scales develop gradually in larval and juvenile fish, the adult number of scales may not be present in very small individuals. See Hubbs and Lagler 1958 or Jenkins and Burkhead 1993 for additional details on meristic counts.

Species Accounts

The species accounts contain general information on the species and, where available, information specific to the MSRB. Individual species accounts appear after the key for each family. Each account provides a description of the species that includes information on its appearance (color and morphological features), similar species (where

PLATE 2. Body forms and anal fin shapes: (a, b) elongate body form, (c) moderately deep body form, (d) deep body form, (e) rounded body cross section, (f, g) moderately compressed body cross section, (h) compressed body cross section, (i) strongly falcate anal fin, (j) slightly falcate anal fin, (k) straight anal fin margin, (l) rounded anal fin.

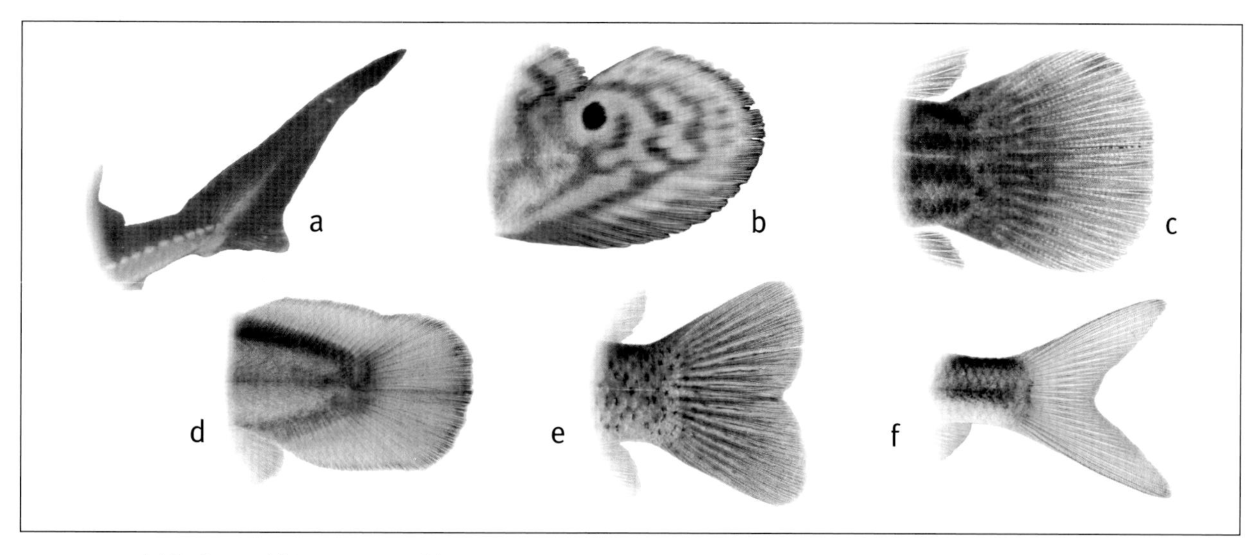

PLATE 3. Caudal fin forms: (a) heterocercal, (b) abbreviate heterocercal, (c–f) homocercal fins: (c) rounded, (d) truncate or straight, (e) emarginate, (f) forked.

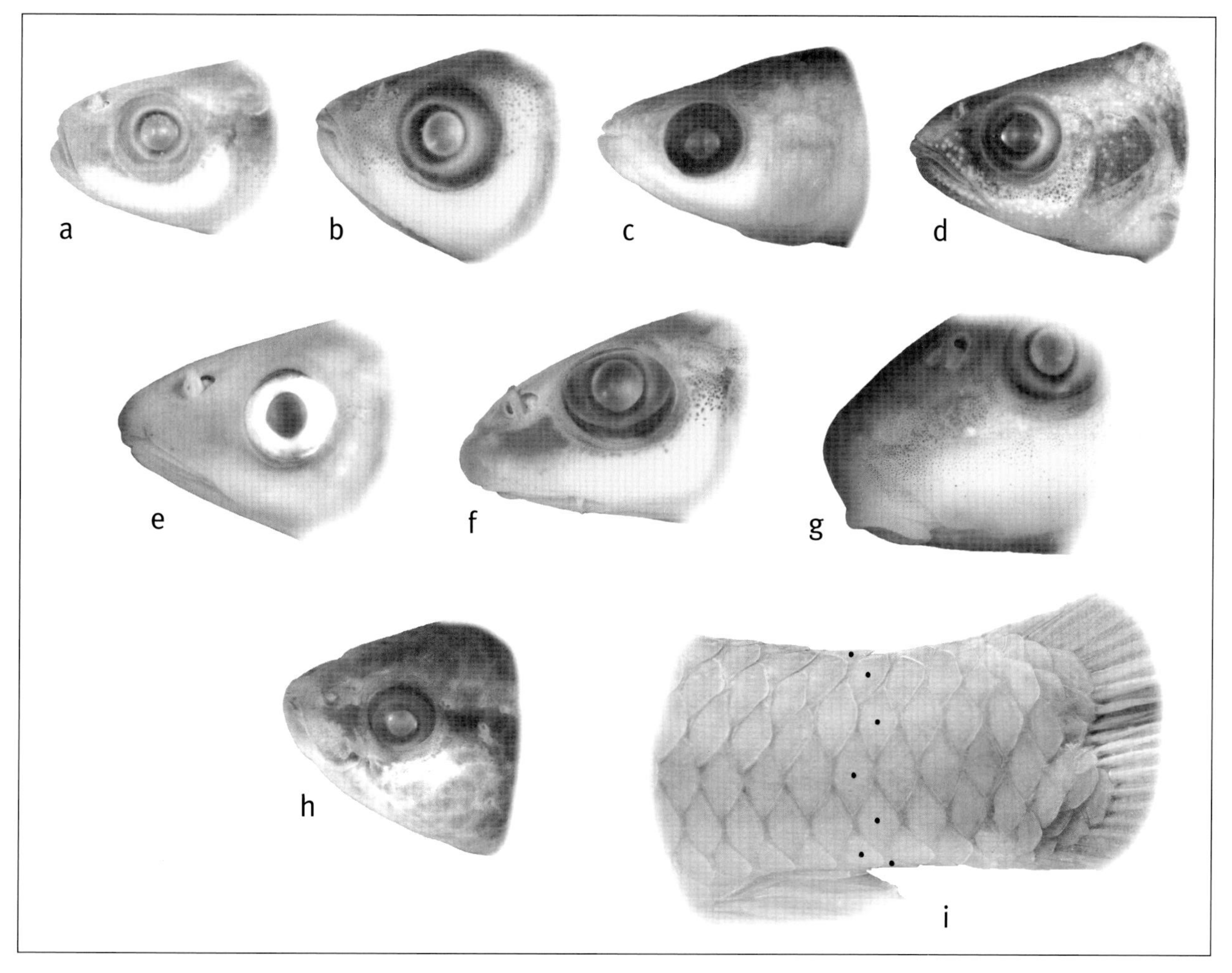

PLATE 4. Mouth shapes and method for counting caudal peduncle scales: (a, b) superior strongly oblique, (c) supraterminal moderately oblique, (d) terminal moderately oblique, (e) subterminal moderately oblique, (f) subterminal slightly oblique, (g) inferior horizontal, (h) terminal nonprotractile, (i) method of counting rows of scales around the caudal peduncle.

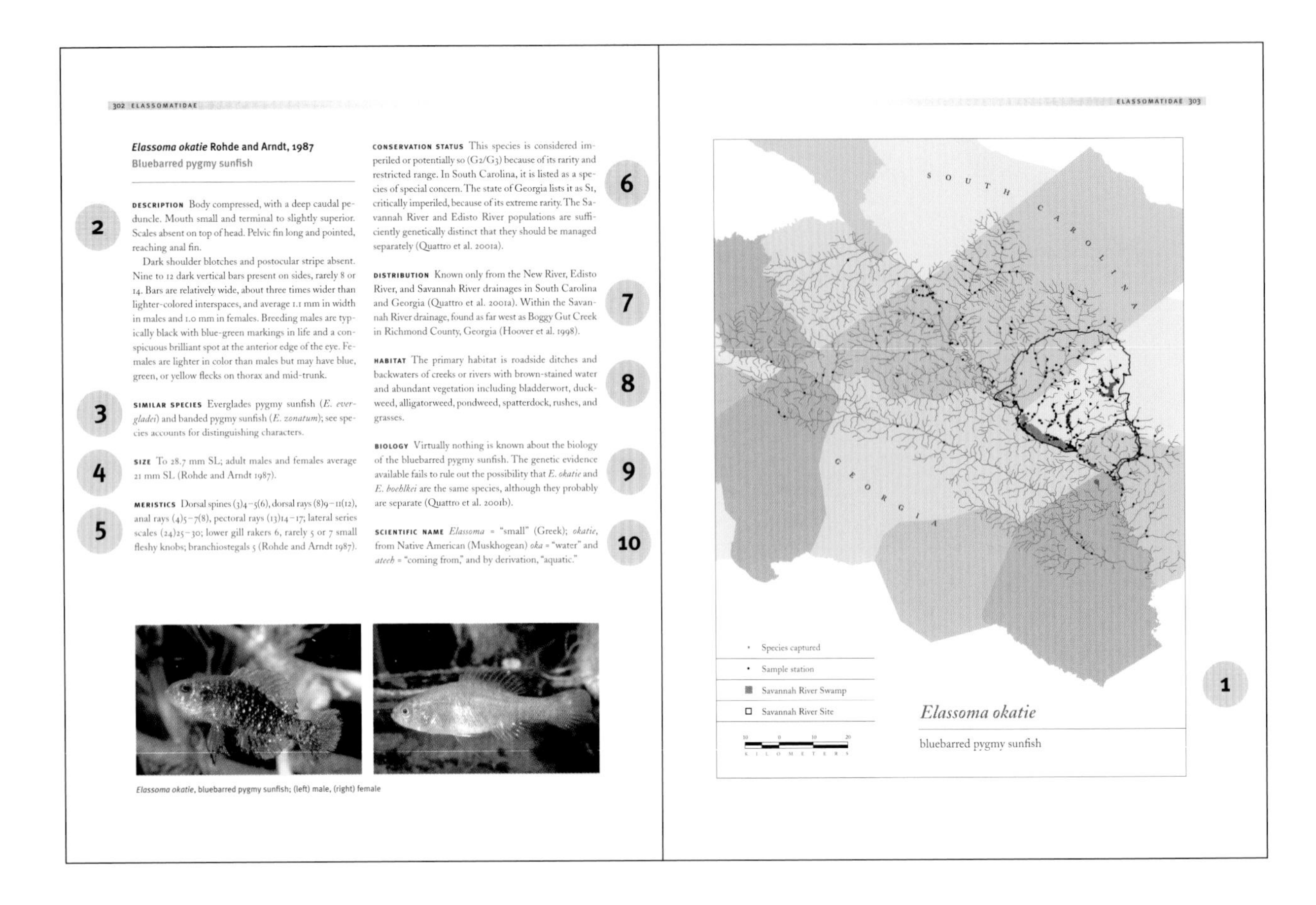

302 ELASSOMATIDAE

***Elassoma okatie* Rohde and Arndt, 1987**
Bluebarred pygmy sunfish

2 **DESCRIPTION** Body compressed, with a deep caudal peduncle. Mouth small and terminal to slightly superior. Scales absent on top of head. Pelvic fin long and pointed, reaching anal fin.

Dark shoulder blotches and postocular stripe absent. Nine to 12 dark vertical bars present on sides, rarely 8 or 14. Bars are relatively wide, about three times wider than lighter-colored interspaces, and average 1.1 mm in width in males and 1.0 mm in females. Breeding males are typically black with blue-green markings in life and a conspicuous brilliant spot at the anterior edge of the eye. Females are lighter in color than males but may have blue, green, or yellow flecks on thorax and mid-trunk.

3 **SIMILAR SPECIES** Everglades pygmy sunfish (*E. evergladei*) and banded pygmy sunfish (*E. zonatum*); see species accounts for distinguishing characters.

4 **SIZE** To 28.7 mm SL; adult males and females average 21 mm SL (Rohde and Arndt 1987).

5 **MERISTICS** Dorsal spines (3)4–5(6), dorsal rays (8)9–11(12), anal rays (4)5–7(8), pectoral rays (13)14–17; lateral series scales (24)25–30; lower gill rakers 6, rarely 5 or 7 small fleshy knobs; branchiostegals 5 (Rohde and Arndt 1987).

6 **CONSERVATION STATUS** This species is considered imperiled or potentially so (G2/G3) because of its rarity and restricted range. In South Carolina, it is listed as a species of special concern. The state of Georgia lists it as S1, critically imperiled, because of its extreme rarity. The Savannah River and Edisto River populations are sufficiently genetically distinct that they should be managed separately (Quattro et al. 2001a).

7 **DISTRIBUTION** Known only from the New River, Edisto River, and Savannah River drainages in South Carolina and Georgia (Quattro et al. 2001a). Within the Savannah River drainage, found as far west as Boggy Gut Creek in Richmond County, Georgia (Hoover et al. 1998).

8 **HABITAT** The primary habitat is roadside ditches and backwaters of creeks or rivers with brown-stained water and abundant vegetation including bladderwort, duckweed, alligatorweed, pondweed, spatterdock, rushes, and grasses.

9 **BIOLOGY** Virtually nothing is known about the biology of the bluebarred pygmy sunfish. The genetic evidence available fails to rule out the possibility that *E. okatie* and *E. boehlkei* are the same species, although they probably are separate (Quattro et al. 2001b).

10 **SCIENTIFIC NAME** *Elassoma* = "small" (Greek); *okatie*, from Native American (Muskhogean) *oka* = "water" and *ateeh* = "coming from," and by derivation, "aquatic."

Elassoma okatie, bluebarred pygmy sunfish; (left) male, (right) female

ELASSOMATIDAE 303

appropriate), size, meristic characters, conservation status (where appropriate), historic distribution and distribution within the MSRB, habitat, biology, and scientific name. One or more photographs of the species are provided with each species account.

The species accounts are organized as follows:

1 DISTRIBUTION MAP Black dots on individual distribution maps represent sites where historical samples were taken and the species in question was not captured. Red dots indicate sites where the species in question was captured.

2 DESCRIPTION Morphological description (body, fins, and other features), coloration, and sexual and seasonal dimorphism.

3 SIMILAR SPECIES Characters that help distinguish the species from others similar in appearance.

4 SIZE Maximum size and most common size in the MSRB and elsewhere. Size, given in millimeters or centimeters, represents total length (TL) unless otherwise specified (10 mm = 1 cm = 0.394 inches, or 1 inch = 2.54 cm = 25.4 mm). Other commonly used length measurements are the standard length (SL) and fork length (FL) (see Plate 1 a).

5 MERISTICS Pertinent number of spines and soft fin rays in various fins, number of scales in various locations, and other information important for identification. The counts of fin rays and other structures are given in the following format: (1)2–3(4). The numbers in parentheses indicate the lowest (left) and highest (right) counts found in the literature. The unbracketed numbers are those most commonly present. If known, specific meristic information for MSRB populations is also given.

6 CONSERVATION STATUS Indicates if the species is introduced and whether it is listed as endangered, a species of concern, etc.

7 DISTRIBUTION Large-scale distribution and distribution and occurrence within the MSRB.

8 HABITAT Stream type, sediment type, vegetation, water quality, and other characteristics of the habitat in which the species is most commonly found.

9 BIOLOGY Information on reproduction (spawning location, season, and behavior), age and growth, diet and feeding, references where descriptions of the eggs and larvae can be found, and other relevant information.

10 SCIENTIFC NAME Etymology.

Occurrence of Fish Species Not Verified or Collected in the MSRB

Some species not included in the keys or the species accounts may occur in the MSRB. They were not included here either because they were not in the collections we examined or because the identification could not be verified. The fathead minnow (*Pimephales promelas*), for example, is common in the live bait trade, has been used for years as a forage fish in a fish hatchery in the Savannah River drainage (Millen in Lincoln County, Ga.), and is approved by the U.S. EPA for use in aquatic toxicological surveys. It is quite possible that this species may be found in the MSRB by continuous introduction, whether or not it is successful in competing with native species to establish a sustaining population (Dahlberg and Scott 1971b). The sailfin molly (*Poecilia latipinna*) occurs in the Savannah River downriver from the MSRB, and could occur here as well. There have been references to river chub (*Nocomis micropogon*) taken near the Savannah Rapids, but no river chubs are available in the museum records from below Clarks Hill/J. Strom Thurmond Lake.

The southern flounder (*Paralichthys lethostigma*) and possibly other flounders migrate up rivers for many miles and may swim up the Savannah River as far as the MSRB. Fishermen have reported catching large adult southern flounders in the area, but we were unable to confirm the species identification. The southern flounder is easily recognized by the fact that both eyes are located on the left side of the body rather than the right, and by the relatively large mouth with visible sharp teeth.

The inland silverside (*Menidia beryllina*) and bluefin killifish (*Lucania goodei*), which occur in the Edisto and Altamaha drainages but have not been reported from the Savannah River, may eventually be found in this area.

The least killifish (*Heterandria formosa*) is generally restricted to the Lower Coastal Plain (Page and Burr 1991). It appears to be relatively common in the Savannah River drainage downstream from our chosen boundary and a small number of individuals have been taken barely within our area of consideration. The sea lamprey (*Petromyzon marinus*) must historically have occurred in the MSRB, but we have found no reliable records. *Lepomis megalotis* reported from Augusta is likely *Lepomis marginatus* (Dahlberg and Scott 1971b). The Carolina pygmy sunfish (*Elassoma boehlkei*) has been reported from the MSRB on the basis of a single specimen.

We will be grateful to receive any information concerning the occurrence of unlisted species.

Use of the Keys

The dichotomous keys used in this book are designed to identify fishes of the MSRB. The key to the families is followed by family and species accounts. Species that are the only representatives of their families are identified in the family key. Others are identified in the keys found at

Petromyzon marinus, sea lamprey. Specimens were collected from above Augusta, Georgia, prior to damming of the Savannah River.

the beginning of the species accounts for each family. The keys were developed for a broad user group ranging from relative novices in the field of fish ecology and morphology to experienced professionals. We tried to limit the use of technical ichthyological terminology where possible; however, it is impossible to escape technical terms entirely when describing fish, and we include a comprehensive glossary to accommodate those unfamiliar with these terms. The illustrations of generalized fishes in the section on general fish morphology (above) and the illustrations in the keys should further aid in recognizing fish features. We recommend reading both sections of each couplet before choosing one and moving on to the next couplet. A user who reaches an obvious "dead end," indicating that a previous couplet was selected in error, should "backtrack" the previous decisions by following the bracketed numbers that accompany each couplet and refer to the previous couplet that led to it.

Many species exhibit sexual dimorphism; males and females may differ in color, body or fin shape, or other features. Also, breeding individuals, especially males, of a number of species display colorations or shapes different from those of nonbreeding fish. Juveniles of many species have a coloration or shape different from that of the adults. The adult features develop over time, and the fish may not resemble the parents until sexual maturity is reached. We tried to limit the use of identification features that change seasonally, differ between the sexes, or develop only at maturity. Special care should be taken while identifying juvenile sunfishes; the identification of smaller sunfishes (<3 cm TL) is known to be problematic. We suggest consulting specialized literature or experts if problems are thought to exist.

Naturally occurring or introduced hybrids of species may also complicate the identification of individual fish. Hybrids of striped bass and white bass from hatcheries have been introduced throughout South Carolina (see details under Moronidae), and bass with features of both species may be found. Naturally occurring hybrids of sunfishes are fairly common, and hybridization within Esocidae, Catostomidae, and other families may also occur. We recommend reading the species description after a fish is identified using this key to confirm the identification. When in doubt, check the key again or consult additional specialized literature.

Collecting and Preserving Specimens

Detailed information on the collection and preservation of fishes is available in Nielsen and Johnson 1983; Etnier and Starnes 1993; Jenkins and Burkhead 1993, and other sources. The scientific collection of fish in the MSRB and elsewhere requires a collecting permit. Other regulations concerning fish collecting may vary among counties; information can be obtained from the Georgia and South Carolina Departments of Natural Resources. Fish can be collected using a variety of techniques including nets, traps, and rod and reel. Fish should be killed only if absolutely necessary and using methods that kill the fish as quickly and humanely as possible (see guidelines of the American Society of Ichthyologists and Herpetologists and American Fisheries Society in Nickum 1988). If questions about the identification of a specimen persist, it should be preserved for further examination. Commonly used preservatives are 50% isopropyl alcohol, 70% ethyl alcohol, and 10% formalin (10% solution of 37% aqueous formaldehyde; note that formaldehyde is a known carcinogen and severe skin irritant). Small fish can be preserved whole, but fish larger than about 200 mm should be carefully opened by making an incision along the right side to allow the preservative to enter the abdominal cavity. Fish can be preserved in glass or plastic containers, which can also be used for storage. Long-term storage may require replacing or refreshing the preservative and monitoring the quality of the preservative. Each sample (jar) should have permanent waterproof labels inside with detailed information on the sampling location (including county and state), date, and the names of the collectors. Individual fish may require individual labels that include the scientific name and the name of the person who identified the specimen. Information should be written in permanent ink. Additional fieldnotes about the habitat and other information are extremely useful for later study. Fish lose color quickly after preservation, and detailed notes of the color of the fish before preservation are often helpful for later identification.

The abbreviations DEF (Dean E. Fletcher) and FDM (F. Douglas Martin) denote observations and unpublished data provided by two of the book's authors.

Key to the Families of Fishes

1a. Jaws absent, buccal funnel present (Plate 5a), pectoral fins absent, 7 gill openings on each side of head, a single median nostril . . . . Petromyzontidae, *Petromyzon marinus* (sea lamprey)—no species account

1b. Jaws present, oral disk absent, pectoral fins present, 1 gill opening on each side of head, paired nostrils . . **2**

[1] **2a.** Body elongate, snakelike or eel-like; scales present, but embedded in skin so that they are obvious only with close inspection . . . . Anguillidae, *Anguilla rostrata* (American eel), p. 89

2b. Body short to elongate but not snakelike or eel-like; scales present or absent but not embedded in skin **3**

[2] **3a.** Both eyes located on same side of head; body strongly compressed side to side and bilaterally asymmetrical **4**

3b. Eyes located one on each side of head; body compressed or not, but bilaterally symmetrical; pectoral fin present on both sides of body **5**

[3] **4a.** Eyes located on right side of head; pectoral fin absent on blind (left) side, and absent (usually) or may be represented by a single ray on eyed side; . . . . Achiridae, *Trinectes maculatus* (hogchoker), p. 406

4b. Eyes located on left side of head; pectoral fin present on both sides of body . . . . Paralichthyidae, *Paralichthys lethostigma* (southern flounder)—no species account

[3] **5a.** Caudal fin with 2 lobes, the upper lobe much larger with an extension of the backbone extending into it (Plate 3 a); 5 rows of enlarged bony scutes running length of body, 1 down center of back (dorsal row), 2 on sides (midlateral rows), and 2 on sides of belly (ventrolateral rows) . . . . . . Acipenseridae (sturgeons), p. 67

5b. Caudal fin rounded, square, or, if with 2 lobes (forked), both lobes about equal or 1 lobe slightly larger (see Plates 3 b–f, 13 d–h for examples); no rows of bony scutes present along sides and back **6**

[5] **6a.** Scales absent; 1 stout spine present in each pectoral fin . . . . Ictaluridae (bullheads, catfishes, and madtoms), p. 205

6b. Scales present, may be thin and flexible or may resemble rows of bony plates; no spines in pectoral fins . . **7**

[6] **7a.** Gular plate present (broad bony plate between the lower jaws, Plate 5 b) **8**

7b. Gular plate absent **9**

[7] **8a.** Caudal fin rounded (Plate 3b); dorsal fin base long and with more than 40 rays . . . . Amiidae, *Amia calva* (bowfin), p. 81

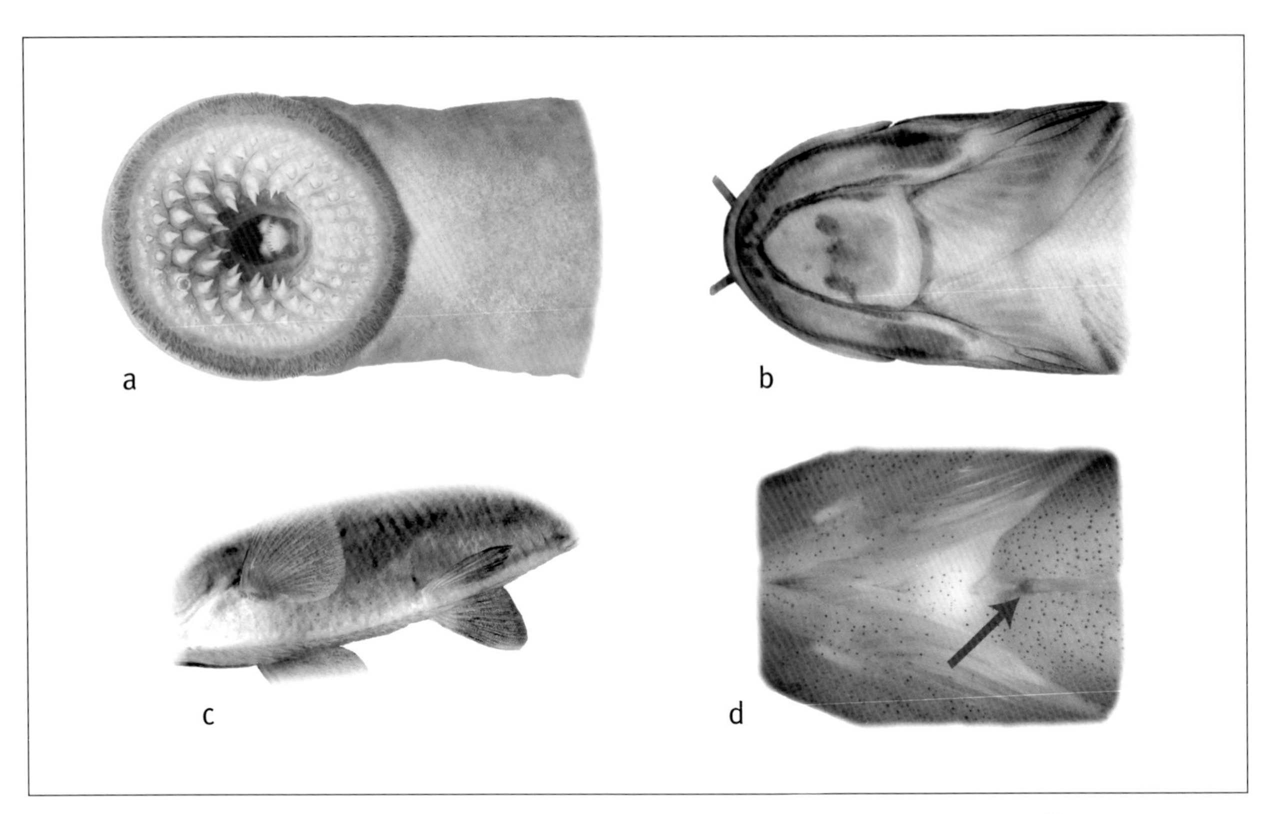

PLATE 5. Morphological characters useful for identification: (a) buccal funnel, (b) gular plate, (c) keeled thorax, (d) anus in jugular position.

8b. Caudal fin forked; dorsal fin base short and with fewer than 25 rays .... Megalopidae, *Megalops atlanticus* (tarpon), p. 85

[7] **9a.** Pelvic fins absent .... Amblyopsidae, *Chologaster cornuta* (swampfish), p. 262
9b. Pelvic fins present **10**

[9] **10a.** Adipose fin present .... Salmonidae (salmons, trouts, and chars), p. 254
10b. Adipose fin absent **11**

[10] **11a.** Body laterally compressed (see Plate 2 g, h for examples) with a prominent keel on belly anterior to pelvic fins (Plate 5 c) .... Clupeidae (herrings and shads), p. 93
11b. Body laterally compressed or not, but without a prominent keel on belly anterior to pelvic fins........ **12**

[11] **12a.** Snout dorsoventrally flattened and shaped like a duck's bill; only 1 dorsal fin present (Plate 15 a, b) .... Esocidae (pickerels), p. 241
12b. Snout not dorsoventrally flattened and not shaped like a duck's bill; 1 or 2 dorsal fins present **13**

[12] **13a.** Jaws long and slender; only 1 dorsal fin present.. **14**
13b. Jaws not long and slender; 1 or 2 dorsal fins present.. **15**

[13] **14a.** Large platelike scales (about 55 along lateral line); anal fin margin rounded and higher than long .... Lepisosteidae (gars), p. 73
14b. Very small scales (about 300 along lateral line); anal fin margin falcate (see Plate 2 i, j for examples) and longer than high .... Belonidae, *Strongylura marina* (Atlantic needlefish), p. 266

[13] **15a.** More than 1 spine in dorsal fin (spines may be thin and flexible but lack branching and joints seen in rays; a hand lens may be required to be sure) **16**
15b. No spines in dorsal fin (may have a single calcified and unbranched ray that looks like a spine) **22**

[15] **16a.** Two widely separated dorsal fins, distance between dorsal fins at least one-half the length of the spinous dorsal fin base. **17**
16b. One or 2 dorsal fins, if 2 the distance between dorsal fins is much less than one-half the length of the spinous dorsal fin base **18**

[16] **17a.** Long anal fin with a single spine and 20 or more rays . . . . Atherinopsidae, *Labidesthes sicculus* (brook silverside), p. 282
17b. Short anal fin with 2 or 3 spines and 10 or fewer rays . . . . Mugilidae (mullets), p. 397

[16] **18a.** Single dorsal fin and opercular spine present; anus located in throat region in larger adults (individuals 50–65 mm SL and larger) (Plate 5 d), at least anterior to pelvic fin origins in intermediate fish (40 mm SL and longer), but located posterior to pelvic origins in small individuals (<25 mm SL) . . . . Aphredoderidae, *Aphredoderus sayanus* (pirate perch), p. 257
18b. One or 2 dorsal fins, and opercular spine may be present or absent, but opercular spine and single dorsal fin not occurring together; anus located closer to anal fin than to pectoral fin at all lengths **19**

[18] **19a.** Lateral line absent; 3–6 dorsal spines . . . . Elassomatidae (pygmy sunfishes), p. 297
19b. Lateral line present; 6 or more dorsal spines **20**

[19] **20a.** Anal fin with 1 or 2 spines . . . . Percidae (darters and perches), p. 369
20b. Anal fin with 3 or more spines **21**

[20] **21a.** Sharp spine at back of gill cover; margin of preopercle strongly serrate . . . . Moronidae (striped basses), p. 285
21b. No sharp spine on back of gill cover; margin of preopercle not strongly serrate. . . . . Centrarchidae (sunfishes), p. 307

[15] **22a.** Caudal fin rounded; dorsal fin located far to rear with more than one-half of its base located posterior to the anal fin origin **23**
22b. Caudal fin forked or emarginate; dorsal fin located farther forward with the entire base located anterior to anal fin origin **25**

[22] **23a.** Mouth terminal; upper jaw not protractile (Plate 4 h) . . . . Umbridae, *Umbra pygmaea* (eastern mudminnow), p. 250
23b. Mouth supraterminal (opens slightly upward, see Plate 4 c for example); upper jaw protractile **24**

[23] **24a.** Dorsal fin located behind anal fin; anal fin of male slender and rodlike . . . . Poeciliidae (livebearers), p. 275
24b. Dorsal fin located nearly over anal fin; anal fin rounded in both sexes. . . Fundulidae (topminnows), p. 269

[22] **25a.** Lips thin and smooth (e.g., Plate 10 a–f); dorsal fin with 8 or 9 rays or, if more, the first principal ray is not jointed like the others and may be calcified, stiff like a spine, and serrate on posterior edge . . . . Cyprinidae (minnows), p. 111
25b. Lips thick and fleshy, often with papillae or ridges (Plate 12 d–h); dorsal fin with 10 or more rays and the first principal ray never calcified, stiff like a spine, and serrate on posterior edge . . . . Catostomidae (suckers), p. 175

Family and Species Accounts

Acipenseridae (sturgeons)

The sturgeons are considered an old and relatively primitive family with 23 living species divided in two subfamilies, Scaphirhynchinae and Acipenserinae. All species are easily recognized by the heterocercal tail, shovel-shaped snout, ventrally located mouth with large fleshy barbels, and large bony plates that cover all or parts of the body. Other than some ossified bones in the head, the skeleton is cartilaginous.

The diet consists primarily of benthic invertebrates. Some sturgeon species are confined to fresh water, but most are anadromous. The sexes are externally indistinguishable except during the reproductive season, when the females are bulging with eggs. Spawning for any individual tends to occur at intervals of several years and takes place in fresh water. The eggs are relatively large (2.5 mm or larger) and are adhesive and demersal. Sturgeons have been harvested for centuries for their eggs to produce caviar; the cured eggs are considered a delicacy in many parts of the world. Sturgeons are among the largest freshwater fishes. The white sturgeon (*Acipenser transmontanus*) is known to reach lengths in excess of 6 m.

Sturgeons grow slowly, mature late (between 5 and 28 years), and have a long life span, possibly more than 100 years. This is probably the reason why populations of almost all commercially important sturgeons have declined considerably due to overfishing and changes in their environment. Many species are now considered endangered, and efforts are being made to restore populations. Sturgeons have been very important economically, mostly because of the caviar, but the meat is highly regarded as well. Aquaculture efforts are currently under way (Etnier and Starnes 1993). A compendium of sturgeon research can be found in Van Winkle et al. 2002. Two species have been reported from the MSRB: the shortnose sturgeon (*Acipenser brevirostrum*) and the Atlantic sturgeon (*A. oxyrinchus*) have been collected near the Savannah River Bluff Lock and Dam (shown above) below Augusta, Georgia.

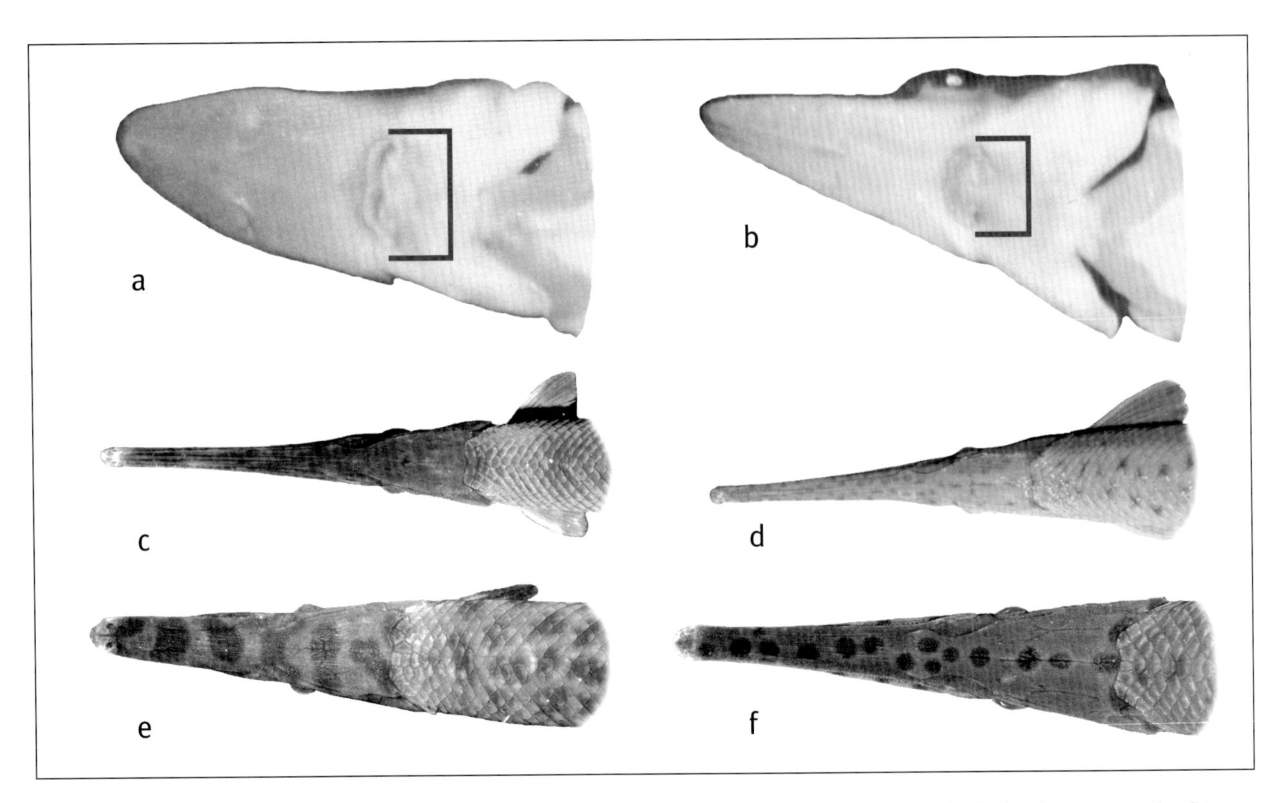

PLATE 6. Sturgeon and gar heads: (a) underside of head of shortnose sturgeon, (b) underside of head of Atlantic sturgeon, (c, d) longnose gar heads, without and with spots, (e) male Florida gar, (f) female Florida gar.

KEY TO THE SPECIES OF ACIPENSERIDAE

1a. Mouth large, gape width more than 62% of the distance between the eyes; snout short and bluntly V shaped (Plate 6 a) . . . . *Acipenser brevirostrum* (shortnose sturgeon), p. 67

1b. Mouth smaller, gape width less than 55% of the distance between the eyes; snout long and narrowly V shaped (Plate 6 b) . . . . *Acipenser oxyrinchus* (Atlantic sturgeon), p. 70

Acipenser brevirostrum Lesueur, 1818
Shortnose sturgeon

DESCRIPTION Tail heterocercal, sharklike; snout long with four sensory barbels on the ventral surface; mouth inferior and protrusible. Body with five rows of enlarged scutes (dorsal and paired midlateral and ventrolateral). Head large, convex, and depressed between the eyes. Snout shorter and more bluntly V shaped and mouth gape wider than that of *A. oxyrinchus* (see ratios in key; Plate 6 a, b); four barbels in ventral transverse row one-third to one-half the distance from snout tip to upper lip (Leim and Scott 1966).

Body brown above, tinged with copper on sides, below lateral scutes reddish mixed with violet shading to a white abdomen. Dorsal scutes with white centers. Iris with a greenish tint (Jones et al. 1978). Pectoral and pelvic fins white edged, anal fin pigmented. Visceral lining black (requires incision to see) (Vladykov and Greeley 1963).

SIZE This is the smallest species of *Acipenser* in North America (Bain 1997); adults attain lengths of about 100 cm and weigh up to 16.5 kg (Dadswell 1976).

MERISTICS Dorsal rays 33–42, anal rays 18–24; dorsal body scutes 8–13, midlateral scutes 22–34, ventrolateral scutes 6–11; upper gill rakers long, triangular, 22–29 on first arch (Vladykov and Greeley 1963).

CONSERVATION STATUS Listed as a federally endangered species; listed as S2 ("imperiled") by Georgia Department of Natural Resources.

DISTRIBUTION Coastal waters, including rivers and lakes, from the St. John River in New Brunswick to the St. Johns River in Florida; may penetrate far up rivers into fresh water (Gruchy and Parker *in* Lee et al. 1980). Commonly caught in gill nets in the Altamaha River, Georgia (Dahlberg 1975). Collins and Smith (1993) collected this species (57–99 cm TL) in the upper Savannah River (rkm 74–298) from February to April, and took 403 fish (26–107 cm TL) in all months in the lower Savannah River (rkm 0–74). Larvae were collected in 1982 and 1983 at two locations near the upper boundary of the SRS and at the mouth of Brier Creek downstream of the SRS in the Savannah River (Muska and Matthews 1983; DOE 1984).

Acipenser brevirostrum, shortnose sturgeon

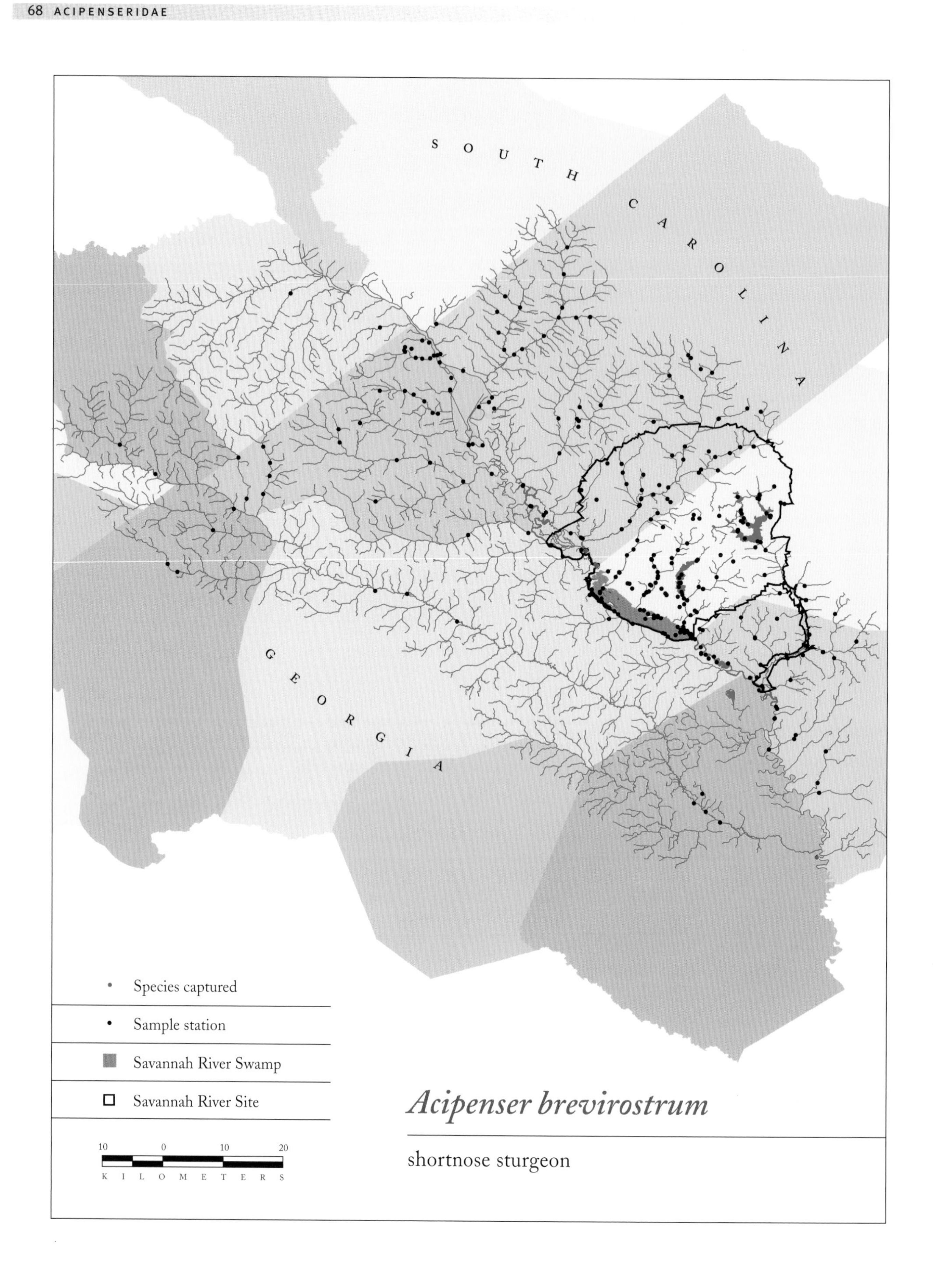

Acipenser brevirostrum

shortnose sturgeon

HABITAT Semianadromous; found in fresh, brackish, and marine waters. Overwintering occurs in estuarine lakes and deeper regions of lower estuaries in salinities up to 20 ppt. In a Georgia study, 12 of 551 tagged shortnose sturgeons were recaptured along the Georgia coast in commercial nets fished for American shad (Collins et al. 1996). During early spring, they move out of overwintering areas and concentrate in river channels. The growth rate is highest in spring and fall, with little growth in the summer. Shortnose sturgeons are amphidromous, utilizing discrete habitats within a freshwater system for both spawning and feeding (Gilbert 1989). They have been caught in drifting gill nets in the Edisto River and in stationary gill nets in the Savannah River (Collins and Smith 1993; McCord 1998). Adults and juveniles in southern rivers aggregate in deep areas near the saltwater-freshwater interface in summer (Hall et al. 1991; Moser and Ross 1995). During 1984–1992, 97,483 shortnose sturgeons were stocked in the Savannah River as part of a state/federal recovery program. Recaptures of marked fish after an average time of 7.2 ± 1.9 years (range 5.9–10.4) indicated that the stocked juveniles made up at least 38% of the population in 2002 (Smith et al. 2002a). Some of the stocked sturgeons did not imprint on the Savannah River and were later found in the Edisto River, the Ogeechee River (Georgia), the Cooper River (South Carolina), and in Winyah Bay (278 km north of the mouth of the Savannah River) (Smith et al. 2002).

BIOLOGY Dadswell (1976) provided the first thorough study of the life history of the shortnose sturgeon. Most shortnose sturgeons live 15–20 years (Gorham and McAllister 1974); the longevity record is 67 years (Dadswell et al. 1984). Dadswell (1976) reported that females spawn every third to fifth year, and males every second year. Adults spawn in the middle of large rivers, usually during peak flood tide in February and March in South Carolina. From mid-January to mid-April 1984–1992, 626 adult shortnose sturgeons were captured in the Savannah River, with significantly more captured in the lower (rkm 42–75) than the upper (rkm 160–299) reaches. Radiotelemetry data indicated that spawning occurred upriver between rkm 179 and rkm 278 (within the MSRB study area, which encompasses rkm 156–rkm 355), at water temperatures of 9.8–16.5 °C (Collins and Smith 1993). Some individuals spawned in consecutive years. Most left the freshwater reaches of the river in spring soon after the January–April spawning season (Collins and Smith 1993). Spawning occurs in fast flows over gravel.

Eggs are demersal, average about 3.0 mm (Dadswell 1976), and are extremely adhesive after fertilization, attaching to hard substrate such as rocks and submerged trees (Collins and Smith 1993) and then becoming nonadhesive after about 2 hours (Vladykov and Greeley 1963). The eggs hatch in 1–2 weeks (Vladykov and Greeley 1963). Larvae and early juveniles are poor swimmers; they stay near the bottom for about 2 weeks and drift with the current, then slowly emigrate downstream (Meehan 1910). See Jones et al. 1978 for larval descriptions. Juveniles migrate to and from fresh water for a number of years. Shortnose sturgeons begin their migration earlier in the season than Atlantic sturgeons (Dadswell et al. 1984).

The age at sexual maturity appears to be 8–15 years in the north and younger in the south depending on coastal location; growth rates vary by region and sex, but all fish mature at approximately the same size throughout the range (50–65 cm TL) (Dadswell et al. 1984). Fecundity ranges from 48,000 to 99,000 (Dadswell 1976). Newly hatched larvae are tadpolelike, dark gray, and have a large yolk sac (Dadswell et al. 1984). At 20 mm TL, shortnose sturgeons in the Hudson River had developed all of the external characteristics of juveniles (Pekovitch 1979).

Juveniles feed on smaller organisms than do adults (Carlson and Simpson 1987); common prey items are aquatic insects (Chironomidae), isopods, and amphipods. Molluscs are an important part of the diet of adults, but not of juveniles (Dadswell 1976). In the Hudson River, the diet probably includes insects and crustaceans, with molluscs dominating (25–50% of the diet) (Curran and Ries 1937).

SCIENTIFIC NAME *Acipenser* = "sturgeon" (Latin); *brevirostrum* = "short snout" (Latin).

Acipenser oxyrinchus Mitchill, 1814

Atlantic sturgeon

DESCRIPTION Tail sharklike (heterocercal), snout long, flattened, and V shaped with four barbels on the lower jaw. Eye small and oval with a pale golden iris. Mouth inferior, protrusible, suckerlike, and lacking teeth. Five rows of enlarged scutes (dorsal and paired midlateral and ventrolateral) present on body; each ventrolateral row of scutes extending past the anus so that two to six bony scutes (Holland and Yelverton 1973) at least as large as the pupil of the eye lie above the anal fin base and below the midlateral row of scutes.

Olive green to bluish gray above and pale below (Jones et al. 1978). Anterior edges of pectoral and pelvic fins, lower caudal lobe, and entire anal fin white. Visceral lining pale, unpigmented (requires an incision to see); preanal scutes in a double row (Vladykov and Greeley 1963).

SIZE Adults may reach 5.49 m TL, but records almost never indicate specimens above 4.27 m (Vladykov and Greeley 1963).

MERISTICS Dorsal rays 30–46, anal rays 22–32; dorsal body scutes 7–13, midlateral scutes 24–35, ventrolateral scutes 8–11 (Jenkins and Burkhead 1993); upper gill rakers 15–24 in specimens below 20 cm, 16–27 in specimens 20–231 cm (Vladykov and Greeley 1963).

CONSERVATION STATUS In May 1999 the National Marine Fisheries Service banned possession and harvest in the exclusive economic zone (EEZ; extending from the state boundary to 200 nautical miles offshore). The Gulf sturgeon (*A. oxyrinchus desotoi*) is federally listed as threatened. The Atlantic sturgeon fishery was closed in South Carolina in 1985 (Collins et al. 2000b).

DISTRIBUTION The Atlantic Coast from Labrador to Florida. The Gulf sturgeon, *A. o. desotoi* (Vladykov 1955), is found along the Gulf Coast from Florida to the Mississippi River in Louisiana and Mississippi; rarely found in French Guiana and Bermuda (Vladykov and Greeley 1963). Credible evidence indicates that *A. oxyrinchus* colonized the Baltic Sea about 1,200 years ago and completely displaced the European sturgeon (*A. sturio*) during the Little Ice Age about 800 years ago. The Atlantic sturgeon was recently extirpated in the Baltic Sea, but

Acipenser oxyrinchus, Atlantic sturgeon (because of difficult lighting conditions, colors are not accurate)

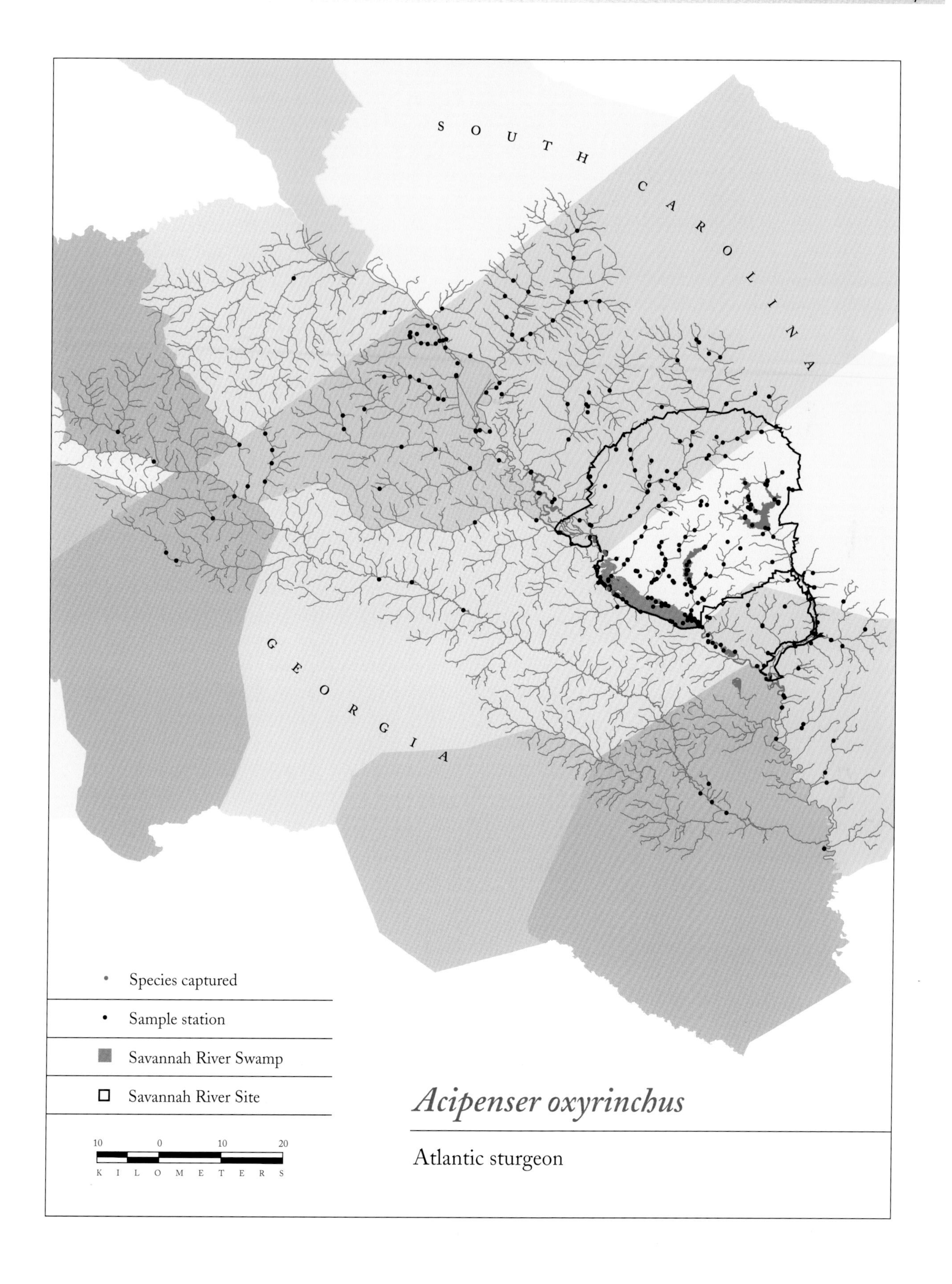

Acipenser oxyrinchus

Atlantic sturgeon

museum specimens verify its former presence there (Ludwig et al. 2002).

Often seen or caught in trawls in saltwater on the Georgia coast and in gill nets in the Altamaha River, Georgia (Dahlberg 1975). Atlantic sturgeons have been collected at depths up to 40 m off the South Carolina coast (Collins and Smith 1997) and have been caught in drifting gill nets in the Edisto and Combahee rivers in South Carolina (Collins et al. 2000a). After the collapse of the more northerly sturgeon fisheries, South Carolina became a major producer of Atlantic sturgeon, producing 55% of the U.S. total in 1976 (Smith et al. 1984).

HABITAT Generally closely associated with estuaries. In South Carolina, adults spend March–November in the river and are widely distributed in terms of salinity and location (M. Collins pers. comm.). They make extensive migrations both north and south of their natal streams. Migrations begin in early March in the Suwannee River in Florida, in February in the St. Marys River in Georgia, and in April in the Chesapeake Bay (Huff 1975). Atlantic sturgeon marked in the Hudson River by Dovel and Berggren (1983) were captured as far south as Cape Hatteras, North Carolina, and as far north as Cape Cod, Massachusetts.

BIOLOGY The Atlantic sturgeon is anadromous. Adults, which may live up to 60 years, reach maturity at an older age in the northern part of the range. Females reach sexual maturity at age 15 and older in the Hudson River (Van Eenennaam et al. 1996) and at 22–34 years in the St. Lawrence River (Scott and Crossman 1973), and males are mature at between 8 and 12 years in the Suwannee River, Florida (Huff 1975). There has been no validated aging study for this species (M. Collins pers. comm.). Individuals spawn only once every 2–6 years (Smith 1985). Fecundity ranges from 1,030,000 to 3,755,000 (Jones et al. 1978).

Adult females begin spawning in mid-May in the Hudson River. They migrate directly to the spawning grounds, which are deep-channel or off-channel habitats (Dovel and Berggren 1983). Paller et al. (1986b) recorded 43 sturgeon larvae, only tentatively identified to species, taken from the upper Savannah River (rkm 113–283) during 1982–1985. Spawning occurs over hard bottoms of clay, rubble, or gravel in running water at temperatures of 14–24 °C (Moser et al. 2000). Males may remain in the river until fall; females typically leave within 4–6 weeks in the northern part of the range. The first artificial propagation of Atlantic sturgeon to the advanced juvenile stage was done in South Carolina (Collins et al. 1999).

The eggs are demersal and adhere to weeds and stones (Vladykov and Greeley 1963). Fertilized eggs range from 2.5 to 2.6 mm (Borodin 1925), have a distinct cross- or star-shaped pigment patch, and hatched in 96 hours at 20 °C (Dean 1895) and in 168 hours at 17.8 °C (Vladykov and Greeley 1963). Larvae are about 7 mm TL at hatching (Smith et al. 1980) and remain on the bottom in deep-channel habitats (Bain 1997). Larvae have teeth (Bigelow and Schroeder 1953). A single larva was found in 1977 in an oxbow channel along the Savannah River (Bennett and McFarlane 1983). Collectors from the University of Georgia took two larvae at rkm 40 in 2000 (M. Collins pers. comm.). Descriptions of the larval stage can be found in Jones et al. 1978. Most juveniles remain in their natal river 3–5 years (some leave earlier) before migrating to the ocean.

Adults are bottom feeders; prey includes mussels, worms, shrimp, and small bottom-dwelling fish (Jones et al. 1978). Juveniles feed on worms, chironomid larvae, isopods, amphipods, and small bivalves (Vladykov and Greeley 1963). More information on Atlantic sturgeon life history can be found in Ryder 1890; Vladykov 1955; Vladykov and Greeley 1963; and Bain 1997.

SCIENTIFIC NAME *Acipenser* = "sturgeon" (Latin); *oxyrinchus* = "sharp snout" (Greek).

Lepisosteidae (gars)

The gars are a small group of relatively primitive fishes. The seven living species in one or two (depending on the authority) genera are restricted to North and Central America and Cuba. Two of the five species known to occur in the United States are found in the MSRB: the longnose gar (*Lepisosteus osseus*) and the Florida gar (*L. platyrhincus*).

The gar's body is long and almost cylindrical, and the family is easily recognized by the long jaws with many clearly visible conical teeth, the abbreviate heterocercal tail, and the ganoid scales (interlocking rhomboid bony scales). Almost all gars are medium to large fish generally found in sluggish waters in rivers, lakes, swamps, brackish estuaries, and coastal marine waters. They can survive periods of low oxygen by using the highly vascularized swimbladder as a lung (Jenkins and Burkhead 1993). Reproduction is during late spring. One female, usually surrounded by several males, disperses the relatively large and adhesive eggs over shallow vegetated areas. No nest building or care for eggs or larvae is known. Eggs of all gars are reported to be toxic to birds and mammals but not to fish (Scott and Crossman 1973; Pflieger 1975). Newly hatched larvae have an adhesive disk on the snout that is used to attach the larva to vegetation until the yolk sac is absorbed and it begins feeding on small zooplankton.

With increasing size, gars become piscivorous ambushers, but crayfish, salamanders, and other food items are frequently included in the diet. Females live longer and grow to a larger size than males (Etnier and Starnes 1993). Although considered a nuisance by many recreational and commercial fishers, gars have not been shown to be damaging to fisheries, and their firm, white, mildly flavored meat is eaten in some areas of Louisiana (Etnier and Starnes 1993).

KEY TO THE SPECIES OF LEPISOSTEIDAE

1a. Snout long and slender, its length 15–20 times the width; no spots on top of head, or less frequently, 1 central row of generally small spots or 2 rows of small spots, 1 on each side of the dorsal surface of the head (Plate 6 c, d) . . . . *Lepisosteus osseus* (longnose gar), p. 74

1b. Snout short and moderately slender, length 5–10 times the width; single central row of large dark spots or blotches on top of head (Plate 6 e, f) . . . . *Lepisosteus platyrhincus* (Florida gar), p. 78

Lepisosteus osseus (Linnaeus, 1758)

Longnose gar

DESCRIPTION Body elongate and nearly cylindrical. Snout extremely long and slender with overhanging upper jaw. Large villiform teeth in upper jaw form a single row on both sides of jaw and on palatine bones and vomer. Gill rakers rudimentary and irregularly arranged. Body covered with nonoverlapping, diamond-shaped, thick, enamel-covered ganoid scales. Thin, bony scutes present on leading edges of unpaired fins and both edges of caudal fin. Tail rounded in posterior outline and abbreviate heterocercal. Dorsal and anal fins located far back on body with origin of anal fin anterior to origin of dorsal fin; pelvic fin midabdominal.

Back and sides brown to olive, shading ventrally to white, yellowish white, or cream; sides sometimes have small spots. Juveniles have a conspicuous dark stripe running from the snout, through the eye, to the base of the caudal fin and may have an interrupted reddish brown stripe or series of spots immediately above the dark lateral stripe. Adults from clear water may have several irregular dark spots on the sides. Top of head most frequently without spots but may have two rows of small spots. Dorsal, caudal, and anal fins frequently have dark blotches. Pectoral and pelvic fins often clear but may have reddish tinge or may be darkened or dusky. Though rare, melanistic (black) (Beecher and Hixson 1982; Mettee et al. 1996; Pigg 1998) and xanthic (yellow or orange) individuals do occur (Mettee et al. 1996).

SIMILAR SPECIES *Lepisosteus osseus* is most similar to *L. platyrhincus* and *Strongylura marina*; both *Lepisosteus* species are easily distinguished from *S. marina* by their larger scales and rounded anal fin. *Lepisosteus platyrhincus* can be distinguished from *L. osseus* by its shorter, broader snout and the larger and more prevalent blotches on top of the head.

SIZE Adults commonly exceed 0.9 m, and may reach lengths of 1.4 m (Robison and Buchanan 1984) or 1.8 m (Mettee et al. 1996; Etnier and Starnes 1993) and weights up to 15.9 kg (Robison and Buchanan 1984) or 22.7 kg (Etnier and Starnes 1993).

MERISTICS Dorsal rays 6–9, anal rays 8–10, pectoral rays 10–13 (Hoese and Moore 1977), pelvic rays 6, caudal rays 11–14 (Jones et al. 1978); total gill rakers 24–28 (Becker 1983) or 14–31 (Suttkus 1963); lateral line scales 57–63 (Robison and Buchanan 1984) or 57–64 (Mettee et al. 1996) or 57–65 (Etnier and Starnes 1993) or 60–66

Lepisosteus osseus, longnose gar; (above) adult, (facing page) juvenile

(Becker 1983); predorsal scales 47–55 (Suttkus 1963; Robison and Buchanan 1984).

DISTRIBUTION From Quebec and the southern tributaries of the Great Lakes south on the Atlantic Coast to Florida; west in the Mississippi drainages to Minnesota, South Dakota, Kansas, and Oklahoma; and west in the Gulf of Mexico drainages to the Mexican tributaries of the Rio Grande and up the Pecos River into New Mexico (Wiley *in* Lee et al. 1980). Believed to have formerly inhabited the Rio Grande in New Mexico but now to be extirpated from that portion of the river (Sublette et al. 1990). Common in the MSRB.

HABITAT Creeks, sloughs, oxbow lakes, swamps, rivers, bayous, and impoundments; most common in sluggish pools and backwaters where there is at least some current. They appear to prefer more current in spawning areas. Catch rates indicate greater abundance in river habitats in Florida than in oxbow lakes (Beecher et al. 1977). In coastal Mississippi, more abundant in deeper waters of large bayous (Goodyear 1967). Because their highly vascularized swimbladder allows them to breathe atmospheric oxygen, longnose gars are capable of surviving in water with very low oxygen concentrations. They are often found in saltwater along the Georgia coast (Dahlberg 1975) and in the Gulf of Mexico (Hoese and Moore 1977; Swift et al. 1977; Etnier and Starnes 1993). Most abundant in the Savannah River and its associated swamp, but often common in tributary streams and some impoundments.

BIOLOGY Adults are gregarious and occur in groups of two to five (Holloway 1954). In Florida, they aggregate in late winter to early spring, often in waters over mud flats with slight current; smaller individuals are found in shallower water than larger ones (Holloway 1954). Spawning aggregations occur in spring, and eggs are scattered over weeds, algal mats, bare rock, or gravel; no parental care is provided to offspring. Longnose gars appear to be frequent symbiotic nest associates of smallmouth bass, *Micropterus dolomieu*; the gar offspring are larger but resemble the host in color (Goff 1984). One of 75 *M. salmoides* nests examined in Lake Weir, Florida, contained eggs of *Lepisosteus* sp. (Chew 1974). Lake populations in Missouri may migrate 10 km or farther up tributary streams to spawn and after spawning may disperse back into lakes, traveling up to 48 km total (Johnson and Noltie 1996).

Eggs are demersal and oval with average dimensions of 4.2 by 3.0 mm (Simon and Wallus 1989). The larvae hatch at 9–10 mm TL and are heavily pigmented with a dark back and light ventral surface to the yolk sac

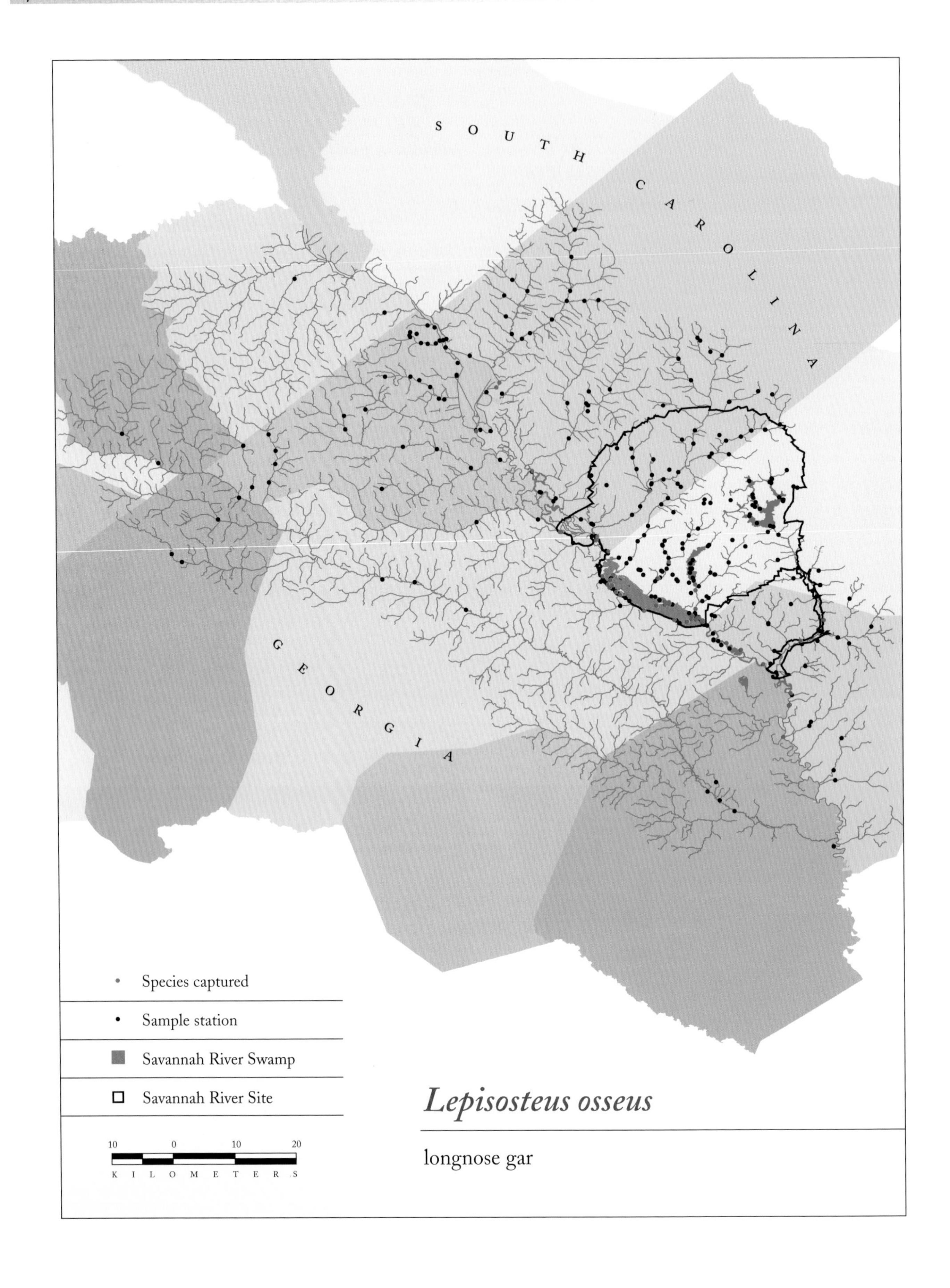

Lepisosteus osseus

longnose gar

(Echelle and Riggs 1972; Yeager and Bryant 1983). Newly hatched larvae attach themselves vertically to objects in the water using an adhesive organ on the snout and stay in place until the yolk is absorbed. Larvae have long dorsal filaments extending from the caudal fin that appear to be extensions of the urostyle (see figure in Becker 1983; or figure 60 in Etnier and Starnes 1993). Simon and Wallus (1989) described egg and larval stages for this species.

In Missouri, females are larger than similar-aged males and attain greater size and age (Netsch and Witt 1962; Klaassen and Morgan 1974; Johnson and Noltie 1997). Males mature at 3–4 years, females at 6 years. Males live up to 11 years while females may reach 22 years (Etnier and Starnes 1993). This is one of the fastest-growing freshwater fish documented, growing up to six times faster than other common large freshwater fishes (Netsch and Witt 1962).

Young longnose gars 17–21 mm TL studied in Lake Texoma, Oklahoma, fed on microcrustaceans, insects, and fish (Echelle 1968). Apparently they must start feeding by 25–26 mm TL, with first feeding items in the Ohio River being minnow or sucker larvae, cladocerans, or insect larvae (Pearson et al. 1979). Young-of-the-year become primarily piscivorous as small as 20–50 mm TL (Mittelbach and Persson 1998). Juveniles and adults typically are opportunistic piscivores but include an occasional shrimp, crayfish, or insect in the diet (Holloway 1954; Goodyear 1967; Netsch 1967; Crumpton 1971; Echelle and Riggs 1972; Becker 1983; Seidensticker 1987; Tyler et al. 1994). Studies indicate that longnose gar particularly prey on available cyprinids and clupeids like menhaden, gizzard shad, and threadfin shad, while centrarchids are mostly absent from the diet (Goodyear 1967; Netsch 1967; Becker 1983; Seidensticker 1987). In Florida and southern Georgia, this species either ceases to feed or feeds very little in the winter (Swift et al. 1977).

SCIENTIFIC NAME *Lepisosteus*, from *lepid* = "scale" and *osteus* = "bony" (Greek), therefore "bony scale"; *osseus* = "bony" (Latin).

Lepisosteus platyrhincus DeKay, 1842
Florida gar

DESCRIPTION Body elongate, slender, almost cylindrical, with thick, nonoverlapping, diamond-shaped ganoid scales. Snout relatively long and slender, the length 5–10 times the width, and with rows of sharp teeth visible. Caudal fin rounded and abbreviate heterocercal. Dorsal fin origin posterior to anal fin origin; both fins located far back on body near caudal fin, with distal edges rounded. Pelvic fins located near the longitudinal middle of the fish.

Suttkus (1963) noted that females have a proportionally longer snout than males. We also observed dimorphism in snout length and width as well as coloration. Individuals with a longer and narrower snout are typically marked with a single row of dark round spots, while those with a shorter and broader snout have a single row of broad dark blotches or bars.

Color is variable (Suttkus 1963) but is generally olive to brown on the dorsum with dark spots or blotches, background color grading into white to tan on the sides and white more ventrally. The ventrum may be black or have black stripes. Sides also with numerous dark spots or blotches. Dorsal, anal, and caudal fins have dark blotches, and pectoral and pelvic fins are often dark or dusky. A specimen orange dorsally with shading to yellow-orange on the sides was collected in midsummer in Florida (Phillips 1958).

SIZE Maximum size 1,330 mm TL (Hammett and Hammett 1939 *in* Suttkus 1963).

MERISTICS Dorsal rays 7–8, anal rays 7–8, pectoral rays 9–11, pelvic rays 6, caudal rays 12–13; gill rakers 25 (19–33); lateral line scales 54–59 (Suttkus 1963).

DISTRIBUTION From the southern tip of Florida to the Savannah River drainage in the north (Suttkus 1963; Page and Burr 1991), and west to the Ochlockonee River drainage of Florida and Georgia (Gilbert *in* Lee et al. 1980). In the MSRB, most commonly collected in the Savannah River and its associated swamp, but also found in tributary streams, especially the lower reaches on the fluvial terraces of the Savannah River.

Lepisosteus platyrhincus, Florida gar; (top) male, (bottom) female.

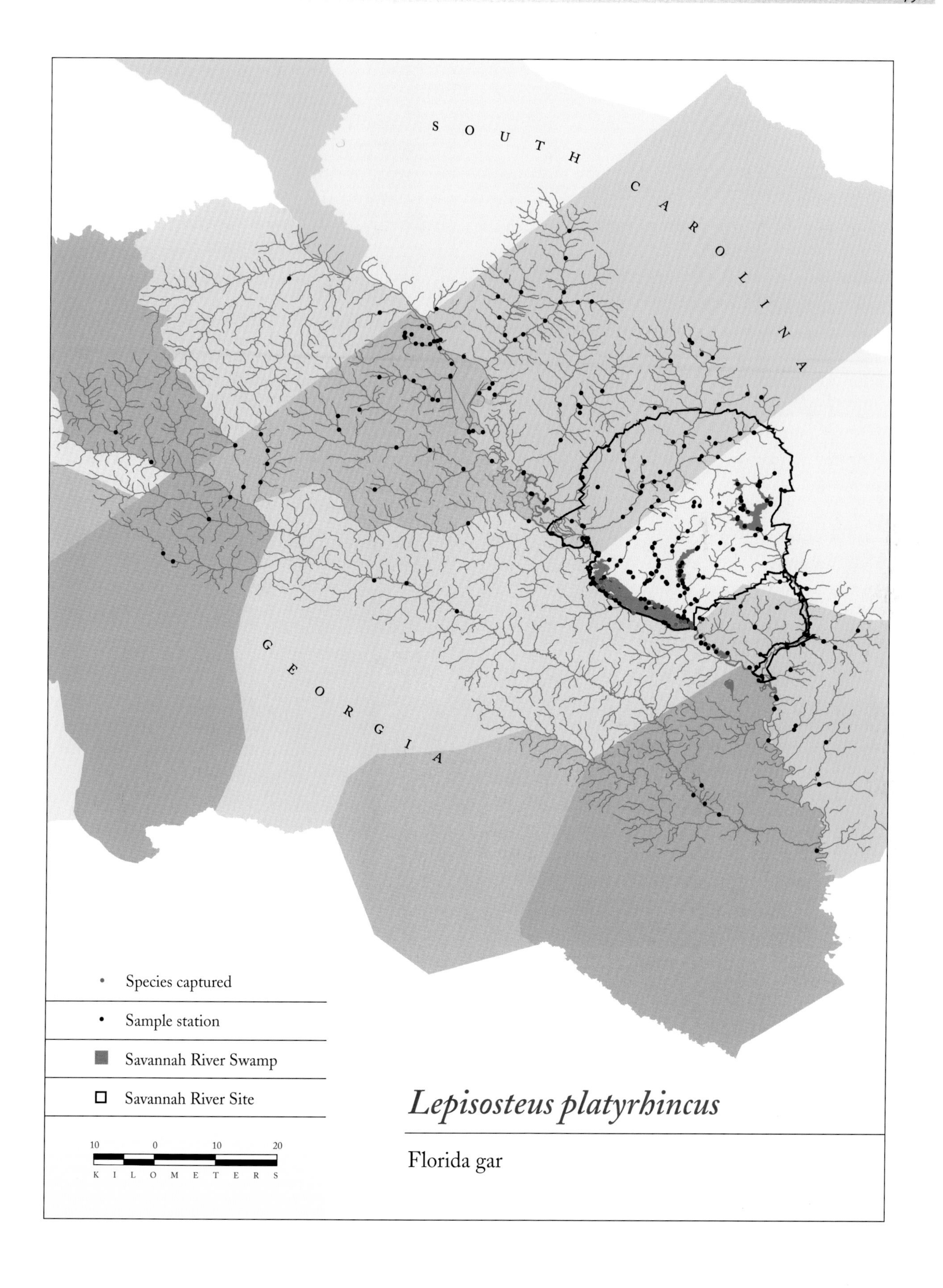

Lepisosteus platyrhincus

Florida gar

HABITAT Sluggish, mud- or sand-bottomed pools with slow or moderate-velocity flow in lowland streams and lakes, often near vegetation. In winter, moving into deeper waters (depths 3.7–7.6 m) near vegetation (Holloway 1954). During the summer, often found almost motionless or slowly cruising at the surface. In the Kissimmee River of Florida, most abundant in slough habitats (Trexler 1995). May be encountered in coastal marine waters. In Florida, Holloway (1954) found them concentrated in shallow mud flats in February; they preferred flats with some current, but had no apparent preference for cover. They were less abundant in smaller streams than in lakes, a trend not directly related to spawning, which occurred later in the year. Florida gars can tolerate very low oxygen conditions because they can use the swimbladder to breathe atmospheric oxygen (Suttkus 1963; McCormack 1967).

BIOLOGY Spawning occurs from March to September (Holloway 1954). Females discharge adhesive eggs among submerged aquatic plants. No parental care is provided to the offspring. Newly hatched larvae have adhesive suckers at the tip of the snout that are used for attachment to vegetation or other structures (Suttkus 1963 describes larvae, in description of gars in general). Females grow larger and mature later than males (Holloway 1954).

Florida gars frequently feed in groups of 2–10 during all hours of the day, but are more active after sunset and on cloudy days (Holloway 1954; Laerm and Freeman 1986). Juveniles eat zooplankton, insects, and small fish; adults prey mostly on fish, shrimp, and crayfish. The concentrations of species in stomachs reflected relative abundance in the habitat sampled by Holloway (1954). Although tilapia are abundant, Florida gars feed primarily on gizzard shad and bluegills in Lake Apopka, Florida (Gu et al. 1996). Crumpton (1971) found similar feeding habits in five central Florida lakes. In the Tamiami Canal of southern Florida, Hunt (1953) found topminnows and palaeomonid shrimp to make up the bulk of the diet (based on average stomach volume), and small sunfish and insects to be only a small proportion of the prey.

SCIENTIFIC NAME *Lepisosteus*, from *lepid* = "scale" and *osteus* = "bony" (Greek), therefore "bony scale"; *platyrhincus* = "flattened snout" (Greek).

Lepisosteus platyrhincus, Florida gar; juvenile

Amiidae (bowfins)

The bowfin, *Amia calva*, is the only living species of the Amiidae, an ancient group that existed as early as the Jurassic period (Nelson 1994) and once shared the earth with dinosaurs. The living bowfin is confined to fresh water, but fossils have been discovered from both freshwater and marine environments. Some species were very large; for example, an extinct African species reached lengths of 2.5–3.5 m (Jenkins and Burkhead 1993). Among living fishes, the bowfin is most closely related to the gars (Lepisosteidae). Primitive characters include a gular plate, an abbreviate heterocercal tail, and a heavily vascularized and subdivided swimbladder (Jenkins and Burkhead 1993).

Amia calva **Linnaeus, 1766**
Bowfin

DESCRIPTION Body elongate, cylindrical, and somewhat compressed near the tail. Mouth large and subterminal with a large gular plate under the lower jaw. The dorsal fin has a straight distal margin and extends two-thirds the length of the body to near the abbreviate heterocercal (hemicercal) tail. Caudal fin rounded; lateral line complete; scales cycloid. The heavily vascularized and subdivided swimbladder is used for air breathing (e.g., Jordan and Evermann 1896; Farmer and Jackson 1998; Hedrick and Jones 1999; Gonzalez et al. 2001).

Males and juveniles with a distinct dark spot (ocellus) on upper part of caudal fin. Larvae black; juveniles up to 80 mm SL with a brownish orange ground color and two to three dark bands on the dorsal, caudal, and anal fins. Dark vertical stripe along posterior edge of operculum; three dark horizontal stripes on head. Dorsalmost and often darkest stripe extending across the snout and past the eye to the posterior margin of the head. Ventralmost stripe extending anteriorly across cheek and around ventral contour of lower jaw. Ground color of larger individuals is variable, ranging from straw to greenish or gray, darkest dorsally and fading to whitish ventrally; stripes on head variably present; stripes or mottling may be visible in dorsal and caudal fins.

Adults are easily distinguished in the breeding season, even from a distance of 3–6 m (Reighard 1903). Breeding males have bright green dorsal, caudal, pectoral, and pelvic fins; the dorsal base of the caudal fin has a large, conspicuous, velvety black ocellus bordered by a broad band of orange or yellow. Stripes on cheek are more distinct in males than females. Fins of breeding females are brownish red, never green, and the ocellus is indistinct and not bordered (Reighard 1903).

SIMILAR SPECIES Mudminnows superficially resemble juvenile bowfins but lack the heterocercal tail and have a much shorter and rounded dorsal fin.

SIZE Females commonly reach 600 mm TL and are larger than males (some reach 450 mm) (Jordan and Evermann 1896). The heaviest specimen on record (9.7 kg) was caught by an angler in January 1980 in Florence, South Carolina (data from the International Gamefish Association [IGFA]). Swift et al. (1977) reported a female from the

Ochlockonee River of Florida and southern Georgia 993 mm in TL.

MERISTICS Dorsal rays 42–53, anal rays 9–12, pectoral rays 16–18, pelvic rays 7, caudal rays 25–28; lateral line scales 62–70 (Jenkins and Burkhead 1993).

DISTRIBUTION In the north, from the St. Lawrence and Ottawa rivers, Lake Champlain, and west through the Great Lakes, south into the Mississippi River basin to Louisiana and west through lower Texas drainages to the Colorado River. In the Atlantic Coastal Plain from Pennsylvania to Florida. Introduced in Iowa, Illinois, North Carolina, and Connecticut (Burgess and Gilbert *in* Lee et al. 1980). In the MSRB, most common in the Savannah River and in backwaters or tributaries on the modern Savannah River floodplain. Also common in some large reservoirs, but generally rare in upland tributaries.

HABITAT Common in sluggish rivers and streams associated with swamps (Swift et al. 1977; Burgess and Gilbert *in* Lee et al. 1980). Most common in shallow, vegetated waters but recorded from relatively deep water in Virginia (Jenkins and Burkhead 1993). Often found around logs and stumps along the banks of the Savannah River or in vegetated margins of backwaters or inlets.

BIOLOGY Spawning occurs from April to early June, peaking in mid-May in Wisconsin (Dean 1899) and in April and May in Michigan. Spawning takes place in quiet bays or inlets with abundant aquatic vegetation and cover such as stumps, bushes, or fallen trees (Reighard 1903). In the Pee Dee River drainage, adults often move upstream in the spring to find suitable spawning habitat and can be found in large numbers below dams and other obstructions. Some anglers await these "runs" and catch appreciable numbers of bowfins by snagging or other methods (Dan Crochet pers. comm.). Males construct a roughly circular nest, 30–90 cm in diameter and 10–20 cm deep, in shallow water (depth about 30–100 cm) among underwater vegetation, tree roots, and logs (Reighard 1901, 1903, 1931). Reighard (1903) found nests in water 12–22 °C, with greatest spawning activity at 16–19 °C. Emergent vegetation may be pushed over and broken off to clear an area for nest construction (Dean 1899). Much of the excavation appears to be done with the snout or even by biting vegetation (Dean 1899; Reighard 1903). Most nest building and much of the courtship/spawning occur at night (Reighard 1903). A nest may contain 2,000–5,000 eggs (Reighard 1931); the fecundity of females is between 2,765 and 64,000 eggs (Jones et al. 1978).

The eggs are adhesive and often attached to plant roots. In Michigan, eggs hatched in 8–10 days (Reighard

Amia calva, bowfin; adult

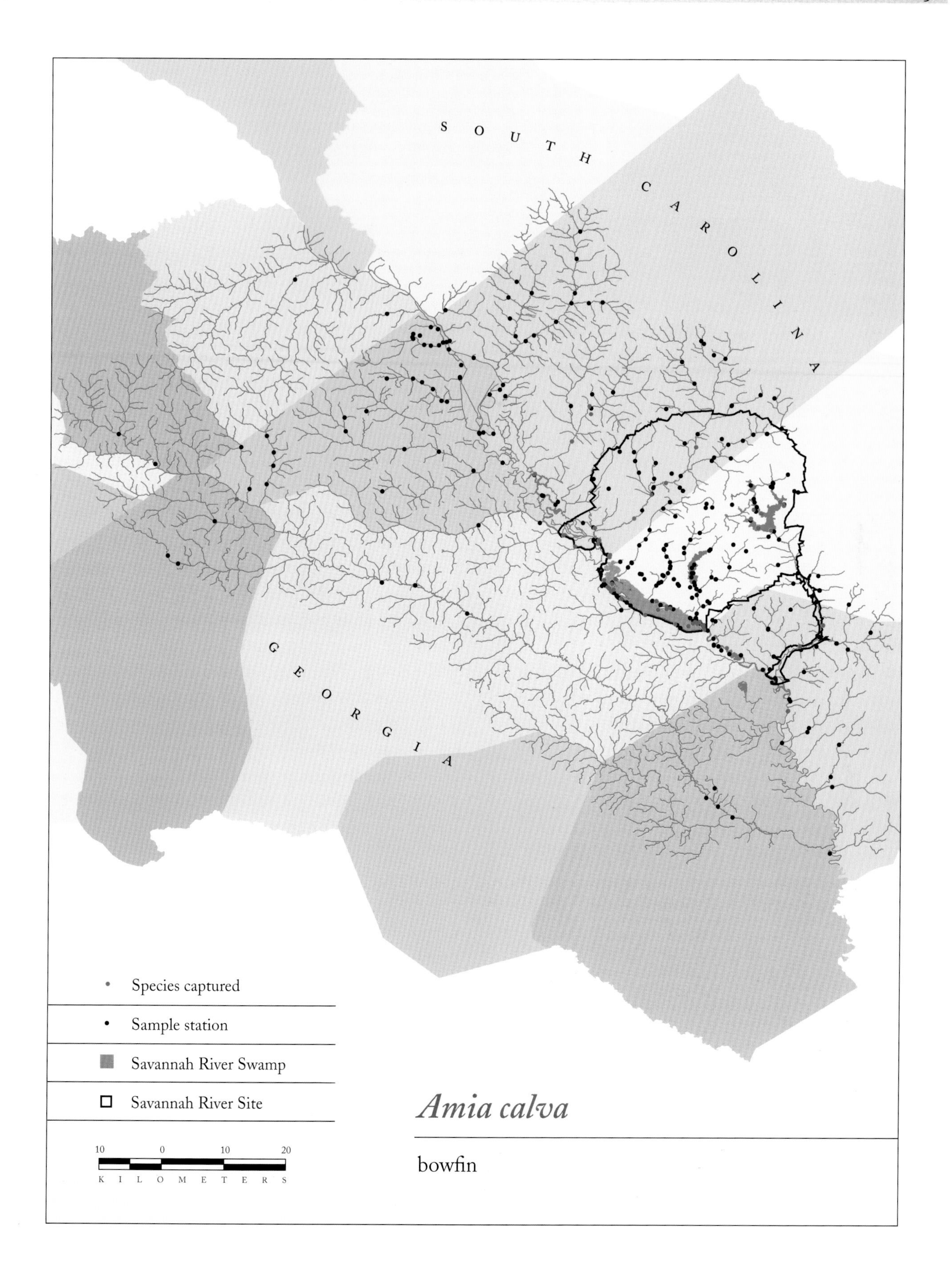

Amia calva

bowfin

Amia calva, bowfin; two color forms of juvenile

1903). Descriptions of larvae are summarized in Jones et al. 1978 and Mansueti and Hardy 1967. Hatchlings have an adhesive organ on the tip of the snout (Reighard and Phelps 1908) that is no longer functional by 12 mm TL (Reighard 1903). After the adhesive organ is absorbed and the blackish young are free swimming, they hover in tightly packed schools (Swift et al. 1977). Larvae begin to feed while still in the nest (Reighard 1903). The male parent continues to guard the swarm of young after they leave the nest; the swarm generally remains in shallow water in the vicinity of the nest (Reighard 1901). The swarming young may drop to the bottom and disperse when disturbed, but the guardian male quickly regathers them. Defense of the eggs and young may last for 1–2 months until the juveniles are 80–100 mm in length (Reighard 1901; Breder and Rosen 1966).

Based on annual marks in scales, the gular plate, and otoliths, Cartier and Magnin (1967) determined that the life span of bowfins in Montreal rarely exceeds 12 years and that there are no differences in growth rates between males and females. Carlander (1969), however, stated that bowfins may live at least 30 years. The recorded average total lengths in the first five year classes in the Montreal area were 228, 388, 455, 528, and 590 mm, respectively, with further growth to a little over 700 mm above 7 years of age (Cartier and Magnin 1967).

Small juveniles eat planktonic microcrustaceans; larger juveniles (>5 cm TL) feed on aquatic insects and small crustaceans such as amphipods (Schneberger 1937). First-year bowfins may become piscivorous at 7–10 cm TL, with up to 90% of their diet consisting of fish (Mittelbach and Persson 1998). In Michigan, bowfins eat crayfish, fish (especially yellow perch, *Lepomis* spp., and bullheads), insects, frogs, and molluscs (Lagler and Hubbs 1940; Lagler and Applegate 1942). In feeding trials, bowfins preferred fathead minnows and crayfish to sunfish (Mundahl et al. 1998). Bowfins may feed infrequently on newts, sirens, and snakes (Jordan and Arrington 2001).

Bowfins studied in a North Carolina swamp were sedentary (Whitehurst 1981) and appeared to be most active at twilight or during the early morning hours (Reighard 1903). In the MSRB, bowfins seasonally migrate into the lower portion of Steel Creek in the spring; most leave during the fall as the temperature of the water flowing from the swamps drops below the water temperature of the river (DEF pers. obs.).

The bowfin is commonly used as a laboratory test animal because it is easy to keep, has interesting physiology and behavior, and is a "living fossil" (Burgess and Gilbert *in* Lee et al. 1980). Neill (1950) described estivation, but northern populations appear to be physiologically incapable of estivation. It is possible that fish of southern populations estivate, but current information is inconclusive (McKenzie and Randall 1990).

SCIENTIFIC NAME *Amia* = ancient name for some kinds of fish (Jordan and Evermann 1896); *calva* = "bare" or "bald," referring to the scaleless head. Local common names include dogfish, mudfish, and grinnel.

Megalopidae (tarpons)

Tarpons are large, predatory marine fishes with a mostly coastal distribution. Late-stage larvae and small juveniles are usually found in estuarine waters, and both juveniles and adults regularly enter fresh water. Tarpons reach lengths in excess of 1 m and have large mouths and eyes, large silvery scales, numerous branchiostegal rays, and elongated posterior rays in the dorsal fins. They have bony gular plates on the throat, a feature shared in the MSRB exclusively with bowfins (Amiidae). The family comprises one genus and two species. Some systematists group the tarpons with the ladyfishes (*Elops* spp.) into the family Elopidae.

Tarpons are found in tropical and subtropical coastal waters of the Indo-Pacific, the western Atlantic, the eastern Atlantic off Africa, and, very rarely, off southern Europe. The vascularized swimbladder can be used as "lungs" for breathing atmospheric oxygen, allowing tarpons to live in waters with very low oxygen content. Like their close relatives the ladyfishes and bonefishes (family Albulidae), they have peculiar flattened, ribbonlike marine larvae called "leptocephali" (singular, leptocephalus). The only other fishes with similar larvae are the eels (order Anguilliformes). Off Florida, tarpons spawn seasonally; off Costa Rica, they spawn all year long, generally near the full moon or near the new moon (Crabtree et al. 1995, 1997). They are highly valued sport fish because of their size, strength, and spectacular leaps when hooked, but they are seldom eaten. *Megalops atlanticus* is the only species representing this family in the MSRB. Young tarpons occasionally wander up the Savannah River into the MSRB for reasons unknown.

Megalops atlanticus **Valenciennes, 1847**

Tarpon

DESCRIPTION Body moderately deep and laterally compressed. Lower jaw projects prominently; maxillary bone large, reaching under the posterior part of the eye in specimens 80–100 mm and far beyond the margin of the eye in larger specimens (Hildebrand 1963). Eyes large, one-fifth to one-third of the head length (Hildebrand 1963). Last dorsal fin ray elongated. Pseudobranchiae undeveloped (look on inside of opercle to see). Caudal fin deeply forked. Lateral line decurved with branched tubes on lateral line scales. Scales large and cycloid with crenulate membranous borders. Gular plate present on the throat between mandibles of lower jaw. Pectoral fins originate very low on sides of thorax. Large axillary scale present at bases of pectoral and pelvic fins.

Back bluish or greenish gray, sides silvery white. Pectoral and pelvic fins unpigmented; other fins more or less dusky.

SIZE Sexually mature at 1.0–1.2 m TL. Males reach 1.6 m FL; females reach 2 m FL; maximum length 2.5 m. Tarpon from Costa Rica tend to be smaller than Florida fish in both mean size and size at maturity (Crabtree et al. 1997).

MERISTICS Dorsal rays 12–16, anal rays 20–25, pectoral rays 13 or 14; lateral line scales 37–48; gill rakers 19–22 upper + 36–40 lower (Jones et al. 1978); branchiostegals about 23 (Hildebrand 1963).

DISTRIBUTION In the western Atlantic, common from Virginia to Brazil, straying as far north as Nova Scotia and as far south as Argentina; in the eastern Atlantic, common from Senegal to Angola, straying as far north as Portugal and France. Individuals tagged in North Carolina have been recaptured as far away as Cuba (Schwartz 2002). Tarpons appear to have entered the Pacific through the Panama Canal. The SCDNR has a photograph of a juvenile captured by a fisherman near Little Hell Landing just below the SRS.

HABITAT Mostly coastal marine waters, but they regularly enter rivers, tidal creeks, canals, and coastal lagoons. During periods of inundation adults move across flooded marshes and are thus found in deeper ponds and borrow pits in the Everglades (Loftus and Kushlan 1987). Adults occur at salinities from 0 to 43 ppt and temperatures from 17 to 37 °C. They are sometimes pelagic in the open ocean, and spawning is restricted to offshore waters. Smaller juveniles are most common in estuarine and other euryhaline situations and are seldom found in marine waters. Larger juveniles frequently penetrate far up rivers and into headwaters of coastal rivers and creeks. Individuals stocked in a Texas reservoir survived and grew for 2–5 years (Howells and Garrett 1992). The swimbladder has a rete mirabile that can be used for aerial respiratory exchange of oxygen and carbon dioxide, so tarpon can tolerate oxygen-deficient waters.

BIOLOGY Spawning occurs in offshore pelagic waters. Individuals may form schools or swim alone. Sexual maturity occurs at 6–7 years. Males have an estimated life span of 32–41 years; females live an estimated 51–78 years (Andrews et al. 2001). The larvae are leptocephali; although teeth are present, there is no evidence of predatory activity. The leptocephali migrate to coastal waters, using their own tissue for energy and thus shrinking during the migration. These shrinking (stage 2) leptocephali enter estuaries or coastal rivers and streams where they undergo a first metamorphosis into small, rodlike, finned larvae that migrate farther up the estuary and undergo a second metamorphosis into juvenile-stage fish. They stay in the upper reaches of estuaries or go up rivers and remain for up to 2 years. Jones et al. 1978 summarizes early development and includes illustrations of the developmental sequence.

The diet is mainly fish and crabs.

SCIENTIFIC NAME *Megalops*, from *mega* = "large" (Greek) and *ops* = "eye" (Greek), or "large eyes"; *atlanticus* = from the Atlantic Ocean.

Megalops atlanticus, tarpon

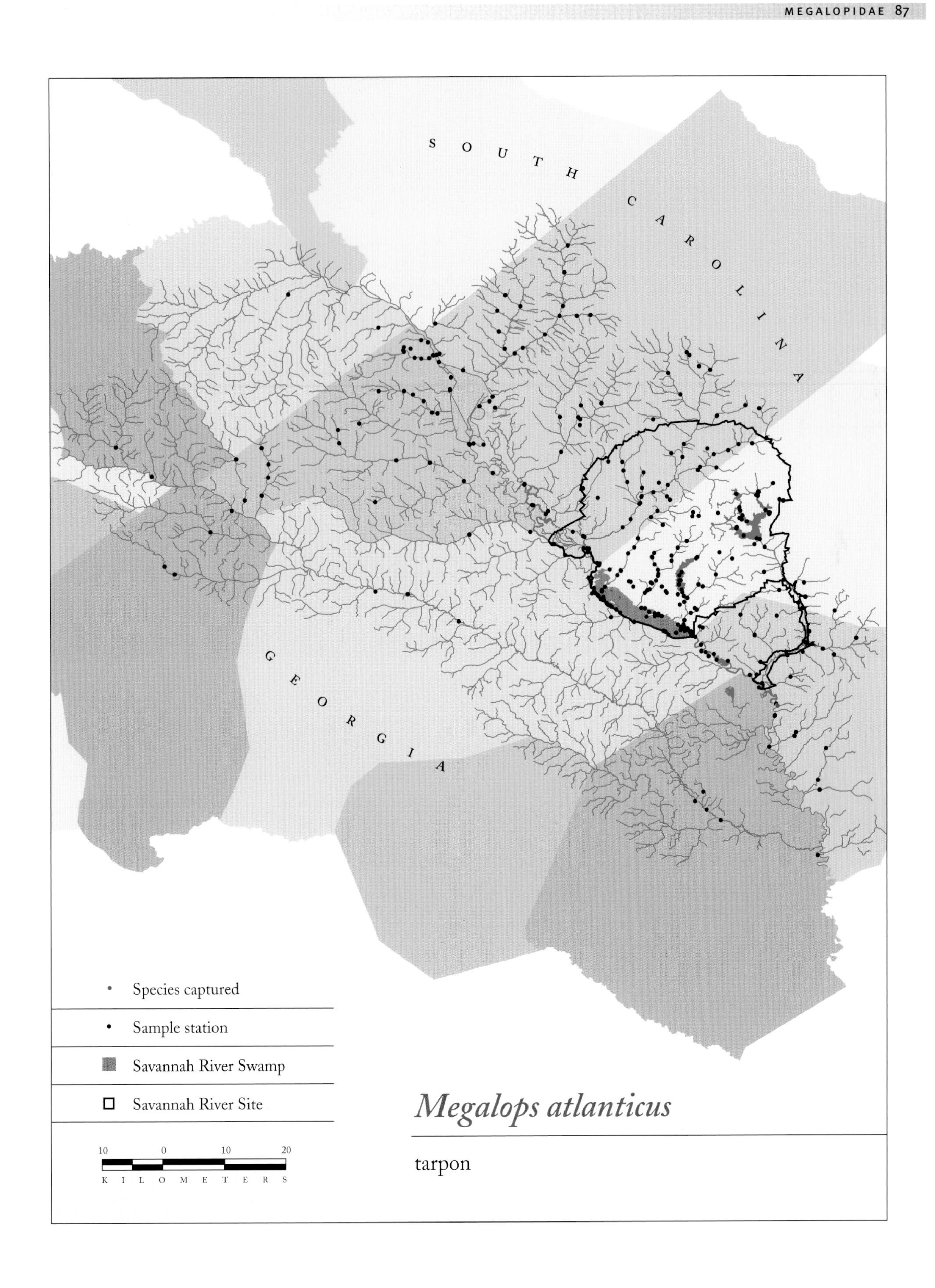

Megalops atlanticus

tarpon

Anguillidae (freshwater eels)

The family consists of one genus, *Anguilla*, with 15 species, including the only North American representative, the American eel (*Anguilla rostrata*). The eels (order Anguilliformes) are most closely related to bonefishes, tarpons, and ladyfishes, and share with them the larval leptocephalus phase. All freshwater eel species are catadromous and semelparous, returning to the ocean to spawn only once after spending the majority of their life in fresh or estuarine water.

Freshwater eels have no pelvic fins, scales that are either very small and embedded or absent, and an anguilliform body shape. This elongate, serpentine body form and the mode of traveling or swimming by means of undulations of the body (anguilliform locomotion) are especially suited for entering and exiting crevices and for burrowing in sediment (Jenkins and Burkhead 1993).

Freshwater eels are found in temperate and tropical waters draining into the Atlantic, western Pacific, and Indian oceans. They are absent from the eastern Pacific. Eels are mostly carnivorous, actively hunting for prey but also scavenging for food. The complex life history features several morphologically very different developmental stages.

Freshwater eel is an important food fish in many parts of the world and is considered a delicacy in many European countries and Japan. Export to foreign markets largely drives the U.S. eel fishery, but some freshwater eel is consumed in the United States, predominantly along the east coast as smoked eel (Etnier and Starnes 1993).

Anguilla rostrata **(Lesueur, 1817)**

American eel

DESCRIPTION Body elongate and cylindrical, compressed toward the tail. Dorsal, caudal, and anal fins continuous; pelvic fins absent. Mouth large and terminal. Scales small and embedded.

Back olive, greenish to yellowish brown, or grayish, becoming lighter on the sides toward the pale yellow or whitish belly. Sexually immature adults (eels are not sexually mature until right before out-migration) tend to have a yellow patina, thus the common name "yellow eel" for this life stage. Out-migrating, sexually mature "silver eels" develop a silver patina on the sides and abdomen.

SIZE Females may reach 120 cm but are typically no more than 75 cm; males reach up to 75 cm but are usually less than 55 cm (Fahay 1978; Jessop 1987). Those found on the SRS are generally no longer than 70 cm (Bennett and McFarlane 1983), and those in small tributaries may reach only 50 cm.

MERISTICS Pectoral rays 14–20; branchiostegal rays 9–13.

DISTRIBUTION The Atlantic Coast of North America and southern Greenland, through the West Indies and along the Caribbean Coast of Central and South America and the Atlantic Coast of South America occasionally south to Brazil. Via the St. Lawrence River into the

Great Lakes, where the species is confined to Lake Ontario (EPRI 1999; Cudmore-Vokey and Crossman 2000). Through navigation canals, throughout the Ohio River and much of the Missouri River drainage, and into New Mexico via the Pecos River (Lee *in* Lee et al. 1980; Van Den Avyle 1982). Introduced into California and Colorado, but the catadromous lifestyle with spawning occurring in the Sargasso Sea precluded establishment. Abundant in the MSRB.

HABITAT Eels occur in a broad diversity of habitats ranging from estuaries to large rivers and small streams, particularly favoring backwaters. Adult females are predominantly found in fresh water unless migrating to the spawning grounds; males remain almost exclusively in salt or brackish water (Fahay 1978). Individuals burrow in mud, hide in holes, or drape themselves over vegetation.

BIOLOGY This catadromous fish has a complex life history (see Fahay 1978; Hardy 1978; and Dixon 2003 for overview, detailed description of various stages, and literature). Spawning takes place in the Atlantic (probably in the Sargasso Sea). Development involves several stages and developmental dimorphisms. Eggs hatch into leptocephali. The larva drifts with the current during its transformation into a transparent free-swimming glass eel, and then into an elver. Glass eels enter Delaware and Indian river estuaries December–May, with the peak in February–April (Wang and Kernehan 1979). During upriver migration the young elvers transform into fully pigmented juveniles when they are about 1 year old and about 65 mm in length; juveniles 80–100 mm are most commonly collected on the SRS. As male eels generally stay in estuarine waters, most or all eels in the MSRB are probably females. Growth to adulthood can take more than 7 years for females and 4–7 years for males; sexual differentiation of the gonads does not occur until the eels are about 200 mm (Fahay 1978). Adults migrate back to sea in fall and midwinter (Vladykov 1964 *in* Jenkins and Burkhead 1993), reaching sexual maturity during their final journey to the spawning grounds. No adult eels are known to return to fresh water. Individuals may live up to 40 years (Jessop 1987), but most live no more than 25 (see Oliviera 1999). Total lengths at 10, 20, and 30 years in a Nova Scotia population were 450–550, 550–600, and 650–715 mm, respectively (Jessop 1987).

Eels may survive drought and low-oxygen conditions via gas exchange through gills and skin (Hughes 1976; Helfman et al. 1997) and are known to leave the water and cross to other water bodies over short distances. They are nocturnal. The diet in tributary streams of the MSRB consists of a large variety of live prey, in particu-

Anguilla rostrata, American eel

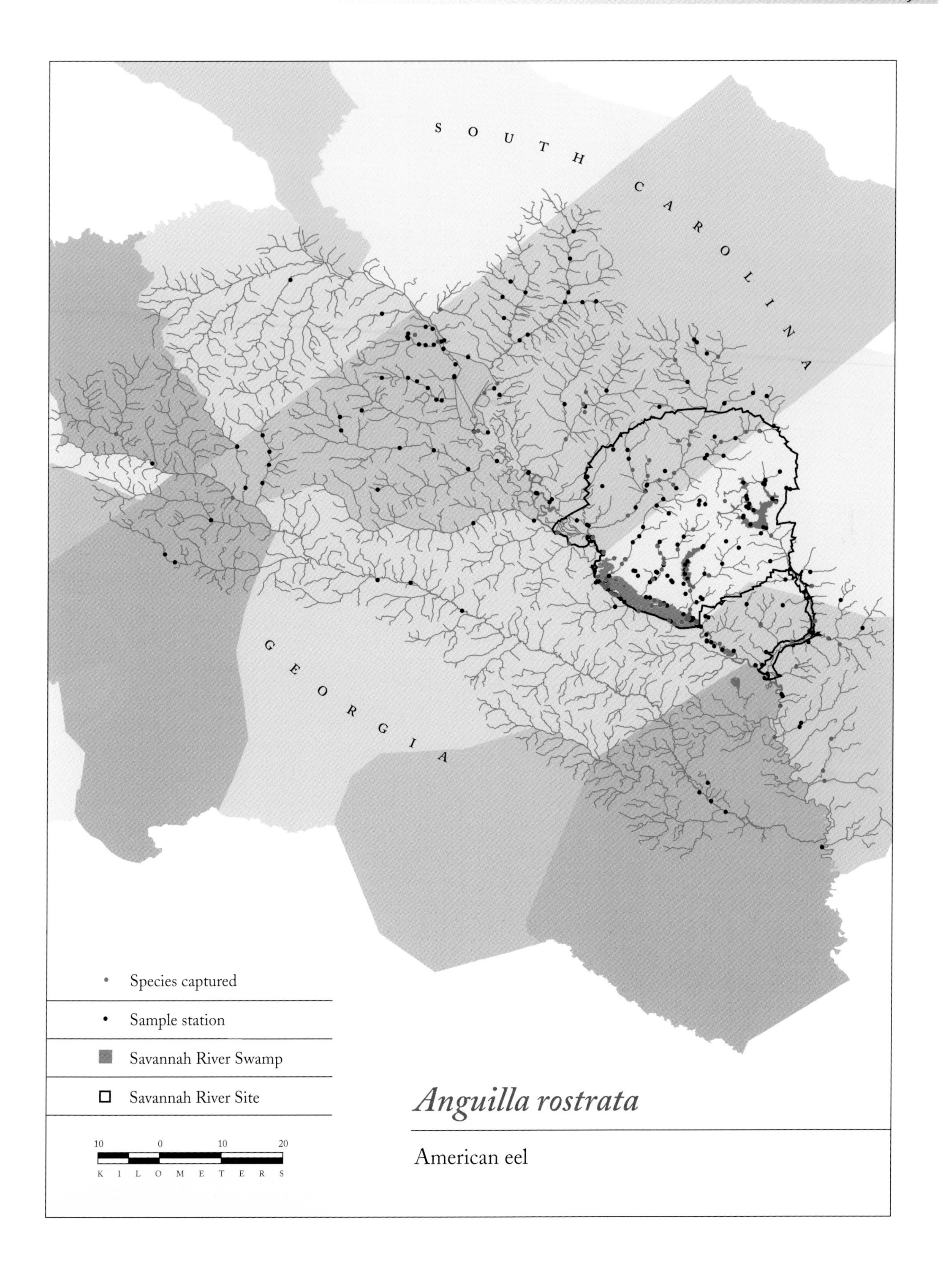

Anguilla rostrata

American eel

lar aquatic insects, crustaceans (especially crayfish), oligochaete worms, and fish (Sheldon and Meffe 1993), but eels also scavenge. Eels are popular baitfish.

Eel populations seem to be declining; barriers to migration, habitat loss and alteration, hydroturbine mortality, oceanic conditions, overfishing, parasitism, and pollution may all be responsible (Castonguay et al. 1994; Haro et al. 2000).

SCIENTIFIC NAME *Anguilla* = "eel" (Latin); *rostrata* = "long nosed" (Latin).

Clupeidae (herrings and shads)

Clupeidae is a large family of anadromous, freshwater or marine fishes including many economically important species like herrings, menhaden, sardines, pilchards, and sprats. The clupeids are characterized by a body shape that is somewhat elongated to deep and moderately to extremely compressed, a silvery color, the presence of an adipose eyelid, large cycloid scales that are easily shed, and a keeled ventral midline composed of scutes (modified scales) that form a serrated belly (Etnier and Starnes 1993). The family has a cosmopolitan distribution and includes 174 species in 63 genera (Grande 1985), 2 of which occur in the MSRB like the Savannah Rapids (shown above) near Augusta, Georgia: *Alosa*, with 3 species, and *Dorosoma*, with 2 species. Some anadromous populations have become landlocked by natural phenomena or by human activity and reproduce in fresh water.

Most clupeids eat zooplankton, swimming with the mouth open and using the long gill rakers to filter food from the water. Growth is generally rapid, and the life span is usually short (2–4 years). Reproductive patterns are diverse in the group, but the eggs are broadcast with no significant parental care given to the eggs or young. Eggs may be pelagic or demersal and adhesive, and are deposited over vegetation or other substrates. The very long, slender larvae look quite different from the adults. See Jacobson et al. 2004 for a 40-year summary of clupeid studies on the Connecticut River with comparisons to other coastal rivers.

The fishery for clupeids is extensive worldwide. Fishing techniques and use of the catch are diverse, from industrial harvesting to use of recreational hook and line, and from mass processing for fish meal and oil to use for human consumption as smoked, pickled, canned, fermented, or salted fish.

KEY TO THE SPECIES OF CLUPEIDAE

1a. Last ray on dorsal fin longer than all other rays (Plate 7 a); scales absent on predorsal median ridge; 10–15 dorsal fin rays; stomach thick walled (gizzardlike). . . . . *Dorosoma* spp. **2**

1b. Last ray on dorsal fin not longer than all other rays; scales present on predorsal median ridge; 14–21 dorsal fin rays; stomach not thick walled . . . . *Alosa* spp. **3**

[1] **2a.** Subterminal mouth with tip of snout projecting beyond tip of lower jaw (Plate 7 b); a notch present on ventral margin of posterior end of each upper jaw (maxilla) (Plate 7 d); 52–70 scales in lateral series; 25–37 anal fin rays . . . . *Dorosoma cepedianum* (gizzard shad), p. 105

2b. Terminal mouth with tip of lower jaw projecting beyond tip of snout (Plate 7 c); without notch in ventral margin of posterior end of each upper jaw (maxilla) (Plate 7 e); 40–48 scales in lateral series; 17–27 anal fin rays. . . . . *Dorosoma petenense* (threadfin shad), p. 108

[1] **3a.** Cheek wider than deep (Plate 7 f); deep mandible with dorsal profile angling steeply up from terminal end when mouth is held open (Plate 7 i); single dark shoulder spot usually present behind operculum, never followed by dusky spots; 44–50 gill rakers on lower limb of first gill arch in large individuals, more than 25 in small individuals; peritoneum sooty or black (requires opening the fish in larger individuals). . . . . *Alosa aestivalis* (blueback herring), p. 96

3b. Cheek width equal to or less than depth (Plate 7 g, h); moderately shallow mandible with dorsal profile gently angling up from terminal end when mouth is held open (Plate 7 j); single dark shoulder spot usually followed by 1–4 dusky spots; more than 55 or fewer than 25 gill rakers on lower limb of first gill arch in large individuals, variable in small individuals; peritoneum pale with scattered melanophores (requires opening the fish) **4**

[3] **4a.** Cheek deeper than wide (Plate 7 g); lower jaw not extending to dorsal profile of head (Plate 7 g); 59–73 crowded gill rakers on lower limb of first gill arch in large individuals, more than 25 in small individuals . . . . *Alosa sapidissima* (American shad), p. 102

4b. Cheek about as deep as wide (Plate 7 h); lower jaw extending to dorsal profile of head (Plate 7 h); 18–23 widely spaced gill rakers on lower limb of first gill arch . . . . *Alosa mediocris* (hickory shad), p. 99

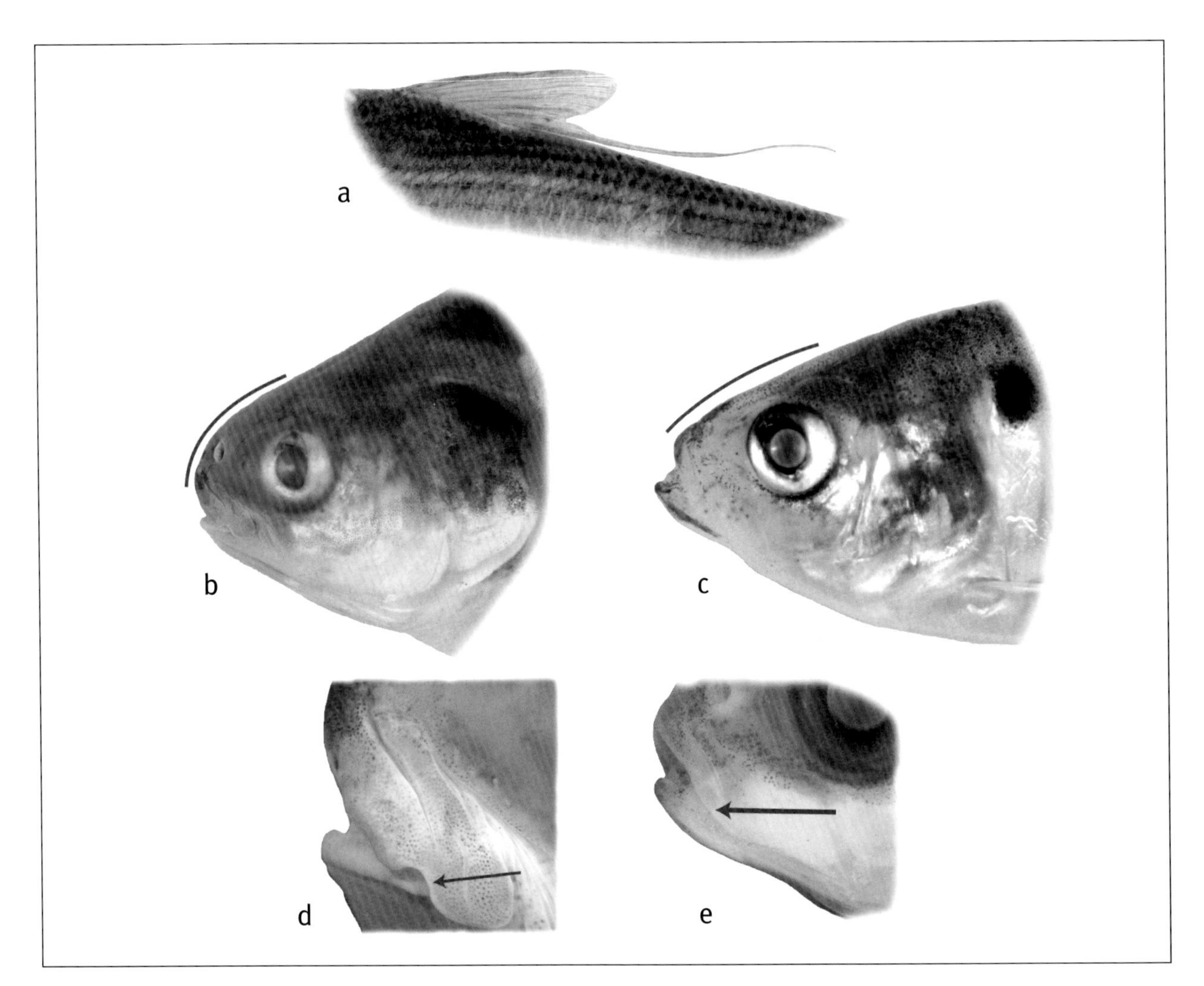

PLATE 7a. Morphological characters of Clupeidae: (a) *Dorosoma* dorsal fin showing elongate final ray, (b) gizzard shad snout, (c) threadfin shad snout, (d) gizzard shad maxillary, (e) threadfin shad maxillary.

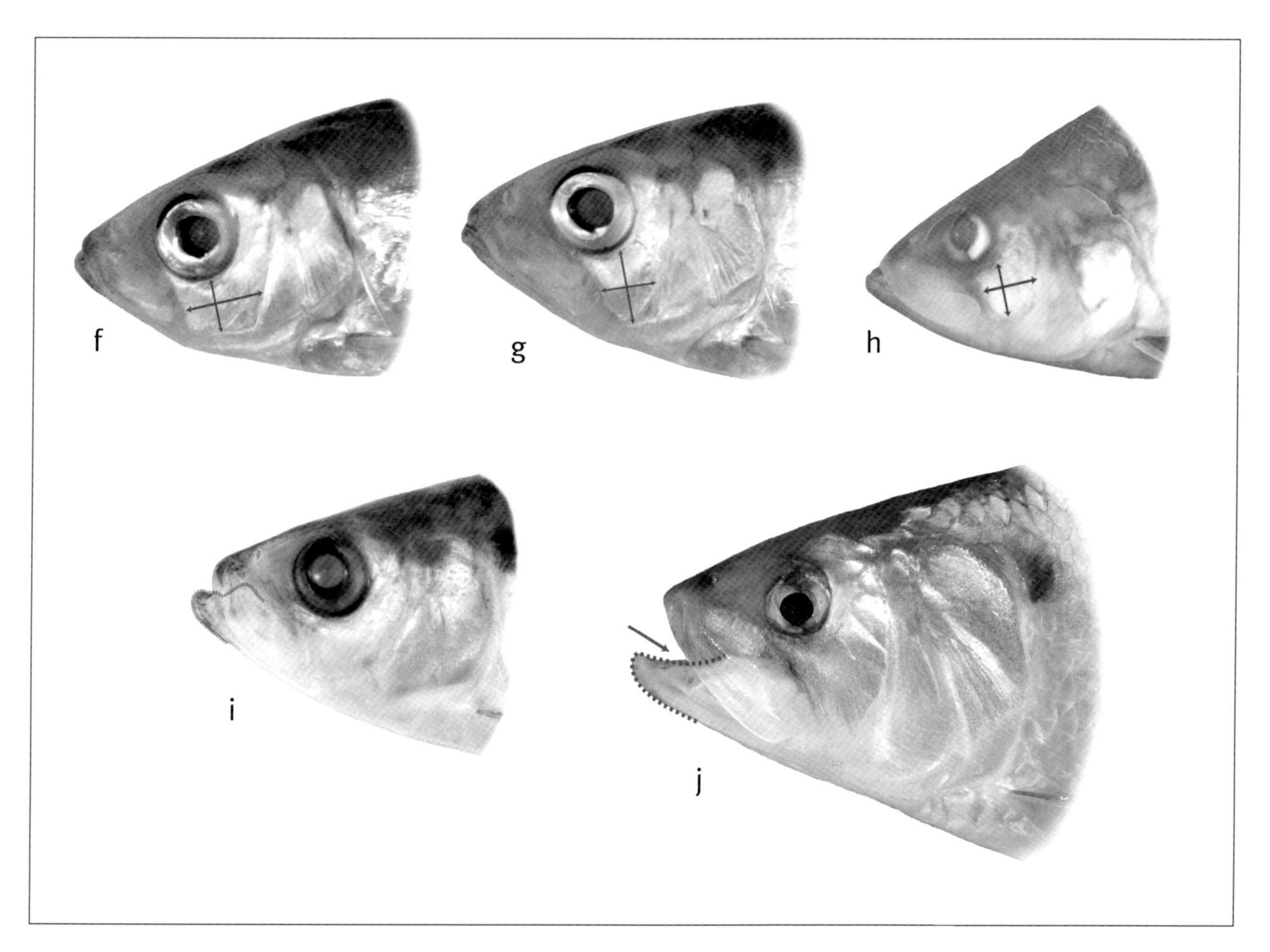

PLATE 7b. Morphological characters of Clupeidae: (f) blueback herring cheek, (g) American shad cheek, (h) hickory shad cheek and mandible, (i) blueback herring mandible profile, (j) American shad mandible profile.

Alosa aestivalis (Mitchill, 1814)

Blueback herring

DESCRIPTION Body deep, depth about half standard length; moderately compressed with a serrate ventral keel. Head short, about as deep as long; eye relatively small in adults and with an adipose eyelid, in diameter equal to or less than snout length. Cheek generally wider than deep. Mouth terminal and oblique; upper jaw with a definite median notch. Maxillary broad with posterior margin rounded, extending to or slightly beyond middle of eye. Teeth present in lower jaw at all ages; no teeth on premaxillaries. No lateral line. Gill rakers long, those at angle of arch 0.75 times eye diameter. Caudal fin forked with lower lobe slightly larger than upper (Carr and Goin 1959).

Back of adult bluish with head and sides silvery. Faint parallel horizontal lines are often present on dorsolateral scale rows, and there is a single faint, dark shoulder spot. Peritoneum black (requires incision to see). Fins plain with green to yellow tint in life.

SIMILAR SPECIES Extremely similar in appearance to both the hickory shad (*A. mediocris*) and the American shad (*A. sapidissima*). To the untrained eye, size at maturity is the principal distinction. Blueback herrings are approximately one-third the size of American shad and one-half the size of hickory shad. Presence of a single shoulder spot, cheek dimensions, and mandible profile may be used to identify a blueback herring. A vertical line through the center of the pupil intersects the subopercular plate of juvenile blueback herrings but not American shad or hickory shad.

SIZE Sexually mature at 250 mm or less. Females reach 290 mm FL (Bozeman and Van Den Avyle 1989); both sexes may reach 400 mm (Rohde et al. 1994).

MERISTICS Dorsal rays 15–20, anal rays 15–21, pectoral rays 14–18, pelvic rays 9–11; lateral series scales 46–54; total scutes 31–36, prepelvic scutes 18–21, postpelvic scutes 12–16 (Bozeman and Van Den Avyle 1989); gill rakers on lower limb of gill arch vary with size, 28–36 in specimens 20–49 mm SL, 30–39 at 60–69 mm SL, 35–41 at 70–

Alosa aestivalis, blueback herring

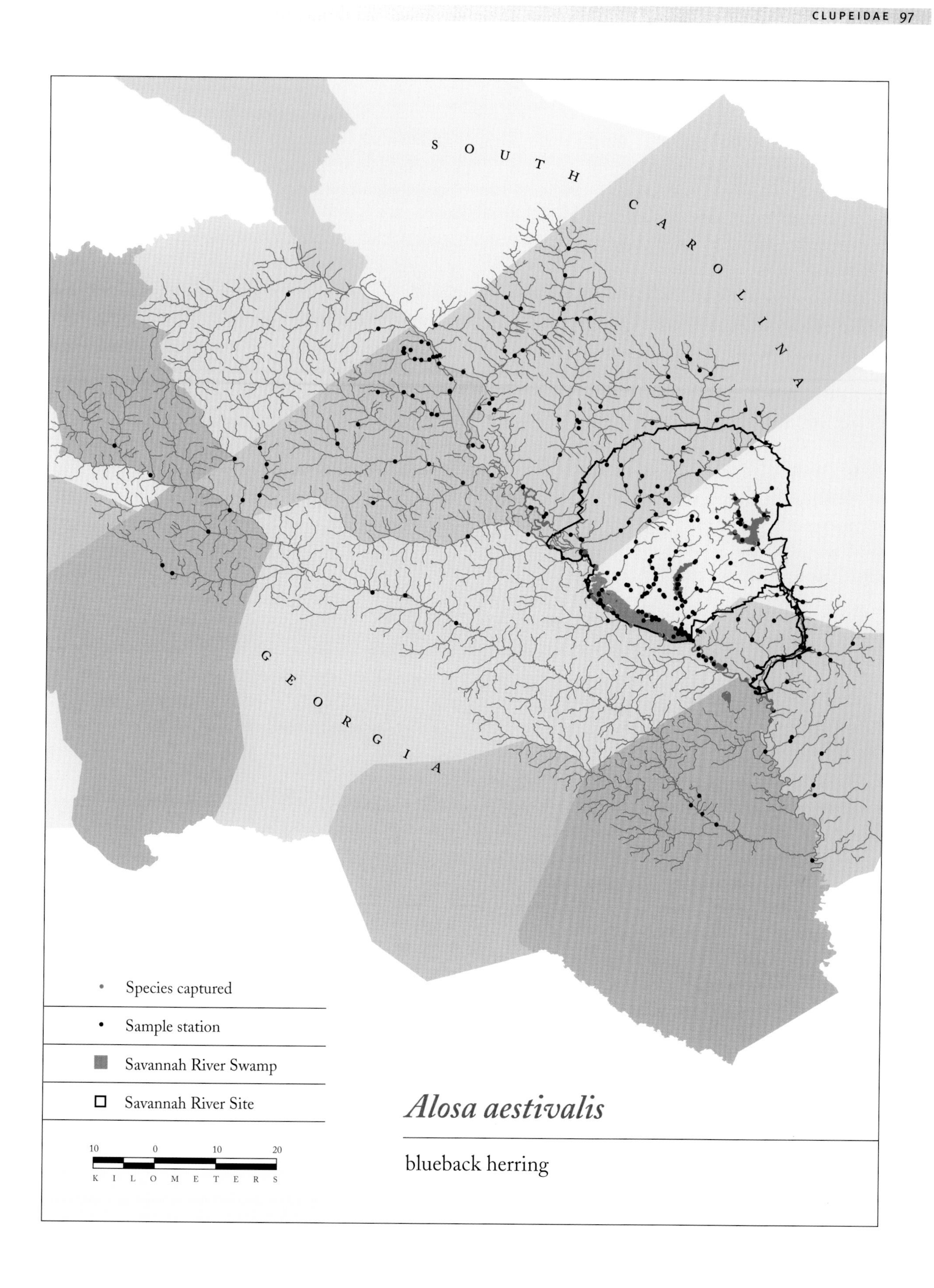

Alosa aestivalis

blueback herring

89 mm SL, 38–44 at 90–109 mm SL, 42–48 at 110–129 mm SL, 42–50 at 130–149 mm SL, and 42–48 at 190–260 mm SL (adults) (Hildebrand 1963); branchiostegals 7,7.

DISTRIBUTION Coastal rivers from Cape Breton, Nova Scotia, to the St. Johns River of Florida (Burgess *in* Lee et al. 1980; Bozeman and Van Den Avyle 1989). Introduced into many reservoirs as forage for game fish (Davis and Foltz 1991). Common in the MSRB in the Savannah River.

HABITAT Adults and larger juveniles of this anadromous species are marine (Bozeman and Van Den Avyle 1989); smaller juveniles remain in the rivers where they hatched until fall. Adults enter freshwater portions of rivers to spawn. Spawning habitat is fresh or slightly brackish water in rivers and ponds with bottoms of sand, gravel, or boulders. In Connecticut, adults spawn in areas with swift stream flow and hard substrate (Loesch and Lund 1977). Females with ripe or nearly ripe eggs are frequently taken in oxbow lakes, sloughs, and swamps in Georgia (Street 1969). Abandoned rice field impoundments in South Carolina are very good spawning habitat (Osteen et al. 1989; Thomas et al. 1992), as are backwater lakes (Meador et al. 1984). The landlocked population in Clarks Hill/J. Strom Thurmond Lake on the Savannah River has very restricted habitat during the summer because the hypolimnion has too little oxygen and the epilimnion is too warm (Nestler et al. 2002).

CONSERVATION STATUS Native but stocked in upstream reservoirs (Davis and Foltz 1991) and now landlocked, so some fish from the MSRB may be from the introduced upstream stock.

BIOLOGY The spawning period is in the spring, starting earliest in the south and proceeding northward. In the Carolinas, spawning occurs in March to early May, but adults may begin migrating into fresh water in late winter. The optimum spawning temperature is thought to lie in the range 21–25.5 °C (Jones et al. 1978; Bozeman and Van Den Avyle 1989). Spent fish return to sea almost immediately. In Connecticut, most spawning is in April–July with small spawning aggregations in September (Loesch and Lund 1977). Immature 1- and 2-year-olds enter rivers shortly after the spawning adults arrive. There is some evidence of stream fidelity; that is, adults probably return to their natal stream to spawn. Spawning is usually along shorelines. Eggs are initially demersal and adhesive in still water or semipelagic in flowing water and may become pelagic after hardening and subsequent release from substrate (Jones et al. 1978). Lippson and Moran (1974) summarized descriptions of blueback herring larvae and their distinguishing features. Juveniles may remain in the lower reaches of rivers or may move upstream in summer (Burbidge 1974) before migrating downstream in late fall (Pardue 1983).

Sexual maturity occurs at 3–4 years. The maximum age in South Carolina is probably 7 or 8 years, but it may be up to 10 farther north (Pardue 1983). The maximum number of spawnings in Connecticut is four (Marcy 1969). Most spawning fish are 4–6 years old. In Georgia, females may mature sexually at age 2, but most spawning females are in age classes 3 and 4 with few surviving to age class 5; males frequently mature at age class 2, but most of those spawning are age class 3 with only a few age class 4 and 5 individuals present in the population (Street and Adams 1969).

Adults feed primarily on zooplankton and sometimes fish, but do not feed extensively in fresh water while on the spawning run. Juveniles eat zooplankton, worms, and midge larvae (Burbidge 1974). Burbidge (1974) observed a direct relationship between standing crops of zooplankton and distribution, growth, and feeding of young-of-the-year blueback herring in the James River of Virginia.

SCIENTIFIC NAME *Alosa*, from Saxon *allis* = old name for the European shad, *Alosa alosa*; *aestivalis* = "of the summer" (Latin).

Alosa mediocris (Mitchill, 1814)

Hickory shad

DESCRIPTION Body compressed with ventral serrate keel. No lateral line. Head long, about four times standard length (Carr and Goin 1959). Eyes of adults relatively small; adipose eyelid present. Scales cycloid, only moderately adherent, and with a definite crenulate membraneous margin (Hildebrand 1963). Lower jaw projects prominently above the dorsal profile (probably the most distinguishing characteristic separating this species from other *Alosa* and evident in both juveniles and adults) (Jenkins and Burkhead 1993). Mouth supraterminal and oblique; upper jaw with definite median notch; maxilla rounded posteriorly, reaching to nearly directly below posterior margin of pupil. Teeth present in lower jaw at all ages; teeth in upper jaw present in all specimens below 150 mm and lacking in specimens larger than 300 mm. Gill rakers on first arch widely spaced. Caudal fin deeply forked.

Sides of the head brassy and tip of the lower jaw dusky in life. Back gray-green shading to silvery white on lower sides. Pectoral fins dusky; pelvic and anal fins plain; caudal and dorsal fins darker, caudal fin with darkened tips. A row of dark spots is present behind the operculum, the anteriormost the most prominent. Narrow dark lines may be present on upper sides in adults. Peritoneum white or silvery (requires an incision to see).

SIMILAR SPECIES Hickory shad adults and juveniles resemble both the blueback herring (*A. aestivalis*) and the American shad (*A. sapidissima*). Gill raker counts in hickory shad approach adult counts in juveniles as small as 18 mm. In *A. aestivalis*, counts in the lower part of the adult range do not occur until juveniles reach sizes greater than 70 mm, and in *A. sapidissima* greater than 125 mm (Jones et al. 1978).

SIZE To 610 mm but usually less than 460 mm.

MERISTICS Dorsal rays 15–20, anal rays 19–23, pectoral rays 15–16, pelvic rays 9; lateral series scales 48–57; total ventral scutes (33)36–37(38), prepelvic scutes (19)20–22(23), postpelvic scutes (12)13–16(17); gill rakers on lower limb of gill arch (18)20–21(23).

DISTRIBUTION Anadromous, ascending coastal streams from the Kenduskeag River in Maine to the St. Johns River in Florida (Burgess *in* Lee et al. 1980). More abundant in southern New England and in Chesapeake Bay than in the rest of the range (Hildebrand 1963; Jenkins

Alosa mediocris, hickory shad

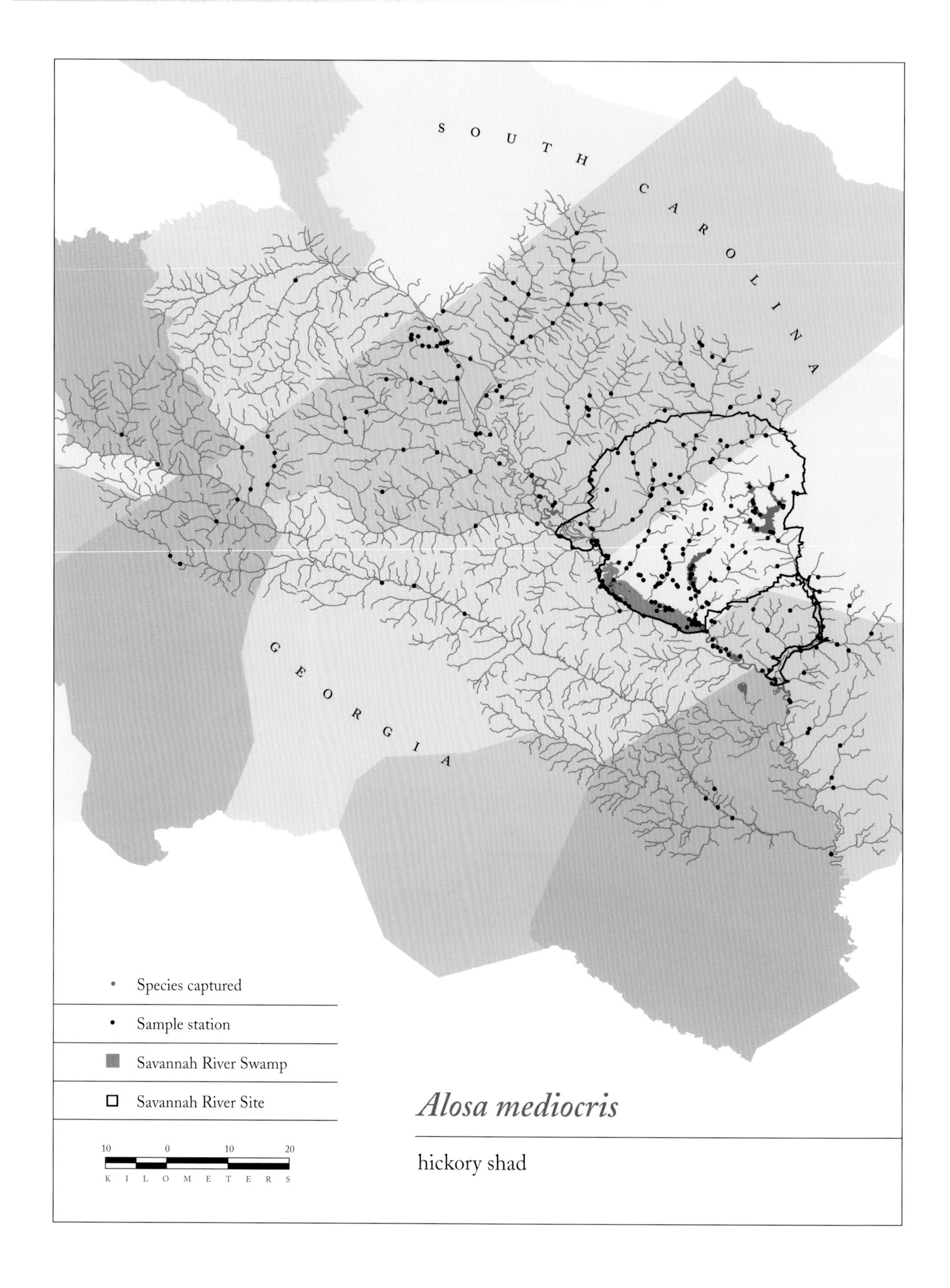

Alosa mediocris

hickory shad

and Burkhead 1993). Present but not abundant in the MSRB.

HABITAT Adults are marine fish. Juveniles probably remain in the freshwater portion of their natal river until early summer. The most frequently used spawning habitat is well up coastal rivers in creeks, ponds, lakes, and backwaters, but there is little in the literature about specific spawning habitats.

BIOLOGY The hickory shad is the least known of the *Alosa* species; Jenkins and Burkhead (1993) summarized the gaps in our knowledge of this species. Adults migrate into rivers in late winter or early spring to spawn. They may also run into fresh water in the fall, but probably not to spawn (Hildebrand 1963). They spawn in March to early May in the southeastern United States at water temperatures ranging from 12.8 to 20.6 °C, with peak spawning between 16.7 and 18.9 °C (Manooch 1984). Spawning is thought to occur at night (Mansueti 1962). Eggs are slightly adhesive but easily break free from the substrate to become pelagic when there is turbulence. Lippson and Moran (1974) summarized descriptions of hickory shad larvae and their distinguishing features.

Adults first spawn at 2 years of age, and in the Altamaha River of Georgia spawn only twice, at ages 2 and 3; a very small number of females and virtually no males survive to age 4 or 5 (Street and Adams 1969), although a life span of up to 8 years is possible (Hildebrand 1963). Juveniles remain in fresh water through the summer and move into more saline water in the fall or early winter (Hildebrand 1963).

The adults are more piscivorous than other *Alosa* species, and also eat fish eggs, small crabs, aquatic insects, and squid.

SCIENTIFIC NAME *Alosa*, from Saxon *allis* = old name for the European shad, *Alosa alosa*; *mediocris* = "mediocre" (Latin), referring to its quality as a food compared with the American shad.

Alosa sapidissima (Wilson, 1811)

American shad

DESCRIPTION Body laterally compressed with a saw-edged ventral keel; body and head deep. Eyes of adults relatively small; adipose eyelid present. Caudal fin forked. No lateral line. Scales cycloid with crenulate membranous margins. No teeth present in lower jaw of specimens more than 20 cm TL. Gill rakers long and slender, longer than eye diameter.

Body silvery white with greenish or bluish back; back becoming brownish in fresh water. Dark shoulder spot sometimes followed by one or more rows of smaller spots; sometimes with dark dorsolateral lines. Fins pale to greenish; dorsal and caudal fins sometimes dusky; anterior pectoral rays, distal tips of dorsal rays, and tips of caudal fin sometimes darkened. Peritoneum pale to silver (requires incision to see).

SIMILAR SPECIES Similar in appearance to both the blueback herring (*A. aestivalis*) and the hickory shad (*A. mediocris*). See species accounts above for distinguishing characters, and see Jenkins and Burkhead 1993 for photographs to easily separate juveniles of the three *Alosa* species.

SIZE To 760 mm. Males are usually 305–477 mm FL; females are usually 383–485 mm FL. Adults average 1.4–1.8 kg but can reach 5.4 kg.

MERISTICS Dorsal rays (14)18–19(21), anal rays (18)21–22(25), pectoral rays (13)16–17(18); lateral series scales 52–65; total ventral scutes 35–38, prepelvic scutes 19–25, postpelvic scutes 12–19; gill rakers on lower limb of gill arch 26–43 in specimens 29–125 mm SL, 59–76 in specimens 300 mm SL and larger; branchiostegals 7,7, rarely 7,6.

DISTRIBUTION The native range is the Atlantic Coast from Labrador to Florida, with introductions producing a Pacific Coast population from the Mexican border through Alaska to the Kamchatka coast of Russia (Stier and Crance 1985). Introduced into most of the Great Lakes but probably not established (Cudmore-Vokey and Crossman 2000). Common in the MSRB, but not usually found in small tributaries.

HABITAT Adults are marine fish (Leggett 1976, 1977; Neves and Depres 1979). They rarely appear in brackish estuaries and fresh water outside the spawning season. Hydrographic characteristics of spawning streams determine habitat suitability. Survival of eggs and larvae is

Alosa sapidissima, American shad

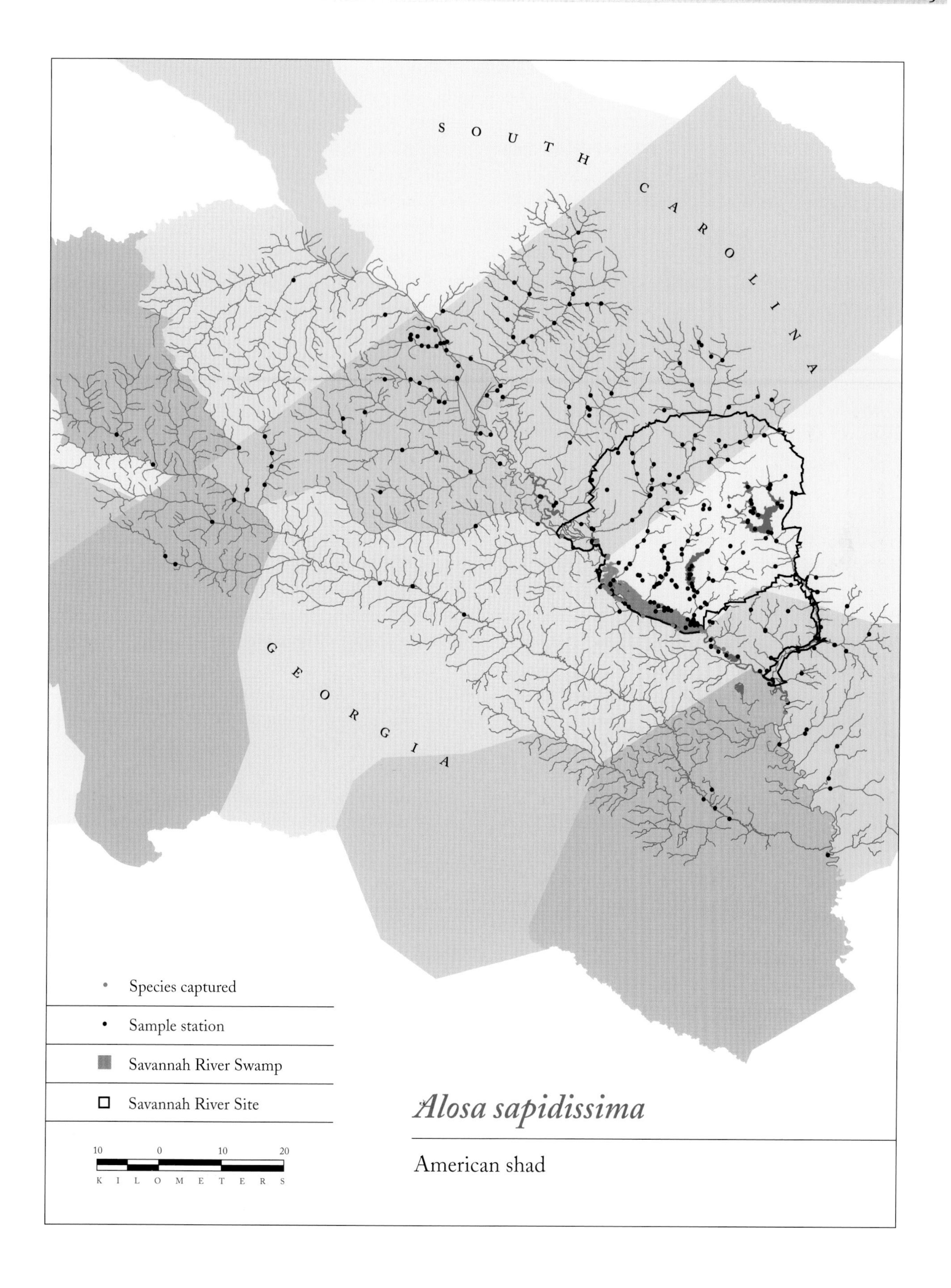

Alosa sapidissima

American shad

highly dependent on river discharge, with optimum discharge correlating directly to river drainage area (Bilkovic 2000). The relationship of drainage area to discharge may be modified in the MSRB by upstream dams. Apparently this species is unable to complete all life stages in fresh water, as all introductions into landlocked situations have failed, though some adult fish survived for several years after introduction (Hildebrand 1963). A recent 40-year summary of early life history and adult population dynamics studies can be found in Leggett et al. 2004 and Savoy et al. 2004, respectively.

CONSERVATION STATUS Fisheries stocks have shown a coastwide decline with some recovery of stock size in the Savannah River (Atlantic States Marine Fisheries Commission 1999).

BIOLOGY Spawning runs may begin in November in Florida and may occur as late as July farther north (Leggett and Whitney 1972; Stier and Crance 1985). Males dominate the earliest runs of the season (Smith 1907). Populations in the Carolinas spawn in March to early May at water temperatures ranging from 10.5 to 21.6 °C (Jones et al. 1978). Southern populations (i.e., south of Cape Hatteras, North Carolina) tend to spawn once and die; northern populations usually are repeat spawners (Leggett and Carscadden 1978). Adults eat little during the 2 months they spend in fresh water on the spawning run, and weight loss and mortality are significant (Leggett 1972; Chittenden 1976; S. M. Davis 1980). Individuals that significantly delay the downstream postspawning migration may lose enough weight to cause mortality (Nichols 1959). Meristic data indicate that northern populations return to their natal streams for spawning (Carscadden and Leggett 1975; Melvin et al. 1986; Bentzen et al. 1989). Tagging experiments support such homing (Hollis 1948), and olfaction and vision appear to aid in homing to natal rivers (Dodson and Leggett 1974). Spawning is usually at night and may occur anywhere in a river, but the most frequented sites have flats or shallow water (Marcy 1972; Stier and Crance 1985). Eggs hatch in 71–86 hours, are demersal or pelagic, and are scattered randomly in the spawning act. Eggs travel 1.6–6.4 km from the point where they were broadcast (Marcy 1972). Larvae have similar survival and growth rates in salinities ranging from 0 to 20 ppt (Limburg and Ross 1995). Lippson and Moran (1974) summarized descriptions of American shad larvae and their distinguishing features.

Larvae and juveniles remain in their tidal freshwater nursery until water temperatures drop in the late fall, when they migrate out to sea or at least migrate downstream to overwinter near the mouth of the river or bay (Milstein 1981). In northern populations, the juveniles may remain in brackish water or inshore for the first year (Hildebrand 1963). In Virginia, stocked juvenile American shad migrated 28 km upstream from the site of release, and the smallest juveniles, both native and stocked, were found farthest upstream (Dixon et al. 1998). This migration cannot be accounted for by swimming alone; they must have made use of selective tidal transport. Juveniles remain in the ocean until they are at least 2 years of age. While at sea they form large schools. Most males mature at 2–3 years; females mature at 3–4 years of age.

Marine adults eat zooplankton, plant material, molluscs, and small fishes. The mysid zooplankters available in freshwater streams are much smaller than marine mysids and are therefore unavailable to adult American shad (Atkinson 1951), which thus feed little in fresh water while on the spawning run. Larvae and juveniles eat crustaceans, insect larvae, and, sometimes, smaller fish (Hildebrand 1963).

SCIENTIFIC NAME *Alosa*, from Saxon *allis* = old name for the European shad, *Alosa alosa*; *sapidissima* = "most delicious" (Latin).

Dorosoma cepedianum (Lesueur, 1818)
Gizzard shad

DESCRIPTION Body depth 2.5 times into standard length (Carr and Goin 1959) and strongly laterally compressed with a serrate ventral keel. Head small with a blunt, rounded snout and subterminal mouth. Ventral edge of upper jaw deeply notched. Adipose eyelid present. Axillary scale present at bases of pectoral and pelvic fins. Digestive tract with a gizzardlike stomach and long, coiled intestine. No lateral line. The last dorsal ray elongate.

Back bluish gray with white to silvery white sides; sometimes with a greenish tinge overall. A large, round dark spot is present on the shoulder immediately behind the opercle, more prominent in young individuals. Six–eight horizontal dark stripes may be present along upper sides above level of shoulder spot. Vertical fins are blackish during spawning and clear with blackish irregular blotches the rest of the time.

SIMILAR SPECIES Very similar in appearance to threadfin shad, *D. petenense*.

SIZE TL 200–529 mm, usually less than 305 mm. Weight up to 1.5 kg, but usually much less (R. R. Miller 1963).

MERISTICS Dorsal rays 10–13(15), anal rays (25)29–35(37); lateral series scales (52)59–67(70); prepelvic scutes 17–20, postpelvic scutes 10–14 (Etnier and Starnes 1993); pectoral rays 12–17, pelvic rays 7–10; gill rakers on lower limb of gill arch as few as 90 at 350 mm TL (Jenkins and Burkhead 1993).

DISTRIBUTION The native range includes most of the United States east of the Rocky Mountains; widely introduced elsewhere as a forage species. Found from the St. Lawrence River and the Great Lakes (where it was introduced, according to Cudmore-Vokey and Crossman 2000) west to the Dakotas and south to mid-peninsula Florida and through Texas and eastern New Mexico into Mexico (Megrey *in* Lee et al. 1980). Loftus and Kushlan (1987) reported gizzard shad as far south as the Everglades. Within the MSRB, most abundant in the main channel of the Savannah River, in backwaters, in swamps, and in reservoirs.

HABITAT Warm, eutrophic bodies of water with soft mud bottoms, high turbidity, and few predators. High bio-

Dorosoma cepedianum, gizzard shad

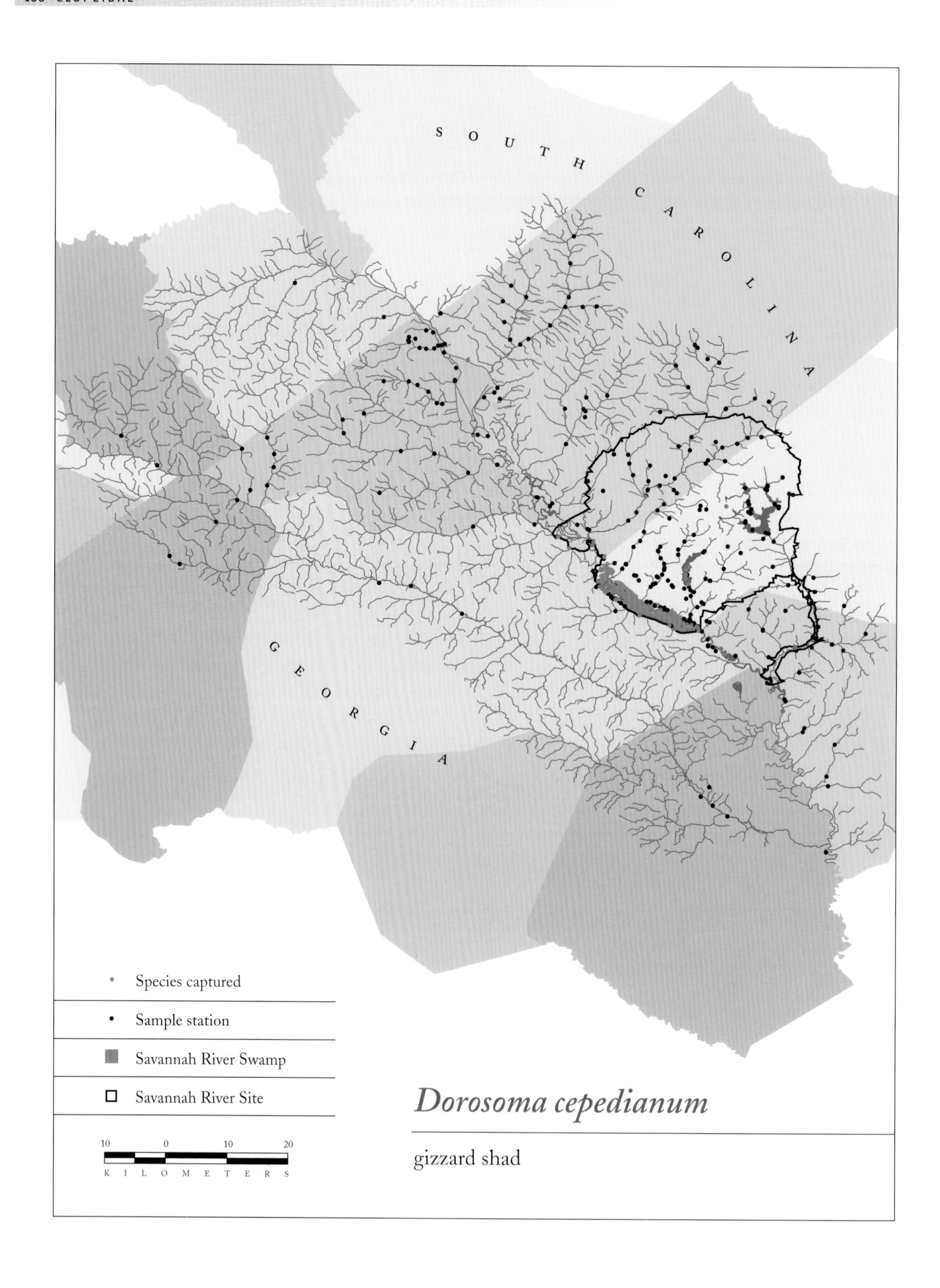

Dorosoma cepedianum

gizzard shad

masses may be found in large sluggish rivers, impoundments, lakes, swamps, bayous, and floodwater pools (Williamson and Nelson 1985). Catch rates in Florida indicated greater abundance in oxbow lakes than in rivers (Beecher et al. 1977). While juveniles will enter brackish water, only full adults occur in salinities nearing seawater (R. R. Miller 1963). They are not found in high-gradient streams or rivers that lack pools (Pflieger 1975). Sloughs, ponds, lakes, and large rivers are the most frequently used spawning habitats. Recently inundated areas appear to be strongly utilized (Storck et al. 1978).

BIOLOGY Gizzard shad form large, active schools. Spawning normally occurs in the spring after the water temperature exceeds 16 °C and with rapidly rising water levels (Williamson and Nelson 1985). Mass spawnings occur at any time of day but are more intense at night (Shelton et al. 1982). Eggs are scattered, usually in shallow water, and are demersal and adhesive, adhering to vegetation, gravel, and detritus. Young reach 150 mm by December of the first year (Lagler and Van Meter 1951). Females may mature as early as year class 1 but typically mature at year class 2 or 3. The usual life span is 4–6 years but may reach 10 years or more (Pflieger 1975). An Illinois population was dominated by year classes 2 and 3, but some lived to 5 years (Lewis 1952).

Feeding, mostly on low-caloric detritus and mud, is restricted to daylight hours with peak feeding activity at 1400–1600 (Pierce et al. 1981). At high biomasses gizzard shad feed mainly on benthic detritus, but at lower biomasses they feed primarily on zooplankton (Schaus et al. 2002). Larvae and juveniles eat zooplankton until they reach 25 mm, at which time they lose their teeth, become slab sided, and take up the adult diet of algae, phytoplankton, mud, and detritus (Smith 1979). Young-of-the-year and year class 1 individuals are capable of filtering particles down to 20 µm (Kutkuhn 1958).

Sudden changes in temperature or oxygen content of the water may result in large die-offs, especially in late summer. More than 10 million shad died in a 6,500 acre impoundment in southern Illinois in October 1952 (Lewis 1952). Because of their abundance and their feeding habits these fish can greatly impact fish, phytoplankton, and zooplankton community structures (Drenner et al. 1982, 1984, 1986, 1996; Guest et al. 1990; Lazzaro et al. 1992). They may occasionally hybridize with threadfin shad (Shelton and Grinstead 1973).

SCIENTIFIC NAME *Doro* = "lanceolate" (Greek), *soma* = "body" (Greek), thus "lanceolate body," referring to shape of juveniles; *cepedianum* = in honor of the French ichthyologist Citoyen Lacépède.

Dorosoma petenense (Günther, 1867)

Threadfin shad

DESCRIPTION Body deep and compressed with a serrate ventral keel and deeply forked caudal fin. No lateral line. Adipose eyelid present. Mouth terminal. Dorsal fin with an elongate posterior ray. Axillary processes present at bases of pectoral and pelvic fins. Gizzardlike stomach with long, coiled intestine.

Back gray or green dorsally, silvery on lower sides with a prominent black spot on each side immediately behind upper part of gill opening. Fins other than the dorsal fin yellowish in life.

SIMILAR SPECIES Gizzard shad, *D. cepedianum*. The threadfin shad's snout is less rounded than the gizzard shad's.

SIZE To 203 mm, but rarely longer than 127 mm (Robison and Buchanan 1984).

MERISTICS Dorsal rays 14–15 (Etnier and Starnes 1993), 11–14 (R. R. Miller 1963), or 10–13 (Robison and Buchanan 1984; Mettee et al. 1996); pectoral rays 12–17; pelvic rays 7–8 (Jenkins and Burkhead 1993); anal rays 17–28; scales in lateral series 40–50; total ventral scutes 23–29, 15–18 prepelvic and 10–12 postpelvic; total gill rakers 300–400 in specimens 66–180 mm SL (Etnier and Starnes 1993).

CONSERVATION STATUS Introduced in the MSRB; the earliest records in the Savannah River are from 1968 (Dahlberg and Scott 1971b).

DISTRIBUTION Probably native to the Ohio River, Tennessee River, and Mississippi River south to the Gulf Coast from Florida through Texas and Oklahoma into Central America. Widely introduced as forage fish and now found from Indiana and Illinois through most of the Southeast including Virginia and in the western United States (Burgess *in* Lee et al. 1980); also successfully introduced in Puerto Rico (Erdman 1972, 1984) and Hawaii (Brock 1960; Hida and Thomson 1962). Prior to 1945 unknown from Illinois, but now found in all Illinois collection stations on the Ohio River during the summer months (Smith 1979). Most abundant in the MSRB in the main channel of the Savannah River and in reservoirs, especially L Lake on the SRS.

HABITAT Primarily found in reservoirs and big rivers; in Florida, found in equal numbers in oxbow lake and river habitats (Beecher et al. 1977). Most common over soft bottoms, but also common over sand.

Dorosoma petenense, threadfin shad

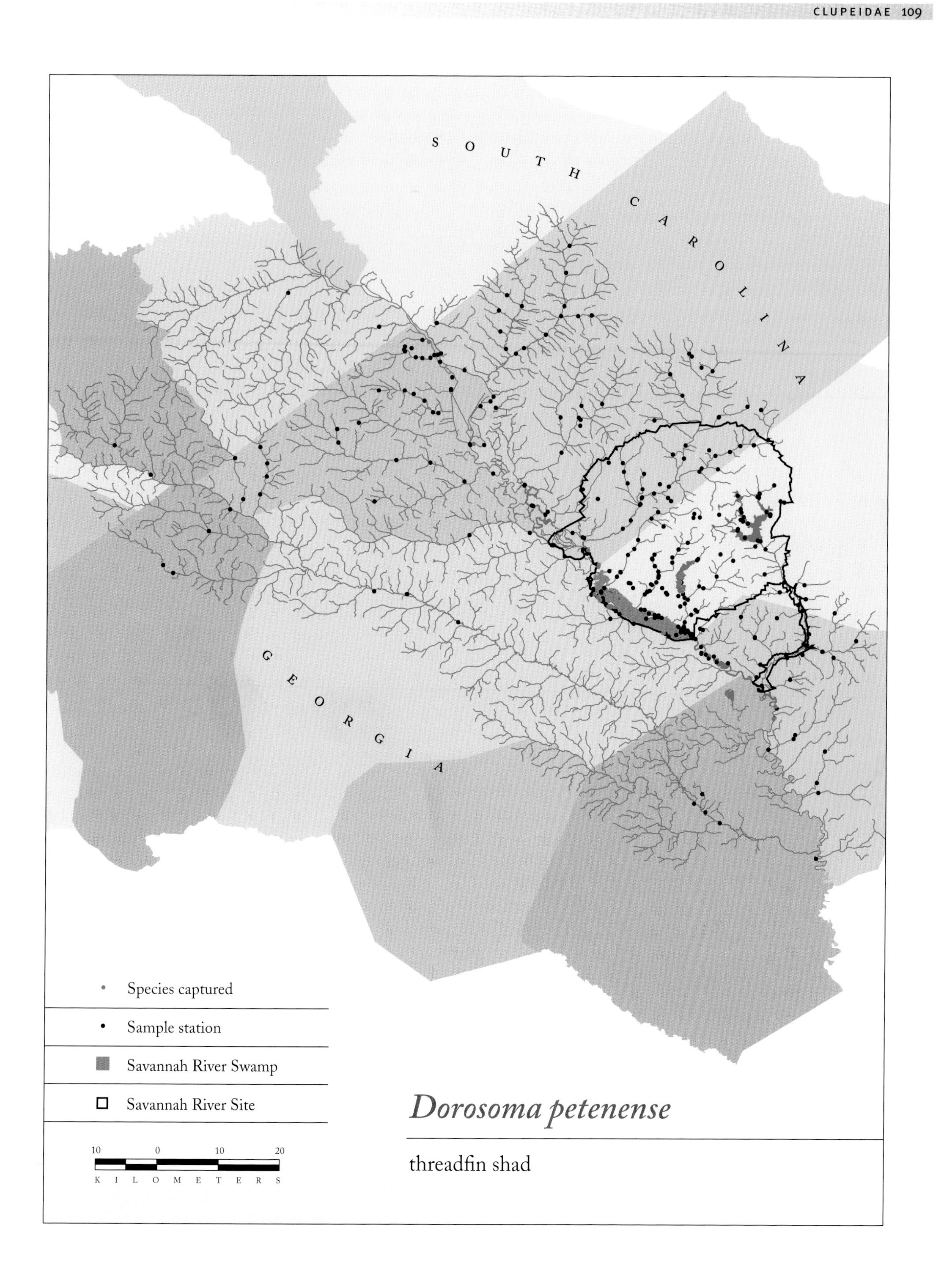

Dorosoma petenense

threadfin shad

BIOLOGY Threadfin shad occur in large pelagic schools usually consisting of similar-sized individuals. During the summer, these schools are confined to the water layer between the surface and the thermocline (Houser and Dunn 1967). They spawn in spring and summer, with perhaps a second spawning peak in the fall (R. R. Miller 1963; Smith 1979). Spawning occurs after sunrise along shorelines. In Tennessee, spawning is nearly restricted to the time period from 0540 to 0730, and this timing may be important in avoiding excess predation (McLean et al. 1982). In Lake Texoma, spawning occurs in the afternoon only at the start of the spawning season and quickly becomes confined to the morning hours (Shelton et al. 1982). Fish may spawn during their second summer and usually live no more than 2–3 years. Individuals spawned in early spring may spawn by the end of their first summer (Robison and Buchanan 1984). Males seldom live past age 3; females not past age 4 (Pflieger 1975).

The diet includes both phytoplankton and zooplankton, with quantities of organic detritus and mud perhaps being ingested as well. Because of its great abundance and feeding habits, this species may greatly affect zooplankton and fish community structures (Guest et al. 1990; Lazzaro et al. 1992).

Threadfin shad are sensitive to sudden changes in temperature and oxygen content, and there are frequent die-offs in summer and winter (Mettee et al. 1996). School cohesion begins to break down at temperatures below 9 °C, and none survive exposure to temperatures below 4 °C (Griffith 1978). In Texas, the minimum temperature tolerated is between 12.2 and 14.2 °C (Hubbs 1951). They may sometimes hybridize with gizzard shad (Shelton and Grinstead 1973).

SCIENTIFIC NAME *Doro* = "lanceolate" (Greek), *soma* = "body" (Greek), thus "lanceolate body," referring to the shape of juveniles; *petenense*, after the type locality, Lake Petén, Yucatán.

Cyprinidae (minnows)

While the name "minnow" is commonly used to refer to any small fish, it specifically refers to members of the family Cyprinidae, the largest fish family in the world, with 1,500 (Etnier and Starnes 1993) to 2,000 (Jenkins and Burkhead 1993) species in about 210 genera. Cyprinids are found in fresh waters of Asia, Africa, North America, and Europe (Jenkins and Burkhead 1993). Close to 300 species have been described from North America, and 15 species from 9 genera are native to the MSRB. Three exotic species—the goldfish (*Carassius auratus*), grass carp (*Ctenopharyngodon idella*), and common carp (*Cyprinus carpio*)—have been introduced from Asia.

Body and fin shapes are diverse in this large family. Most species native to the MSRB are relatively small (less than 15 cm in length), except for golden shiners (*Notemigonus crysoleucas*), which may reach 25 cm. Some of the exotic species like the grass carp, however, can exceed 1 m. The general body shape of North American cyprinids is moderate to elongate, and all fins are soft rayed. However, note that although not strictly a spine, the first ray of the dorsal and anal fins in the goldfish and common carp is thick, stiff, and serrated (Plate 8 a). The scales are cycloid and the lateral line is well developed. Teeth are absent from the jaws, but pharyngeal teeth are present in the throat. Pharyngeal teeth vary in structure and may be highly specialized for a particular diet. The shape of the mouth is diverse and sometimes reflects the depth of the water column utilized. Some species have barbels, although these may be inconspicuous.

While some species are largely monomorphic, sexual dimorphism in color and morphology is common, especially during the reproductive season. Males within the same population may exhibit different degrees of sexual dimorphism. Dimorphic features include the development of greatly enlarged fins and deeper bodies. Conical keratinized breeding tubercles are another striking breeding feature. In males these are most developed on the head but also may be present on the fins and body, particularly in areas where the male and female have contact during spawning. These tubercles are most often small and numerous, but may be developed into conspicuous horny structures as in the *Nocomis* and *Semotilus* species. The females of some species also develop small tubercles, but these are less pronounced than the males' tubercles. Cyprinid species exhibit a broad range of colors that can be quite spectacular, especially in the breeding season, and are one reason why some species are extensively cultured for the aquarium trade.

Cyprinids are polygamous, and their reproductive behavior varies greatly (see review in Johnston and Page 1992). Broadcasting, the most primitive and common breeding strategy, involves scattering eggs with no prior modification of the substrate; it may be done pelagically, over vegetation, or over a variety of bottom substrates. Broadcasters often gather in very large schools to spawn. Crevice spawners hide their eggs within crevices of logs or rocks. Nest building includes the construction of pits or saucer-shaped depressions in the sediment and the construction of pebble mounds or ridges. Egg clusterers lay their eggs in a single layer on the flat underside of a submerged object. Some species, including several local ones, spawn on the nests of other nest-building species such as bluehead chubs (*Nocomis leptocephalus*) and sunfish.

As can be expected of such a large family, food habits are very diverse. Most species eat insects and aquatic crustaceans, but quite a few eat plant material and fish. The length of the intestine is correlated with the diet. Herbivores have a long, coiled intestine, and the lining of the body cavity (the peritoneum) is black; carnivores have a relatively short, S-shaped intestine and usually a white peritoneum; omnivores such as bluehead chubs and golden shiners have an intestine of intermediate length.

In other parts of the world, especially in Asia, cyprinids are important sport and commercial fish. In the United States, cyprinids are important bait fishes, and several species (e.g., red shiners [*Cyprinella lutrensis*], golden shiners, and fathead minnows [*Pimepheles promelas*]) have been widely introduced as bait bucket escapees. Some carps have been introduced to control undesirable aquatic vegetation or molluscs, and in many areas these species have become pests themselves. Sterile (triploid) grass carp have been introduced extensively in the MSRB to control vegetation.

Table 17 lists characters useful in identifying native minnows.

KEY TO THE SPECIES OF CYPRINIDAE

1a. Long dorsal fin, more than 13 rays; first principal ray of dorsal and anal fins thick, serrated, stiff, and spine-like (Plate 8 a) **2**

1b. Short dorsal fin, less than 12 rays, usually 8 or 9; first principal ray of dorsal and anal fins thin and flexible **3**

[1] **2a.** Two pairs of conspicuous barbels on each side of upper jaw (Plate 8 b) . . . . *Cyprinus carpio* (common carp), p. 131

2b. No barbels at corners of mouth . . . . *Carassius auratus* (goldfish), p. 119

[1] **3a.** Anal fin located far to the rear, distance from anal fin origin to caudal fin base will go 2.5 times or more into distance from tip of snout to anal fin origin; similarly, the distance from the anal fin origin to the caudal fin base is equal to the distance from the anal fin origin to the pelvic fin origins; pharyngeal teeth with prominent parallel grooves. . . . . *Ctenopharyngodon idella* (grass carp), p. 122

3b. Anal fin origin further forward, distance from anal fin origin to caudal fin base will go less than 2.5 times into distance from tip of snout to anal fin origin; similarly, the distance from the anal fin origin to the caudal fin base is longer than the distance from the anal fin origin to the pelvic fin origins; pharyngeal teeth without grooves **4**

[3] **4a.** Single round, dark spot on anterior base of dorsal fin (Plate 8 f); 1 pair of small, flaplike preterminal maxillary barbels, each located in a groove above the upper lip, may be difficult to see in juveniles (Plate 8 c); pharyngeal teeth 2,5–4,2 (Plate 11 p, q) . . . . *Semotilus atromaculatus* (creek chub), p. 171

4b. No dark spot at base of dorsal fin; no barbels *or* terminal maxillary barbels present in corner of mouth (Plate 8 d, e); pharyngeal teeth with greater row counts symmetrical (see Table 17) **5**

[4] **5a.** Body compressed, sharp scaleless keel along belly between pelvic fins and anal fin (Plate 8 g); usually more than 12 anal fin rays; lateral line strongly downcurved anterior to pelvic fins (Plate 8 h) . . . . *Notemigonus crysoleucas* (golden shiner), p. 144

5b. Body not compressed; no scaleless keel posterior to pelvic fins; usually fewer than 12 anal fin rays; lateral line not strongly downcurved anterior to pelvic fins, moderately downcurved in *P. stonei* **6**

TABLE 17. Characters useful for identifying native minnows in the MSRB.

	Dorsal rays[1]	Anal rays[1]	Pharyngeal teeth[1]	Mouth position	Body profile	Postanal stripe	Lateral stripe[2]
Cyprinella leedsi	8	8	0,4–4,0	Subterminal, moderately oblique	Laterally compressed	Variable, absent/sometimes anus to caudal peduncle	Anterior one-fourth distinctly faded or absent; narrow; upper edge sharper than lower edge
C. nivea	8	8	1,4–4,1	Subterminal, moderately oblique	Laterally compressed	Variable, absent/sometimes anus to caudal peduncle	Anterior one-fourth distinctly faded or absent; narrow; upper edge sharper than lower edge
Hybognathus regius	8	8	0,4–4,0	Subterminal or inferior	Cylindrical/slightly compressed	Absent	Absent/diffuse
Hybopsis rubrifrons	8	8	1,4–4,1	Inferior, slightly oblique/ near horizontal	Cylindrical, ventrally flattened	Variable	Narrow; anterior and sometimes posterior punctate; upper edge sharper than lower edge
Nocomis leptocephalus	8	7	0,4–4,0	Slightly subterminal, slightly oblique	Cylindrical	Absent	Indistinct/variable
Notropis chalybaeus	8	8	2,4–4,2	Terminal, moderately oblique	Cylindrical	Anus to caudal peduncle	Narrow, upper and lower edges sharp
N. cummingsae	8	10–11	1,4–4,1	Terminal, moderately oblique	Cylindrical	Anus to anal fin base, sometimes onto caudal peduncle	Intermediate depth; upper edge sharper than lower edge
N. hudsonius	8	8	2,4–4,2 or 1,4–4,1	Distinctly subterminal or slightly inferior	Cylindrical	Absent or faint along anal fin base	Narrow, variable; darkest posteriorly; punctate anteriorly & less distinctively posteriorly
N. lutipinnis	8	8	2,4–4,2	Terminal, moderately oblique	Cylindrical	Absent	Intermediate depth; upper edge sharper than lower edge
N. maculatus	8	8	0,4–4,0	Subterminal, moderately oblique	Cylindrical	Anus to caudal peduncle	Narrow with large caudal spot (spot deeper than stripe)
N. petersoni	8	7	2,4–4,2	Terminal, moderately oblique	Cylindrical	Anus to caudal peduncle	Narrow, punctate especially anteriorly; distinctly triangular caudal spot
Notemigonus crysoleucas	(7)8(9)	12–15	0,5–5,0	Superior, strongly oblique	Laterally compressed	Absent	Present in juveniles but fades with age; narrow
Opsopoeodus emiliae	9	8	0,5–5,0	Superior, strongly oblique	Cylindrical/slightly compressed	Anus to caudal peduncle	Variable, narrow, anterior often punctate
Pteronotropis stonei	8	10–11	2,4–4,2	Terminal, moderately oblique	Laterally compressed	Absent	Deep; upper edge sharper than lower edge
Semotilus atromaculatus	8	8	2,5–4,2	Terminal, moderately oblique	Cylindrical	Absent or faint along anal fin base	Usually indistinct/variable

[1] Modes are provided for meristic features.
[2] Narrow = stripe depth less than diameter of eye; intermediate depth = stripe depth about equal to diameter of eye; deep = stripe depth greater than diameter of eye.

PLATE 8. Morphological characters of Cyprinidae: (a) serrate dorsal ray of common carp, (b) barbels of common carp, (c) creek chub barbels, (d) bluehead chub barbels, (e) rosyface chub barbels, (f) creek chub dorsal fin spot, (g) golden shiner keel, (h) golden shiner lateral line, (i) taillight shiner caudal spot, (j) bluehead chub mouth, (k) rosyface chub mouth, (l) pugnose minnow mouth.

[5] **6a.** Mouth very small, superior, and strongly oblique (Plate 4 a; Plate 8 l); dorsal rays 9; pharyngeal teeth 0,5–5,0 (Plate 11 n) . . . . *Opsopoeodus emiliae* (pugnose minnow), p. 165

6b. Mouth larger, terminal or subterminal and horizontal or slightly to moderately oblique (see Plate 4 d–f for examples); dorsal rays 8; pharyngeal teeth with fewer than 5 teeth on greater rows **7**

[6] **7a.** Mouth with terminal maxillary barbels (Plate 8 d, e; note that barbels may be small and inconspicuous, and it may be necessary to open the mouth to find them) **8**

7b. Mouth without barbels **9**

[7] **8a.** Short, round snout with large, slightly subterminal mouth (Plate 8 d, j); stocky caudal peduncle with slightly forked caudal fin; 7 anal rays; caudal fin red or reddish orange in life; pharyngeal teeth 0,4–4,0 (Plate 11 f) . . . . *Nocomis leptocephalus* (bluehead chub), p. 140

8b. Long snout overhanging a small inferior mouth (Plate 8 e, k); slender caudal peduncle with deeply forked caudal fin; 8 anal rays; caudal fin without color in life; pharyngeal teeth 1,4–4,1 (Plate 11 e) . . . . *Hybopsis rubrifrons* (rosyface chub), p. 137

[7] **9a.** Prominent large black basicaudal spot, larger than pupil of eye and wider than width of lateral stripe (Plate 8 i); possessing all three of the following: pharyngeal teeth 0,4–4,0 (Plate 11 l), body slender and elongate, and pigment on membranes of dorsal fin . . . . *Notropis maculatus* (taillight shiner), p. 159

9b. No black caudal spot, or if present less distinct and narrower than depth of the lateral stripe and always smaller than pupil of eye; either pharyngeal teeth not 0,4–4,0 *or* if 0,4–4,0, then deep bodied with pigment in membranes of dorsal fin *or* narrow elongate body without pigment in membranes of dorsal fin, although rays may be outlined with melanophores **10**

[9] **10a.** Dorsal fin usually with distinct black pigment on membranes, an anterior blotch or posterior blotch may be present, or the distal one-third to one-half of the fin may be black (Plate 9 a–f); body deep and laterally compressed; lateral profile of head and anterior body broadly conical (Plate 9 g, h) **11**

10b. Dorsal fin membranes clear or rays outlined with melanophores; body shallow with a more rounded cross section, and elongate; lateral profile of head and anterior body more narrowly cylindrical (Plate 9 i, j). **13**

[10] **11a.** Long anal fin, 10–11 rays; prominent wide (deep) bluish black lateral stripe (Plate 9 h); pharyngeal teeth 2,4–4,2 (Plate 11 o) . . . . *Pteronotropis stonei* (lowland shiner), p. 168

11b. Short anal fin, 8 rays; narrower, less distinct lateral stripe that becomes fainter anteriorly (Plate 9 g); pharyngeal teeth 0,4–4,0 or 1,4–4,1. . . . . **12**

[11] **12a.** Black blotch on anterior portion of dorsal fin (anterior to first principal ray), distal two-thirds entirely black in expanded fin of nuptial males (Plate 9 c, d); dorsolateral sides marked with a blurred crosshatched pattern (Plate 9 k); pharyngeal teeth 0,4–4,0 (Plate 11 a). . . . . *Cyprinella leedsi* (bannerfin shiner), p. 125

12b. Black blotch on posterior portion of dorsal fin (posterior 2 membranes), expanded dorsal fin of nuptial males milky white with black pigment on most membranes (Plate 9 e, f); dorsolateral sides marked with a sharp, clear crosshatched pattern (Plate 9 l); pharyngeal teeth 1,4–4,1 (Plate 11 b) . . . . *Cyprinella nivea* (whitefin shiner), p. 128

[10] **13a.** Either no pigment on chin or only small, scattered melanophores visible only with strong magnification when viewed from ventral side (Plate 10 a, b); mouth distinctly subterminal or inferior (see Plate 4 e, f and Plate 8 k for examples). **14**

13b. Pigment on chin shaped as terminal bar, thin- or thick-lined U, or terminal spot when viewed from ventral side (Plate 10 c–f); mouth terminal or nearly terminal (see Plate 4 d; Plate 9 i for examples), snout of *N. petersoni* and *N. lutipinnis* sometimes slightly overhanging mouth **15**

PLATE 9. Morphological characters of Cyprinidae: (a) male lowland shiner dorsal fin, (b) female lowland shiner dorsal fin, (c) breeding male bannerfin shiner dorsal fin, (d) female bannerfin shiner dorsal fin, (e) breeding male whitefin shiner dorsal fin, (f) female whitefin shiner dorsal fin, (g) bannerfin shiner body profile, (h) lowland shiner body profile, (i) yellowfin shiner body profile, (j) spottail shiner body profile, (k) bannerfin shiner lateral scale color pattern, (l) whitefin shiner lateral scale color pattern.

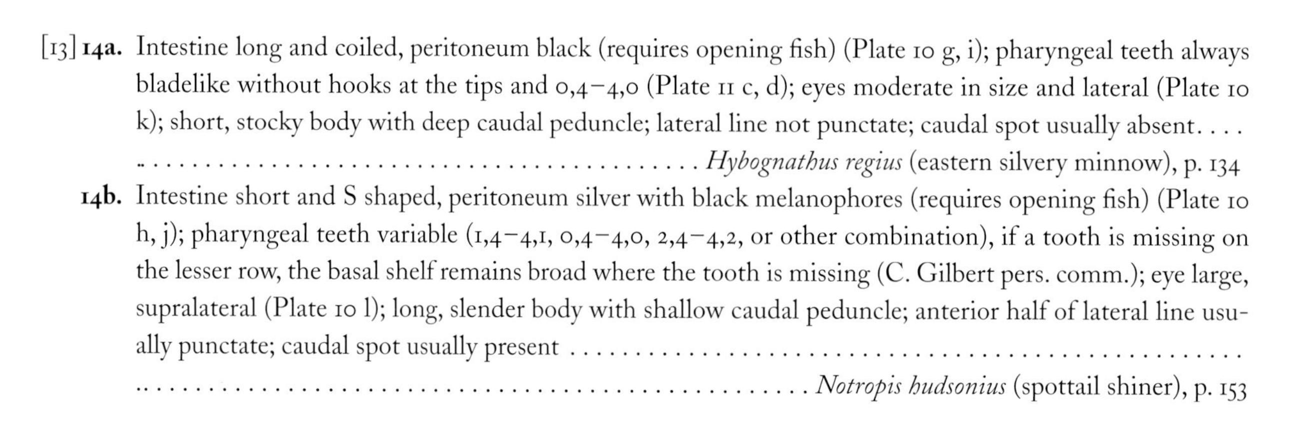

[13] **14a.** Intestine long and coiled, peritoneum black (requires opening fish) (Plate 10 g, i); pharyngeal teeth always bladelike without hooks at the tips and 0,4–4,0 (Plate 11 c, d); eyes moderate in size and lateral (Plate 10 k); short, stocky body with deep caudal peduncle; lateral line not punctate; caudal spot usually absent. . . . . *Hybognathus regius* (eastern silvery minnow), p. 134

14b. Intestine short and S shaped, peritoneum silver with black melanophores (requires opening fish) (Plate 10 h, j); pharyngeal teeth variable (1,4–4,1, 0,4–4,0, 2,4–4,2, or other combination), if a tooth is missing on the lesser row, the basal shelf remains broad where the tooth is missing (C. Gilbert pers. comm.); eye large, supralateral (Plate 10 l); long, slender body with shallow caudal peduncle; anterior half of lateral line usually punctate; caudal spot usually present . . . . *Notropis hudsonius* (spottail shiner), p. 153

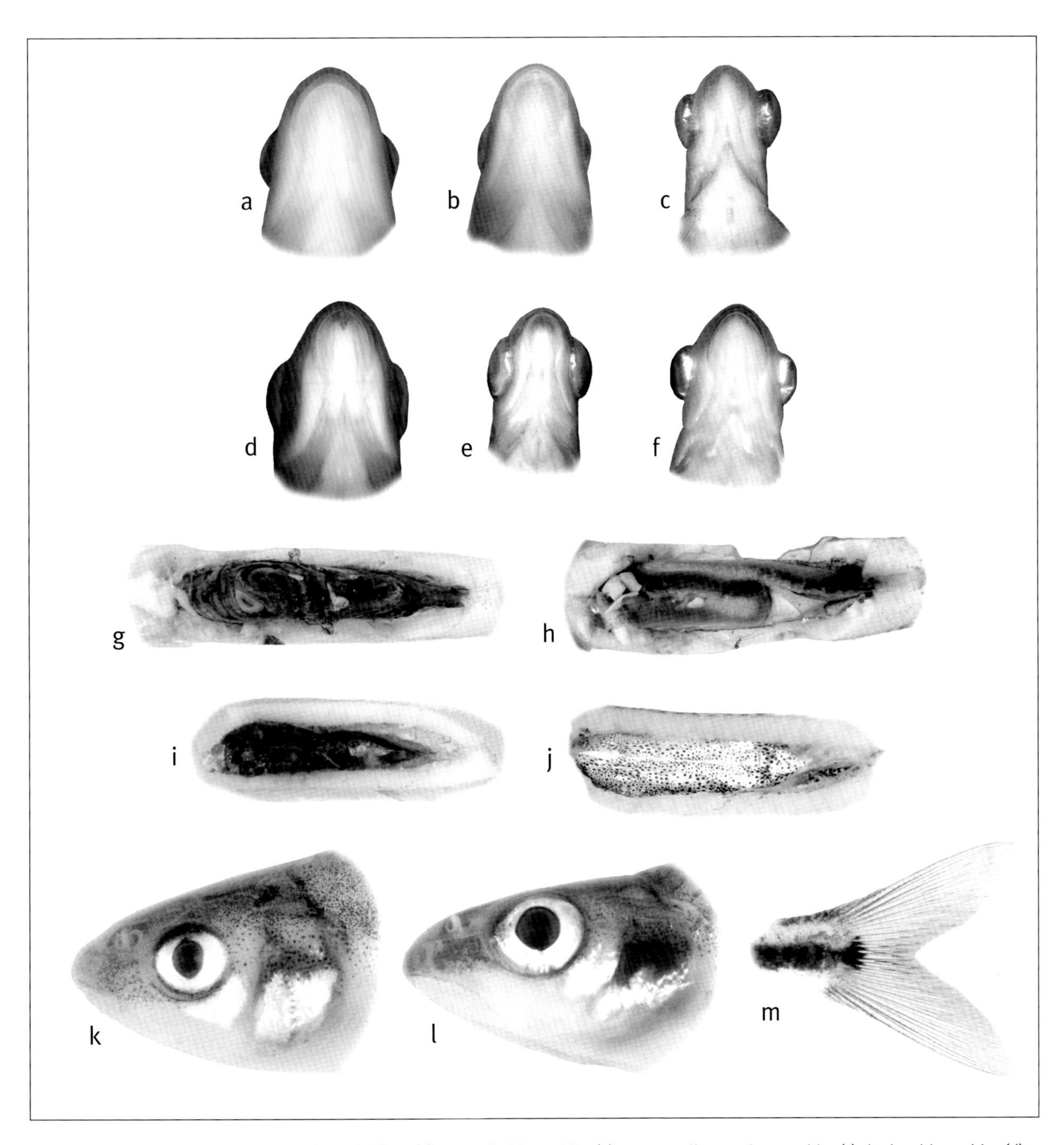

PLATE 10. Morphological characters of Cyprinidae: (a) spottail shiner chin, (b) eastern silvery minnow chin, (c) dusky shiner chin, (d) coastal shiner chin, (e) ironcolor shiner chin, (f) yellowfin shiner chin, (g) long, convoluted intestine, (h) short, S-curved intestine, (i) dusky or black peritoneum, (j) white or silvery peritoneum, (k) smaller, lateral eye of eastern silvery minnow, (l) larger, dorsolateral eye of spottail shiner, (m) triangular caudal spot of coastal shiner.

[13] **15a.** Long anal fin, rays (9)10–11(13); pigment on chin shaped as a distinct, thick-lined horseshoe by a line outlining the lower jaw, entire ventral side of lower jawbone (dentary) black (Plate 10 c), visible when viewed from the ventral side; pharyngeal teeth 1,4–4,1 (Plate 11 i) . . . . . *Notropis cummingsae* (dusky shiner), p. 150

15b. Short anal fin, rays 7–8; pigment on chin shaped as terminal spot, terminal bar, or thin-lined U when viewed from ventral side (Plate 10 d–f); pharyngeal teeth 2,4–4,2 (Plate 11 h, k, m). **16**

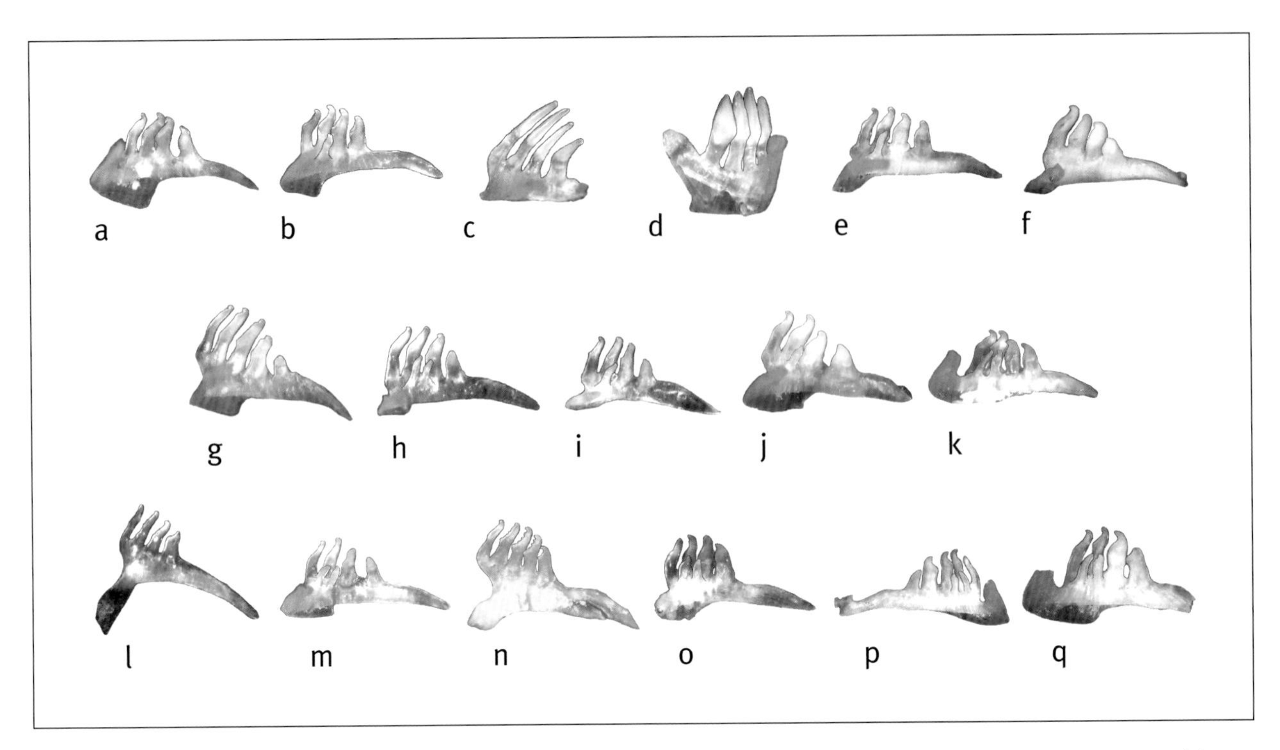

PLATE 11. Pharyngeal teeth: (a) bannerfin shiner, (b) whitefin shiner, (c, d) ventral and dorsal views of eastern silvery minnow, (e) rosyface chub, (f) bluehead chub, (g) golden shiner, (h) ironcolor shiner, (i) dusky shiner, (j) spottail shiner, (k) yellowfin shiner, (l) taillight shiner, (m) coastal shiner, (n) pugnose minnow, (o) lowland shiner, (p, q) left and right arches of creek chub.

[15] **16a.** Terminal black spot on chin when viewed from ventral side (Plate 10 d); anal fin rays 7; distinct triangular basicaudal spot (Plate 10 m) . . . . *Notropis petersoni* (coastal shiner), p. 162

16b. Pigment on chin shaped as a terminal bar or thin-lined U when viewed from ventral side (Plate 10 e, f), anal fin rays 8; well-defined basicaudal spot absent or with poorly defined margins and not distinctly triangular **17**

[16] **17a.** Postanal stripe extending from anus across base of anal fin onto caudal peduncle; thick, prominent terminal black bar across chin, some pigment extending onto anterior gular region (Plate 10 e) when viewed from ventral side; roof of mouth black pigmented. . . . . *Notropis chalybaeus* (ironcolor shiner), p. 147

17b. No postanal stripe, scattered melanophores may be visible under magnification; pigment on chin a thin-lined U along outer margin of ventral side of lower jawbone (Plate 10 f) when viewed from ventral side; roof of mouth not black pigmented . . . . *Notropis lutipinnis* (yellowfin shiner), p. 156

Note: Observation of pharyngeal teeth arrangement requires removal of pharyngeal arch. For details, see Introduction to Fish Identification, above, under Fish Morphology.

Carassius auratus (Linnaeus, 1758)
Goldfish

DESCRIPTION Body deep and stoutly compressed with a short, deep caudal peduncle. Dorsal and ventral profiles distinctly convex. Head scaleless (Scott 1954); snout longer than eye diameter (Okada 1959–1960). Mouth terminal and oblique. Barbels absent. Dorsal fin long, wedge-shaped (sometimes slightly falcate), highest anteriorly. Both dorsal and anal fins with long anterior stiff, serrate spine formed from fused rays. Relatively short pelvic fins. Large cycloid scales. Lateral line complete. Paired fins longer in males (Breder and Rosen 1966). Nuptial males have tubercles on operculum, and sometimes on back and pectoral fins.

Coloration in the wild is variable, but frequently with brassy olive brown dorsum and sides, white vent, and mostly translucent fins. Body coloration may vary from olive brown (Khan 1929) to silvery to gold with black blotches, and yellowish to white below (Jones et al. 1978).

SIZE Commonly reaches 120–220 mm SL, with a maximum of 410–457 mm SL reported (Carlander 1953).

MERISTICS Dorsal spines 2–3 (1 large one), dorsal rays (14)15–18 (Jenkins and Burkhead 1993) or 15–21 (Scott and Crossman 1973), anal spines 2–3 (1 large one), anal rays 5–6, pectoral rays 14–17, pelvic rays 8–10; pharyngeal teeth 0,4–4,0; lateral line scales 28–32 (Jenkins and Burkhead 1993) or 25–31 (Scott and Crossman 1973).

CONSERVATION STATUS The first exotic fish known to be introduced in North America; released as early as the late 1600s (Courtenay et al. 1984).

DISTRIBUTION The native distribution is eastern Europe to China; introduced into all 48 contiguous U.S. states; the Canadian provinces of British Columbia, Alberta, and Ontario; and Mexico, where it is rare (Hensley and Courtenay *in* Lee et al. 1980). Introduced into Puerto Rico, where it successfully breeds in artificial pools but shows no sign of establishing self-sustaining populations in natural waters (Erdman 1984). Collected in every U.S. state except Alaska (Courtenay et al. 1991). In the MSRB, reported from only two locations on the SRS—a pond and a stream—but probably present in many garden ponds.

Carassius auratus, goldfish

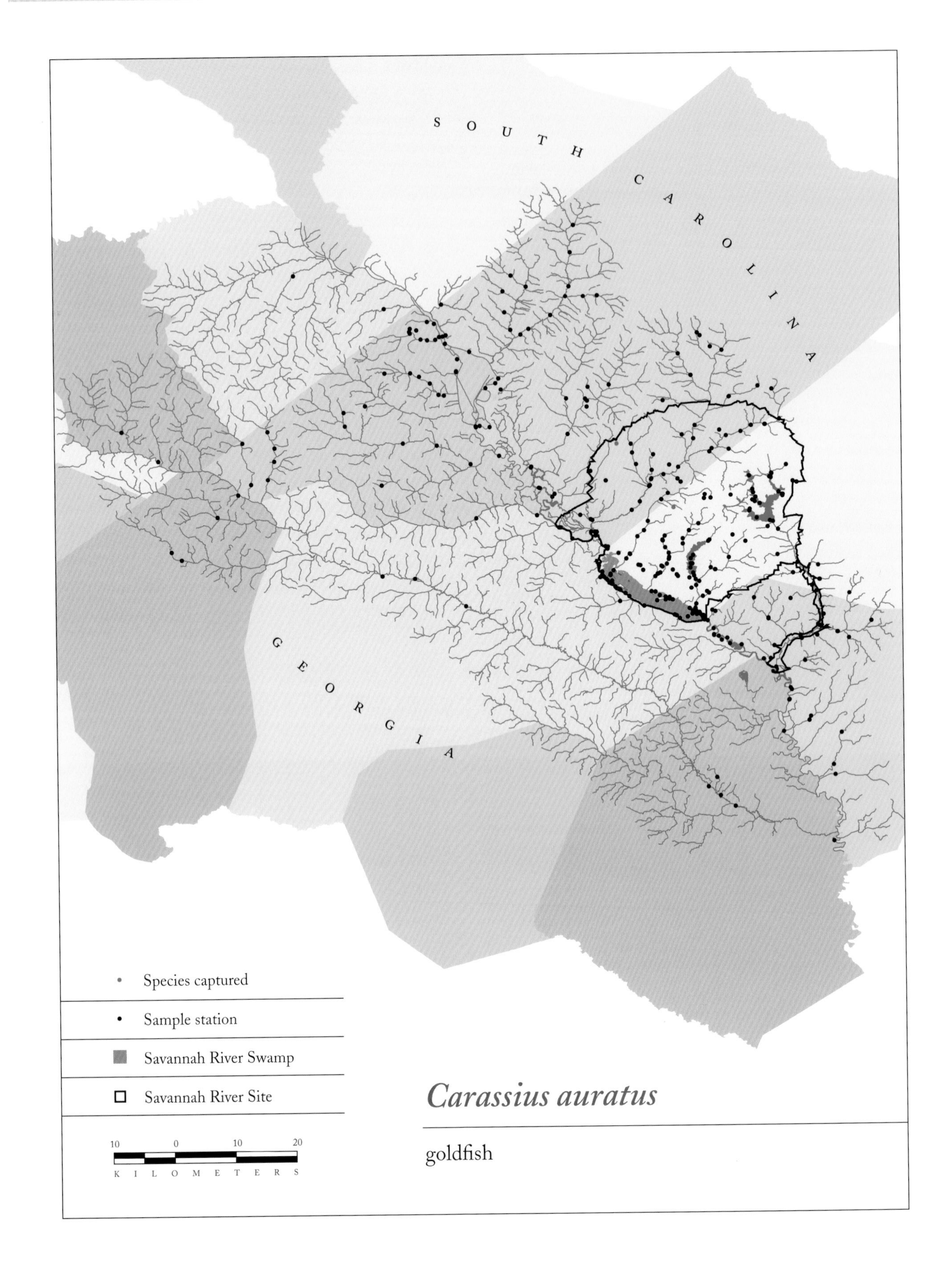

Carassius auratus

goldfish

HABITAT Shallow water in warm lakes, reservoirs, rivers, and quiet streams with dense vegetation. Goldfish can tolerate turbid water and often thrive in degraded waters (Rohde et al. 1994). They have been found in salinities up to 17 ppt, but mainly below 15 ppt (Jones et al. 1978). They may be benthic or may school at the surface.

BIOLOGY Goldfish have a prolonged spawning season (late March–mid-August) in North America (Okada 1959–1960). Individual fish spawn 3–10 lots of eggs at intervals of 8–10 days (Smith 1909). Several thousand eggs are broadcast over submerged vegetation at water temperatures between 15.5 and 23 °C. Adults spawn in pairs and do not care for the developing offspring. Under crowded conditions this species produces a secretion that suppresses further spawning (Swingle 1953 *in* Carlander 1969). Fecundity varies from 2,000 (Smith 1909) to 400,000 eggs; eggs are about 1 mm in diameter (Slastenenko 1958) and hatch in 3–6 days. Larvae often cling to plants (Sterba 1962) and become free swimming in 1–2 days (Chen 1926). Spawning, egg, and larval descriptions can be found in Grimm 1937; Battle 1940; Okada 1959–1960; and Berg 1964. Larval descriptions are summarized in Mansueti and Hardy 1967; Lippson and Moran 1974; Wang and Kernehan 1979; and Auer 1982. Domestic goldfish reach sexual maturity at 2–4 years and around 75 mm TL (Fearnow 1925; Nikolsky 1963). The usual life span is 6–7 years, with a maximum of 30 years reported (Robison and Buchanan 1984).

Juveniles feed mostly on zooplankton and insect larvae. Adults are omnivores; they are commonly used to control aquatic weeds and algae in small ponds and are among the most popular aquarium fish. They hybridize with *Cyprinus carpio*, the common carp, in Lake Erie (Trautman 1957) and in the Chicago area (Smith 1979).

SCIENTIFIC NAME *Carassius*, from *karass* = the vernacular name for European crucian carp; *auratus* = "golden" or "gilded" (Latin).

Ctenopharyngodon idella (Valenciennes, 1844)
Grass carp

DESCRIPTION Body elongated and chubby, with a short, deep caudal peduncle. Subterminal mouth slightly oblique with nonfleshy, firm lips and no barbels. Head lacks scales (Chilton and Muoneke 1992). Crowns of pharyngeal teeth in the row with the most teeth have parallel grooves (Jenkins and Burkhead 1993). All fins lack spines. Dorsal fin short. Anal fin set closer to the tail than in most cyprinids. Lateral line complete.

Darkest dorsally, with color ranging from olive to brownish olive shading into brownish yellow or silvery white on the sides. White ventrally. Scales on sides large with dark outlines forming a crosshatched pattern. Top and sides of head generally darker than body. All fins dark or dusky and may have reddish tinge.

MERISTICS Dorsal rays 8–9 (Jenkins and Burkhead 1993) or 8–10 (Chilton and Muoneke 1992), anal rays 8–10, pectoral rays 19–22, pelvic rays 8–9; pharyngeal teeth 2,4–4,0 (Jenkins and Burkhead 1993) or 2,4–4,2 (Chilton and Muoneke 1992); lateral line scales 35–42 (Jenkins and Burkhead 1993) or 40–42 (Chilton and Muoneke 1992).

SIZE Length can exceed 1 m. Weight commonly 29–36 kg, but may reach 45 kg (Lopinot 1972). Young stocked at 20 cm can reach 46 cm by fall (Rohde et al. 1994).

DISTRIBUTION Native to the Pacific drainages of Asia from the Amur River of China and Siberia south to the West River in southern China and Thailand; stocked in at least 35 U.S. states (Greenfield 1973), but usually fails to become established at site of stocking (Guillory *in* Lee et al. 1980). Reproducing established populations are found in the Mississippi River from Illinois and Missouri to Louisiana (Conner et al. 1980; Burr and Warren 1986; Zimpfer et al. 1987; Courtenay et al. 1991; Etnier and Starnes 1993), the Illinois River in Illinois (Raibley et al. 1995), the lower Missouri River (Raibley et al. 1995), the Ohio River, the Trinity River of Texas (Waldrip 1992; Webb et al. 1994; Elder and Murphy 1997), and in Minnesota (Courtenay et al. 1991). Imported into Mexico in 1960 to control water hyacinths (Sutton 1977). In the MSRB, introduced into a number of ponds and reservoirs and found in the Savannah River.

HABITAT Within the native range, large rivers; introduced into private ponds and a few public waters as biological control agents for aquatic macrophytes (Cross 1969). High tolerance for salinity.

Ctenopharyngodon idella, grass carp

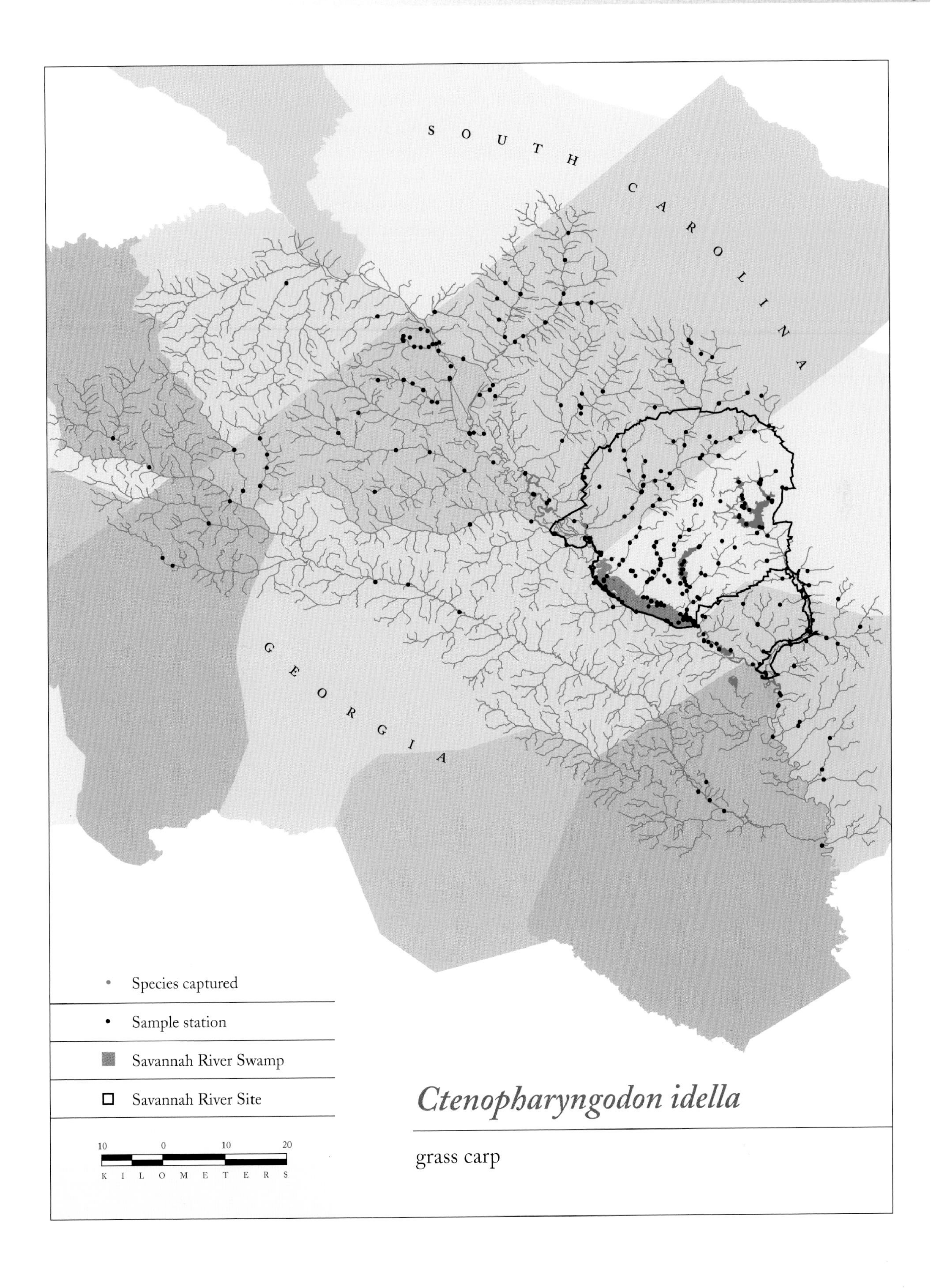

Ctenopharyngodon idella

grass carp

BIOLOGY Grass carp spawn in flowing water and do not reproduce in lakes and ponds. Spawning in the eastern United States occurs in the early spring in rising water at about 15–17 °C (Stanley et al. 1978). The eggs are semipelagic and remain suspended during the 20–40 hour incubation period (Berg 1964; Konradt 1968). Reproductive success depends largely on an adequate flow of oxygen-rich water to suspend the eggs until they hatch.

Young become herbivorous at about 73 mm. Grass carp consume 50–60% of their body weight per day (Wiley and Wike 1986) and sometimes up to 100% (Shireman and Smith 1983). When macrophyte abundance is low or when competitors are few, the diet may shift to include animal material such as insects and small fish (Laird and Page 1996). Reduction or elimination of aquatic macrophytes by grass carp may have serious effects on the fish and waterfowl that depend on them (Laird and Page 1996).

Most introduced grass carp are triploid (sterile), including those found in Georgia and South Carolina (Chilton and Muoneke 1992). Triploid grass carp were developed in 1983 (Cassani and Maloney 1991). The hybrid grass carp *Ctenopharyngodon idella* × *Hypoptalmichthys nobilis* proved a disappointment due to its poor growth performance and consumption rate (Wike 1987).

SCIENTIFIC NAME *Ctenopharyngodon* = "comblike," "throat," and "teeth" (Greek), referring to the gill rakers; *idella* = "distinct" (Greek).

Cyprinella leedsi (Fowler, 1942)

Bannerfin shiner

DESCRIPTION Body deep and laterally compressed; relative body depth increasing with size. Lateral profile of head and anterior body conical, but ventral profile tends to be more flattened. Eyes moderate in size and laterally placed. Mouth subterminal and moderately oblique. Barbels absent. Lateral line complete.

Narrow lateral stripe present, steel blue in life and dark in preservative, but anterior one-fourth distinctly faded or absent. Upper edge sharper than lower edge, particularly anteriorly. Dorsum olive tan, more tan in nuptial males. Ventral head and chin white, but upper and lower lips lightly pigmented. Belly immaculate, white; flanks silvery gray. Postanal stripe variable, ranging from absent to extending onto caudal peduncle. Black blotches present on dorsal fin anterior to first principal ray. Other fins translucent. Scales on dorsolateral sides lined with black, forming a blurred and indistinct crosshatched pattern. Roof of mouth unpigmented.

Breeding males are distinctive: dorsal fin greatly expanded with a rather rounded distal edge, the lower one-third milky white and the distal two-thirds evenly colored black with white tip on anterior three or four membranes; pectoral, pelvic, and particularly anal fins expanded (Rabito and Heins 1985); anal fin milky white and brightest along the distal edge, with irregular blotches across the central to distal portion that are most intense anteriorly; pelvic fin milky white, especially distal tips, with irregular gray pigment on rest of the fin, most intense on distal portion of anterior rays; caudal fin with white tips; lateral stripe becomes more prominent. Nuptial males have tubercles, especially on the dorsal head; those on the dorsal midline are generally larger than those in bordering rows (see photos in Mayden 1989). Enlarged tubercles, 1 per scale, are present on second and/or third scale rows above anal fin and on caudal peduncle.

SIMILAR SPECIES The bannerfin shiner is most likely to be confused with the whitefin shiner (*C. nivea*) but can be distinguished by its blurred rather than clear dorsolateral crosshatch pattern. Also, the black blotches on the dorsal fin of the bannerfin shiner lie anterior to the first principal ray but are on the posterior two membranes of the whitefin shiner's dorsal fin. Pharyngeal teeth counts differ and are particularly diagnostic. Tuberculation of the dorsum of nuptial males also differs.

SIZE Adult size is up to 80 mm SL (Mayden 1985); males (average about 50 mm SL) are larger than females (average 41–46 mm SL) (Rabito and Heins 1985).

MERISTICS Dorsal rays 8, anal rays 8(9), pectoral rays (12)14–15(16), pelvic rays 8; pharyngeal teeth 0,4–4,0; lateral line scales (35)36–37(39) (Mayden 1989).

Cyprinella leedsi, bannerfin shiner; breeding male

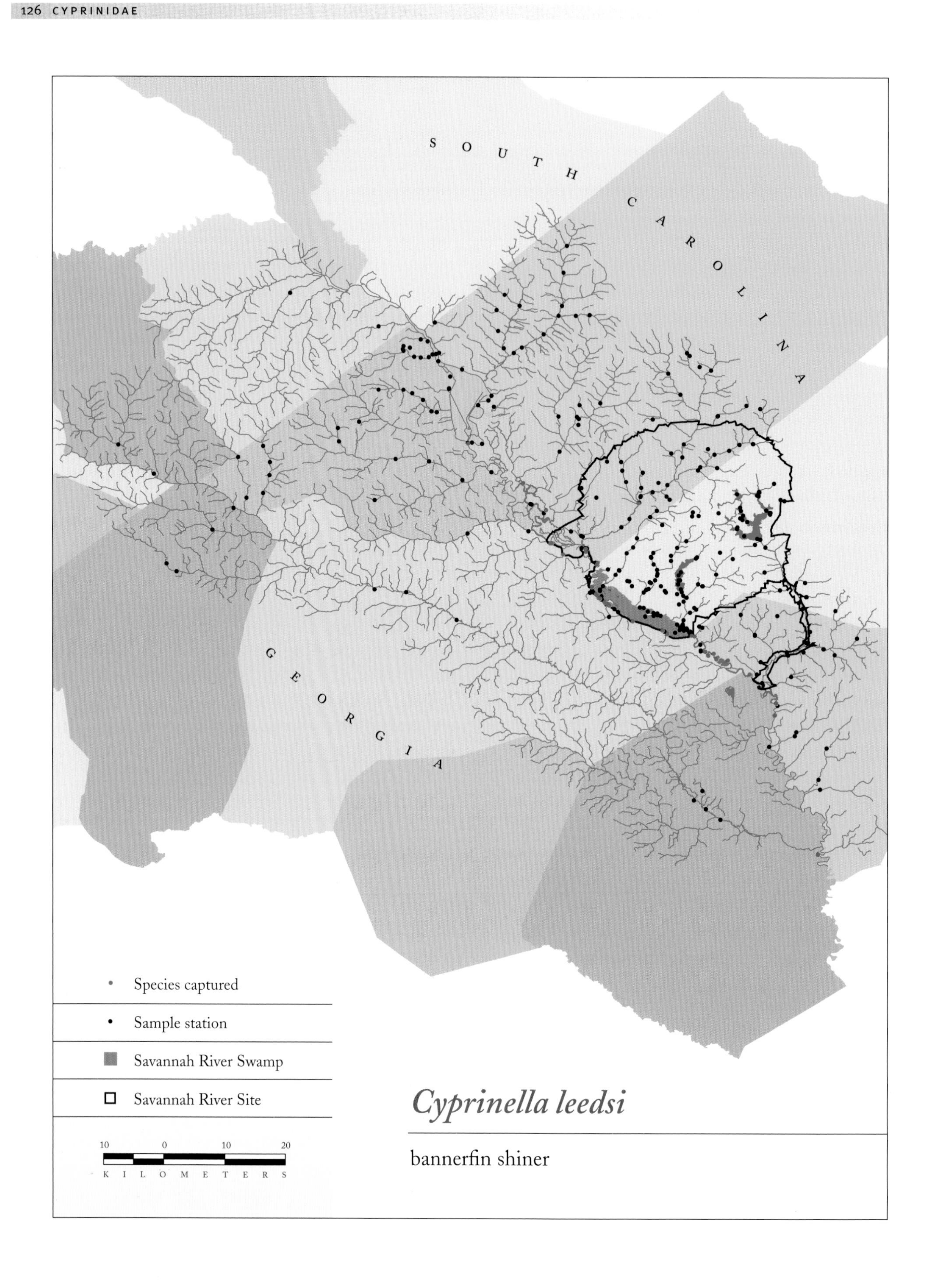

Cyprinella leedsi

bannerfin shiner

Cyprinella leedsi, bannerfin shiner; nonbreeding individual

DISTRIBUTION From the Savannah River of Georgia and South Carolina south to the Ochlockonee and Suwannee river drainages of Florida (Gilbert *in* Lee et al. 1980).

HABITAT In the MSRB, bannerfin shiners are found almost exclusively in the Savannah River. Schools often gather in eddies behind woody debris or other large flow obstructions such as wooden wing dams. They rarely enter tributary streams further upstream than on the modern floodplain of the Savannah River.

BIOLOGY Bannerfin shiners, classified as crevice spawners by Johnston and Page (1992), are well suited for laboratory study because they readily spawn in aquaria in predictable and controllable locations. Rabito and Heins (1985) described territorial defense and spawning in bannerfin shiners in aquaria. Breeding males aggressively defend the area around the crevices in which eggs are deposited. Females deposit a clutch of eggs by repeatedly returning to the spawning site and laying small batches of 5–15 eggs. An entire spawning episode can last 15–45 minutes, and clutches contain 26–228 eggs. Females can spawn multiple clutches at 3–10 (generally 4–6) day intervals by having groups of oocytes synchronously develop from a reservoir of immature oocytes (Heins and Rabito 1986). In laboratory experiments, narrow crevices (2–4 mm wide) were most frequently chosen for spawning; about 90% of the deposited eggs were successfully fertilized and began development. Rabito and Heins 1985 describes fertilized eggs; Loos and Fuiman 1978 includes a cursory comparison of the larvae with those of other *Cyprinella* species. Ritualistic displays include head-to-tail circling and parallel swims. The protracted spawning season in Georgia and Florida extends at least from May to September (Heins and Rabito 1986).

The bannerfin shiner is an invertivore. Wiltz (1993) examined 25 specimens collected in the MSRB and found large numbers of pelycypods (bivalve mollusks), a variety of aquatic insects (especially chironomids), and microcrustaceans in their digestive tracts.

SCIENTIFIC NAME *Cyprinella* = "small carp" (Greek); *leedsi* = named after Arthur N. Leeds, who helped collect the type series (Mayden 1989).

Cyprinella nivea (Cope, 1870)
Whitefin shiner

DESCRIPTION Body laterally compressed and deep; relative body depth increasing with size. Head and anterior body conical in lateral profile. Eyes moderate in size and laterally placed. Mouth subterminal and moderately oblique. Barbels absent. Lateral line complete.

Narrow lateral stripe steel blue in life (dark in preservative), but anterior one-fourth distinctly faded or absent; upper edge sharper than lower edge, particularly anteriorly. Stripe more prominent in nuptial males. Dorsum olive tan, ventral head and chin white; upper and lower lips lightly pigmented, with generally lighter pigment on anterior terminal end. Belly immaculate white, flanks white in large males, more gray in smaller individuals. Postanal stripe variable, ranging from absent to extending onto caudal peduncle. Black blotches on posterior two membranes of dorsal fin; other fins translucent, but anterior pectoral ray may be outlined with black. Scales on dorsolateral sides with black linings forming a sharp and clear crosshatched pattern. Roof of mouth unpigmented except for a few scattered melanophores along the anterior.

Nuptial males have distinctive characteristics: dorsal fin expanded with white edging, most intense anteriorly; black present on most membranes, but most intense on posterior three membranes; lower portion of dorsal fin marked with milky white that is most intense in posterior rays between black blotches; pectoral, pelvic, and anal fins white, brightest on distal edges; caudal fin with white tips; tubercles present, especially on the dorsal head, tubercles on dorsal midline generally about same size as tubercles in bordering rows (see photos in Mayden 1989).

SIMILAR SPECIES See the bannerfin shiner (*C. leedsi*) account.

SIZE Large for a shiner, reaching 91 mm SL, 113 mm TL (Cloutman and Harrell 1987).

MERISTICS Dorsal rays 8, anal rays (7)8(9), pectoral rays (13)15(16), pelvic rays 8; pharyngeal teeth 1,4–4,1; lateral line scales 35–41 (Mayden 1989).

DISTRIBUTION From the Neuse River drainage of North Carolina south to the Savannah River drainage of South Carolina and Georgia (Gilbert and Burgess *in* Lee et al. 1980). In the MSRB, primarily in the Savannah River, less common in middle and lower parts of tributaries.

HABITAT In the Savannah River, schools of whitefin shiners gather in eddies behind woody debris. Particularly large schools sometimes appear below wooden wing dams.

Cyprinella nivea, whitefin shiner; breeding male

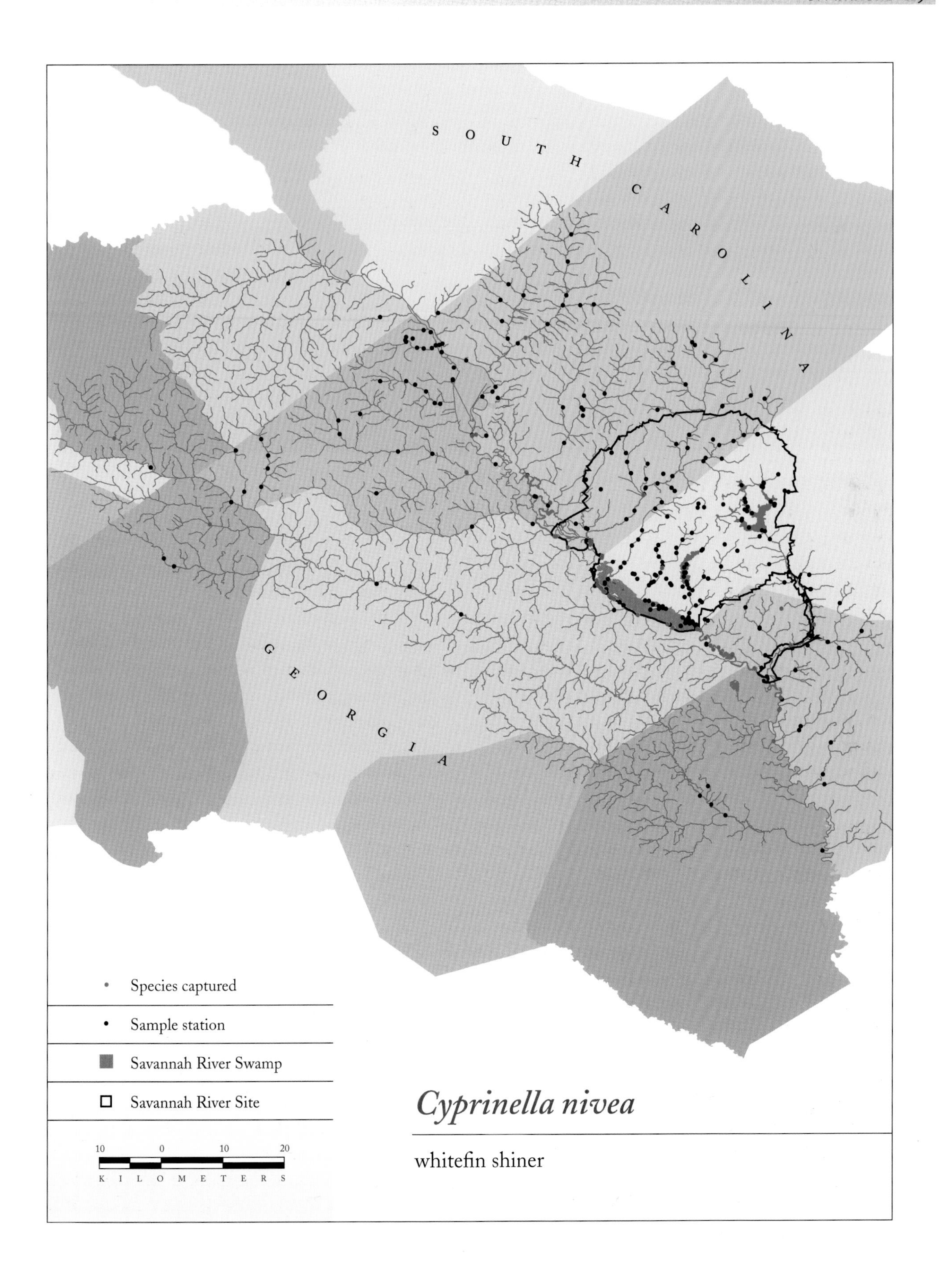

Cyprinella nivea

whitefin shiner

Cyprinella nivea, whitefin shiner; nonbreeding individual

They also frequent the lower parts of tributary streams, particularly the sections lying on the fluvial terraces along the Savannah River Valley. In the tributary streams, they frequent scour pools behind stumps or woody debris.

BIOLOGY Spawning in the Broad River (South Carolina) occurs from June to August at water temperatures of 23–26 °C (Cloutman and Harrell 1987). Spawning has not been observed, but we suspect that whitefin shiners are crevice spawners like other *Cyprinella* species. Females of 52 and 72 mm TL contained 112 and 545 mature eggs (clutch size), respectively; ovary morphology indicates that females lay multiple clutches by spawning repeatedly throughout the protracted spawning season (Cloutman and Harrell 1987). Loos and Fuiman 1978 includes a cursory comparison of the larvae with those of other *Cyprinella* species.

Cloutman and Harrell (1987) conducted life history studies of whitefin shiners in the Broad River, Cherokee County, South Carolina, and reported that annulus formation on scales occurred in late June; males grew faster than females; and both sexes lived up to 4 years.

Insects dominate the diet, but plant material including vascular plants and algae has also been found in whitefin shiner guts (Cloutman and Harrell 1987); it is not known if ingestion was intended or incidental or whether the material was digested. Whitefin shiners have a relatively short S-shaped intestine typical of many carnivorous shiners. Feeding rates were reduced in cold winter months when 42% of the guts were empty (Cloutman and Harrell 1987). Bain and Helfrich (1983) reported that whitefin shiners preyed on offspring in nests of bluegill (*L. macrochirus*).

SCIENTIFIC NAME *Cyprinella* = "small carp" (Greek); *nivea* = "snow" or "snowy" (Latin), referring to white on fins of breeding males (Mayden 1989).

Cyprinus carpio Linneaus, 1758

Common carp

DESCRIPTION Body deep and stoutly compressed. Dorsal profile distinctly convex with a crest behind the head; ventral profile more flattened. Mouth subterminal, a pair of long, conspicuous barbels near each corner of upper jaw. Mouth lacking teeth, but pharyngeal arch with three rows of teeth. Dorsal fin long and falcate anteriorly. Dorsal and anal fins with a long, stiff, serrate anterior spine formed from fused rays. Lateral line complete.

Color may be golden or brassy olive, slate gray, or olive green above grading into golden yellow ventrally. Scales on sides and back with a black basal spot. Lower half of caudal and anal fins often with a reddish hue (Trautman 1957; Slastenenko 1958). Spawning males develop pearl organs on the head and pectoral fins (Swee and McCrimmon 1966).

MERISTICS Dorsal spine 1 (large serrated), dorsal rays (15)18–20(23), anal spine 1 (large serrated), anal rays (4)5(6), pectoral rays (14)15–16(17), pelvic rays 8–9; pharyngeal teeth 1,1,3–3,1,1; lateral line scales usually 35–39 (Jenkins and Burkhead 1993) or 32–41 (Berg 1964).

CONSERVATION STATUS Exotic in North America; first released in 1830 in the Hudson River (Courtenay et al. 1984).

SIZE Adults can reach 1.2 m TL and can weigh up to about 41 kg (Migdalski 1962).

DISTRIBUTION Native to temperate portions of Eurasia from southern Norway to Spain, east to Siberia and south into China and northern India; widely introduced in North America and established in most drainages south of 50 degrees north latitude (Allen *in* Lee et al. 1980). In the MSRB, found mostly along the main stem of the Savannah River.

HABITAT Moderately warm and generally shallow waters of rivers, lakes, and reservoirs, usually in association with aquatic vegetation; rocky shoal areas; and protected bays, over sand, clay, or mud bottoms (Trautman 1957). Mostly found in areas of slow current (Pflieger 1975). The common carp adapts to a wider range of conditions (e.g., habitat disturbance, especially siltation and high turbidity) than almost any native North American fish (Becker 1983). Carp migrate into lake shallows, stream tributaries, marshes, and floodplains prior to spawning (Sigler 1955). Carp can adapt to brackish water (exceeding 17 ppt) (Schwartz 1964), but larvae usually die at salinities greater than 4 ppt (Askerov 1975).

BIOLOGY Often found in schools. In North America, common carp spawn from late March to August at water temperatures between 18 and 23 °C (Sigler 1958), sometimes up to 28 °C near the surface in areas of dense veg-

Cyprinus carpio, common carp

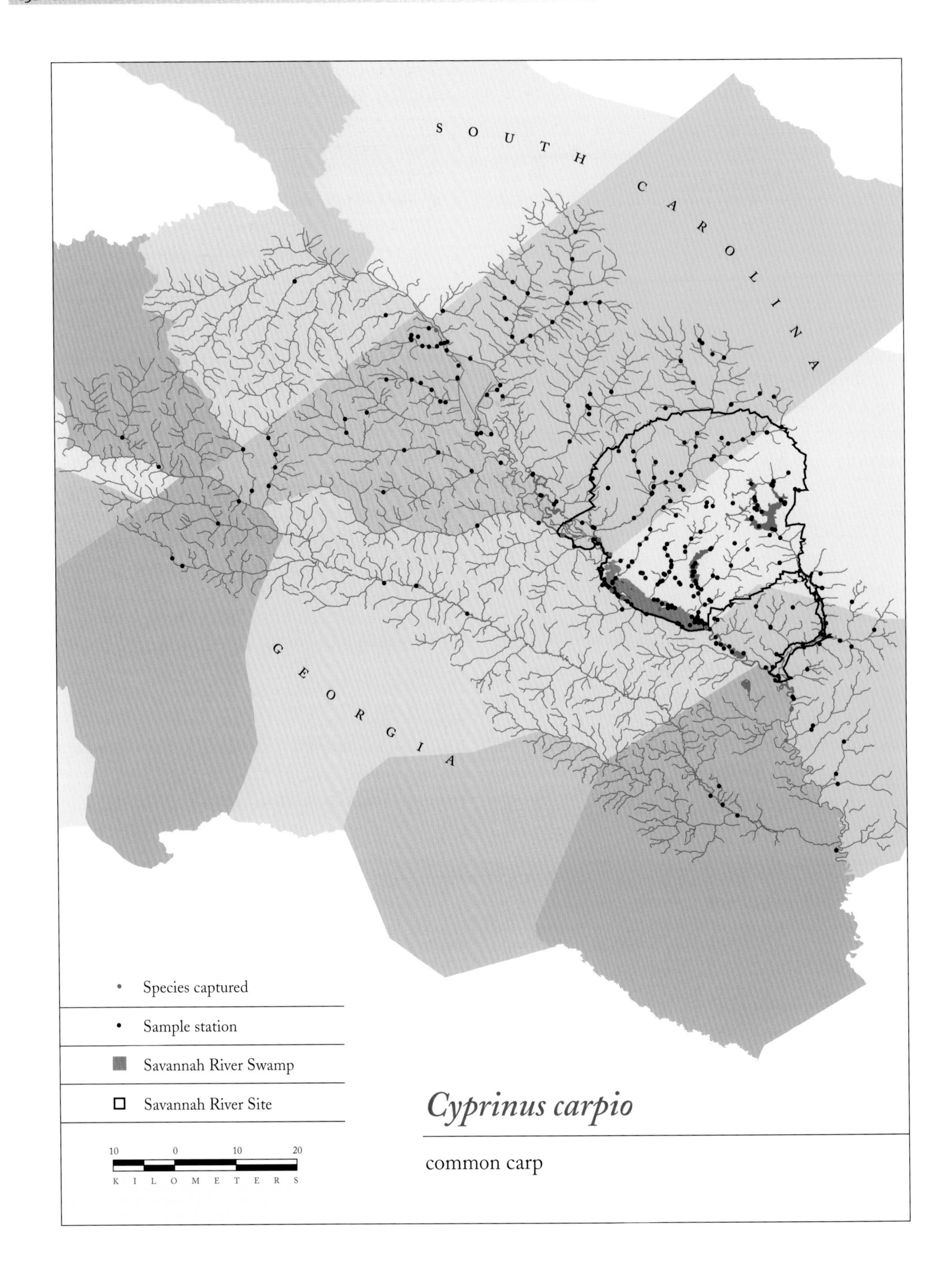

Cyprinus carpio

common carp

etation. Spawning can occur two or three times over several days or weeks. Mating groups include one female and several males, and no nest is prepared. Eggs are deposited on substrate such as grass blades and are not guarded. Eggs vary from 1.24 to 1.42 mm and are yellowish green (Nahamura 1969). Hatching occurs in 78 hours at 19–23 °C (Verma 1970). Fecundity estimates range from 479,000 to 600,000 (Moroz 1968). For descriptions of larvae and juveniles see Fish 1932; Smallwood and Derrickson 1933; Brogensky 1960; Mansueti and Hardy 1967; Hoda and Tsukahari 1971; Lippson and Moran 1974; Wang and Kernehan 1979; and Auer 1982. Larvae smaller than 5.0 mm attach to aquatic vegetation. Congregation and schooling begin at about 9.5 mm. Juveniles begin to move into deeper water at about 25 mm. Females mature at 2–5 years, and males at 1–2 years of age (Becker 1983). Common carp live up to 15 years, and ages up to 24 years have been reported (Slastenenko 1958).

Common carp are omnivorous, showing preferences for chironomids, cladocerans, oligochaetes, and plankton (Slastenenko 1958; Scott and Crossman 1973). Juveniles feed on larval fish when invertebrates are scarce (Astanin and Trofimova 1969). Carp root in the bottom sediments when feeding, stirring up silt and increasing water turbidity, which can have a deleterious effect on other fishes.

SCIENTIFIC NAME *Cyprinus* = "carp" (Greek); *carpio* = "carp" (Latin).

Hybognathus regius Girard, 1856

Eastern silvery minnow

DESCRIPTION Body stocky to slightly compressed, shallow. Caudal peduncle deep. Eyes moderate in size and laterally placed. Mouth small and subterminal or inferior. Barbels absent. Lateral line complete and gently downcurved anteriorly. Intestine long and coiled; peritoneum black.

Black lateral stripe absent or indistinct in life; sometimes with a relatively narrow, diffuse stripe extending from posterior edge of operculum to caudal peduncle. Stripe wider and more distinct posteriorly; entire stripe intensifies in preservative. Dorsum pale tan/olive, scales peppered with tiny melanophores. Ventral head and sides silvery, lighter ventrally. Chin without pigment or with only very small scattered melanophores (visible with magnification). Upper lip with small scattered melanophores. Roof of mouth largely unpigmented with only tiny scattered melanophores. Postanal stripe absent. Fins translucent, at least dorsal and anterior pectoral rays outlined lightly in black (may require magnification to see).

Breeding males have light yellow on sides of body and head (Raney 1939); all fins tinged with yellow, lightest in dorsal fins. Midline of belly silvery; dorsum of male darker than that of female. Breeding females remain silvery (Raney 1939). Breeding tubercles present in males on head, fins, and posterior edges of scales of the dorsum and sides, more abundant anteriorly (Raney 1939).

SIZE Up to 120 mm TL; females (usually up to 100 mm SL) are larger than males (usually up to 83 mm SL) (Raney 1939, 1942).

MERISTICS Dorsal rays 8, anal rays 8, pectoral rays (14)15 (16), pelvic rays (7)8; pharyngeal teeth 0,4–4,0; lateral line scales (32)34–37(38) (Jenkins and Burkhead 1993).

Hybognathus regius, eastern silvery minnow

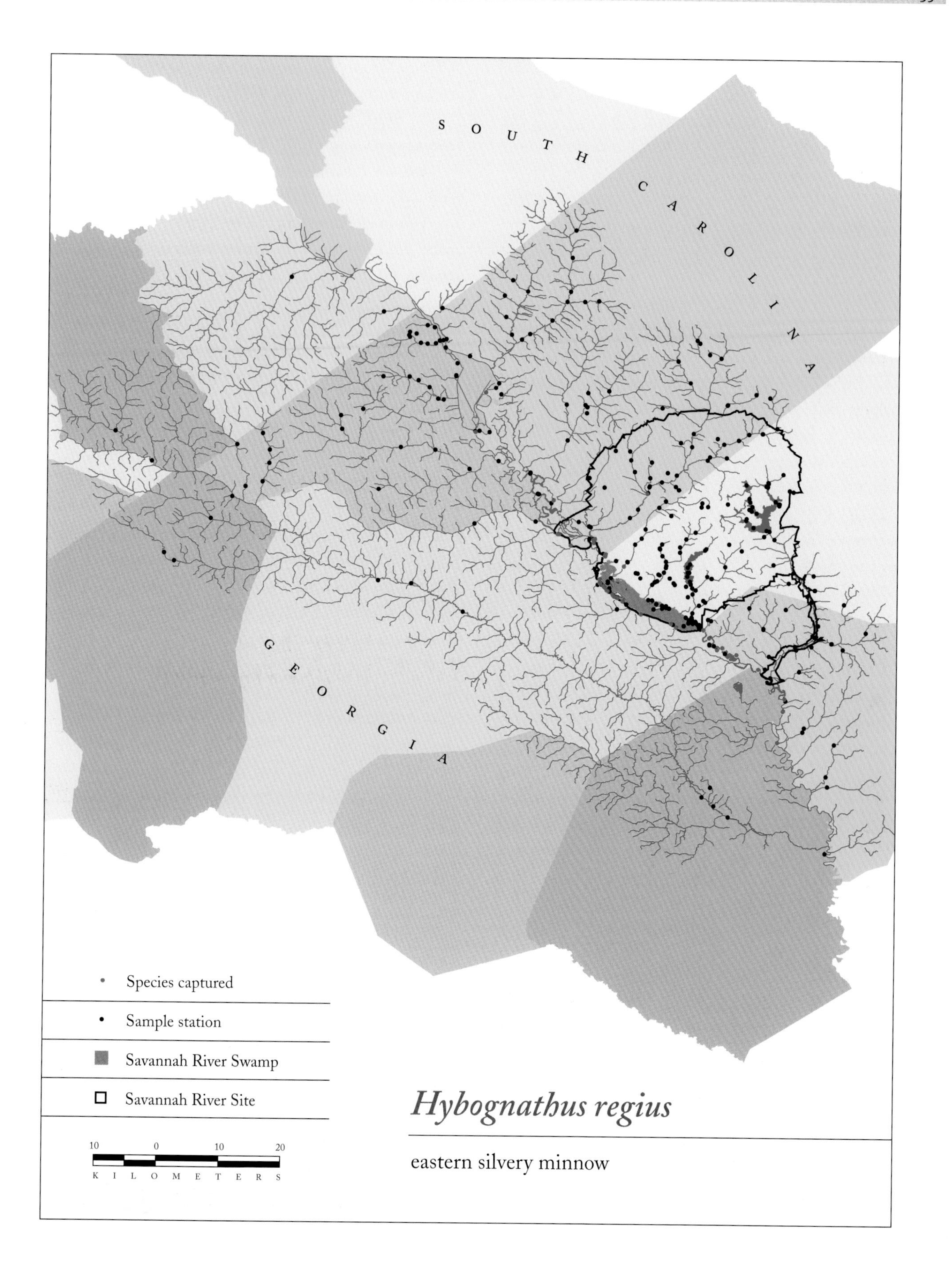

Hybognathus regius

eastern silvery minnow

DISTRIBUTION On the Atlantic slope from the St. Lawrence River drainage, Quebec, down to the Altamaha River drainage, Georgia (Page and Burr 1991). Apparently uncommon in the MSRB, which is near the southern extent of the range, but most commonly found in the Savannah River or oxbows, sloughs, or sections of tributary streams on the modern Savannah River floodplain; sporadically ascends tributary streams, especially portions on fluvial terraces.

HABITAT Typically pools and backwaters of low-gradient streams, lakes, and small to large sluggish rivers (Wang and Kernehan 1979; Page and Burr 1991). Spawning habitat includes upper portions of tidal creeks, inlet streams of lakes, small coves, and other slow-flowing water bodies (Raney 1939, 1942; Wang and Kernehan 1979).

BIOLOGY The eastern silvery minnow is one of the earliest to spawn of the northeastern cyprinids. Spawning occurs before aquatic macrophytes become abundant (Raney 1939), from April through May at water temperatures between 10 and 21 °C in Delaware and New York. Males migrate onto the spawning grounds first, with their numbers peaking in April, and remain there longer than females (Raney 1939). Eggs are broadcast freely, without prior modification of the substrate and without parental care (Johnston and Page 1992). Spawning peaks around midday and occurs generally in trios consisting of a female accompanied by a male on each side, although an additional one to seven males may be present (Raney 1939). Males sometimes spawn repeatedly with different females and do not appear to be territorial (Raney 1939). Spawning substrate includes sand and gravel in tidal creeks of Delaware, and in nontidal waters spawning habitat overlaps with that of the spottail shiner (*N. hudsonius*) (Wang and Kernehan 1979). The demersal and apparently nonadhesive or only briefly adhesive eggs are distributed on the soft bottom or at the base of newly sprouted grass in water less than 30 cm deep, commonly less than 15 cm. Eggs hatch in 6–7 days at 13–21 °C, and larvae swim up from the bottom to form small schools about 2 weeks after hatching (Raney 1939). Larval descriptions are provided in Raney 1939; and Wang and Kernehan 1979. Fuiman et al. (1983) discussed larval characteristics useful for identification.

In females measuring 60–90 mm SL, clutch size increased with female size; 2,000–6,600 mature ova were found (Raney 1939). In a New York culture pond, spawning occurred at age 2 (Raney 1942), but age at maturity varied with growth rate, and only individuals with high growth rates matured in 1 year (Forney 1957). Fish in the culture pond reached 41–61 mm TL at the end of the first growth season, 65–89 mm TL at the end of the second growth season, and 76–87 mm (males) and 78–96 mm (females) at the end of the third (Raney 1942).

The diet consists primarily of microscopic materials found in the bottom ooze, especially diatoms and other algae (Raney 1939; Wiltz 1993), although it may include some insects (Schwartz 1963). The long, coiled intestine (Raney 1939) and modifications of the pharyngeal epithelium into fine papillae that filter out diatoms and other small items make this species well suited for such a feeding strategy (Hlohowsky et al. 1989).

SCIENTIFIC NAME *Hybognathus* = "humped jaw" (Greek); *regius* = "royal" (Latin).

Hybopsis rubrifrons (Jordan, 1877)

Rosyface chub

For a brief time after 1991 this species was referred to as *Notropis rubescens*, a name change necessitated by synonymy when *Hybopsis* was downgraded to a subgenus (Robins et al. 1991). The name *Hybopsis rubrifrons* was restored when *Hybopsis* was reelevated to generic level.

DESCRIPTION Body long, cylindrical, and ventrally flattened. Eyes large and dorsolaterally placed. Mouth small, inferior, slightly oblique or nearly horizontal, distinctly overhung by long snout. Maxillary barbels long and slender, sometimes densely ornamented with taste buds (Dimmick 1988). Lateral line complete.

A narrow black lateral stripe, upper edge sharper than lower edge, extends from tip of snout to caudal fin and terminates in a moderately distinct but variable caudal spot that streaks onto the caudal fin. Lateral line entirely punctate, but punctation variably visible above black lateral stripe. In life, dorsal edge of black stripe bordered by thin iridescent coppery stripe. Dorsum tan olive, ventral head white, belly white to goldish yellow, and flanks pearly silver. Chin and both lips without pigment. Lips noticeably fleshy and almost suckerlike. Roof of mouth without pigment. Postanal stripe variable, but often extending from anus to along anal fin base. Fins translucent, but some rays, particularly first pectoral, dorsal, and caudal, may be lightly outlined with melanophores.

Breeding males have small tubercles on the paired and median fins as well as on the head, including the snout, dorsal head, face, cheeks, chin, and throat. Dorsal scales, especially above the black stripe, are also outlined with tubercles. Breeding females also have tubercles, especially on the head, but they are less pronounced and abundant than on males. Red develops on anterior one-third of the body during spawning (Page and Burr 1991).

SIMILAR SPECIES The bluehead chub (*N. leptocephalus*) and sometimes the pugnose minnow (*O. emiliae*) also have single maxillary barbels at each corner of the mouth. The rosyface chub has a small inferior mouth and pharyngeal teeth 1,4–4,1 versus 0,4–4,0 in the other two species. The pugnose minnow can easily be distinguished by its small superior and strongly oblique mouth, slightly compressed body, and nine dorsal rays (versus eight in the rosyface chub). The bluehead chub has a large, only slightly subterminal mouth, red in the caudal fin in life, and seven anal rays (versus eight in the other species). See the description of the spottail shiner (*N. hudsonius*) for comparison with a similar species that lacks barbels.

Hybopsis rubrifrons, rosyface chub

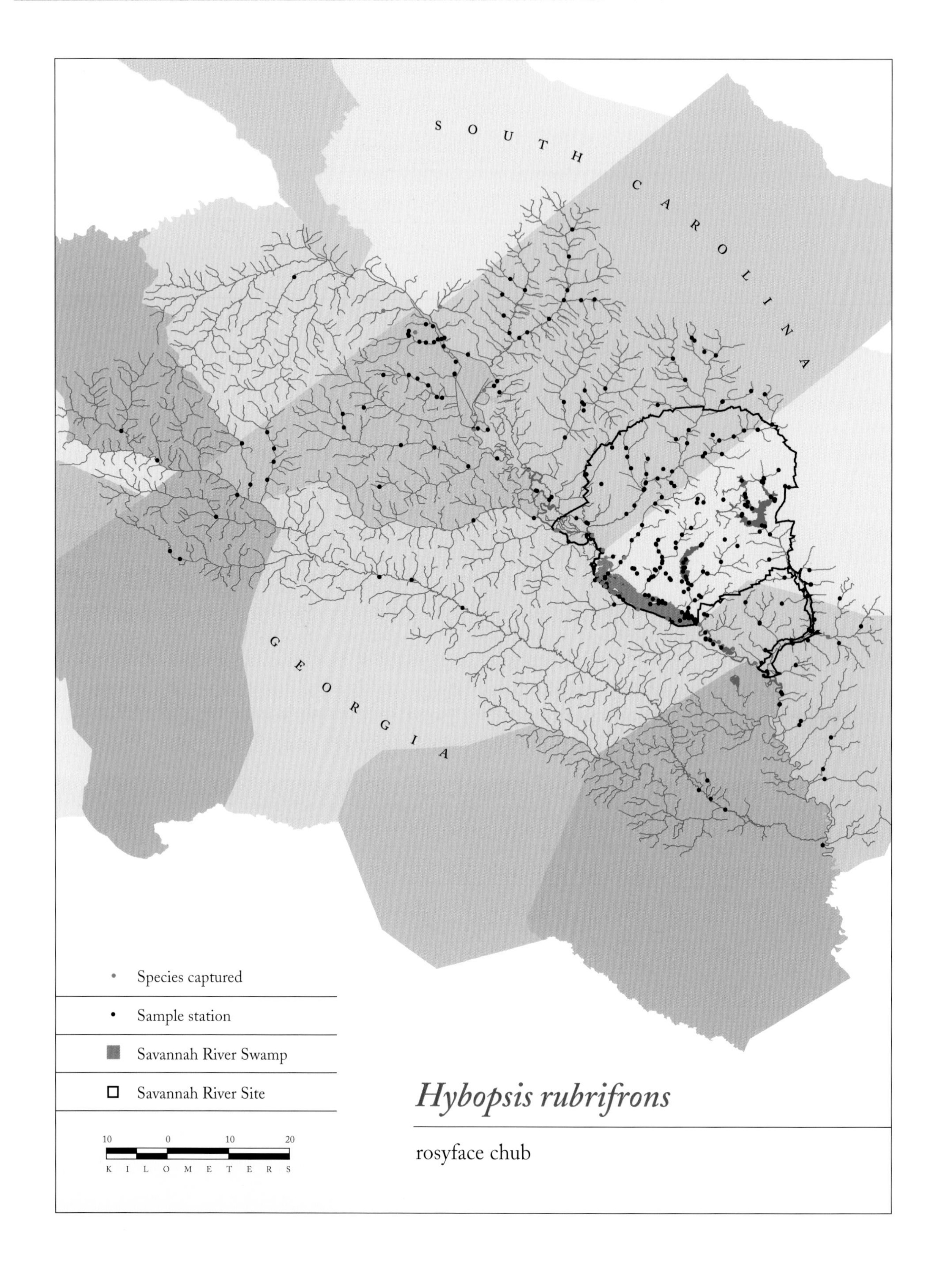
S O U T H C A R O L I N A
G E O R G I A
Species captured
Sample station
Savannah River Swamp
Savannah River Site
10 0 10 20
K I L O M E T E R S
Hybopsis rubrifrons
rosyface chub

SIZE Large for a minnow native to the MSRB, up to 84 mm TL (Page and Burr 1991).

MERISTICS Dorsal rays 8, anal rays 8; pharyngeal teeth 1,4–4,1; lateral line scales 35–39 (Page and Burr 1991).

DISTRIBUTION Piedmont and upper edge of Upper Coastal Plain in the Saluda, Savannah, and Altamaha river drainages of South Carolina and Georgia (Page and Burr 1991) and Blue Ridge province in the Savannah River drainage (Clemmer *in* Lee et al. 1980). In the MSRB, generally found in the Savannah River proper and sporadically in portions of tributaries on the modern Savannah River floodplain and its associated fluvial terraces. In Fourmile Branch and Pen Branch on the SRS, at least in the spring, frequently found on the uppermost fluvial terrace near the terrace-upland border.

HABITAT Generally found over sand and gravel substrates in flowing pools or runs of tributary streams, usually occupying the lower portion of the water column, near the bottom, even in deep (1.5–2 m) pools.

BIOLOGY Little is known about the biology of the rosyface chub. The adults appear to spawn on the nests of bluehead chubs; large numbers of brightly colored rosyface chubs have been observed schooling over nests along with yellowfin shiners (*Notropis lutipinnis*) in Georgia (Brady Porter pers. comm.). This is consistent with the observation that North American cyprinids with red chromatic coloration are generally nest-associating species (Carol Johnston pers. comm.). Facultative nest association is a possibility in view of Wiltz's (1993) report of this species spawning over clean gravel in fast riffles. Spawning in the MSRB has been reported from April through June at water temperatures of 19–24 °C (Wiltz 1993).

The diet has not been studied, but two specimens from Pen Branch collected in May contained aquatic insects, cladocerans, and copepods (DEF pers. obs). The mouth morphology and short S-shaped intestine also indicate benthic invertivory. Davis and Miller (1967) found few taste buds on the barbels and suggested that the barbels may be of limited use for feeding; however, Dimmick (1988) showed that barbels are densely ornamented with taste buds. Rosyface chubs have both the large optic lobes typical of sight feeders and the well-developed vagal lobes typical of taste feeders (Davis and Miller 1967). These factors, along with the variable barbels, small subterminal mouth, and large eyes, may indicate varied methods of feeding.

SCIENTIFIC NAME *Hybopsis* = "rounded/bulging face" (Greek); *rubrifrons* = "red forehead" (Latin).

Nocomis leptocephalus (Girard, 1857)
Bluehead chub

DESCRIPTION Body stout, cylindrical, and ventrally flattened anterior to the pelvic fins. Eyes relatively small and laterally placed. Mouth large, slightly subterminal, and slightly oblique. Terminal maxillary barbels short and rather flat, heavily ornamented with taste buds (Dimmick 1988). Barbels relatively small, especially compared with those of the rosyface chub (*H. rubrifrons*). Lateral line complete.

Deep black lateral stripe, variable and often indistinct, with diffuse upper and lower edges. Scales on stripe heavily outlined in black. Stripe extends from snout (may be narrow or indistinct on head) to base of caudal fin, then streaks onto the caudal fin; in life, a narrower tan (light in preservative) stripe lies above it. Dorsum tan olive. Scales on dorsum dusky with thick black lining on posterior edge that produces an uneven crosshatched scale pattern. Ventral head and belly white, flanks silvery and iridescent. Scales on belly and flanks with light posterior dark edging formed by dark melanophores. Chin, both lips, and roof of mouth without pigment. Postanal stripe absent. Fins tinged with reddish orange in life, particularly the dorsal, caudal, and pectoral fins. Outer branched portions of rays in dorsal, anal, and pelvic fins of breeding males dusky and lined with black.

Breeding males develop blue on side of head, especially the operculum and cheek, and sometimes a blue lateral stripe, and dusky areas of fins become darker. Breeding males also develop a nuptial crest: the forehead dorsally between the eyes becomes swollen in a distinctly convex bulge that has (7)8(9) large, conspicuous horny tubercles; smaller males generally have fewer cephalic tubercles than larger males (Lachner and Wiley 1971). Small tubercles are present on the head and fins, the pectoral fins with distinctive rows of tubercles. Breeding females also have small tubercles on the head and all fins.

SIMILAR SPECIES See the rosyface chub (*H. rubrifrons*) account for characters useful for distinguishing fishes in the MSRB with single maxillary barbels. The creek chub (*S. atromaculatus*) differs in having a terminal mouth with black pigment on the lower lip, a spot at the dorsal fin base, modally eight anal rays, and pharyngeal teeth 2, 5–4,2.

SIZE Males grow faster and larger than females; males reach at least up to 210 mm SL (Lachner and Wiley 1971).

MERISTICS Dorsal rays 8(9), anal rays 7, pectoral rays (15)16–18(19), pelvic rays (7)8; pharyngeal teeth 0,4–4,0, may rarely lack 1 tooth on either side; lateral line scales (38)39–41(42) (Lachner and Wiley 1971; Jenkins and Burkhead 1993) or 36–43 (Page and Burr 1991).

Nocomis leptocephalus, bluehead chub; (above) male, (facing page) female

DISTRIBUTION East of the Appalachian Mountains in the Blue Ridge province, Piedmont, and Upper Coastal Plain from Virginia to Georgia and west through Alabama to southern Mississippi and lowermost Mississippi tributaries in Louisiana and Mississippi; apparently absent from Lake Ponchartrain drainages (Jenkins and Lachner *in* Lee et al. 1980). Widely distributed across the MSRB in small to intermediate streams of the upland area where pebbles for nest building are available. Very rarely collected from the Savannah River near the mouths of tributaries that enter the river without being filtered through an extensive swamp on the modern Savannah River floodplain (e.g., some of the small tributaries on the Georgia side of the river bordering the SRS).

HABITAT Stream mesohabitat characterized by deep, swift water (Meffe and Sheldon 1988), with the fish often residing deep within woody debris or undercut banks. Large males can also frequently be found in swift-flowing water within riprap placed in streams at areas such as beneath bridge crossings.

BIOLOGY Bluehead chub males are mound builders (Johnston and Page 1992): they excavate a depression in the stream bottom that is then filled with pebbles to form a platform on which a dome nest is constructed (Raney 1947; Maurakis et al. 1991). Spawning occurs from May through July at water temperatures from 16 to 28 °C, but generally between 19 and 23 °C (Wallin 1992; DEF unpub. data). During the spawning season bluehead chubs spawn episodically. Nest areas are reused repeatedly, but the males may move the nest or construct an entirely new nest (Wallin 1989). During a given spawning cycle, 1 day of spawning is typically preceded by 2–3 days of nest building (Wallin 1989). Wallin observed multiple males contributing to nest construction, but most of the work was done by the largest male; the nests averaged 50 cm diameter, were about 14 cm high, and contained an average of 14,000 or more stones 6–12 mm in diameter (Wallin 1989). Males defend territories with behavioral displays and aggression at the upstream edge of the nest near the pit where spawning occurs (Maurakis et al. 1991, 1997). Females gather behind the nest and swim forward to the pit when they are ready to spawn (Wallin 1989, 1992). Reduced water velocities in the pit at the upstream edge of the nest may enhance fertilization of eggs (Maurakis et al. 1992). The male clasps the female in a strong grip, perhaps to reduce cuckoldry by other males (see details in Sabaj et al. 2000). Males continue to move stones onto or around the nest for 5–15 days after spawning, gradually building the nest higher and keeping it free of silt (Wallin 1992). Wallin (1992) found that hatching occurred in 3 days and larvae were found in the nest for 4–12 days at 19–29 °C. Larvae are illustrated in Loos et

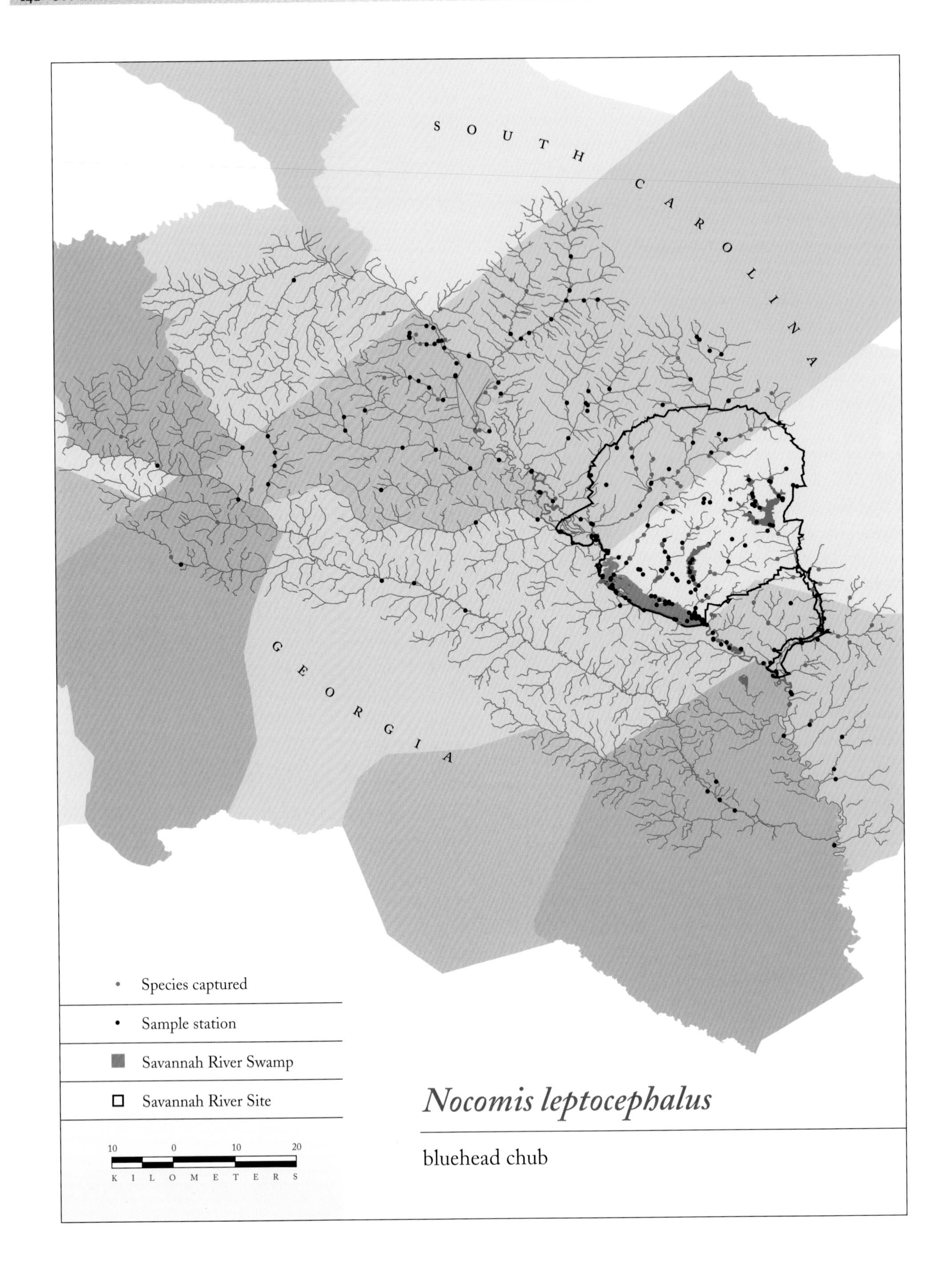
SOUTH CAROLINA
GEORGIA
Species captured
Sample station
Savannah River Swamp
Savannah River Site
10 0 10 20
KILOMETERS
Nocomis leptocephalus
bluehead chub

Nocomis leptocephalus, bluehead chub; closeup of breeding male head

al. 1979; myomere counts are provided in Green and Maurakis 2000; and Fuiman et al. 1983 discusses larval characteristics useful for identification.

Members of the nest-building genus *Nocomis* are often keystone species within a stream because other species use their nests for spawning (Lachner 1952; Johnston and Birkhead 1988; Vives 1990; Johnston and Page 1992; Johnston 1994). In the MSRB, such species include yellowfin shiner (*Notropis lutipinnis*) (McAuliffe and Bennett 1981; Wallin 1989, 1992), blackbanded darter (*Percina nigrofasciata*) (Johnston and Kleiner 1994), and rosyface chub (*Hybopsis rubrifrons*)(Brady Porter pers. comm.). Bluehead chubs showed no aggression toward the nest-associating yellowfin shiners.

Lachner (1952) recorded males that were tuberculate at 115 mm SL and reported that many bluehead chubs live to 2 years but few live to 3 years (Lachner 1952); however, Jenkins and Burkhead (1993) pointed out that Lachner did not examine any large males (>130 mm SL). Lachner (1952) reported a fecundity of 710–800, based on four females ranging in size from 82 to 92 mm SL.

The bluehead chub is omnivorous and has an intermediate-length gut longer than those of the other local native carnivorous cyprinids but shorter than that of the herbivorous eastern silvery minnow (*H. regius*). The longer intestine and increased number of taste buds on the lips are morphological adaptations for the ingestion and digestion of plant material (Davis and Miller 1967). In the MSRB, their diet consists of large amounts of aquatic and terrestrial insects, followed by crustaceans (crayfish, shrimp, and cladocerans), clams, and algae and detritus (Sheldon and Meffe 1993). Gatz (1979) observed a similar diet of invertebrates with some algae in North Carolina Piedmont streams, but Flemer and Woolcott (1966) reported larger amounts of algae, including filamentous algae, diatoms, and desmids, in the diet of bluehead chubs from Virginia.

SCIENTIFIC NAME *Nocomis* = after Nokomis, "Daughter of the Moon" in Longfellow's *Song of Hiawatha* (Jenkins and Burkhead 1993); *leptocephalus* = "small/slender head" (Greek).

Notemigonus crysoleucas **(Mitchill, 1814)**
Golden shiner

DESCRIPTION Body deep and strongly laterally compressed; not as streamlined as minnows that inhabit swift water (Hubbs and Cooper 1936). Body of adults deeper than that of juveniles. Head very small, especially in large adults. Eyes large and laterally placed. Mouth superior, strongly oblique. Barbels absent. Lateral line complete and arched strongly downward anterior to pelvic fins. Sharp, scaleless median keel present along belly between pelvic fins and anus.

Juveniles have a black lateral stripe from the operculum to the base of the caudal fin; most distinct in small individuals, indistinct in fish over 80 mm SL, and lacking in adults. Dorsum ridge yellow-olive. Entire lateral sides of body, operculum, and sides of head flashy silver to gold; ventral head, breast, and belly generally lighter. Olive iridescent stripe sometimes present on dorsolateral sides. In preservative, body scales, especially ventrolaterally, of large adults may be marked by rows of spots formed by distinct basal spots on scales. Upper and lower lips with pigment, darker medially. Chin pigment absent or light, only on terminal arch bordering lower lips. Floor of mouth mostly black; roof of mouth lacks pigment. Postanal stripe absent. Fins translucent with rays variably lined with melanophores.

Breeding adults (at least males) may have red-tinged pectoral, pelvic, and anal fins, and the distal (branched) portion of the anal fin marked with black, darker anteriorly. Dorsal fin, although generally not as dark as anal fin, pigmented with similar pattern. Caudal fin black distally. Small breeding tubercles present on posterior margin of body scales, fins, and head, with largest tubercles on chin.

SIZE At least up to 255 mm TL (209 mm SL in Par Pond).

MERISTICS Dorsal rays (7)8(9), anal rays (8)12–15(19), pectoral rays 16–17(18), pelvic rays (8)9; pharyngeal teeth 0,5–5,0 (Plate 11 g); lateral line scales (39)44–56(57) (Jenkins and Burkhead 1993).

DISTRIBUTION The Atlantic slope from the Maritime Provinces of Canada south to Florida, west to Texas, and north to Saskatchewan; introduced in Arizona, New Mexico, and California (Lee *in* Lee et al. 1980). In the MSRB, widely distributed and locally common in lentic and sluggish lotic waters.

HABITAT Widely distributed in fluvial systems, including upland streams. Most commonly found in still or sluggish backwaters on the modern Savannah River floodplain, but they regularly ascend the downstream portions of tributaries crossing the fluvial terraces of the Savannah River, where they inhabit the slow deeper side braids as well as upland streams. Local habitat also includes upland reservoirs and ponds, and ditches. Although rare or absent in flowing sections of headwater streams, golden shiners inhabit beaver ponds (Snodgrass and Meffe 1999). Felley and Hill (1983) reported similar habitats for the Illinois River in Oklahoma, while Hubbs and Cooper (1936) indicated a preference for weedy lakes.

Notemigonus crysoleucas**, golden shiner; (left) adult, (right) small individual**

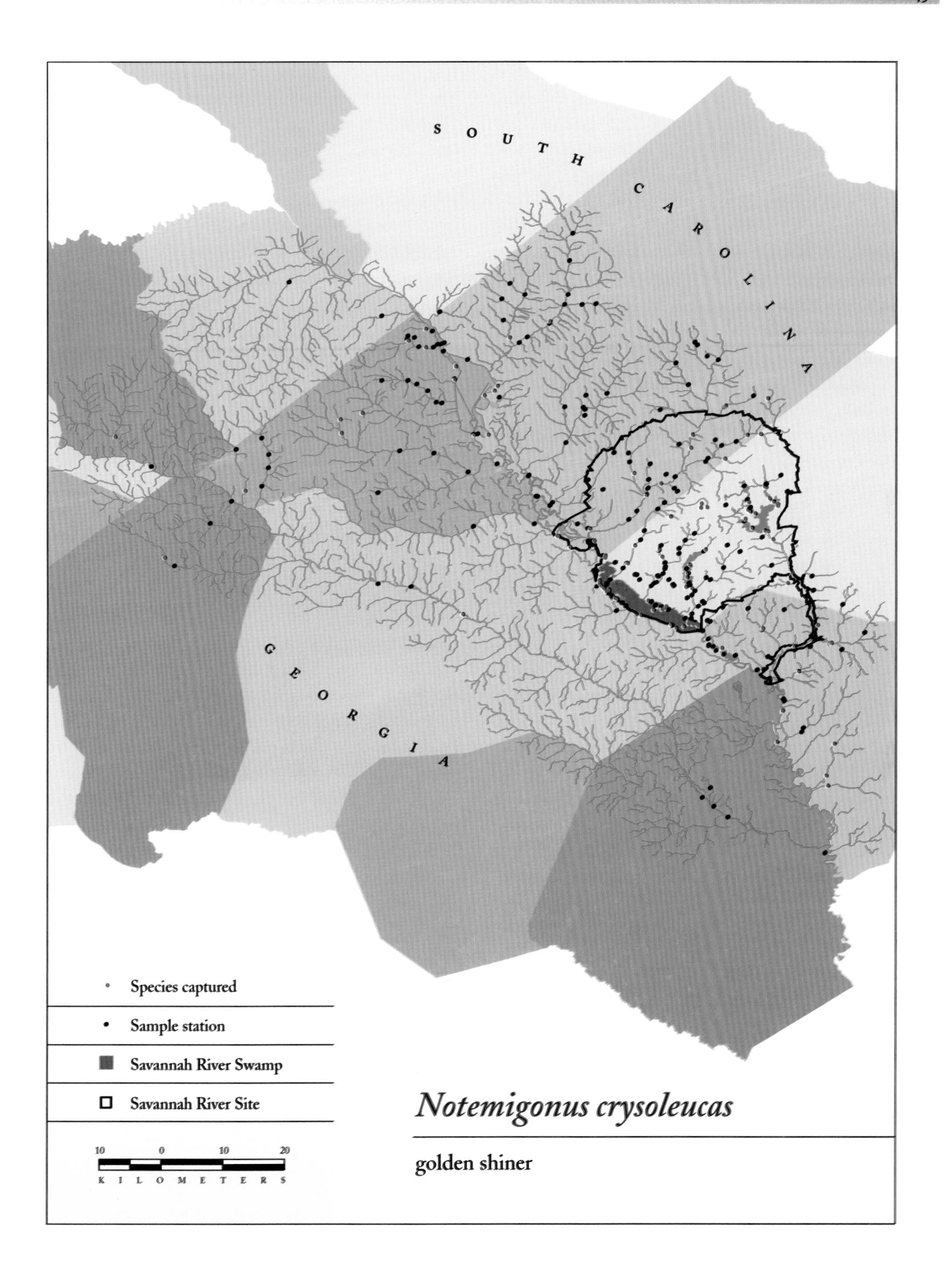

Notemigonus crysoleucas

golden shiner

BIOLOGY The golden shiner is a broadcast spawner that will also spawn in the nests of other species. Spawning occurs at 18–27 °C (Piper et al. 1982). Larvae were collected in the Savannah River in March and April (Wiltz 1993). Cahn (1927) found gravid females in June and July in Wisconsin. The eggs are adhesive and may be broadcast over a variety of objects, including filamentous algae, rooted aquatic plants (Piper et al. 1982), and small pebbles (Loos et al. 1979). Larvae are described in Snyder et al. 1977; and Loos et al. 1979; and larval descriptions are summarized in Jones et al. 1978. Fuiman et al. 1983 describes characteristics useful for larval identification.

Golden shiners have been observed to spawn with several centrarchid species, including largemouth bass (*Micropterus salmoides*) (Kramer and Smith 1960; Chew 1974), spotted sunfish (*Lepomis punctatus*) (Carr 1946), green sunfish (*L. cyanellus*) (Pflieger 1975), bluegill (*L. macrochirus*) (DeMont 1982), and pumpkinseed (*L. gibbosus*) (Shao 1997). Golden shiners spawned in 4–75% of the largemouth nests in a Minnesota lake (Kramer and Smith 1960). In a New York pond, shiners spawned in about one-third of the pumpkinseed nests, but nests near the shore were used less frequently (Shao 1997). Golden shiners have also been observed to spawn with bowfins (*Amia calva*) (Katula and Page 1998). Responses of the nest-guarding sunfish to golden shiners vary. The spotted sunfish Carr (1946) observed tried to prevent nest entry; in contrast, guarding male bluegills attacked conspecific juveniles but displayed no aggression toward the shiners (DeMont 1982). Nest-guarding bass sometimes display aggression toward the shiners (Kramer and Smith 1960). The number of shiners that may gather over a nest is variable. Kramer and Smith (1960) reported 5–100 spawning golden shiners over bass nests, while DeMont (1982) observed small groups of 2–5 shiners swimming over bluegill nests. Spawning is generally not synchronized with the host, but golden shiner larvae leave the nest at the same time as or before spotted sunfish, largemouth bass, and bluegill host larvae (Carr 1946; Kramer and Smith 1960; Shao 1997). Reproductive success is influenced by the success of the guarding host. Survival rates of shiner larvae were better in successful largemouth bass nests than in failed or unused nests (Kramer and Smith 1960). Similarly, golden shiners selected pumpkinseed nests that had already been spawned in by the host, and spawning in guarded nests enhanced the survival of shiner offspring (Shao 1997). Interactions between host bluegills and golden shiners included schooling of larvae and early juveniles of similar size (DeMont 1982). Cultured eggs hatched in 2–3 days while held at 21–24 °C (Snyder et al. 1977); development takes longer at higher or lower temperatures (Shao 1997).

Most females mature at 1 year of age in southern Michigan and at age 2 in northern Michigan (Cooper 1936), and at 1–3 years of age in New York (Forney 1957). In Michigan, the growth rate also varies with latitude; fish at ages 1, 3, and 6 years had average TLs of 43, 81, and 125 mm, respectively in the northern part, and 68, 106, and 184 mm, respectively, in the southern part (Cooper 1936). Females grow more rapidly than males, and growth slows after maturation. Maturity may be determined by size rather than age. The number of ova in females increases with size (Cooper 1936). Golden shiners may live up to 7 years (Cooper 1936).

Golden shiners are omnivorous, and like many cyprinids appear to be opportunistic feeders. Like the bluehead chub, this species has an intermediate gut length—longer than that of the local native carnivorous cyprinids but shorter than that of the herbivorous eastern silvery minnow. The gut of the golden shiner has an extra coil in addition to the typical S-shaped loop of carnivorous cyprinids. In a study in the MSRB, golden shiners ate mainly cladocerans and bryozoans, along with a variety of other aquatic invertebrates, fish, ants, algae, and higher plants (Wiltz 1993). Gatz (1979) reported the diet to consist primarily of diatoms, aquatic and terrestrial insects, and filamentous algae. Golden shiners may sometimes eat primarily crustaceans and insects (Cahn 1927) and may eat large numbers of molluscs when these are available (Forbes and Richardson 1908 *in* Carlander 1969).

SCIENTIFIC NAME *Notemigonus* = "angled back" (Greek); *crysoleucas* = "golden white" (Greek), referring to the body color.

Notropis chalybaeus **(Cope, 1869)**
Ironcolor shiner

DESCRIPTION Body cylindrical and slightly compressed; large individuals are generally deeper bodied than smaller ones. Eyes large and laterally placed. Mouth large, terminal, and moderately oblique. Barbels absent. Lateral line complete.

Narrow, intensely black lateral stripe with sharp upper and lower edges extending from tip of snout to base of caudal fin, then streaking onto caudal fin fork. Indistinct caudal spot present as continuation of stripe. In life, a narrower iridescent straw to yellow line (light in preservative) lies above the black stripe. Dorsum tan olive, and scales on dorsum dusky and lined with black (posterior edges noticeably darker), forming crosshatched scale pattern. Ventral head and belly white, flanks whitish. Chin marked with thick, prominent terminal bar; some pigment extending onto anterior gular region. Roof of mouth mostly black with dense large melanophores. Postanal stripe extends from anus onto caudal peduncle. Fins translucent, but dorsal, anal, caudal, and anterior pelvic and pectoral rays outlined in black.

Breeding males with a bright orange to red iridescent stripe above the black lateral stripe, gold belly, and silver to coppery flanks. Rosy gold patches sometimes present at base of caudal fin aligned above and below black stripe. Small breeding tubercles on body (especially dorsolaterally), fins (most prominent on pectoral fins), and head, with enlarged tubercles around mouth. Breeding colors extend onto dorsal and caudal fins, and tend to be orange in brown water and rosy in habitats with clear water (Marshall 1947).

SIMILAR SPECIES The ironcolor shiner is most likely to be confused with the dusky shiner (*N. cummingsae*), yellowfin shiner (*N. lutipinnis*), or coastal shiner (*N. petersoni*), its closest relative within the MSRB. The ironcolor shiner has 8 (mode) anal rays and a postanal stripe, and is unique in having a dark black roof of the mouth and an intensely black terminal bar on the chin. The dusky shiner has more than 8 (mode 10–11) anal rays, a thick horseshoe-shaped line on the chin, and a postanal stripe; the pharyngeal teeth are 1,4–4,1 as opposed to 2,4–4,2 in the other species. The coastal shiner has 7 (mode) anal rays, a black terminal spot on the chin, a postanal stripe, and a distinct triangular caudal spot. The yellowfin shiner has 8 (mode) anal rays and is the only species of this group with yellow on the paired and anal fins. It also has a thin horseshoe-shaped line on the chin and lacks the postanal stripe. See Table 17 for a comparison of useful identification characters for the Cyprinidae.

Notropis chalybaeus, ironcolor shiner, male

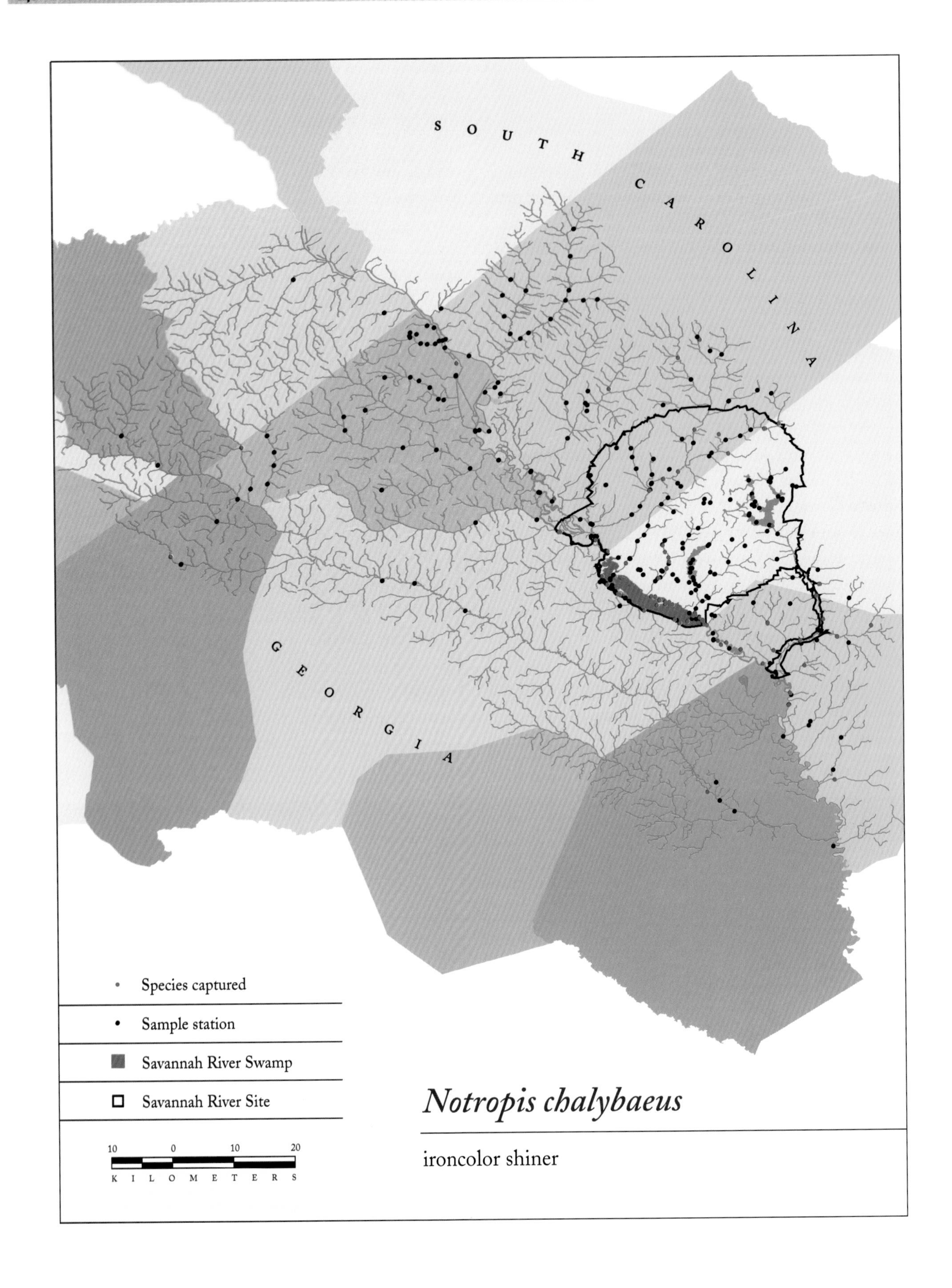
SOUTH CAROLINA
GEORGIA
Species captured
Sample station
Savannah River Swamp
Savannah River Site
10 0 10 20
KILOMETERS
Notropis chalybaeus
ironcolor shiner

SIZE Adults measure 32–44 mm SL (Warren et al. 1991); females are generally larger than males (Marshall 1947).

MERISTICS Dorsal rays (7)8, anal rays 8, pectoral rays 11–13, pelvic rays 7–8(9); pharyngeal teeth 2,4–4,2; lateral scale series 32–34(36) (Jenkins and Burkhead 1993).

DISTRIBUTION The Atlantic Coastal Plain from the Hudson River in New York to tributaries of Lake Okeechobee, Florida, west to the Sabine River of Texas and Louisiana, and north in the Mississippi Valley to Wisconsin; a disjunct population is present in the San Marcos River of Texas (Swift *in* Lee et al. 1980). Common throughout the MRSB in appropriate habitat.

HABITAT In the MSRB, found in slower portions of the sandy-bottomed tributary streams of the upland fluvial terraces and modern floodplain of the Savannah River. Inhabited sections of streams are generally slow enough that association with flow obstructions is unnecessary. Robison (1977) found ironcolor shiners in the midwater portion of a stream, and Snodgrass and Meffe (1999) determined them to be more abundant in beaver ponds than in the streams above. Beaver ponds are likely important for spawning and represent a source population for many stream reaches. Similarly, in braided systems such as those common on the Savannah River fluvial terraces, ironcolor shiners are typically found in slow side channels and backwaters.

BIOLOGY Females broadcast their eggs without prior modification of the substrate, and without parental care. During spawning a male and female streak across a pool with their ventral surfaces pressed together, distributing the adhesive eggs in the water column (Marshall 1947); the eggs probably settle on the sandy bottom (Marshall 1947). The eggs hatch in 54 hours at an average air temperature of 17 °C. Larvae are described and illustrated in Marshall 1947; and Wang and Kernehan 1979. Additional characteristics and drawings of a mesolarva are available in Loos and Fuiman 1978. Fuiman et al. 1983 discusses characters useful for larval identification.

The spawning season extends from April to September in Florida (Marshall 1947) and Alabama (Mettee et al. 1996), and from late May through mid-August at water temperatures of 24–33 °C in Illinois culture ponds, with peak spawning in June and July (Burr et al. 1989; Warren et al. 1991). In the MSRB, ironcolor shiners spawn in cattle tanks in June and July, but not in August; May spawning is suspected. Females spawn multiple clutches, and groups of ova are synchronously recruited from a reservoir of immature oocytes (Warren et al. 1991). Clutch size averages 109 (70–147) eggs. Both sexes mature in 1 year at about 32 mm SL (Warren et al. 1991).

Ironcolor shiners are visual invertivores (Marshall 1947). In the MSRB, the diet consists of aquatic insects, microcrustaceans, diatoms, and algae/detritus (Sheldon and Meffe 1993). In Louisiana, large amounts of microcrustaceans—especially cladocerans, insects, ostracods, and water mites—are eaten (DEF unpub. data). Although plant material is ingested, it is generally in the same condition in the anterior and posterior ends of the digestive tract; this and the short S-shaped gut suggest that plant material is not digested (Marshall 1947).

SCIENTIFIC NAME *Notropis* = "keeled back" (Greek) (note: the keel in Rafinesque's type specimen was the result of drying or shriveling of the fish [Jordan and Evermann 1896]); *chalybaeus* = "iron colored" (Greek).

Notropis cummingsae Myers, 1925
Dusky shiner

DESCRIPTION Body cylindrical to slightly compressed. Eyes large and laterally placed. Mouth terminal, moderately oblique. Barbels absent. Lateral line complete. Dusky shiners in the MSRB are generally smaller and paler with a deeper, frailer body than those in the neighboring Combahee drainage in South Carolina (Hubbs and Raney 1951).

A moderately deep, dusky to black lateral stripe extends from snout to caudal fin (hence the name dusky shiner), the upper edge sharper than the lower edge. Stripe terminates in an indistinctive caudal spot that streaks onto caudal fin. A narrow dark line runs around tip of snout. In life, dorsal edge of black stripe bordered by thin iridescent coppery stripe (light in preservative). Dorsum olive to gold. Ventral head and belly white, flanks grayish silver. Chin pigmented with thick black line forming a horseshoe outlining lower jaw. Entire ventral side of lower jawbone (dentary) black. Upper lip black. Anterior portion of roof of mouth sometimes with scattered melanophores. Postanal stripe present from anus across anal fin base, sometimes onto caudal peduncle. Fins translucent with dorsal, caudal, some pelvic, pectoral, and anal rays edged with black, especially anteriorly.

Breeding males have small tubercles on the head, body, and paired fins. Breeding females also have tubercles, but less widely distributed—mostly on the face, chin, and snout.

SIMILAR SPECIES The dusky shiner may be confused with the ironcolor shiner (*N. chalybaeus*), coastal shiner (*N. petersoni*), and yellowfin shiner (*N. lutipinnis*). See the species account for the ironcolor shiner for characters useful in distinguishing these four species.

SIZE Adults 28–60 mm SL; females (usually 30–50 mm SL) are larger than males (usually 28–43 mm SL) (DEF unpub. data).

MERISTICS Dorsal rays (7)8(9), anal rays (9)10(12), pectoral rays (12)15(17), pelvic rays (8)9(10); pharyngeal teeth 1,4–4,1; lateral line scales (34)36–37(40) (DEF unpub. data).

DISTRIBUTION Mostly on the Atlantic Coastal Plain from the Tar River drainage of North Carolina to the Altamaha River drainage of Georgia, plus three disjunct

Notropis cummingsae, dusky shiner

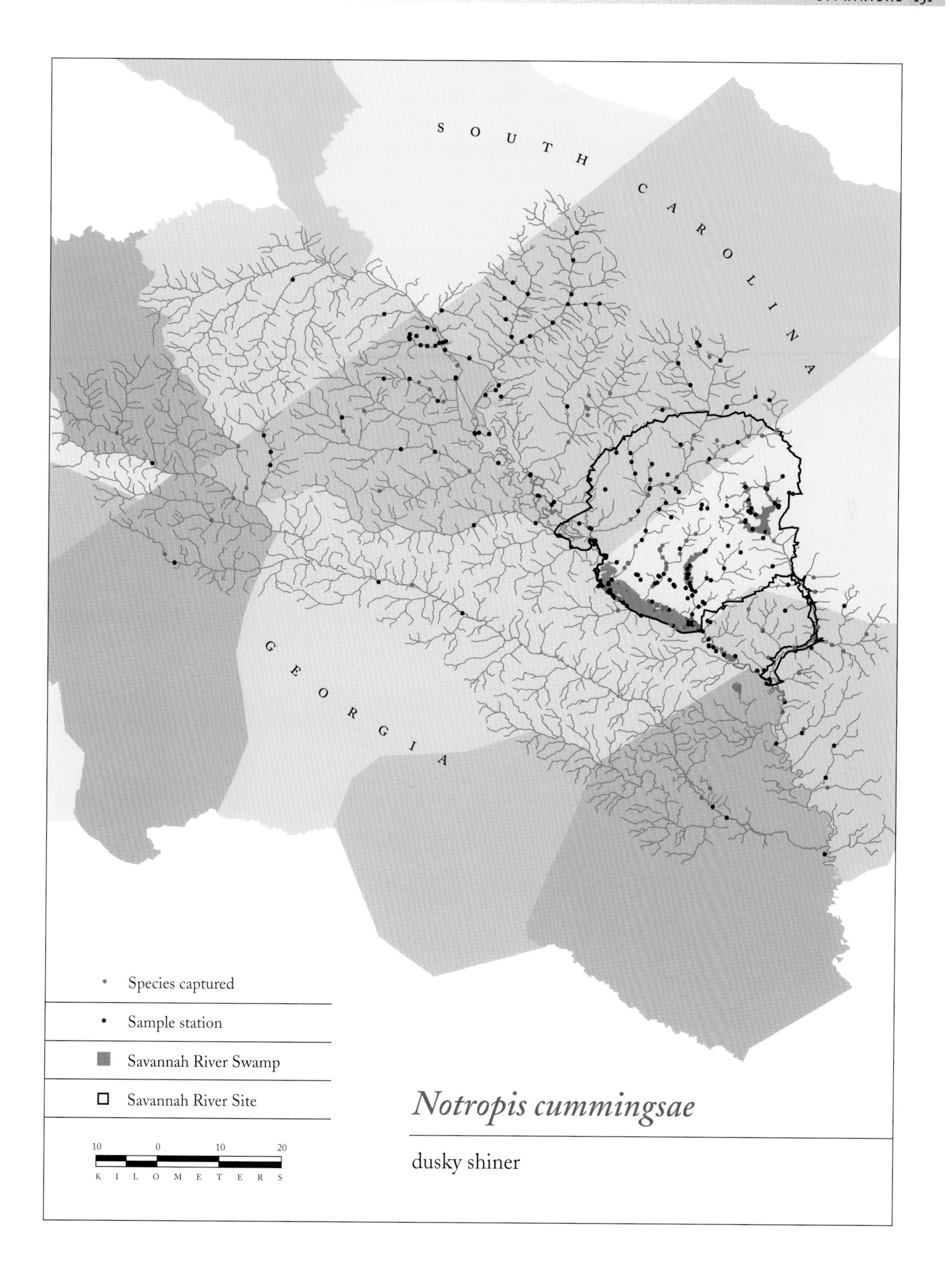

Notropis cummingsae

dusky shiner

populations: one in the middle Chattahoochee drainage of Georgia and Alabama, another in the Choctawhatchee drainage east to the Aucilla drainage of Florida and Georgia, and the third in the lower St. Johns drainage, Florida (Burgess et al. 1977; Gilbert and Burgess *in* Lee et al. 1980). Common and abundant in the MSRB, especially in intermediate-sized tributary streams, particularly those sections flowing across the fluvial terraces of the Savannah River, but may be found in large streams and small headwaters.

HABITAT Dusky shiners are frequently found in the middle of the water column, above yellowfin shiners and coastal shiners, in eddies behind flow obstructions in larger streams or in scour pools in smaller streams. During warm months they are found farther from brush and flow obstructions than the lowland shiner (*Pteronotropis stonei*) (DEF pers. obs.). In the winter, dusky shiners often take cover in—rather than behind—dense brush or vegetation that is creating a flow obstruction. They are generally more abundant at road crossings, clear-cuts, and in the thermally impacted streams on the SRS.

BIOLOGY Dusky shiners typically spawn from late May through the end of July, episodically in 1–3-week intervals, and at water temperatures between 20 and 30 °C, with peak spawning activity occurring around 25 °C (DEF unpub. data). The release of warm or cold water from reservoirs to tributary streams can expedite or delay the time of spawning, respectively. Although dusky shiners are typically found in habitats with a noticeable current, Fletcher (1993) noted localized spawning migrations to large, still pools or backwaters that were also used for nursery areas; juveniles subsequently migrated out of the pools to join the adults. Within the pools, dusky shiners spawned on the nests of redbreast sunfish (*Lepomis auritus*), feeding voraciously on sunfish eggs and larvae while spawning (Fletcher 1993). Even when offspring of both species were present in the nest, the shiners fed almost exclusively on sunfish young.

Schultz (1999b) compared the life histories of dusky shiners in two streams in the MSRB. In both streams, females and to a lesser extent males accumulated lipid reserves in the winter and early spring before their gonads began to enlarge. These lipid reserves were subsequently depleted during the spawning season, reaching a low point in late summer when large adults began to die off and became rare. Dusky shiners grew faster and had larger ovaries, a longer spawning season, and greater maximum lipid storage reserves in the more productive Fourmile Branch than in Upper Three Runs.

Dusky shiners are frequently observed in areas with noticeable current, facing upstream while they feed on drifting food items. They are invertivorous: much of the adult diet consists of terrestrial insects, along with aquatic insects, microcrustaceans, water mites, and smaller amounts of oligochaetes, detritus, and algae (Sheldon and Meffe 1993). The relatively short S-shaped gut is consistent with a carnivorous diet.

SCIENTIFIC NAME *Notropis* = "keeled back" (Greek) (note: the keel in Rafinesque's type specimen was the result of drying or shriveling of the fish [Jordan and Evermann 1896]); *cummingsae* = named after Mrs. J. H. Cummings.

Notropis hudsonius (Clinton, 1824)

Spottail shiner

DESCRIPTION Body elongate, cylindrical, and slightly compressed. Eyes large and dorsolaterally placed. Mouth moderate in size, subterminal or slightly inferior, and moderately oblique. Barbels absent. Lateral line complete.

Narrow black lateral stripe present but variable, extending from tip of snout to caudal fin, darkest posteriorly, terminating in a moderately distinct, but variable, caudal spot. Lateral line punctate, punctation generally less distinctive posteriorly. In life, dorsal edge of black stripe bordered by thin, iridescent, straw-colored stripe (light in preservative). Dorsum tan olive. Dorsal scales dusky with dark posterior lining forming a rounded crosshatch pattern. Ventral head white, belly white to straw dorsally, flanks gray to pearly silver. Chin and lower lip without pigment, but upper lip with scattered melanophores visible under magnification. Roof of mouth without pigment. Postanal stripe absent or faint along anal fin base. Fins translucent; dorsal, caudal, and anterior pectoral fin rays may be lightly outlined with melanophores.

Breeding males with minute, difficult to detect tubercles on the head, including the lips and chin. Anterior body, anterior pectoral fins, and sometimes other fins may also have tubercles (Jenkins and Burkhead 1993). Breeding females have tubercles on the head and nape (Jenkins and Burkhead 1993).

SIMILAR SPECIES Spottail shiners are most likely to be confused with rosyface chubs (*Hybopsis rubrifrons*), but lack the barbels and fleshy lips. The upper lip of the spottail shiner is sparsely sprinkled with melanophores that are distinctly visible with magnification, but both lips of the rosyface chub entirely lack pigment. The snout of the rosyface chub also overhangs the mouth more distinctly so that when viewed ventrally, it makes up a significant portion of the anterior profile. In contrast, the lips of the spottail shiner are at or near the edge of the anterior profile. The rosyface chub is also noticeably more ventrally flattened.

SIZE Large for a shiner, reaching at least 118 mm TL (100 mm SL); adults are commonly 100 mm TL (80 mm SL).

MERISTICS Dorsal rays 8, anal rays (7)8(9), pectoral rays 13–16(17), pelvic rays (7)8–9(10) (Jenkins and Burkhead 1993); pharyngeal teeth modally 1,4–4,1, rarely 4–4, 2,4–4,2, or some other combination (C. Gilbert pers. comm.); lateral line scales (34)35–38(40) (Jenkins and Burkhead 1993).

Notropis hudsonius, spottail shiner

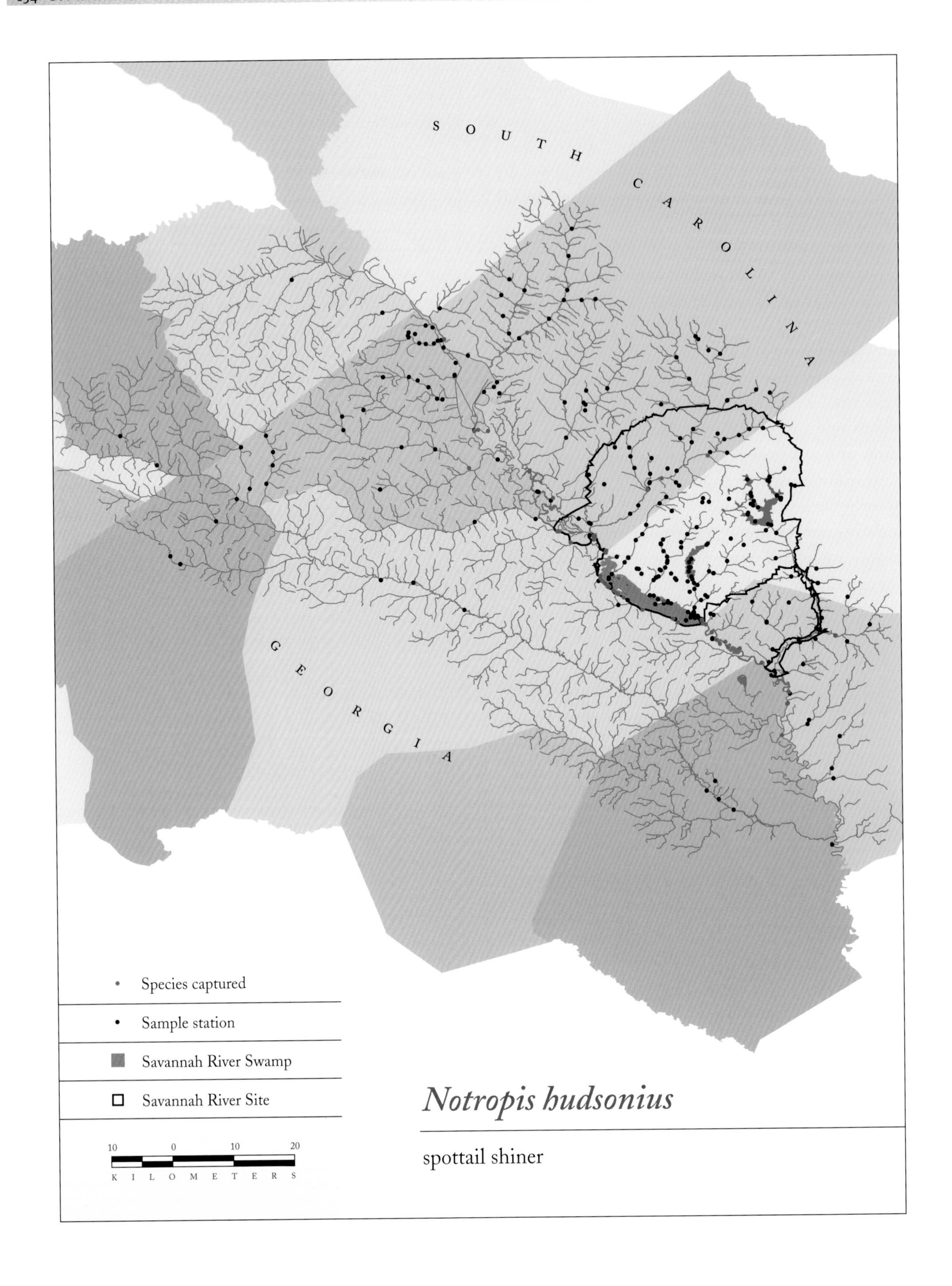

Notropis hudsonius

spottail shiner

DISTRIBUTION From the Altamaha drainage of Georgia on the Atlantic slope north to the Connecticut and Thames river drainages of Connecticut, Massachusetts, and New Hampshire; through most of the St. Lawrence River drainage and Great Lakes basin; in the upper Mississippi River drainage and in northwestern Canada east of the Rocky Mountains to near the mouth of the McKenzie River. This species appears to have entered the Chattahoochee River (Gulf Coast) drainage via stream capture from the Savannah River (Gilbert and Burgess *in* Lee et al. 1980). In the MSRB, most common in the Savannah River.

HABITAT Typically found in large bodies of water, but may ascend larger streams (Fish 1932). In the Savannah River, spottail shiners frequently occur in large schools ranging in size from 50 to several hundred individuals on sandbars of the inside bank of curves (particularly in warmer months) or in eddies behind flow obstructions. This species is occasionally found in lower portions of tributary streams where it is sporadic, spatially and temporally, in sections crossing fluvial terraces of the Savannah River; more frequent on portions crossing the modern Savannah River floodplain. Commonly found in slower water of rivers over gravel substrate (Cahn 1927). May be abundant in areas of lakes with abundant submergent vegetation (Griswold 1963).

BIOLOGY Spawning occurs from May into June or July. Northern populations may spawn later than southern populations, and local variations occur (Fish 1932; McCann 1959; Peer 1966; Wells and House 1974; Loos et al. 1979; Mansfield 1984). In Lake Michigan, spawning commences at a water temperature of 18 °C (Mansfield 1984). Spottail shiners broadcast their eggs without prior modification of the substrate and without providing parental care (Johnston and Page 1992). Riverine populations spawn most often over sand or gravel in shallow riffles (Wright and Allen 1913; Loos and Fuiman 1978; Loos et al. 1979). In Lake Erie, spawning takes place near the shore (Fish 1932), with individuals moving from deeper water onto sandy shoals during spawning episodes (Peer 1966; Wells and House 1974). In Lake Michigan, Mansfield (1984) observed them to migrate 100 m up a tributary stream and spawn on broken concrete and rocks. They may also spawn on underwater structures such as intake cribs or patches of *Cladophora* in lentic waters (Wells and House 1974). In riverine habitats, high densities of larvae were collected drifting in the current soon after hatching (Mansfield 1984). Eggs and larvae are described in Loos and Fuiman 1978; Fish 1932; Mansueti and Hardy 1967; Lippson and Moran 1974; and Jones et al. 1978. For larval characters useful for identification, see Fuiman et al. 1983.

Sexual maturity is reached in 1 or 2 years (McCann 1959; Smith and Kramer 1964; Peer 1966; Wells and House 1974) and may be influenced by latitude and the fish's size (Peer 1966). Ovaries of 1-year-old females (70–90 mm TL) in Iowa contained 1,400 eggs; older females contained 1,300–2,600 mature eggs, and no correlation was found between fish size and egg number (McCann 1959). Wells and House (1974) reported 915–3,709 eggs in Lake Erie females between 97 and 131 mm TL.

Spottail shiners may live up to 4 (McCann 1959; Peer 1966) or 5 years (Smith and Kramer 1964; Wells and House 1974), but fish over 3 years of age are rare (McCann 1959; Smith and Kramer 1964; Peer 1966; Wells and House 1974). In a sample taken along a gradient of northern latitudes (Iowa to Saskatchewan), growth rates decreased in higher latitudes. Reported total lengths were 39–77 mm, 62–98 mm, 83–108 mm, and 105 mm (from north to south), at 1, 2, 3, and 4 years, respectively (Peer 1966).

Spottail shiners are opportunistic invertivores that feed on insects (especially chironomids), bivalve molluscs, and microcrustaceans in the Savannah River (Wiltz 1993). Researchers working in other areas have found a similar diet (Cahn 1927; McCann 1959), but have also reported inclusion of some plant material (Griswold 1963). Vadas (1990) reported that 26% of the diet consists of green algae. Based on the presence of benthic, planktonic, and surface-dwelling organisms in the digestive systems he examined, McCann (1959) concluded that the spottail shiner feeds at multiple levels in the water column. The intestine is a long S in shape. Opportunistic feeding aggregations have been found preying on eggs of alewives (*Alosa pseudoharengus*) on clupeid spawning grounds (Edsall 1964) and on mayflies during mayfly hatches (Adams and Hankinson 1932). The invasion of Asian clams may be influencing the diet in the MSRB.

Schooling may reduce time and energy spent avoiding predators, and may allow for more efficient feeding (Seghers 1981).

SCIENTIFIC NAME *Notropis* (Greek) = "keeled back" (Greek) (the keel in Rafinesque's type specimen was the result of drying or shriveling of the fish [Jordan and Evermann 1896]); *hudsonius* = "of the Hudson" (described from the Hudson River).

Notropis lutipinnis (Jordan and Brayton, 1878)
Yellowfin shiner

DESCRIPTION Body stout and cylindrical, more arched dorsally (peaking at or in front of dorsal fin origin) than the slightly flattened ventral profile. Eyes large and laterally placed. Mouth large, terminal, and moderately oblique, with a long snout. Barbels absent. Lateral line complete.

An intense black lateral stripe of intermediate depth extends from tip of snout to caudal fin base and streaks onto caudal fin fork, but no caudal spot present; upper edge of stripe sharper than lower edge. In life, dorsal edge of black stripe bordered by narrower iridescent yellow to gold stripe (light in preservative); rosy patches above and below black stripe on caudal fin base. Dorsum tan olive. Ventral head and belly white, flanks silvery gray. A thin-lined U runs along outer margin of ventral side of lower jawbone (dentary). Roof of mouth lacks pigment except for small scattered melanophores along anterior curve. Postanal stripe absent. Pectoral, dorsal, and particularly pelvic and anal fins yellow tinged.

Breeding males and some breeding females develop bright coloration: ventral body below the black lateral stripe becomes bright red; iridescent stripe above the black lateral stripe intensifies and becomes deeper; dorsum becomes red with a narrow iridescent yellow to gold median stripe; ventral head yellow gold and dorsal head whitish; paired and median fins intense whitish yellow-gold, particularly the proximal two-thirds. Tubercles prevalent on head, body, and all fins of at least males.

SIMILAR SPECIES Ironcolor shiner (*N. chalybaeus*), dusky shiner (*N. cummingsae*), and coastal shiner (*N. petersoni*). See description of ironcolor shiner above for characters to distinguish these species.

SIZE Maximum at least 78 mm SL; adults commonly reach 60 mm SL.

MERISTICS Dorsal rays 8(9), anal rays (9)8(12); pharyngeal teeth 2,4–4,2; lateral line scales (36)37–39(40).

Notropis lutipinnis, yellowfin shiner; breeding colors

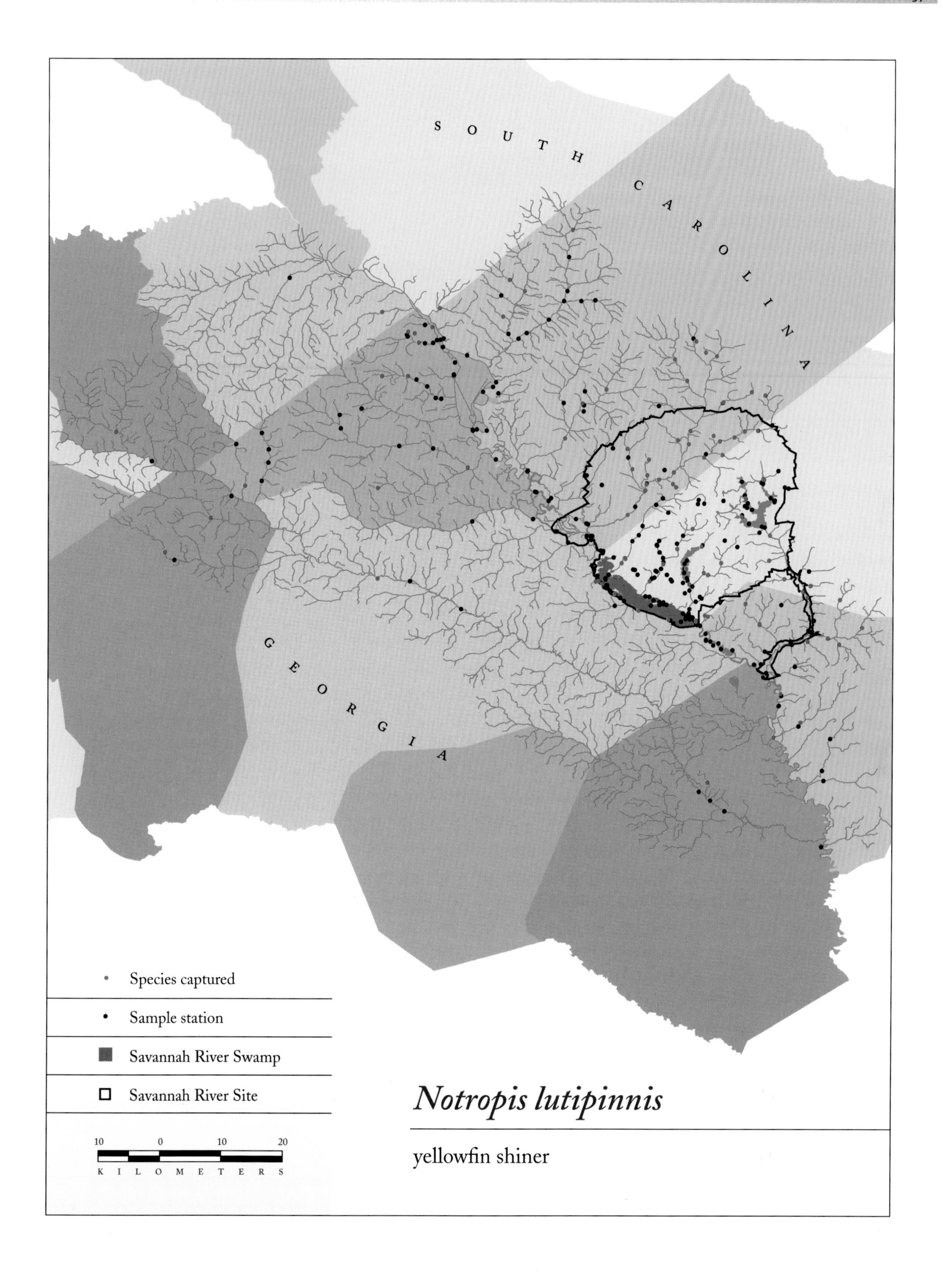

Notropis lutipinnis

yellowfin shiner

Notropis lutipinnis, yellowfin shiner; nonbreeding colors

DISTRIBUTION The Santee River drainage in North and South Carolina to the Altamaha River drainage of Georgia; also in the headwaters of the Coosa and Chattahoochee rivers in Georgia, where it may be native (Gilbert and Burgess *in* Lee et al. 1980). In the MSRB, abundant and widely distributed in tributary streams.

HABITAT Large adults are found in faster mesohabitats than are the other shiners typically found in headwater streams of the MSRB. Small individuals generally inhabit slower water (Meffe and Sheldon 1988). In general, most abundant in sandy-bottomed runs and flowing pools of headwater streams where at least some gravel is present, but also found in intermediate-size streams; generally in lower portions of the water column.

BIOLOGY Spawning occurs from May through July in the MSRB at water temperatures between 16 and 28 °C, with most spawning activity taking place between 19 and 23 °C (Wallin 1992; DEF unpub. data). McAuliffe and Bennett (1981) first reported spawning aggregations of yellowfin shiners over the nests of bluehead chubs (*Nocomis leptocephalus*). Wallin (1989, 1992) further studied this relationship and found the yellowfin shiner to be an obligate nest associate of nest-building minnows. Yellowfin shiners begin gathering on the nests while the male chubs are building them. Spawning begins simultaneously with the chubs; the latter spawn for 1 day, but the yellowfin shiners continue to spawn for 1–2 following days. Several hundred shiners may gather over a nest on the peak day, but numbers dwindle with time (DEF pers. obs.). The brightly colored males jockey for position at the front of the nest, where spawning occurs; drab females gather at the downstream edge and move forward to spawn. The behavior of brightly colored females is undescribed. Wallin (1992) found that hatching occurred in 3 days at 19–20 °C, and larvae remained in the nest for 4–12 days. Eggs and larvae are described in Loos and Fuiman 1978; Fuiman et al. 1983 discusses characters useful for identifying larvae.

Meffe et al. (1988) noted postspawning die-offs of adults, and Sheldon and Meffe (1995) observed migrations of adults during spawning season. Beaver ponds may interrupt downstream migrations of young-of-the-year, which accumulate upstream of the ponds in the winter (Snodgrass and Meffe 1999).

Yellowfin shiners are invertivores that feed primarily on terrestrial insects, followed by aquatic insects, microcrustaceans, and small amounts of clams, oligochaetes, detritus, and algae (Sheldon and Meffe 1993). The short, S-shaped intestine is consistant with this diet.

SCIENTIFIC NAME *Notropis* = "keeled back" (Greek) (the keel in Rafinesque's type specimen was the result of drying or shriveling [Jordan and Evermann 1896]); *lutipinnis* = "yellow fin" (Latin).

Notropis maculatus (Hay, 1881)
Taillight shiner

DESCRIPTION Body elongate, slender, and slightly compressed. Eyes moderately large and laterally placed. Snout rounded; mouth small, subterminal, and moderately oblique. Barbels absent. Lateral line incomplete.

A narrow black lateral stripe extends from lachrymal groove on face to caudal fin base; pored scales of lateral line that overlie black stripe frequently punctate. Large, round, prominent median basicaudal spot present (larger than pupil of eye) with elongate black blotches above and below it in the secondary caudal rays. In life, narrower iridescent yellow to gold stripe above black lateral stripe (light in preservative). Dorsum tan olive. Ventral head and belly white, flanks silvery gray, sides silvery. Scales of dorsum peppered with melanophores with dark posterior edges forming a crosshatched pattern. Scales on sides variably outlined in black forming a crosshatched pattern. Upper lip with black pigment, especially on lateral sides. Tip of snout with weak or no pigment. Chin variable, with black along lower jaw and sometimes with a line streaking the midline of the gular region. Roof of mouth lacking pigment. Postanal stripe extending from anus to caudal peduncle. Fins translucent; dorsal, anal, caudal, and anterior pelvic and pectoral rays generally outlined in black, especially distal portions.

Breeding males have red over much of the body and head, red in iris of eye, and red distally on dorsal, pelvic, anal, and caudal fins (Burr and Page 1975). Basal portion of the dorsal, pelvic, and anal fins remains translucent, but black is present on leading edge and distal (branched) portion of the posterior dorsal, pelvic, and anal rays, and in fork of tail. Breeding males also have small tubercles on chin, sides of head, and along dorsal side of anterior pectoral fin rays. Breeding females generally lack red, except for occasional pale red snouts, and also lack tubercles. Nonreproductive males larger than 30 mm TL may be identified by a band of dusky spots along the anterior margin of the dorsal fin (Cowell and Barnett 1974).

SIZE Probably up to 70 mm TL, but most adults are 40–60 mm TL.

MERISTICS Dorsal rays 8, anal rays (7)8; pharyngeal teeth 0,4–4,0; lateral scale series (34)35–37(39), 8–15 pored scales (Douglas 1974; Robison and Buchanan 1984; Etnier and Starnes 1993).

DISTRIBUTION The Atlantic and Gulf coastal plains below the Fall Line (Robison 1978), from the Cape Fear River south along the Atlantic Coast, across the Gulf drainages, and as far up the Mississippi River drainage as southern Illinois (Burr et al. 1988) and western Kentucky (Burr and Page 1975). Taillight shiners appear to be relatively rare in the MSRB, where they are primarily found in swamps, oxbows, and sloughs on the modern floodplain of the Savannah River but can also be found in backwaters of the braided channels of tributary streams on the fluvial terraces.

HABITAT Still or nearly still waters with mud or silt bottoms, sometimes with dense vegetation. Brenneman (1992) found taillight shiners in far downstream reaches but not in headwaters of a tributary of the Pearl River in Missis-

Notropis maculatus, taillight shiner; (left) breeding male, (right) nonbreeding colors

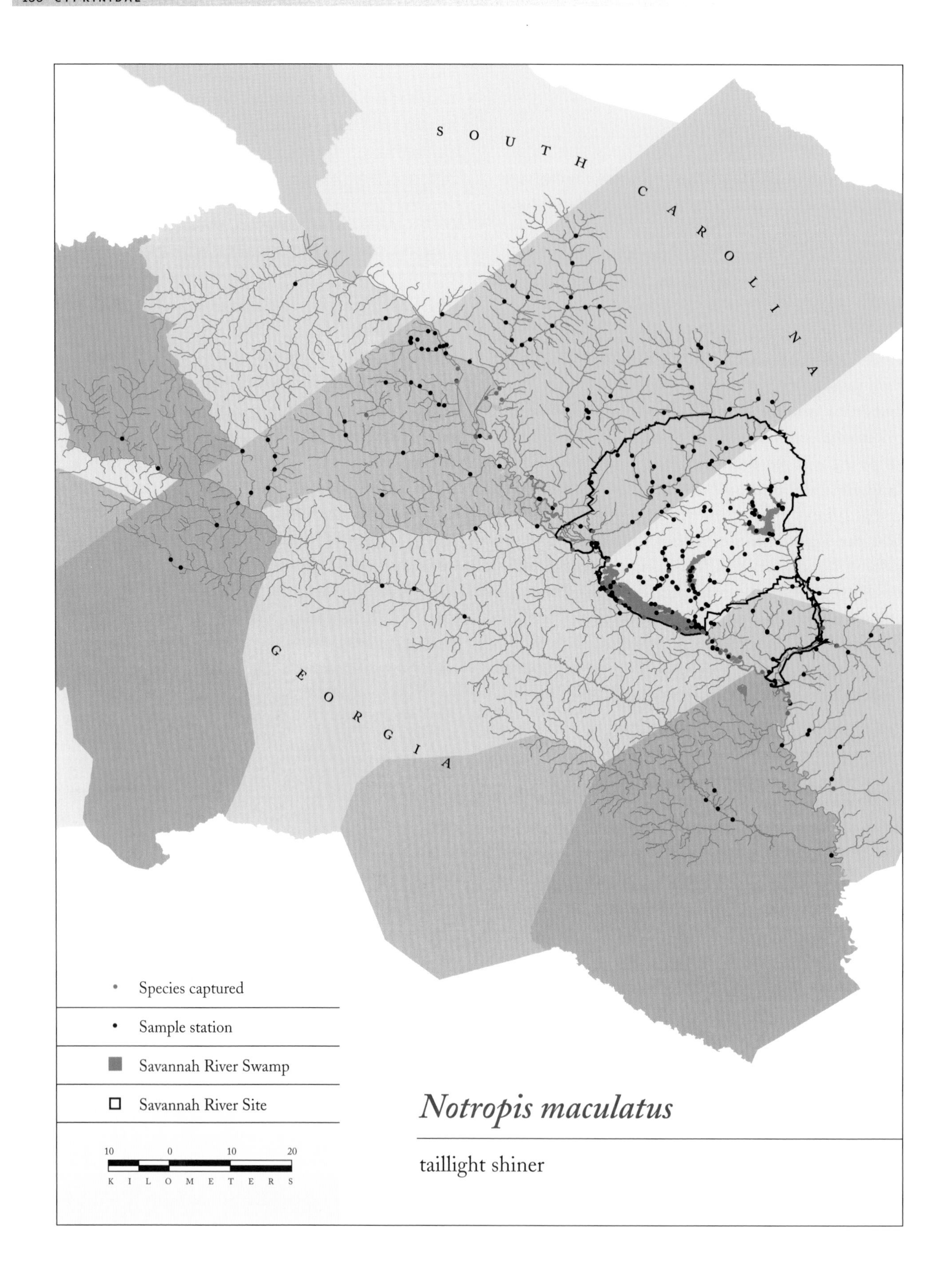

Notropis maculatus

taillight shiner

sippi. Habitats include nearshore areas in oxbows with cypress swamp margins, backwaters, sloughs, and densely vegetated margins of lakes (Burr and Page 1975; Robison 1978). Cowell and Barnett (1974) reported finding them generally close to the bottom on the lakeward side of shoreline vegetation. Females were more common in open water, and males were found more often along the shoreline.

BIOLOGY Based on the reported presence of young-of-the-year and adults in breeding condition, spawning occurs from March to May in Kentucky (Burr and Page 1975), and from March through June in Alabama (Mettee et al. 1996). In a Florida lake, they spawned from March to early October at water temperatures of 23–32 °C in 1970–1971 (Cowell and Barnett 1974), and from March through September at 20–34.5 °C in 1969 (Beach 1974). Males in breeding colors have been found from April to mid-June in Arkansas (Robison and Buchanan 1984) and in spring and summer (through July) in the MSRB. The taillight shiners Chew (1974) studied broadcast their eggs in nests of largemouth bass (*Micropterus salmoides*), which showed no aggression toward them; more than half of the eggs in such largemouth bass nests were from shiners (Chew 1974). Burr and Page (1975) reported spawning beneath or adjacent to a large log in 15–30 cm of water. The eggs are adhesive and demersal, and hatch in 60–72 hours (Cowell and Barnett 1974). Larvae are described in Millard 1981. Loos and Fuiman 1978; and Fuiman et al. 1983 provide characteristics useful for larval identification.

Cowell and Barnett (1974) reported postspawning mortality of 1-year-old fish in Florida and, on the basis of the absence of annual increments in scales and otoliths, suggested the maximum life span to be 1 year; in Kentucky, most individuals probably live less than 2 years (Burr and Page 1975). See Cowell and Barnett 1974 for information on growth rates of larvae in aquaria. Fecundity increases with size. Ova counts (ova > 0.8 mm diameter) averaged 246 (25–431) in females from Kentucky (Burr and Page 1975) and 78–408 in females measuring 42–59 mm TL from Florida (Cowell and Barnett 1974).

Taillight shiners are invertivorous. Individuals collected in the MSRB had fed on large numbers of cladocerans and copepods, but also on insects; plant material was found in only 1 of 21 individuals examined (Wiltz 1993). The diet is similar in Florida, and based on diet and prey availability, Cowell and Barnett (1974) concluded that feeding occurs in littoral vegetation. The short S-shaped intestine indicates a carnivorous diet.

SCIENTIFIC NAME *Notropis* = "keeled back" (Greek) (the keel in Rafinesque's type specimen was the result of drying or shriveling [Jordan and Evermann 1896]); *maculatus* = "spotted" (Latin).

Notropis petersoni Fowler, 1942
Coastal shiner

DESCRIPTION Body robust, cylindrical, but slightly compressed. Eyes large and laterally placed. Mouth large and terminal or slightly subterminal. Barbels absent. Lateral line complete.

A narrow black lateral stripe, upper edge sharper than lower edge, extends from snout to caudal fin base, narrowing and fading immediately anterior to caudal spot. Darkness of lateral stripe variable (generally fades in preservative). Stripe overlain by a distinctly punctate lateral line. Caudal spot distinctly triangular, the point of the triangle pointing anteriorly; spot same width as or narrower than lateral stripe, with light tan spots above and below. In life, dorsal edge of black stripe bordered by tan coppery stripe (light in preservative). Dorsum tan olive. Dorsal scales dusky with dark posterior lining that forms a rounded crosshatch pattern. Ventral head white. Belly fading from white to straw dorsally. Flanks silvery gray or pearly iridescent. Upper lip black. Chin marked with a laterally expanding terminal spot formed by black on anterior lower lip and anteriormost gular region. Roof of mouth generally with only very small scattered melanophores. Postanal stripe extending from anus to caudal peduncle. Fins translucent. Dorsal, anal, caudal, and anterior pectoral and pelvic rays outlined in black.

Breeding males have small breeding tubercles on head, body, and fins, especially pectoral fins; and enlarged tubercles around dorsal edge of mouth and ventral curve of lower jaw. Tubercles present but smaller on heads of breeding females.

SIMILAR SPECIES Ironcolor shiner (*N. chalybaeus*), dusky shiner (*N. cummingsae*), and yellowfin shiner (*N. lutipinnis*). See ironcolor shiner species account for characters to distinguish these species.

SIZE At least up to 79 mm SL, but adults in the MSRB more commonly reach 64 mm SL.

MERISTICS Dorsal rays 8, anal rays (6)7(8) (Bailey et al. 1954); pharyngeal teeth 2,4–4,2; lateral line scales 35–38 (Davis and Louder 1971).

DISTRIBUTION From the Cape Fear and Waccamaw river drainages of North Carolina south to the Miami area in Florida, west along the Gulf Coast to the Jordan River of Mississippi; apparently absent from the Pascagoula drainage (Swift *in* Lee et al. 1980). Widely distributed in the MSRB, but most abundant in the Savannah River or its

Notropis petersoni, coastal shiner

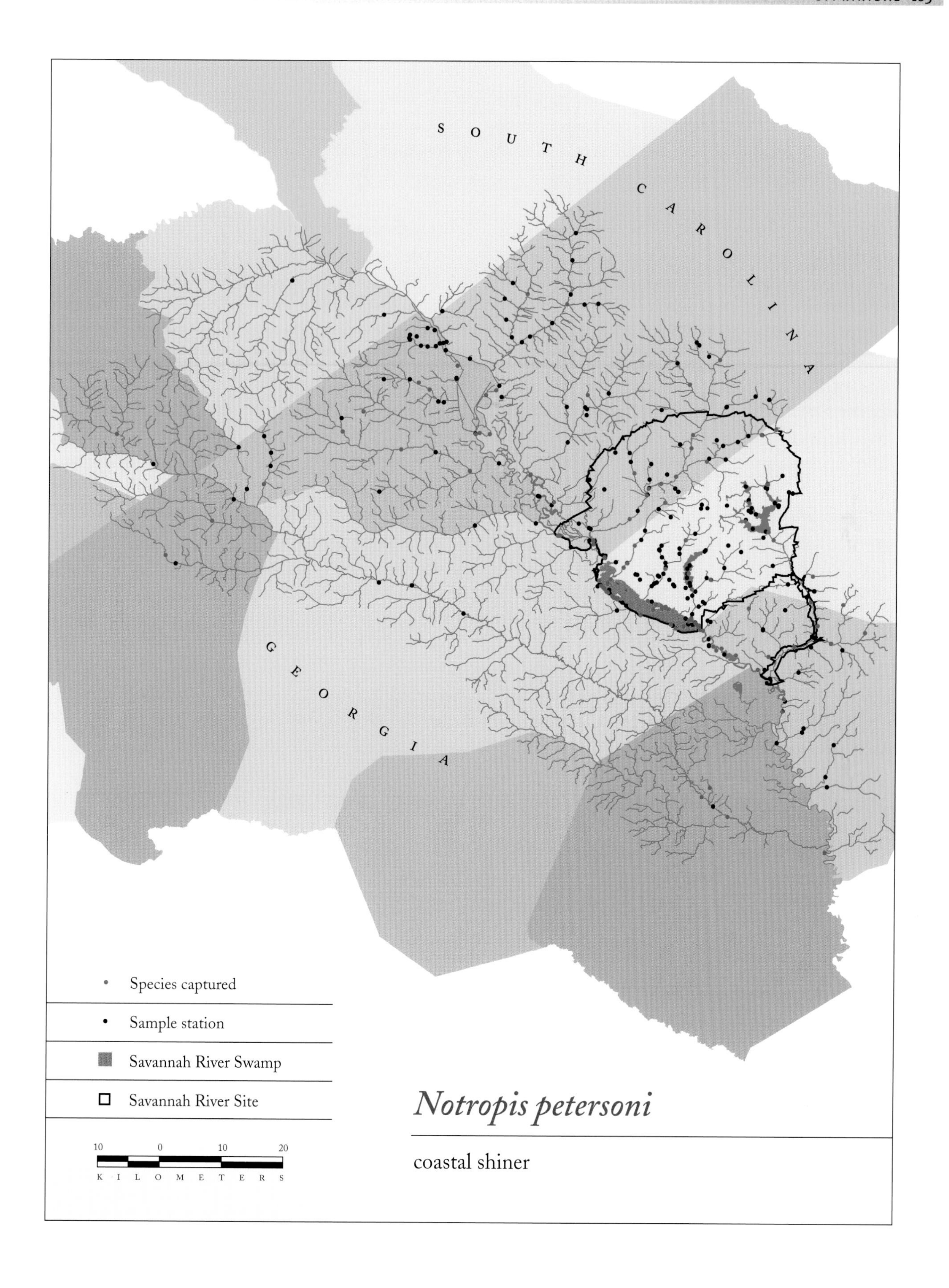

Notropis petersoni

coastal shiner

tributaries on the modern Savannah River floodplain; also common in sections of tributaries flowing across the Savannah River fluvial terraces, especially postthermal streams, and ascends large to intermediate tributaries of the upland area. Relatively common in reservoirs such as Par Pond on the SRS.

HABITAT Deeper scour pools or runs of tributary streams, where they orient facing upstream, deep in the water column near the bottom. Davis and Louder (1971) reported coastal shiners inhabiting shorelines of lakes and streams with a water velocity less than 30 cm/sec.

BIOLOGY Details of the life history vary with latitude along the Atlantic seaboard as indicated by studies done in North Carolina and Florida (Davis and Louder 1971; Cowell and Resico 1975, respectively). Spawning occurs over a protracted season. In North Carolina, with a temperate climate comparable to that of the MSRB, the coastal shiner spawns from late April through mid-July at water temperatures of 17–26 °C in Lake Waccamaw, but from May through August in streams where the water is cooler; in both habitats peak spawning occurs in June (Davis and Louder 1971). In south Florida, coastal shiners spawn from March through late August or early September at water temperatures of 19–27 °C, primarily at downstream locations (Cowell and Resico 1975). For larval identification characters, see Loos and Fuiman 1978.

In the Florida population, coastal shiners matured in 1 year, the females at average lengths of 59 (range 43–73) mm TL, the males at 56 (range 39–67) mm TL (Cowell and Resico 1975); in North Carolina, they were sexually mature at 2 years and 45 mm TL (Davis and Louder 1971). Both studies found sex ratios skewed toward females: between 1.8 and 2.3 females to 1 male in Florida, and 1.4 females to 1 male in North Carolina. Egg counts from ovaries increased with the female's size, with averages ranging from 328 to 854 in females measuring 40–50 to 70–81 mm TL (Davis and Louder 1971). Growth was fast in the Florida population, with all young-of-the-year reaching a length of 38 mm TL by December. Cowell and Resico (1975) suggested the Florida population to be an annual species based on lack of annuli on the otolith. Large postspawning die-offs occurred in the summer months (Cowell and Resico 1975). In contrast, North Carolina coastal shiners reportedly live through age 3, averaging 24, 46, and 59 mm TL at ages 1, 2, and 3, respectively (Davis and Louder 1971). Larvae grew to 8 mm in 1 month and to 14 mm in 2 months, and young-of-the-year joined schools of adults at about 18 mm in their third month after hatching (Davis and Louder 1971).

Coastal shiners are invertivores. They feed on a variety of organisms, including microcrustaceans, especially cladocerans, and insects (Davis and Louder 1971; Wiltz 1993). Feeding is apparently opportunistic; bivalve molluscs dominated the diet in an MSRB study (Wiltz 1993). Some populations make occasional feeding migrations, as demonstrated in Lake Waccamaw, where shiners migrated from the lake into inflowing streams after heavy rainfall to feed (Davis and Louder 1971).

SCIENTIFIC NAME *Notropis* = "keeled back" (Greek) (the keel in Rafinesque's type specimen was the result of drying or shriveling [Jordan and Evermann 1896]); *petersoni* = after C. Bernard Peterson, who assisted with the collection of the type specimens.

Opsopoeodus emiliae Hay, 1881

Pugnose minnow

DESCRIPTION Body slender, cylindrical, but slightly compressed. Eyes large and laterally placed. Mouth very small and strongly oblique. Small maxillary barbels sometimes present on one or both sides of mouth, but only a few very small taste buds present (Gilbert and Bailey 1972; Dimmick 1988). Lateral line appears complete in MSRB specimens; reported as usually complete by Gilbert and Bailey (1972), and as incomplete by Robison and Buchanan (1984).

A variable but generally narrow black lateral stripe extends from tip of snout to caudal fin base, expanding as it streaks onto fork of caudal fin. No distinct black caudal spot. Light tan patches on caudal fin base above and below black lateral stripe. In life, with narrow coppery stripe above the black lateral stripe (light in preservative). Upper and lower lips lined with black. Chin largely unpigmented, but scattered melanophores are visible at high magnification. Roof of mouth with small scattered melanophores, especially anteriorly. Dorsum olive tan. Ventral head and belly white, flanks silvery gray. Scales above black lateral stripe outlined in black forming a distinctive diamond-shaped crosshatch pattern. Predorsal scales crowded (Gilbert and Bailey 1972). Postanal stripe present from anus onto caudal peduncle. Black pigment in dorsal and caudal fins.

Nuptial coloration of males includes dark silver-blue body, white tips on anal and pelvic fins, dusky dorsal fin with intensified black in anterior and posterior portions and white in the middle. Small white knobs develop on first three dorsal rays (Page and Johnston 1990b). Tubercles develop on head, especially on chin and around upper lip, and on dorsal surface of pectoral rays (Gilbert and Bailey 1972). Scales on midlateral and lower sides with large, flashy iridescent patches.

SIZE At least up to 65 mm TL, but usually less than 50 mm SL in the MSRB.

MERISTICS Dorsal rays 9, anal rays 8, pectoral rays (13)14–15(16) (Etnier and Starnes 1993); pharyngeal teeth 0,5–

Opsopoeodus emiliae, pugnose minnow

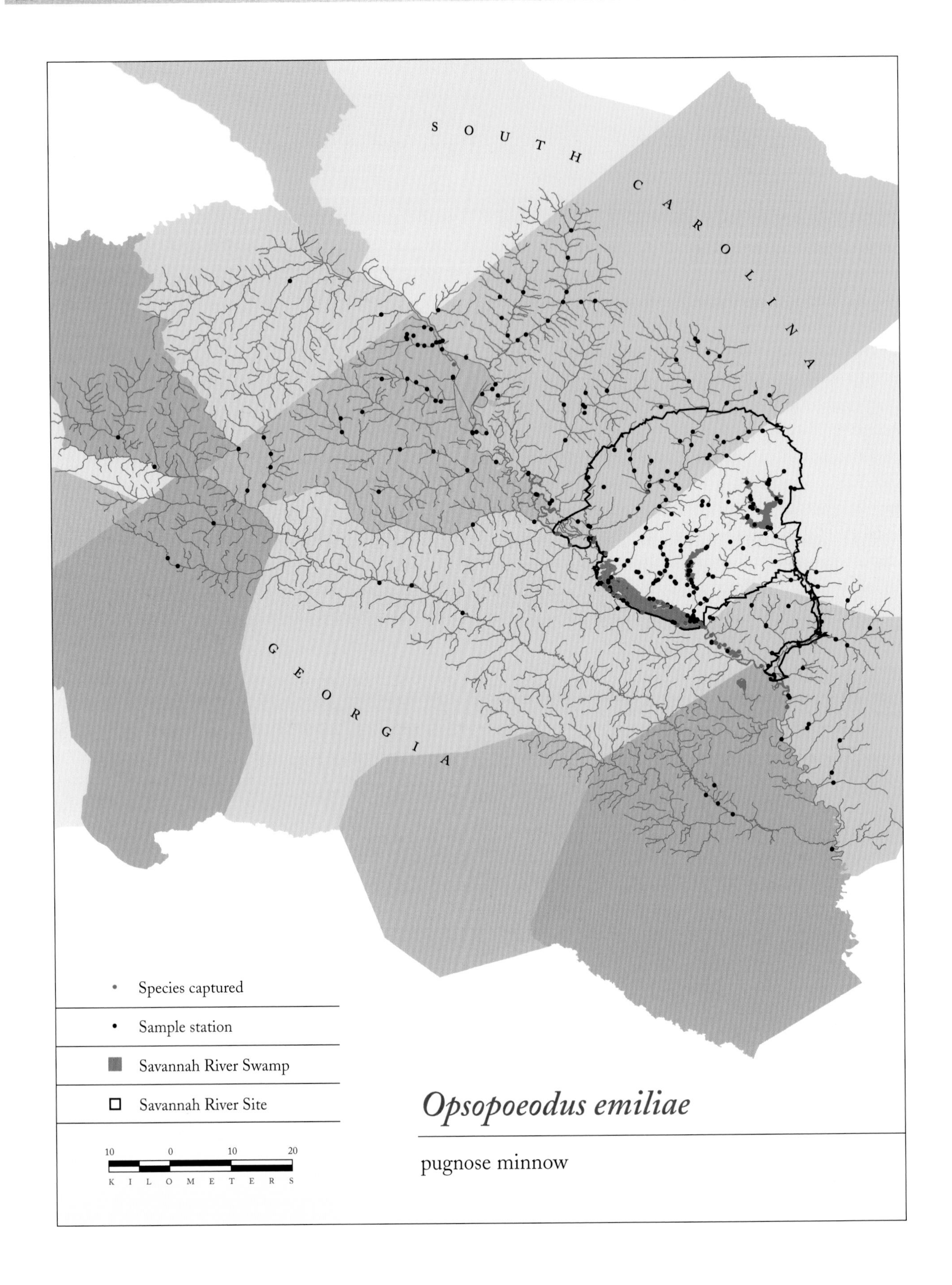

Opsopoeodus emiliae

pugnose minnow

5,0; lateral line scales 36–39 (Gilbert and Bailey 1972; Page and Burr 1991).

DISTRIBUTION From the Edisto River of South Carolina south to Lake Okeechobee, Florida; west along the Gulf Coast to the Nueces River of Texas, and north through the Mississippi Valley to Wisconsin, Michigan, Ohio, and West Virginia (Gilbert *in* Lee et al. 1980). Apparently uncommon, and rarely abundant at any single location in the MSRB. Most commonly collected in small numbers in the Savannah River, at least in the late winter and spring. Wiltz (1993) collected this species from the Savannah River and the mouth of Beaverdam Creek, Georgia. It occurs sporadically in tributary streams on the fluvial terraces of the Savannah River and is rarely reported in the upland streams.

HABITAT Sluggish, well-vegetated backwaters, sloughs, and oxbows (Smith 1979; Robison and Buchanan 1984; Etnier and Starnes 1993; Mettee et al. 1996), or rivers in deeper holes over soft bottoms (Cahn 1927). In the Savannah River, most commonly collected in eddies behind flow obstructions. More abundant in oxbow than in river habitats in Florida (Beecher et al. 1977), but found in similar frequencies in the main channel and floodplain habitats of the Ochlockonee River, Florida (Leitman et al. 1991).

BIOLOGY The spawning period is reported to be early summer in Missouri (Pflieger 1975) and Arkansas (Robison and Buchanan 1984), late spring and early summer in Tennessee (Etnier and Starnes 1993), and May–July in the southern United States (Mettee et al. 1996). Except for the spawning behavior documented in detail by Page and Johnston (1990b), the biology is not well known. Pugnose minnows are egg-clusterers (Johnston and Page 1992); that is, they lay their eggs in a single layer on the underside of a flat object (Page and Johnston 1990b). A single layer of eggs attached to the underside of an object suspended above the substrate will be better oxygenated, less susceptible to siltation, and well hidden from predators. Males defend territories to which they attract females. The males regularly perform circling swimming displays while on the nest. Spawning occurs with the pair inverted, and the female presses her abdomen against the overhanging substrate and deposits the adhesive eggs. Eggs are laid in small batches during repeated brief spawning events that last for about 1 second. Eggs are deposited singly or in strings. The larvae are described in Millard 1981. Additional larval and juvenile characters are presented in Loos and Fuiman 1978; Auer 1982; Fuiman et al. 1983; and Leslie and Timmins 2002. Males provide parental care to developing young, including defense against predators, but the males studied by Page and Johnston (1990b) frequently left the nest while offspring were present. Fertilized eggs hatched successfully at 21–27 °C, but died at cooler and warmer temperatures (see details in Page and Johnston 1990b).

The pugnose minnow is likely a carnivore, and the oblique mouth position suggests that it feeds in mid-water or near the surface (Gilbert and Bailey 1972). The diet includes insects such as chironomids, microcrustaceans, larvae (unpublished dissertation by William McLane cited in Gilbert and Bailey 1972), and small crustaceans such as amphipods (Cahn 1927).

SCIENTIFIC NAME *Opsopoeodus* = "mouth" and "dainty feeding" (Greek), referring to the upturned mouth; *emiliae* = named after Mrs. Emily Hay, wife of the author of the species.

Pteronotropis stonei (Fowler, 1921)
Lowland shiner

In the past, this species was known as the sailfin shiner, *Pteronotropis hypselopterus*. Recent taxonomic investigations revealed that the sailfin shiner is actually part of a complex of several species (Suttkus and Mettee 2001; Suttkus et al. 2003). The latter study elevated the lowland shiner, *Pteronotropis stonei*, to species status. Although these studies downgraded *Pteronotropis* to a subgenus of *Notropis*, we follow Mayden (1989) in its recognition at the level of genus.

DESCRIPTION Body laterally compressed, males deeper bodied than females. Eyes large; head small. Mouth small, nearly horizontal, and terminal. Barbels absent. Lateral line complete and moderately downcurved anterior to origin of pelvic fin.

The most distinctive color feature is the broad black lateral stripe with a bluish tinge that extends from the tip of the chin and upper jaw to the caudal fin base; a thin burnt orange stripe (light in preservative) borders its upper margin. Dorsum olive gold above, belly and flanks white. Rosy red spots present at base of caudal fin. Black lateral stripe continues across chin, on which pigment is darkest anteriorly. Inside of upper lip darkened by melanophores. Postanal stripe absent. Pectoral and pelvic fins tinged with yellow, as are distal portions of the large, falcate anal fin and the caudal fin. Pelvic and anal fins also may be dusky. Dorsal fin mostly black with a clear basal portion and clear patch at the anterior distal corner, clear patches larger on females.

Breeding males have small tubercles at least on chin, sides of head, anal fin, and posterior edges of scales, especially anteriorly.

SIMILAR SPECIES The body form is similar to the two *Cyprinella* species found in the MSRB (the bannerfin shiner, *C. leedsi*, and the whitefin shiner, *C. nivea*), but the lowland shiner is easily distinguished by the long anal fin with 10–11 (versus 8) rays, the prominent deep bluish black lateral stripe that runs along the entire length of the body, and pharyngeal teeth count of 2,4–4,2 (versus 1,4–4,1 or 0,4–4,0 in the other species).

SIZE Adults are commonly 30–45 mm SL; males are larger than females.

MERISTICS Dorsal rays 8, anal rays 10–11; pharyngeal teeth 2,4–4,2.

Pteronotropis stonei, lowland shiner; male

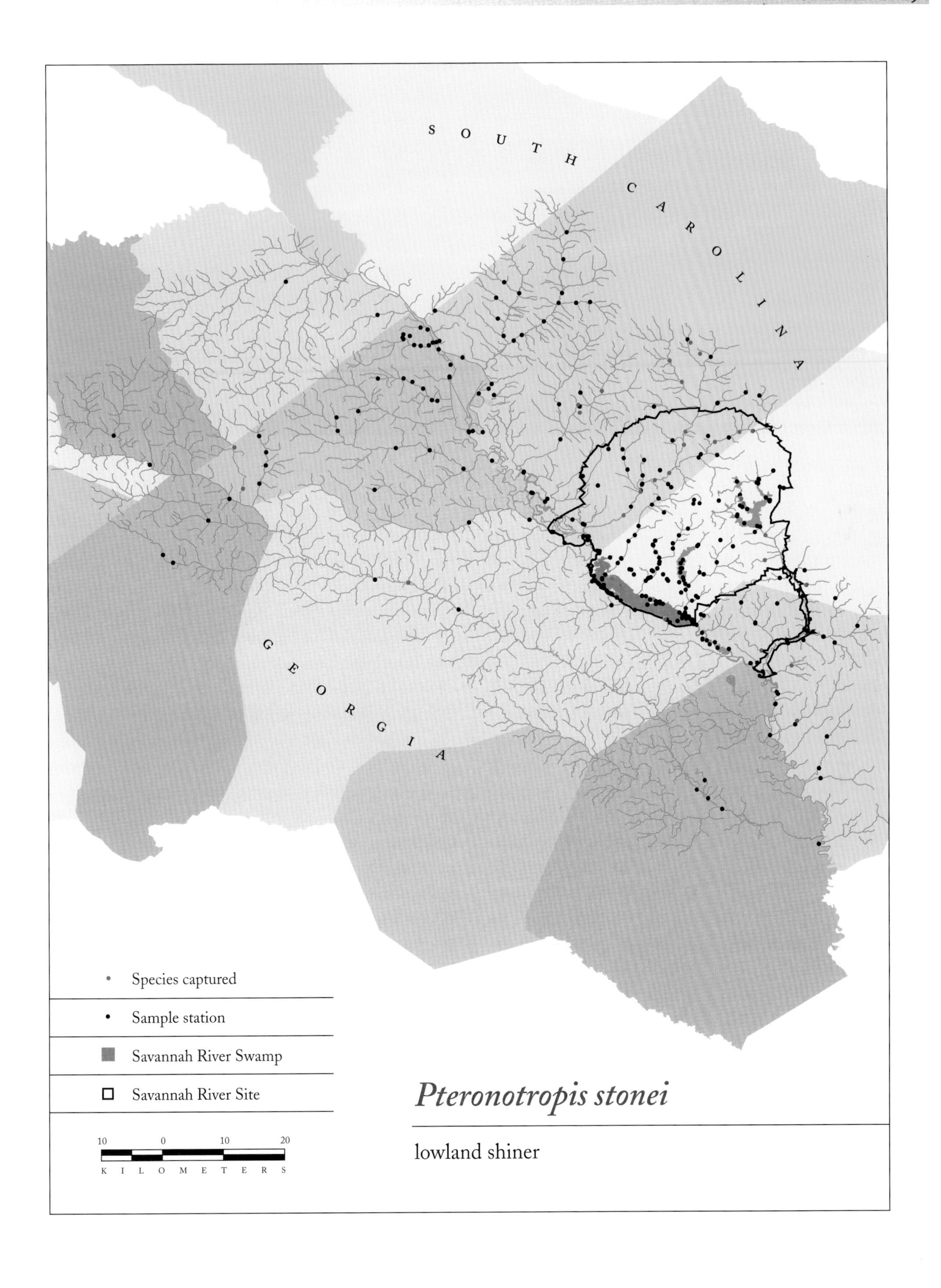

Pteronotropis stonei

lowland shiner

Pteronotropis stonei, lowland shiner; female

DISTRIBUTION On Atlantic Coastal Plain and lower Piedmont from the Pee Dee River of South Carolina south to the Satilla River of south Georgia (Suttkus et al. 2003). In the MSRB, restricted to upland streams, where it is sometimes locally abundant.

HABITAT Small to intermediate—and occasionally large—streams with moderate to swift flow and sandy substrates. The headwater streams where lowland shiners are found generally have moderate flow with alternating runs, scour pools, and often pools created by root dams, in contrast to swifter streams with some gravel in the substrate where the yellowfin shiner (*N. lutipinnis*) is the dominant shiner. In headwaters, lowland shiners are generally found in scour or plunge pools, but in larger streams they are frequently observed in eddies immediately downstream of woody debris, overhanging woody vegetation, or aquatic macrophytes. In the winter, lowland shiners often move from behind the debris up into it. They can also be found in small ephemeral braids on the floodplain during floods, particularly in the winter. In all habitats, schools of 10 to more than 75 individuals are generally found facing upstream in the middle of the water column.

BIOLOGY These broadcast spawners scatter their demersal and adhesive eggs across the bottom. Individual females spawn multiple clutches throughout a protracted spawning season that lasts from late May through July in the MSRB, commonly at water temperatures of 20–25 °C (DEF pers. obs.). Spawning likely occurs in runs with aquatic macrophytes, as suggested for the flagfin shiner (*P. signipinnis*) by Albanese (2000a), but this has not been confirmed. Loos and Fuiman (1978) reported on egg size; and Fletcher and Wilkins (1999) examined larval characters. For the first 3–5 days after hatching, while still nourished by yolk supplies and not yet actively feeding, the larvae avoid being swept away by the current by attaching themselves to stationary objects with a glue secreted from glands in the head (Fletcher and Wilkins 1999). Subsequently, the larvae swim into the water column and begin feeding. Lowland shiners likely live 1–2 years, as also suggested for *P. signipinnis* (Albanese 2000b). They readily spawn in aquaria.

Lowland shiners are frequently observed feeding on invertebrates drifting in the current in mid-water and occasionally at the surface. In the Hillsborough River of Florida, lowland shiners ate aquatic insects, rotifers, crustaceans, pelecypods, and algae; ingestion of adult chironomids indicates at least some surface feeding (Hoover 1980).

SCIENTIFIC NAME *Pteronotropis* = "winged" and "keeled back" (Greek); *stonei* = named in honor of Witmer Stone, who collected the holotype in 1917.

Semotilus atromaculatus (Mitchill, 1818)
Creek chub

DESCRIPTION Body stout and cylindrical; head and anterior body broad. Eyes relatively small and laterally placed. Mouth large, terminal, and moderately oblique. A single pair of short, broad-based, flaplike preterminal maxillary barbels is present, each barbel located in a groove above the upper lip and ornamented with taste buds at the terminal end (Dimmick 1988). Lateral line complete (Jenkins and Burkhead 1993).

A moderately deep black lateral stripe with diffuse upper and lower edges extends from snout (may be narrow or indistinct on head) to base of caudal fin, then streaks onto caudal fin. A narrower goldish tan stripe is above it (light in preservative). Lateral stripe variable and often indistinct in large adults, more distinct in juveniles and in preservative. Large but often indistinct caudal spot usually present, more distinct in preservative and in juveniles.

Dorsum dusky olive with heavily pigmented scales. Posterior margin of scales on dorsum and sides outlined in black; particularly in juveniles forming a crosshatched pattern; intensity of black sometimes variable, producing an uneven spotted appearance. Ventral head and belly mostly white, flanks silvery gray. Chin has pigment on anterior tip of lower lip. Also, black on anterior portion of upper lip. Roof of mouth without pigment. Postanal stripe absent or faint along anal fin base. Small black spot present at anterior base of dorsal fin, on base of first three to four rays; spot often surrounded immediately by reddish orange in adults (light in preservative). Anal, pectoral, and pelvic fins often reddish orange or yellowish orange.

Breeding males develop olive or apple green tints above the diffuse brown or olive midlateral stripe, which is overcast with violet iridescence; a rosy pink or reddish cast is present ventrally on lower body, cheek, opercle, and lower fins (Reighard 1910; Jenkins and Burkhead 1993). Small tubercles are abundant on areas used to grasp females during spawning (i.e., opercular region, sides of body and tail, posterior edge of dorsal fin, and upper edge of pectoral fins). The most distinctive feature of nuptial males is the four or five large, hornlike tubercles on the dorsal surface of each side of the head in the interorbital area (Reighard 1910). Breeding females lack bright colors and tubercles on the head.

SIZE Males reach 250 mm (Hubbs and Cooper 1936) or up to 263 mm SL (Greeley 1930a) and are larger than females, which reach 245 mm SL (Greeley 1930a).

Semotilus atromaculatus, creek chub

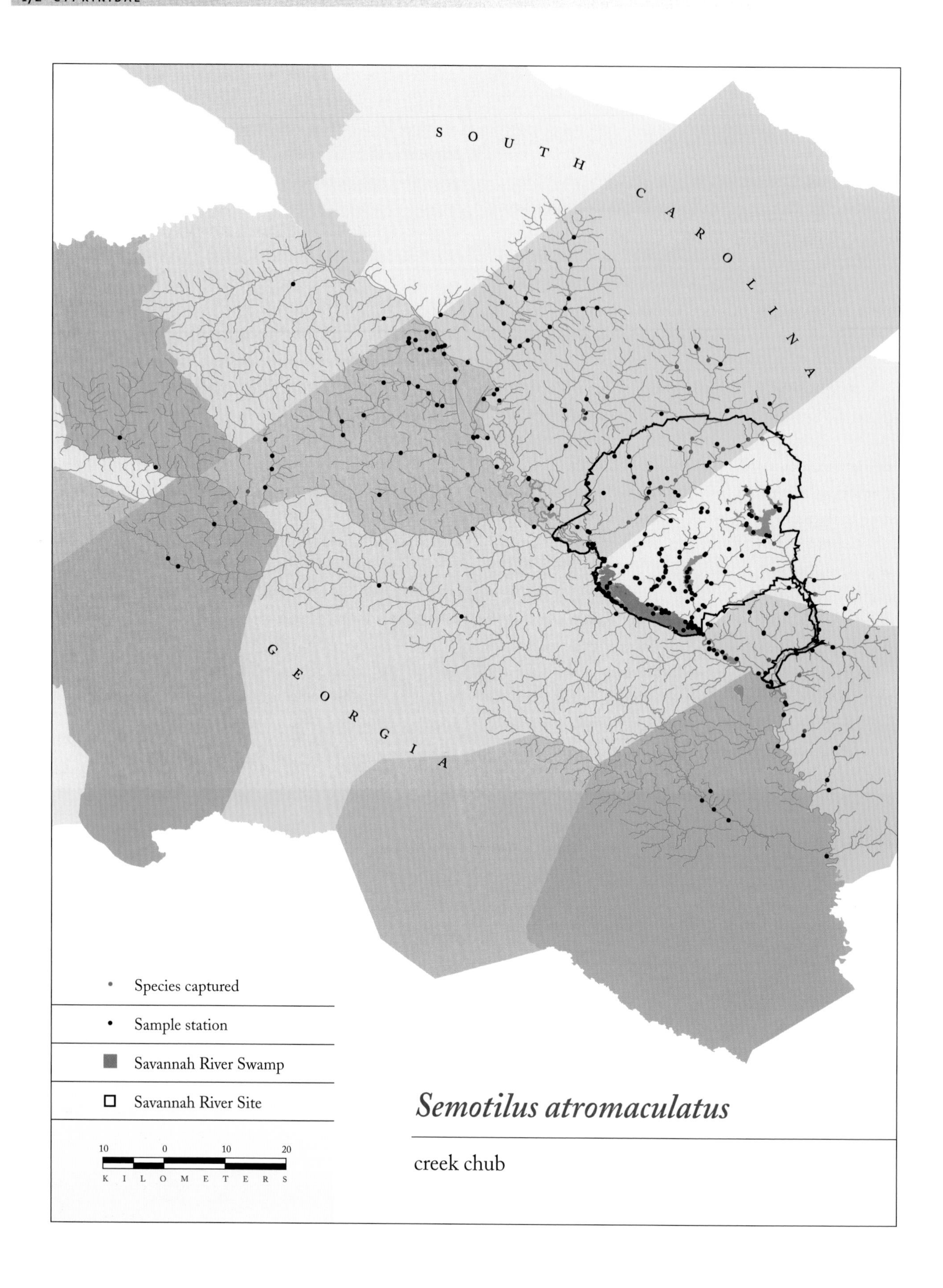

Semotilus atromaculatus

creek chub

MERISTICS Dorsal rays 8, anal rays 8, pectoral rays (14)15–18(20), pelvic rays (7)8(9); pharyngeal teeth 2,5–4,2; lateral line scales (49)52–58(62) (Jenkins and Burkhead 1993).

DISTRIBUTION Widely distributed through most of North America east of the Rocky Mountains; absent from most of Florida, Texas, and Oklahoma, with an apparently disjunct population in the Pecos River drainage of New Mexico (Lee and Platania *in* Lee et al. 1980). Common and widely distributed in the MSRB, primarily in the upper headwaters and very small tributaries of the upland areas.

HABITAT Although its distribution overlaps with that of the bluehead chub (*N. leptocephalus*), the creek chub is generally found in smaller streams barely 1–1.5 m wide and less than 1 m deep. Creek chubs are generally part of fish assemblages in headwater streams (Felley and Hill 1983) with a small upstream drainage area (Larimore and Smith 1963) and in narrow, shallow channels (Meffe and Sheldon 1988; McMahon 1982). Although rare in larger rivers, they sometimes occur near the mouth of small tributary streams (Starrett 1951). A microhabitat analysis indicated that creek chubs spend more time near the stream edge than do many other cyprinids (McNeely 1987), and, although their seasonal habitat shifts, often take refuge in the shelter of roots and overhanging banks of deeper pools (Greeley 1930a; Moshenko and Gee 1973).

BIOLOGY Creek chubs typically spawn earlier than most other nest-building minnows such as the *Nocomis* species (Maurakis et al. 1990), and in the MSRB only the tail end (late April or May) of the creek chub spawning season appears to overlap with that of the bluehead chub. Spawning was observed in April and May in Missouri and Ohio (Pflieger 1975; Ross 1977), early June in Wisconsin (Cahn 1927), and May in Manitoba (Moshenko and Gee 1973).

The creek chub employs the pit-ridge-building spawning strategy (Johnston and Page 1992). Nest building and spawning behavior, described in great detail by Reighard (1910), are summarized here. Nest building is initiated by the male, which constructs a pit about 30 cm wide and 4–6 cm deep by picking up substrate—often gravel or a mixture of gravel and sand—with its mouth and dropping it in a small pile at the upstream edge of the pit. Spawning occurs in the pit, which the male then fills in and covers by moving substrate from the pit's downstream edge, extending the mound into an ever-lengthening ridge. Reighard (1910) reported larger stones being placed on the bottom of the pit, and eggs and larvae developing within these large interstitial spaces; smaller stones put on the top prevent entry of potential egg predators and filter out sediment before it reaches the offspring. However, this may be influenced by the available substrate. Nest ridges are typically about 25 cm wide and 0.5–2 m long but can be as long as 5.5 m. Ross (1976) examined stream dimensions and water velocities of spawning areas and determined that nests are built solitarily in small runs, or in groups of three or four, sometimes less than 50 cm apart, at the downstream edge of larger pools. Maurakis et al. (1990) compared nest structures of different *Semotilus* species. Larvae were described by Embody (1914 *in* Greeley 1930a) and Loos et al. (1979); Fuiman et al. (1983) discussed characters useful for identifying larvae.

The peak spawning period lasts 3–4 days (Ross 1976), during which multiple females spawn in each nest. The spawning act includes a strong clasp of the female by the male that may reduce cuckoldry by other males. The male wraps his body in a U around the female while pushing her into a vertical position with her head up and abdomen against the substrate. Each spawning event lasts less than a second, and a female spawns her entire clutch by repeatedly visiting the nest and releasing about 25–50 eggs at a time. Females may enter the spawning area of a nest by employing one of three behaviors determined by habitat and number of active nests. Downstream tail-first drifting and a circle swim (about 60 cm diameter) prior to entry less frequently provoked a defensive response by the guardian male than did direct head-first entry (Ross 1976).

A guardian male defends its nest from rivals by head butting smaller males with its large cephalic tubercles, or by displays such as parallel swims when the males are of similar size. Fighting may be rare, with rival males more often being chased away without contact. Miller 1964 describes parallel swims and provides additional information on nesting behavior. The spawning success of a male is reduced when it has to use more intensive aggression and more lengthy parallel swims to defend its nest (Ross 1977). Potential egg predators are not consistently driven away from nests (Ross 1977), and it seems that males may defend the spawning site more to avoid disruption of spawning than to deter egg predators.

Small "nest watching" males employ a parasitic spawning strategy. This type of male lurks behind the nest until the guardian male leaves it during defensive interactions with another large male, then enters the spawning pit and attempts to spawn with the females there until the guardian male returns and chases it away (Ross 1977).

Creek chubsuckers (*Erimyzon oblongus*) have been observed spawning in creek chub nests. Creek chub males usually drive the creek chubsuckers away from the spawning pits, and the suckers may successfully spawn only in inactive nests (Page and Johnston 1990b).

Female creek chubs in a central Illinois population matured in 2 years at 63 mm SL, and fecundity increased with female size, with mature ova numbers ranging from 438 to 7,154 (Schemske 1974). In northern populations, females usually mature at 3 years, and males at 4 years of age (Greeley 1930a). Males and females in the Ontario population studied by Powles et al. (1977) matured at 60 and 65 mm TL, respectively.

Creek chubs have been reported to live up to 3 years in Illinois populations (Lewis and Elder 1956; Schemske 1974), and up to 7 years in Ontario (Powles et al. 1977). Males generally grow faster than females (Greeley 1930a). In a Missouri population, creek chubs reached 38–64 mm by the end of their first summer, and adults ranged from 100 to 200 mm (Pfleiger 1975). In tributaries of Clear Creek in Illinois, fish averaged 61, 87, 138, and 168 mm TL in August at ages 0, 1, 2, and 3, respectively (Lewis and Elder 1956).

The large terminal mouth, hooked pharyngeal teeth specialized for grasping and holding prey, and short gut of the creek chub are well suited for its predaceous, carnivorous feeding habits (Hubbs and Cooper 1936). In the MSRB they feed largely on terrestrial insects along with some aquatic insects (Sheldon and Meffe 1993). In other geographical areas creek chubs prey on similar organisms (Minckley 1963; Gatz 1979; McNeely 1987), and in Spring Creek, Oklahoma, on oligochaetes, fish, and crayfish as well (McNeely 1987). The diet shifts seasonally (Starrett 1950; Felley and Hill 1983), most likely based on the availability of different prey types. In Iowa, for instance, creek chubs eat terrestrial insects in large numbers only in the spring and summer, and eat large numbers of fish in the fall when young-of-the-year abundance is highest for most species; aquatic insects, while eaten year-round, dominate the winter diet (Starrett 1950). Terrestrial insects dominated the diet in summer and fall in an Oklahoma stream (Felley and Hill 1983). The diet of the creek chub may also vary according to habitat type. Angermeier (1985) found that the relative amounts of terrestrial insects and the diversity of prey taxa consumed varied among sites in age 0 fish. Diet also changes with size, and fish become a more important food item by age 3; by age 4 about 70% of the creek chub's gut volume can consist of fish (Keast 1985b).

SCIENTIFIC NAME *Semotilus* = "spotted banner" (Greek), referring to the dorsal fin with the basal spot; *atromaculatus* = "black spot" (Latin), describing the spot at the anterior base of the dorsal fin.

Catostomidae (suckers)

Suckers are a widespread family of fishes. Except for 2 species, the 75 known sucker species (there are probably a number of undescribed species) are restricted to North American fresh waters from the Arctic Circle to Mexico, with the greatest diversity found in the southeastern United States (Jenkins and Burkhead 1993). Nine species are found in the MSRB. It is thought that the longnose sucker, *Catostomus catostomus* (northern North America and Siberia), recently invaded Siberia, and that the distribution of *Myxocyprinus asiaticus* (rivers in China), a species with many ancestral characters, is an indication that suckers were more widely distributed in the past (Etnier and Starnes 1993).

All suckers have cycloid scales. The origin of the dorsal fin is in front of the pelvic fins. The anal fin is placed well back on the body. The mouth is inferior with protractile, fleshy lips that are highly sensitive and well adapted to bottom feeding. The diet generally consists of small epibenthic and benthic invertebrates such as insect larvae and crustaceans, and plant material. A few species forage on mid-water zooplankton, and some others are specialized for feeding on molluscs (Jenkins and Burkhead 1993). Although suckers ingest detritus along with living food, they are able to sort and eject unwanted material via the mouth and the gill openings (Jenkins and Burkhead 1993).

Reproduction in suckers was reviewed by Page and Johnston (1990a). Spawning typically occurs in the spring, and the demersal, adhesive or nonadhesive eggs, 1.5–3.5 mm in diameter, are randomly scattered across the spawning substrate. The males have well-developed nuptial tubercles on the anal fin and the lower part of the caudal fin. Adult males can usually be recognized by their longer anal fin and longer inner pelvic rays relative to the outer ones (Jenkins and Burkhead 1993).

Suckers occupy a wide range of freshwater habitats but seem to avoid silted areas, a contradiction to the common belief that they are "mud-loving" (Jenkins and Burkhead 1993). Suckers can dominate the fish biomass in streams and are recreationally and commercially harvested for human consumption or fishing bait (Jenkins and Burkhead 1993). Genetically suckers are tetraploid; that is, they have double the usual number of chromosomes.

We here recognize *Moxostoma* and *Scartomyzon* as valid genera based on several recent accounts; however, the status of the genus *Scartomyzon* may be in question. Harris et al. (2002) have suggested subsuming *Scartomyzon* into the genus *Moxostoma* based on the fact that the species of both genera appear in the same clades. If *Scartomyzon* is shown to be invalid, only the brassy jumprock in our area will require a change in genus name.

KEY TO THE SPECIES OF CATOSTOMIDAE

1a. Long-based dorsal fin, more than 20 rays (Plate 12 a) **2**
1b. Relatively short-based dorsal fin, 18 or fewer rays (Plate 12 b, c) **3**

[1] **2a.** No nipple projection on mid-anterior edge of lower lip (Plate 12 d); relatively elongate snout with corner of mouth not extending to below anterior edge of eye . . . . *Carpiodes* sp. cf. *cyprinus* (quillback), p. 178
2b. Nipple projection present on mid-anterior edge of lower lip (Plate 12 e); short snout with corner of mouth extending to below anterior edge of eye . . . . *Carpiodes* sp. cf. *velifer* (highfin carpsucker), p. 181

[1] **3a.** Dorsal surface of head flat or concave between eyes (Plate 12 i); 4–6 prominent dark saddles present . . . . *Hypentelium nigricans* (northern hogsucker), p. 190
3b. Dorsal surface of head convex between eyes; prominent dark saddles usually absent (small juvenile *Minytrema*, *Erimyzon*, *Moxostoma*, and *Scartomyzon* often have 3 or 4 moderately dark saddles) **4**

[3] **4a.** Scales on sides of body marked with black spots or dashes (shape of spots variable, often small rectangles or thick lines) at their bases forming 10–11 parallel lines . . . . . *Minytrema melanops* (spotted sucker), p. 193
4b. Scales on sides of body not marked with black spots forming parallel lines (sides of small *Scartomyzon* may be marked with continuous lines, but scales are not spotted) **5**

[4] **5a.** Lateral line absent; swimbladder with 2 chambers . . . . *Erimyzon* **6**
5b. Lateral line complete; swimbladder with 3 chambers **7**

[5] **6a.** Smaller scales, (39)40–43(45) scales in midlateral series; adults with 5–8 dark (frequently faint) dorsolateral/midlateral blotches, but young with only a narrow, dark midlateral stripe . . . . *Erimyzon oblongus* (creek chubsucker), p. 184
6b. Large scales, 34–38(39) scales in lateral series; adults without dark blotches, but young with a narrow, dark midlateral stripe . . . . *Erimyzon sucetta* (lake chubsucker), p. 187

[5] **7a.** Lower lip with large V-shaped indentation in posterior margin of lower lips (Plate 12 f) . . . . *Moxostoma collapsum* (notchlip redhorse), p. 196
7b. Posterior edge of lower lip straight or nearly straight (Plate 12 g, h) **8**

[7] **8a.** Number of scales around caudal peduncle 15–16 (see Plate 4 i for counting method); pelvic rays usually 9 . . . . *Scartomyzon* sp. (brassy jumprock), p. 202
8b. Twelve scales around caudal peduncle; pelvic rays usually 10 . . . . *Moxostoma robustum* (robust redhorse), p. 199

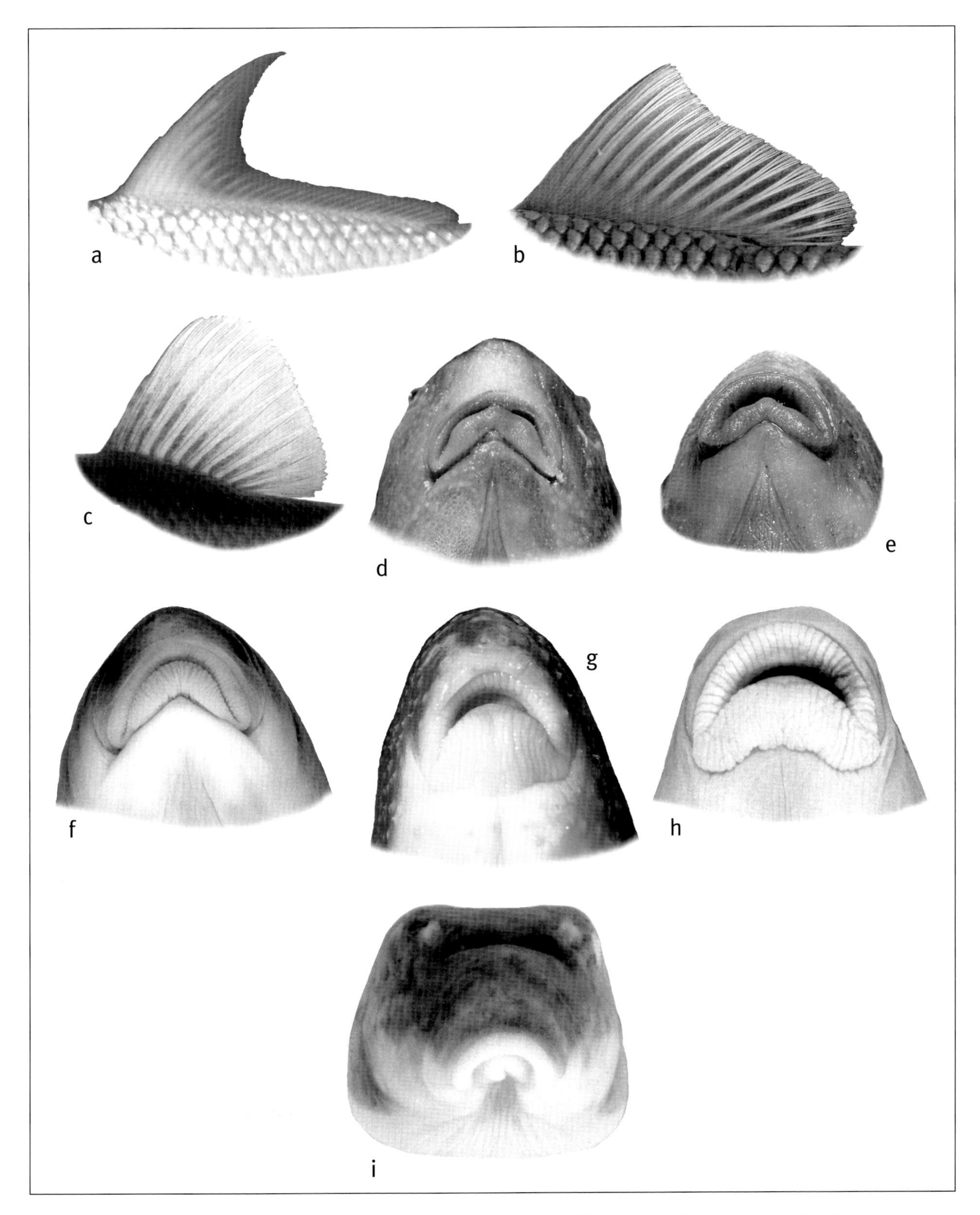

PLATE 12. Morphological characters of Catostomidae: (a) *Carpiodes* dorsal fin, (b) notchlip redhorse dorsal fin, (c) *Erimyzon* dorsal fin, (d) quillback lips, (e) highfin carpsucker lips, (f) notchlip redhorse lips, (g) robust redhorse lips, (h) brassy jumprock lips, (i) northern hogsucker showing concavity between the eyes.

Carpiodes cyprinus (Lesueur, 1817)
Quillback

The taxonomic status of the quillback is uncertain. Southeastern populations, including those in the MSRB, differ from those in the Mississippi and Great Lakes basins and may represent a different species (R. E. Jenkins pers. comm.).

DESCRIPTION Body deep and laterally compressed, with an arched back and relatively flat ventral profile. Base of the dorsal fin long, with elongate anterior rays extending to approximately one-half the length of the dorsal fin base when depressed. Mouth small and inferior, lips relatively large. Eyes comparatively large. Scales large and conspicuous.

Color generally silvery to bronze or olive with a darker back and lighter underside. Nuptial tubercles of breeding males are relatively small and cover most of the head (except for the top), also occurring on pectoral fins and, to a lesser extent, pelvic fins (Huntsman 1967).

SIMILAR SPECIES The quillback resembles the highfin carpsucker (*Carpiodes velifer*). They can be distinguished by the lower lip: the highfin carpsucker has a nipplelike projection at the middle of the anterior edge and the quillback does not (although Beecher [1979] noted an anomalous median lip projection on two quillbacks from the Apalachicola River, Florida). Also, the anterior rays of the highfin carpsucker's dorsal fin are considerably longer than those of the quillback, often extending nearly to the end of the dorsal fin base in the highfin but only approximately half the length of the dorsal fin base in the quillback. The quillback's snout is slightly longer than the highfin's; the quillback's mouth does not extend to the anterior edge of the eye, while the mouth does reach the anterior edge of the eye in the highfin carpsucker.

SIZE Adults are generally 30–50 cm TL (Mettee et al. 1996) but are known to reach 66 cm (Trautman 1981).

MERISTICS Dorsal rays 25–33, anal rays 7–8, pectoral rays 14–19, pelvic rays 8–10; lateral line scales 35–41; gill rakers 27–51 based on specimens collected in Tennessee (Etnier and Starnes 1993). However, meristics of quillback populations from the MSRB may differ from those of Tennessee populations.

DISTRIBUTION Distribution not continuous; on the Atlantic slope found from the St. Lawrence River and the Delaware River drainage of New Jersey and Pennsylva-

Carpiodes cyprinus, quillback, male

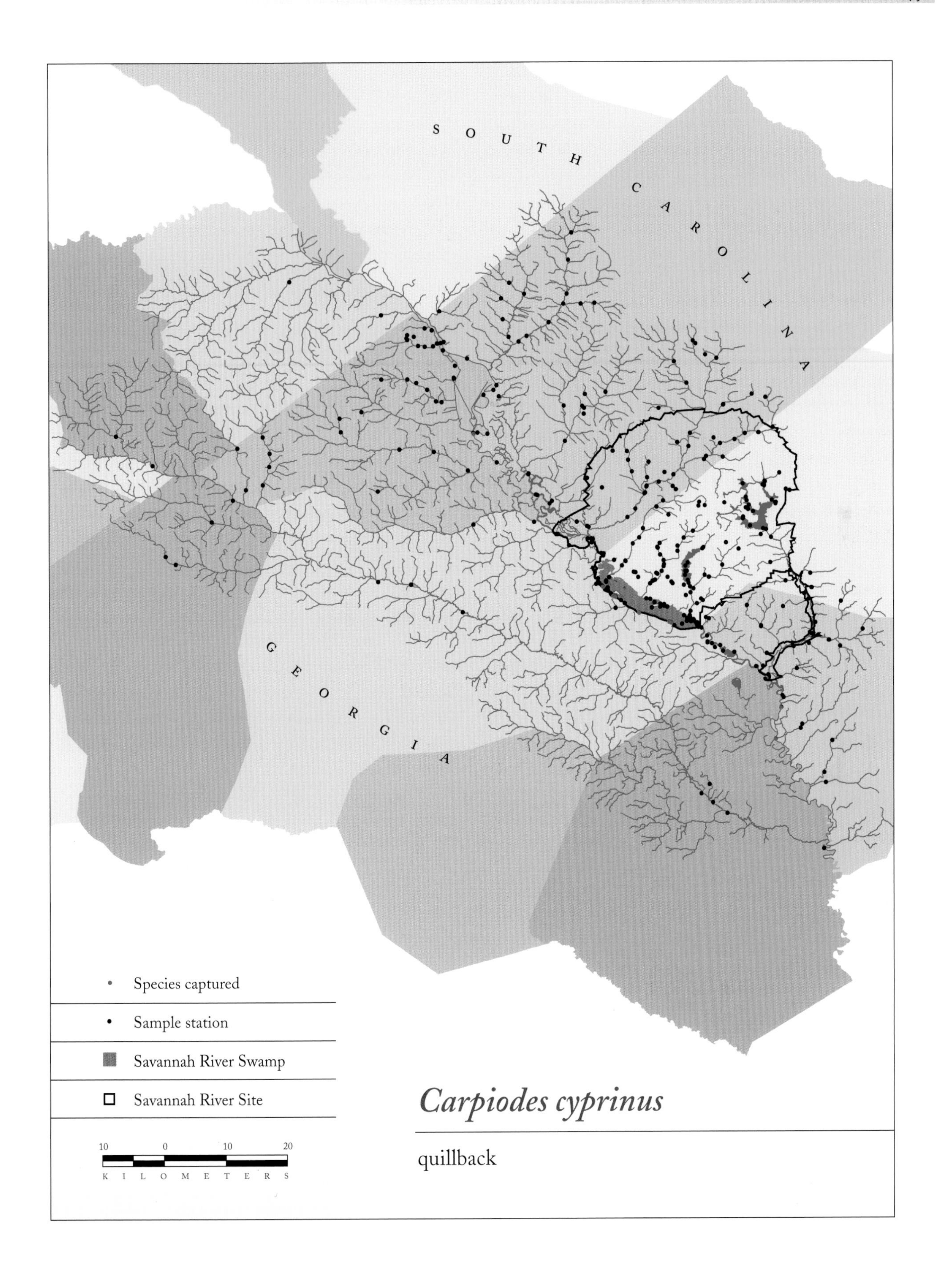

Carpiodes cyprinus

quillback

nia south to the Altamaha River of Georgia, but apparently absent from the Rappahannock and York rivers of Virginia and the Neuse and Tar rivers of North Carolina. Occurs in Gulf Coast rivers in Mobile and Apalachicola drainages; in the Mississippi, Ohio, and Missouri river basins from Louisiana north to the southern Great Lakes tributaries; and north and west into Manitoba, Saskatchewan, and Alberta (Platania and Jenkins *in* Lee et al. 1980). Found in all of the Great Lakes except Lake Superior (Cudmore-Vokey and Crossman 2000). In South Carolina, found primarily in the northern portion of the state (Rohde et al. 1994). Within the MSRB, this species appears to be largely restricted to the Savannah River.

HABITAT Typically large streams, rivers, and reservoirs; sometimes found in smaller streams with large, permanent pools (Pflieger 1975). Beecher and Hixson (1982) noted that quillbacks concentrated 500–1,000 m downstream from dams in Florida. Except when spawning, quillbacks generally prefer areas of fairly slow current, although they may move to swifter areas when water levels are low (Beecher 1980). In reservoirs, they are generally more common in littoral and offshore bottom areas than in open water during the summer (Hubert and O'Shea 1992).

BIOLOGY Spawning occurs in the spring at water temperatures ranging from 7 to 25 °C (Gale and Mohr 1978; Curry and Spacie 1984; Parker and Franzin 1991). Quillback spawning was intermittent and prolonged in a Quebec river (D'Amours et al. 2001). Pfleiger (1975) reported that quillback spawn over gravel near the ends of deep riffles or runs, although they may also spawn in slow-flowing waters over soft substrates (Robison and Buchanan 1984). Gale and Mohr (1978) reported spawning in a wide variety of habitats. In Virginia, some adults leave rivers and spawn in tributaries (Jenkins and Burkhead 1993). Spawning migrations up to 32 km upstream were observed during high-water periods in Manitoba, although migrations were much shorter (2–3 km) when discharges were low (Parker and Franzin 1991). Fecundity ranges from approximately 15,000 to 360,000 ova per female and is related to size (Woodward and Wissing 1976; Parker and Franzin 1991).

Larval quillbacks can be abundant and may constitute an important component of the assemblage of drifting larval fish in rivers (Gale and Mohr 1978; D'Amours et al. 2001). In Virginia, young spawned in tributaries may enter rivers to mature (Jenkins and Burkhead 1993). Growth in a Manitoba population was initially rapid but began to decline around age 4 (Parker and Franzin 1994). Males grew slightly faster than females in an Ohio population, where some fish were mature by age 4 and the maximum age was 11 years (Woodward and Wissing 1976).

This bottom feeder consumes aquatic insect larvae, small molluscs, and other small invertebrates along with plant material and organic debris in bottom sediments (Robison and Buchanan 1984; Rohde et al. 1994). Beecher (1980) considered it a trophic generalist that feeds primarily on small particles.

SCIENTIFIC NAME *Carpiodes* = "carplike" (Latin); *cyprinus* (Greek) alludes to the generic name of the carp.

Carpiodes velifer (Rafinesque, 1820)

Highfin carpsucker

Like that of the quillback, the highfin carpsucker's taxonomic status is uncertain. Southeastern populations, such as those in the MSRB, may be a different species from those in the interior United States (R. E. Jenkins pers. comm.).

DESCRIPTION Body deep and laterally compressed, back highly arched. Base of dorsal fin long, and anterior rays of dorsal fin much elongated. Mouth small and inferior. Lips medium to small, the anterior edge of the lower lip with a nipplelike projection. Eyes comparatively large. Scales fairly large and conspicuous.

Color generally silvery to bronze with a darker back and lighter underside. Breeding tubercles in spawning males cover the fins, most of the body, and the head except for the cheeks and opercles (Huntsman 1967).

SIMILAR SPECIES The highfin carpsucker can be confused with its close relative, the quillback. However, the quillback lacks a nipplelike projection on the anterior edge of the lower lip and has shorter anterior dorsal fin rays and a slightly longer snout (see the quillback account for more details).

SIZE The smallest of the carpsuckers; adults are generally 23–41 cm (Rohde et al. 1994; Mettee et al. 1996).

MERISTICS Dorsal rays 21–27, anal rays 7–8, pectoral rays 14–17, pelvic rays 8–10; lateral line scales 33–37; gill rakers 40–70 based on Tennessee specimens (Etnier and Starnes 1993); meristics of MSRB specimens may differ.

CONSERVATION STATUS Highfin carpsucker populations have declined in many areas as a result of siltation, pollution, and impoundment (Rohde et al. 1994). Pflieger (1975) reported this species to be less tolerant of turbidity and siltation than other carpsuckers and noted that its abundance in Missouri has declined as a result. Etnier and Starnes (1993) reported that it has likewise been adversely affected by environmental changes in Tennessee. Reduction in pollution and siltation in the Ohio River has led to an increase in highfin carpsuckers and other species intolerant of pollution (Pearson and Pearson 1989).

DISTRIBUTION Range not continuous; found in Lake Michigan drainages near Chicago and in the upper Mississippi River drainages, including the Missouri, Ohio, and Tennessee river drainages, south to the Arkansas River and Red River drainages in Oklahoma, Arkansas, and Louisiana; Gulf of Mexico drainages from the Pearl River of Mississippi and Louisiana to the Choctawhatchee

Carpiodes velifer, highfin carpsucker

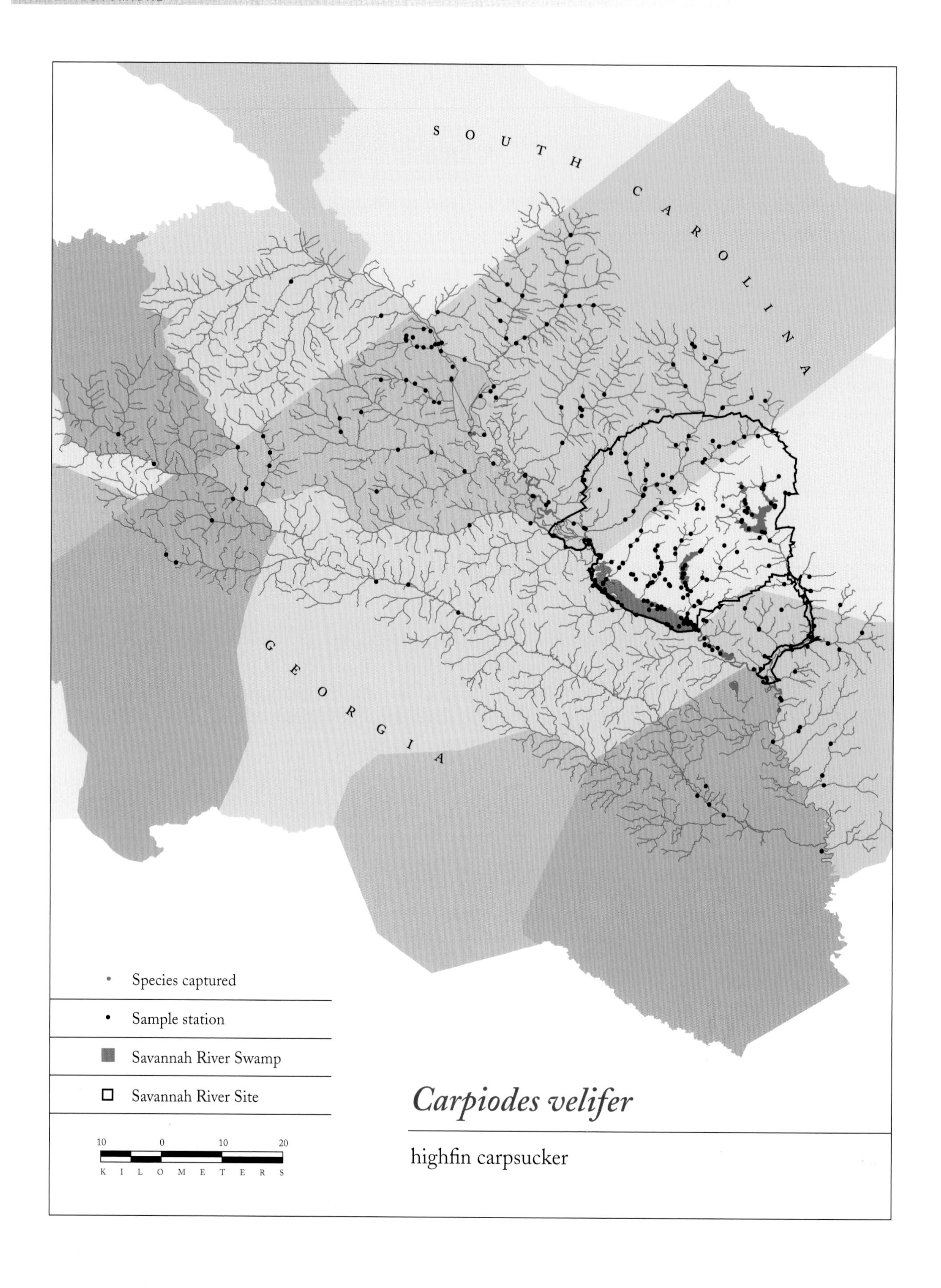

Carpiodes velifer

highfin carpsucker

drainage of Alabama and Georgia; and Atlantic slope drainages from the Altamaha River of Georgia to the Cape Fear River of North Carolina (Lee and Platania *in* Lee et al. 1980). Populations in the lower Cape Fear, Catawba, and Yadkin river drainages of North Carolina may represent a taxonomically distinct native species (R. E. Jenkins pers. comm.). Uncommon in the mid-Atlantic region (Rohde et al. 1994). This species appears to be rare in the MSRB; there are only a few records from the Savannah River.

HABITAT Relatively clear, medium to large rivers with gravel substrates (Etnier and Starnes 1993), in this respect differing somewhat from the quillback, which is sometimes found in smaller and more turbid streams. In the Missouri Ozarks, highfin carpsuckers are more common in large reservoirs than in streams (Pflieger 1975).

BIOLOGY The biology of the highfin carpsucker is poorly known. It is a schooling fish and presumably a bottom feeder like the other carpsuckers, and it has the habit of skimming the surface with its back and dorsal fin exposed and leaping clear of the water (Trautman 1981). In northwest Florida rivers, this species appears to be more abundant in October and November (Beecher and Hixson 1982).

In Missouri, the highfin carpsucker apparently spawns in the summer over deep gravel riffles (Pflieger 1975). Fecundity estimates for females collected in Ohio ranged from 41,644 to 62,355 ova (Woodward and Wissing 1976). The young closely resemble young quillbacks (Rohde et al. 1994) but grow more slowly (Vanicek 1961; Woodward and Wissing 1976). Male highfin carpsuckers in Ohio grow slightly faster than females (Woodward and Wissing 1976). The maximum life span is about 8 years (Vanicek 1961).

SCIENTIFIC NAME *Carpiodes* = "carplike" (Latin); *velifer* = "sail-bearing" (Latin), referring to the elongate rays of the dorsal fin.

Erimyzon oblongus (Mitchill, 1814)
Creek chubsucker

DESCRIPTION Small, with a chubby body and short dorsal fin. No lateral line. Mouth oblique, nearly terminal, with the rear margin of the suckerlike lips forming an acute V-shaped angle.

Young have mostly whitish sides with a continuous black midlateral stripe extending from the snout to the base of the tail. With maturation, this stripe breaks into five to eight dark blotches on an olive to brown background that grades to white on the underside. Blotches may be faded in larger individuals.

The anal fin of the male is bilobed, the membranes of the posterior lobe often highly incised; the anal fin of the female is not lobed (Jenkins and Burkhead 1993). Breeding males have three large tubercles on each side of the snout.

SIMILAR SPECIES The creek chubsucker can easily be confused with the similar lake chubsucker. The two species can be separated by counting midlateral series scales (where the lateral line would be if a lateral line were present): the creek chubsucker has 40–45, and the lake chubsucker has 34–39 lateral series scales.

SIZE Adults are generally 114–360 mm in length (Rohde et al. 1994), and the sexes are equal in size, in contrast to most other suckers, in which the females are larger (Page and Johnston 1990).

MERISTICS Dorsal rays 9–14, anal rays 7–8, pelvic rays 8–10; gill rakers 7–10 on lower limb of first arch (Jones et al. 1978); lateral series scales (39)40–43(45).

DISTRIBUTION *Erimyzon o. oblongus* is found from Maine and southern Lake Erie drainages to the Savannah River; *E. o. clavifrons* occurs in the Gulf of Mexico drainages from the Escambia River of Florida and the Chattahoochee River of Alabama west to the San Jacinto River system of Texas and north through the Mississippi Valley to southern Great Lakes drainages in Wisconsin, Illinois, Indiana, Michigan, and Ohio; individuals intermediate between these subspecies occur in the Altamaha and Ogeechee river systems of Georgia (Wall and Gilbert *in* Lee et al. 1980). The creek chubsucker occurs throughout South Carolina except for the mountainous northwestern part of the state and is common in the MSRB.

HABITAT Usually found in pools of clear, slow-flowing streams, but also occurs in some lakes and river backwaters (Wagner and Cooper 1963); often found in associa-

Erimyzon oblongus, creek chubsucker; adult and juvenile

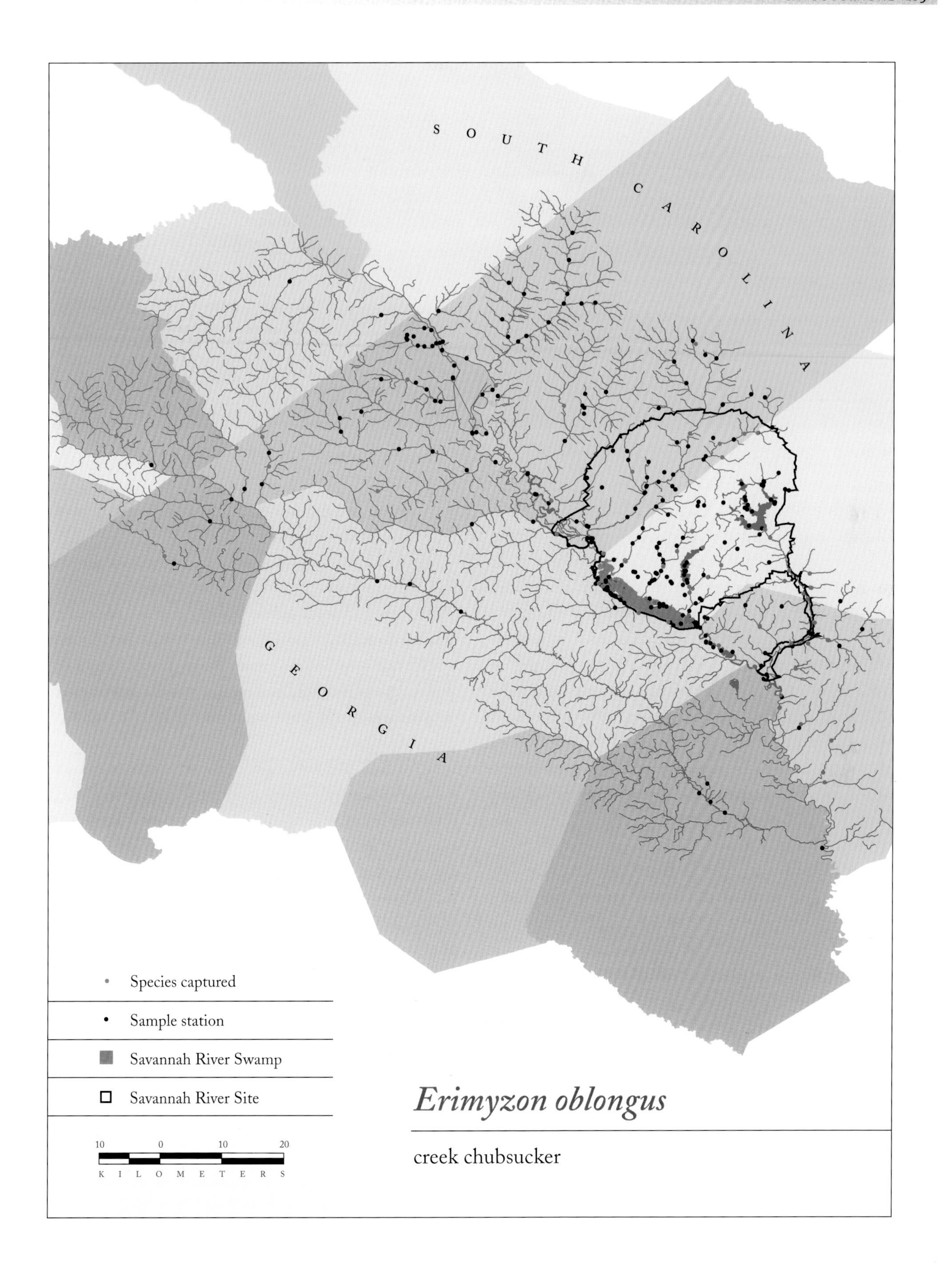

Erimyzon oblongus

creek chubsucker

Erimyzon oblongus, creek chubsucker; small juvenile

tion with soft bottoms and abundant aquatic vegetation (Rohde et al. 1994). In Virginia, most commonly found in low- to moderate-gradient streams with a variety of bottom types, and less often associated with aquatic vegetation than the lake chubsucker. Within these habitats, the young are frequently found in calm shallows and the adults in deeper pools (Jenkins and Burkhead 1993). The creek chubsucker is well represented in first–fourth-order streams on the Savannah River Site (Paller 1994) and is more typically found in pools than in runs or riffles (Meffe and Sheldon 1988). Its relatively small size enables this fish to inhabit smaller streams than the larger suckers (Page and Johnston 1990a).

BIOLOGY Spawning occurs in North Carolina from late March to late April (Carnes 1958 *in* Jenkins and Burkhead 1993), typically at water temperatures ranging from approximately 14 to 24 °C (Jenkins and Burkhead 1993). Page and Johnston (1990a) described spawning behavior. Males defend territories over gravel beds or near spawning pits dug by other fish. The large breeding tubercles on their heads play a role in aggressive interactions between territorial males. A female drifts downstream into the territory of a male and digs into the gravel with her snout to indicate her readiness to spawn. Spawning usually involves single pairs, unlike most suckers, in which there are typically two or more males for each female. No nest is prepared, and the eggs settle into the substrate downstream from the spawning pair. Jones et al. (1978) summarized descriptions of creek chubsucker larvae and their distinguishing features.

Creek chubsuckers spawned over finer substrates and in slower currents than other types of suckers in an Indiana stream, selecting fine gravels in areas 50–75 cm deep with current velocities of 0.1–0.24 m/sec (Curry and Spacie 1984). Wagner and Cooper (1963) reported that fecundity is related to size, and that the largest creek chubsuckers in a Pennsylvania lake produced an average of 29,000 eggs.

Growth is relatively rapid the first 2 years and then slows (Wagner and Cooper 1963). In Virginia, some creek chubsuckers mature by age 2 (Jenkins and Burkhead 1993). The maximum life span is approximately 7 years, and males tend to be shorter-lived than females (Wagner and Cooper 1963; Jenkins and Burkhead 1993).

Aquatic insect larvae, small crustaceans, and a variety of other small invertebrates make up the diet. This species probably feeds near the bottom like other suckers, although its nearly terminal mouth suggests that it may feed higher in the water column as well (Pflieger 1975). On the SRS, creek chubsuckers consume primarily detritus, algae, diatoms, copepods, cladocerans, and chironomid larvae (Sheldon and Meffe 1993). Larger individuals also eat small clams.

SCIENTIFIC NAME *Erimyzon* = "to suck" (Greek); *oblongus* = "oblong" (Latin), referring to the body shape.

Erimyzon sucetta (Lacépède, 1803)
Lake chubsucker

DESCRIPTION Body chubby, small to moderate in size. Scales large; dorsal fin short; no lateral line. Mouth oblique, nearly terminal, with the rear margin of the suckerlike lips forming an acute V-shaped angle.

Adults are generally brown to olive dorsally, grading to yellow or white ventrally. Young have a continuous black stripe that extends from the snout to the base of the tail; the stripe may persist in adults but often becomes indistinct. Young may have orange caudal and dorsal fins (Jenkins and Burkhead 1993).

The anal fin of the breeding male is bilobed, and membranes of the posterior lobe are often highly incised; the anal fin of the female is not lobed (Jenkins and Burkhead 1993). Breeding males have three to four large tubercles on each side of the snout and smaller tubercles on the anal fin.

SIMILAR SPECIES The lake chubsucker can be confused with the creek chubsucker. The two can be separated by counting midlateral series scales (where the lateral line would be if one were present): the lake chubsucker has 34–38, and the creek chubsucker has 40–45.

SIZE Adults are approximately 100–410 mm in TL (Rohde et al. 1994).

MERISTICS Dorsal rays 10–13, anal rays usually 7–8, pectoral rays 14–16, pelvic rays 7–8 (Robison and Buchanan 1984; Jenkins and Burkhead 1993); lateral series scales 34–38(39).

DISTRIBUTION Native to Atlantic slope drainages from southern Virginia south to Lake Okeechobee, Florida, and most abundant on the Coastal Plain; west along the Gulf Coast to the Guadalupe River of Texas, and north in the Mississippi Valley to southern tributaries of the Great Lakes and the Canadian tributaries to Lake Erie (Wall and Gilbert *in* Lee et al. 1980). Common on the Coastal Plain of South Carolina and extending into the lower Piedmont, but absent from higher-elevation areas in the northwestern part of the state (Rohde et al. 1994). Collected from many locations in the MSRB; common in Par Pond on the SRS.

HABITAT Lakes, ponds, impoundments, river backwaters, sloughs, and sluggish portions of streams (Swift et al. 1977). Commonly associated with dense aquatic vegetation, generally in clear water (Pflieger 1975; Trautman 1981). A study in Florida found this species to be less common in more eutrophic lakes (Bachman et al. 1996). Seldom found in small streams on the SRS, but sometimes occurs in the lower stream reaches and swamps on the Savannah River floodplain.

BIOLOGY Spawning occurs primarily during March and April in South Carolina (Rohde et al. 1994). Tubercu-

Erimyzon sucetta, lake chubsucker; adult and juvenile

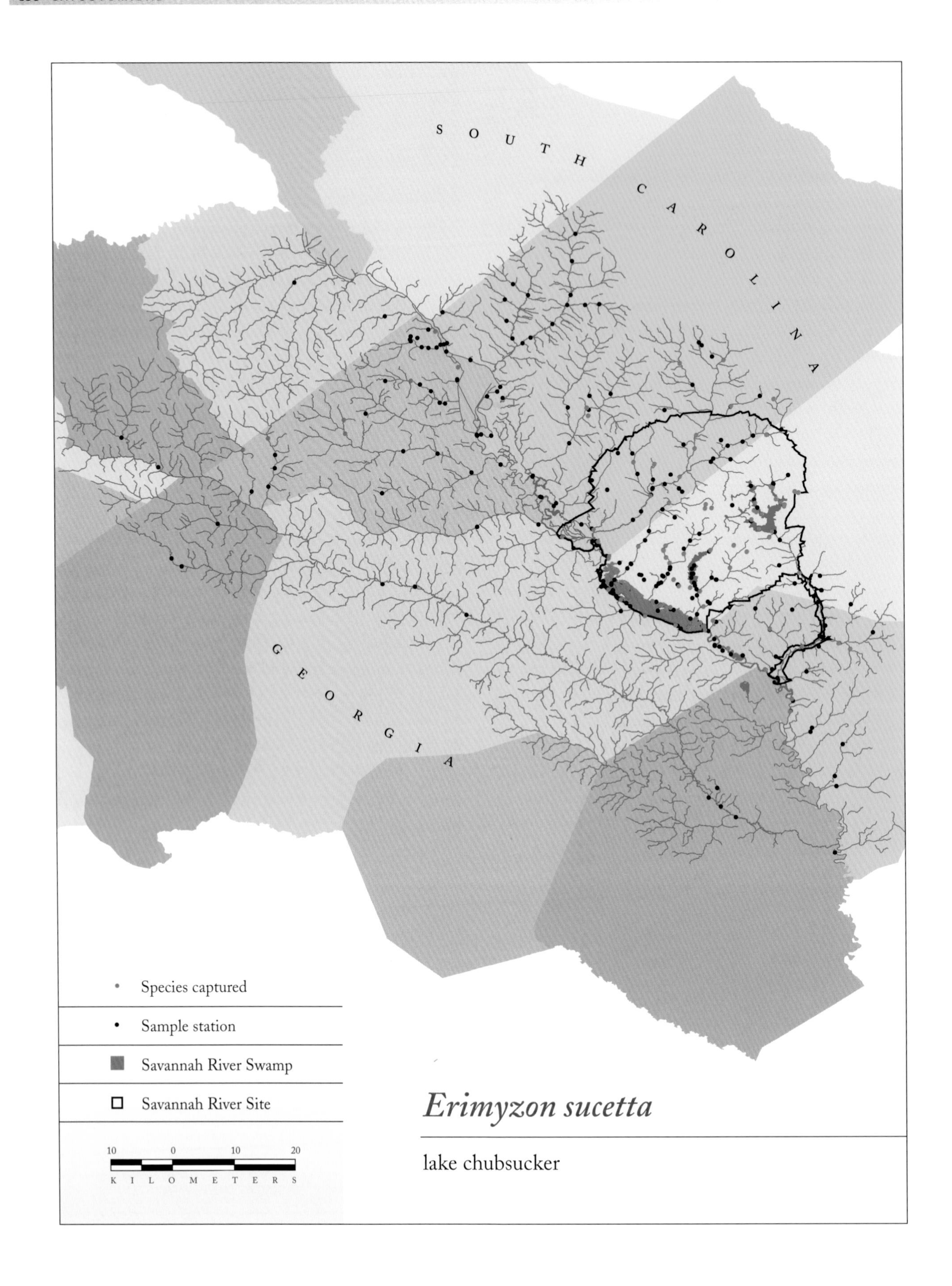

Erimyzon sucetta

lake chubsucker

Erimyzon sucetta, lake chubsucker; small juvenile

lated males and females distended with eggs have been collected as early as late February and early March in northern Florida (Swift et al. 1977). The males prepare a gravel nest within streams (Scott and Crossman 1973), but in ponds and lakes, eggs may be randomly scattered over vegetation (Pflieger 1975) or deposited in the nests of largemouth bass, *Micropterus salmoides* (Carr 1942 *in* Page and Johnston 1990a). Females in Florida ponds produced an average of 18,478 eggs (Shireman et al. 1978). The eggs hatch in 6–7 days at water temperatures of 22.2–29.4 °C (Rohde et al. 1994). The young often school with similar-sized individuals of other species that also have a dark lateral stripe (Rohde et al. 1994). Florida fish reach sexual maturity by age 3 (Shireman et al. 1978). Maximum longevity is approximately 8 years (Carlander 1969), although a Nebraska study indicated that few individuals reach age 4 (Winter 1983 *in* Jenkins and Burkhead 1993).

Lake chubsuckers between 83 and 152 mm long studied in Florida ponds fed on copepods, filamentous algae, chironomid larvae, cladocerans, ostracods, and algae (Shireman et al. 1978). The intestines of all fish in this study contained sand grains and detritus indicating bottom feeding. Jenkins and Burkhead (1993) summarized the foods of lake chubsuckers as microcrustaceans, aquatic insects, molluscs, algae, and detritus. Eberts et al. (1998) suggested lake chubsuckers as a good forage species for largemouth bass because they do not typically consume fish eggs or larvae.

SCIENTIFIC NAME *Erimyzon* = "to suck" (Greek); *sucetta* = "sucker" (from French *sucet* [Jordan and Evermann 1896]).

Hypentelium nigricans (Lesueur, 1817)
Northern hogsucker

DESCRIPTION Body slender and tapering; head large, bony, and squarish in cross section. Eyes located near the top and rear of the head, with the space between the eyes forming a shallow concavity. Snout long; protrusible mouth ventral; lips large, fleshy, and papillose. Dorsal fin short; pectoral fins relatively large and broad.

Color tan to olive, grading to white on the underside. Generally with four to six darker crossbars or mottling extending across the back and sides.

Males develop medium to large breeding tubercles on the anal fin, lower lobe of the caudal fin, and underside of the caudal peduncle; small tubercles occur elsewhere (Jenkins and Burkhead 1993).

SIZE Adults are typically 20–38 cm long (Rohde et al. 1994). Northern hogsuckers tend to be smaller in small streams, where sexual maturity may be attained at 100 mm (Etnier and Starnes 1993). The largest fish are typically females.

MERISTICS Dorsal rays 9–12, anal rays 6–8, pectoral rays 15–18; lateral line scales (42)46–50(55); gill rakers 21–26 (Pflieger 1975; Robison and Buchanan 1984; Etnier and Starnes 1993; Jenkins and Burkhead 1993).

DISTRIBUTION Widespread throughout the Mississippi and Ohio river basins, through Ontario Great Lakes drainages, and upper reaches of most Atlantic slope rivers from New York to northern Georgia (Buth and Murphy *in* Lee et al. 1980). In South Carolina, more common in the mountain and Piedmont regions in the northwestern part of the state (Rohde et al. 1994), but also found in clear streams in the upper Sand Hills ecoregion. In the MSRB, this species has been collected in a number of locations, including the Savannah River and tributary streams, although it is seldom abundant in any of them.

HABITAT Usually found in medium to large streams with clear water, substantial flow, and gravel or cobble bottoms, but occurs sometimes over finer substrates as well. Runs, deep riffles, and pools with substantial current are apparently preferred (Robison and Buchanan 1984), but northern hogsuckers are observed about equally in slow, moderate, and fast currents in Virginia (Jenkins and Burkhead 1993). In SRS streams, this species prefers relatively deep and wide sites with medium to strong currents (Meffe and Sheldon 1988). Northern hogsuckers are relatively intolerant of siltation, pollution, and other types of an-

Hypentelium nigricans, northern hogsucker

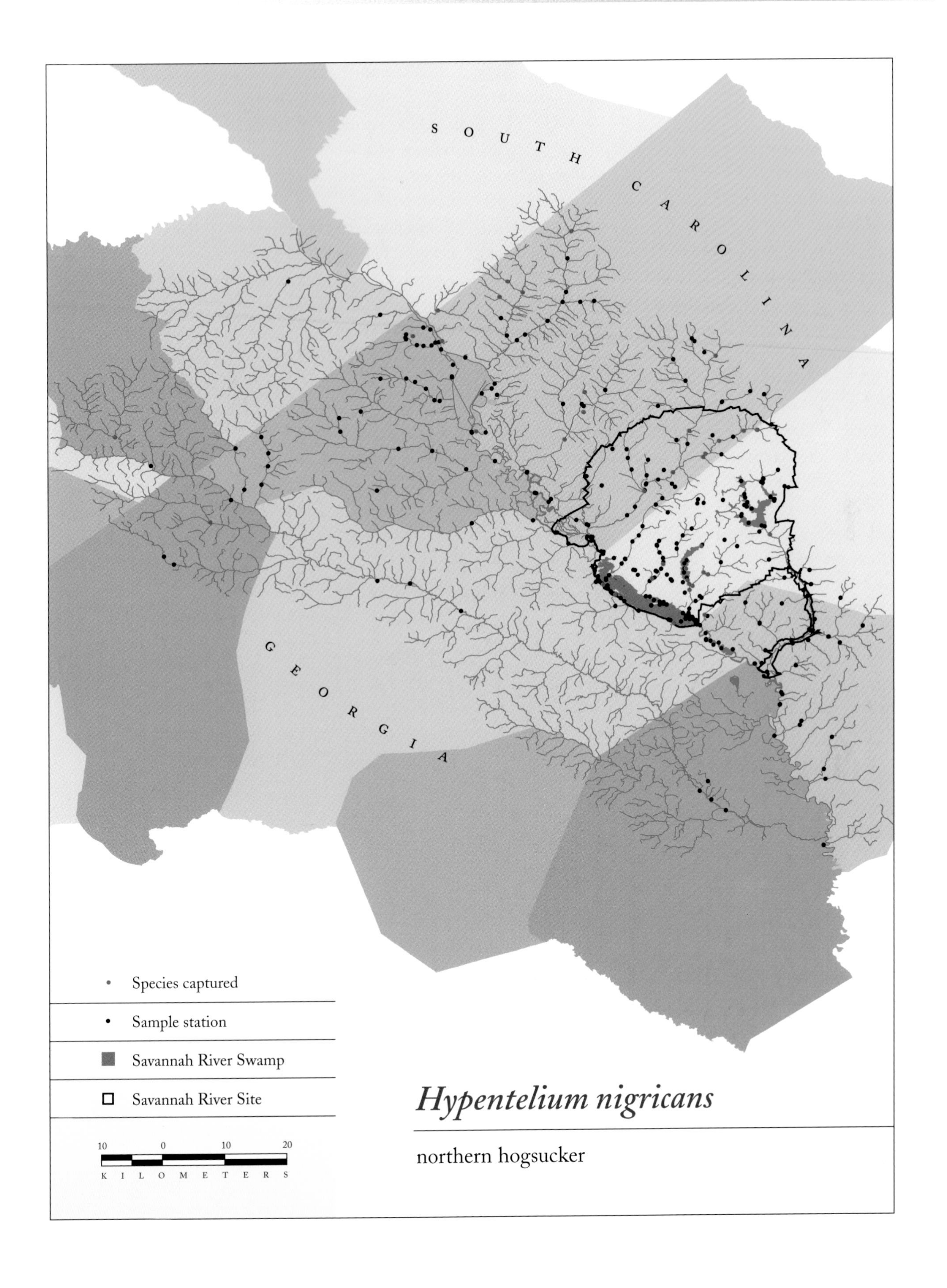

Hypentelium nigricans

northern hogsucker

thropogenic disturbance (Smith 1979) and may be a good sentinel species for assessing environmental perturbation.

BIOLOGY Spawning occurs during April and May at water temperatures of 11–23 °C, often in shallow riffles at the ends of pools (Jenkins and Burkhead 1993). Curry and Spacie (1984) reported that northern hogsuckers in an Indiana stream preferred to spawn over medium gravel in riffles at a depth of 30–60 cm and current velocities of 0.4–0.9 m/sec. Jenkins and Burkhead (1993) reported that spawning adults form a mobile troupe consisting of a single relatively large female and 3–11 males. This type of arrangement is unusual in suckers, which generally spawn in trios consisting of 2 males and a single female. No nest is built, but the gravel may be swept clean by the vigorous swimming of the spawning fish (Etnier and Starnes 1993). The nonadhesive eggs settle onto the substrate and are abandoned by the parents (Robison and Buchanan 1984; Etnier and Starnes 1993). Males generally reach sexual maturity in 2 years and females in 3; the life span is up to 10 years (Raney and Lachner 1946 *in* Jenkins and Burkhead 1993).

This unusual sucker is well adapted for life on the bottom in relatively fast waters. Its large head, sloping snout, and large pectoral fins enable it to hold a position on the bottom with relatively little expenditure of energy, and its mottled coloration blends well with rocks, gravel, and bottom debris.

The northern hogsucker is an active feeder that roots into the bottom in search of insect larvae, small crustaceans, other benthic invertebrates, and algae (Pflieger 1975; Jenkins and Burkhead 1993). In its search for edible material it vacuums up large amounts of sand and detritus, which it voids through the gill openings (Rohde et al. 1994). It may overturn small stones in its search for food, and it scrapes attached algae and associated invertebrates from rocks. These activities may attract minnows and other fishes to exposed invertebrates (Pflieger 1975). The northern hogsucker is more likely to be found singly or in small groups than in the large schools characteristic of many other suckers (Pflieger 1975).

SCIENTIFIC NAME *Hypentelium* = "five lobed" (Latin), referring to the lips; *nigricans* = "blackish" (Latin).

Minytrema melanops (Rafinesque, 1820)
Spotted sucker

DESCRIPTION Body medium sized and moderately slender. Parallel rows of dark spots along the sides. Mouth subterminal and horizontal, with lips forming an acute V-shaped angle. Lateral line absent or incomplete (developed on only a few scales).

Body generally olive to brown turning to brassy or yellow on the sides and silvery to white below. Breeding males may have a grayish pink streak above a dark lateral stripe and another dark area with lavender highlights above that (McSwain and Gennings 1972; Etnier and Starnes 1993).

Spawning males have medium to large tubercles on the snout, cheek, anal fin, and caudal fin (McSwain and Gennings 1972; Etnier and Starnes 1993).

SIZE Adults are 150–495 mm TL (Trautman 1981).

MERISTICS Dorsal rays 10–13, anal rays 7, pectoral rays 16–18, pelvic rays 9–10; lateral scale series 42–46; gill rakers 20–26 (Robison and Buchanan 1984; Etnier and Starnes 1993).

DISTRIBUTION The lower Great Lakes basin (Lakes Erie, Huron, and Michigan); the upper Mississippi Valley of Minnesota, Wisconsin, and Iowa, south to Louisiana; west along the Gulf Coast to the Colorado River of Texas and east to the Suwannee River drainage of Florida; north on the Atlantic slope to the Cape Fear drainage of North Carolina (Gilbert and Burgess *in* Lee et al. 1980). In South Carolina, common throughout most of the state except the mountainous northern areas and the Lower Coastal Plain (Rohde et al. 1994). This species has been collected at many locations in the MSRB. It is one of the most abundant of the larger fishes found in the middle portion of the Savannah River and is common in larger streams on the SRS. Small individuals are sometimes found in smaller streams.

HABITAT Low-gradient rivers and large creeks with silt, sand, gravel, or clay bottoms, typically in pools, backwaters, and channel margins, especially near sunken timber; seldom found in lakes (Swift et al. 1977). Prefers submerged aquatic vegetation, lots of detritus, soft substrates, and quiet backwaters away from current (Robison and Buchanan 1984). Spotted suckers studied in the upper Mississippi River actively selected habitats with submerged snags for foraging or protection (Lehtinen et al.

Minytrema melanops, spotted sucker; adult

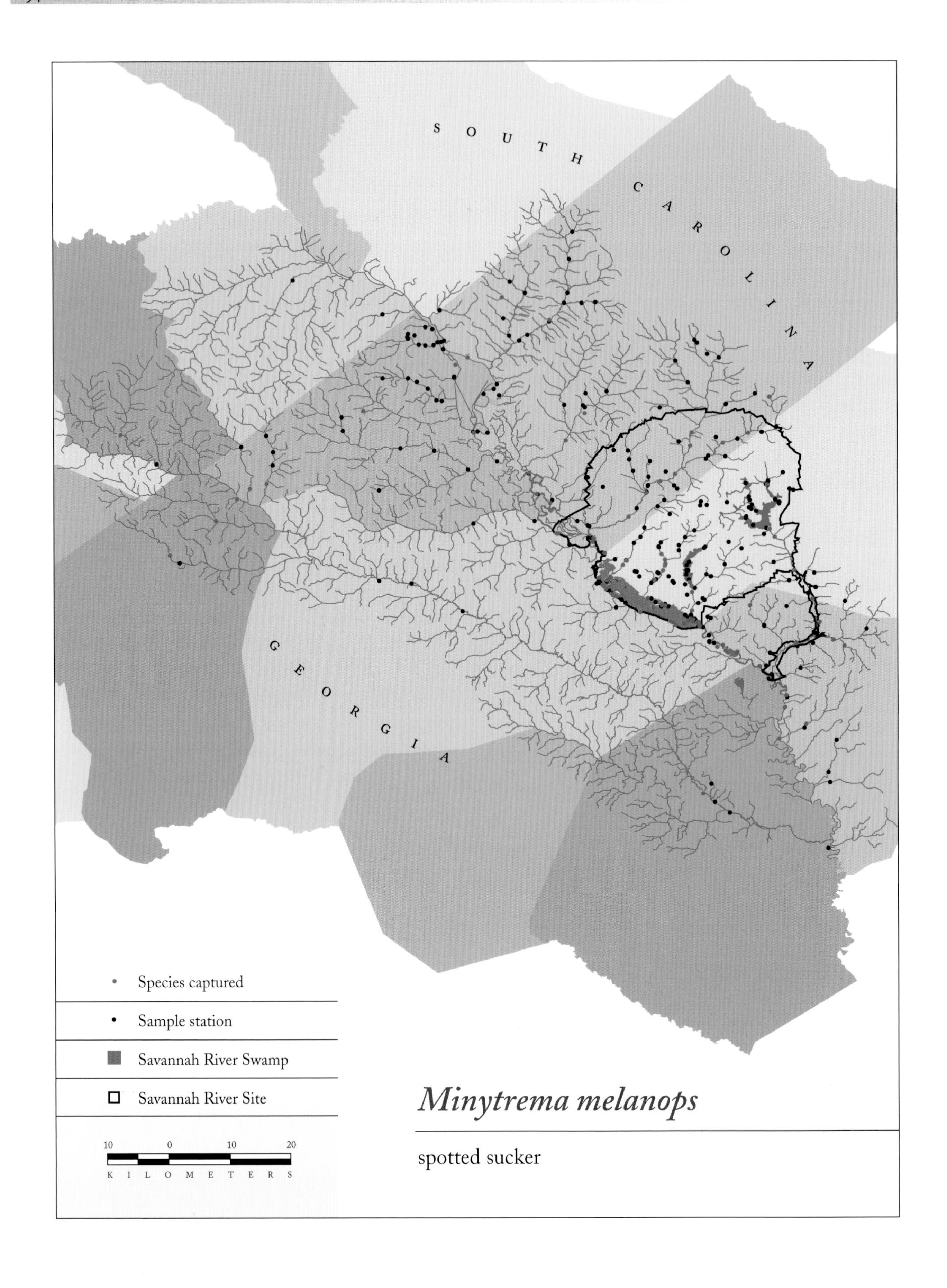

Minytrema melanops

spotted sucker

Minytrema melanops, spotted sucker; juvenile

1997); Florida fish showed no clear preference between oxbow and river habitats (Beecher et al. 1977).

BIOLOGY In Georgia, spawning occurs from mid-March to early May at water temperatures between 12.2 and 19.4 °C (McSwain and Gennings 1972). Males actively defend breeding territories over coarse rubble in riffles less than 0.5 m deep with moderate current. Prespawning courtship activities involve bumping and prodding of the female by the male. Spawning usually involves two males and one female, with the female positioned between the males. The demersal, adhesive eggs hatched in 108–156 hours at 16.1–20.0 °C (Robison and Buchanan 1984). Larval spotted suckers may form schools with other species when feeding (White and Hagg 1977). Sexual maturity is usually attained at 3 years in Missouri (Pflieger 1975), and the maximum life span is at least 6 years (Rohde et al. 1994).

The larvae are selective mid-water zooplanktivores, but juveniles and adults become generalist bottom feeders (White and Hagg 1977). Adults consume copepods, cladocerans, chironomids, organic detritus, and sand, feeding primarily at dawn and dusk. The diet shifts seasonally, and they may subsist largely on organic detritus during the winter. Finely spaced gill rakers make spotted suckers efficient at selecting and concentrating small food particles in quiet waters (White and Hagg 1977). In the Satilla River of Georgia, spotted suckers are largely insectivores with a preference for chironimid larvae (Coomer et al. 1977; Henry 1979). This species generally consumes smaller aquatic insects than are eaten by sunfishes (Henry 1979).

SCIENTIFIC NAME *Minytrema* = "reduced aperture" (Greek), referring to the reduced number of lateral line pores; *melanops* = "black in appearance" (Greek).

Moxostoma collapsum (Cope, 1870)
Notchlip redhorse

The notchlip redhorse is a recently recognized southeastern Atlantic slope species that was formerly subsumed under the silver redhorse, *Moxostoma anisurum* (R. E. Jenkins pers. comm.). Therefore, all previous accounts of southeastern Atlantic slope populations of silver redhorse actually describe the notchlip redhorse. The following information was taken mostly from accounts of the silver redhorse that, based on location, would now be classified as notchlip redhorse.

DESCRIPTION Body robust to elongate; scales large; head moderate in size; lateral line complete. Mouth horizontal; lips fleshy with many fine ridges, most of which are finely divided. Lower lip bilobed, its rear margin forming an acute V-shaped angle. Pharyngeal teeth comblike. Swimbladder with three chambers (R. E. Jenkins pers. comm.).

Back generally tan or olive grading to silvery or brassy on the sides and to white on the undersurface. Dorsal and caudal fins generally olive to light gray; lower fins light gray to white, usually with an orange or pink tint (Jenkins and Burkhead 1993).

Spawning males have well-developed tubercles on the anal fin and ventral lobe of the caudal fin. Spawning males and females usually have a dusky to blackish mid-side stripe (R. E. Jenkins pers. comm.).

SIMILAR SPECIES Several features separate the notchlip redhorse from other large suckers of similar appearance in the MSRB. Its V-shaped lower lip distinguishes it from the brassy jumprock and the robust redhorse, both of which have a lower lip with a straight or nearly straight edge. The notchlip redhorse can also be distinguished from the brassy jumprock by its dorsal fin with 14–16 rays (versus 12–13) (Rohde et al. 1994). Its complete lateral line and lack of rows of spots distinguish it from the spotted sucker. Like the robust redhorse, the notchlip redhorse generally has 12 scales around the caudal peduncle (versus 16 in the brassy jumprock and spotted sucker) (R. E. Jenkins pers. comm.)

SIZE Adults are approximately 250–420 mm long (Rohde et al. 1994), although some may exceed 500 mm (Jenkins and Burkhead 1993).

MERISTICS Dorsal rays (12)14–15(17), caudal rays (17)18 (19), anal rays 7, pelvic rays (8)9(10), pectoral rays (16)17–19(20); lateral line scales (38)40–42(46); circum peduncle 12(15) (Jenkins and Burkhead 1993).

Moxostoma collapsum, notchlip redhorse

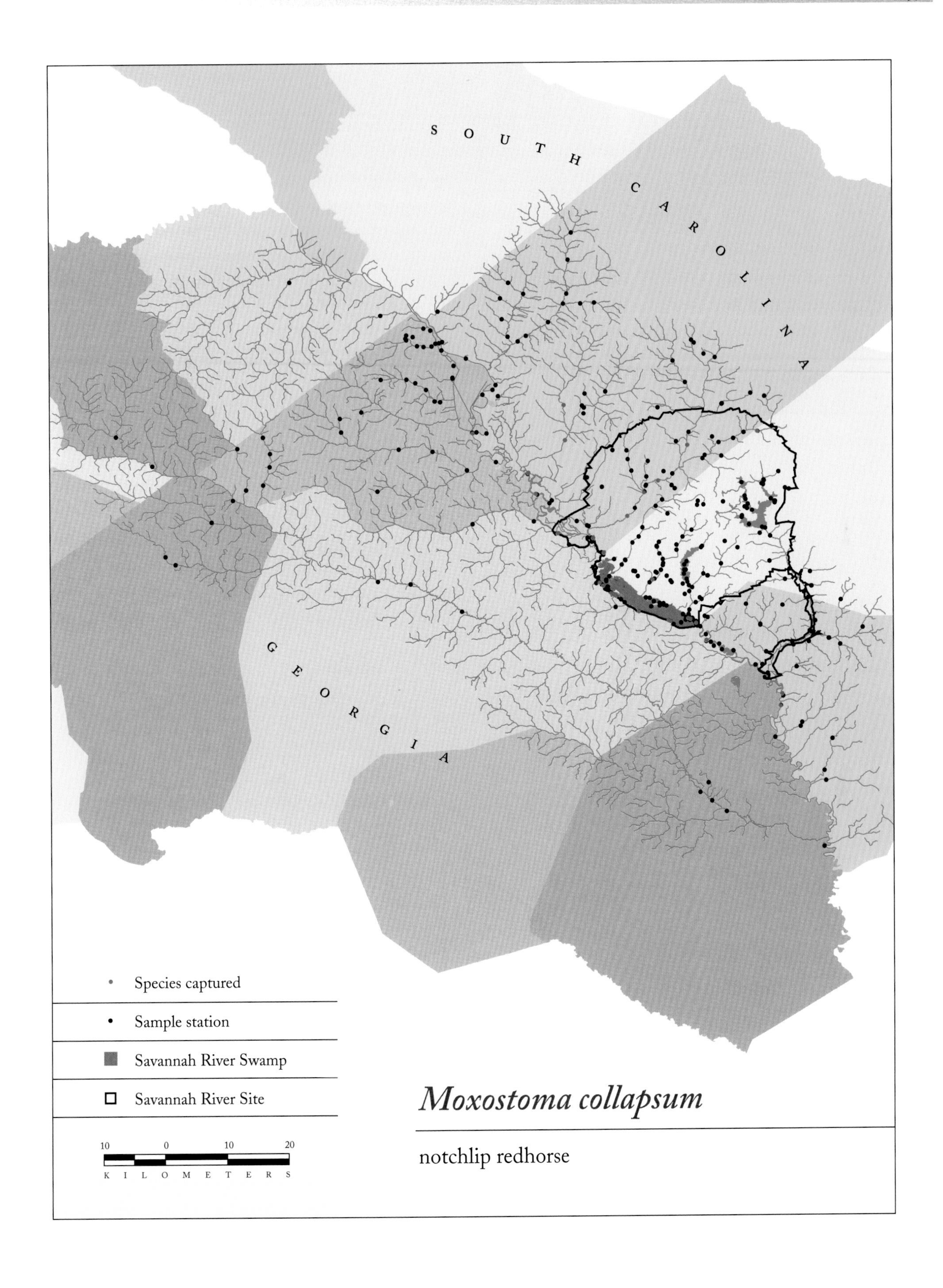

Moxostoma collapsum

notchlip redhorse

DISTRIBUTION Southeastern Atlantic slope drainages from Virginia to Georgia (R. E. Jenkins pers. comm.). In South Carolina, most common in the Piedmont region (Rohde et al. 1994). In the MSRB, this species has been collected from the Savannah River and its larger tributaries.

HABITAT Typically rivers and larger streams, but also medium-sized streams of moderate gradient, and natural and artificial lakes (Jenkins and Burkhead 1993). Usually found in deeper pools over silt, sand, gravel, or rock bottoms, although migrating adults may occur in runs and riffles. Juveniles often inhabit small pools and relatively shallow backwaters (Jenkins and Burkhead 1993).

BIOLOGY Although this is a fairly common species, its biology is poorly known (R. E. Jenkins pers. comm.). In Virginia, the notchlip redhorse spawns in groups during mid-April to mid-May in runs and riffles with gravel and small cobble substrates (R. E. Jenkins pers. comm.). If its diet is like that of the silver redhorse, the notchlip redhorse probably subsists largely on immature aquatic insects, microcrustaceans, other small benthic invertebrates, algae, and detritus (Pflieger 1975; Etnier and Starnes 1993; Jenkins and Burkhead 1993).

SCIENTIFIC NAME *Moxostoma* = "sucking mouth" (Greek); *collapsum* = "flattened sideways" (Latin).

Moxostoma robustum (Cope, 1870)
Robust redhorse

DESCRIPTION The largest and heaviest sucker found in the MSRB. Body thick; scales large; snout blunt; mouth horizontal with thick fleshy lips and molariform pharyngeal teeth.

Upper body bronze or coppery golden, lower sides brassy mixed with dusky, grading to white ventrally. Caudal fin reddish, brightly so in juveniles and small adults. Adult males develop medium to large tubercles on the head during the spawning season (R. E. Jenkins pers. comm.).

SIMILAR SPECIES The robust redhorse can be distinguished from the notchlip redhorse by the posterior margin of its lower lip, which is relatively straight and medially sometimes has a flaplike posterior extension (versus the V-shaped lip of the notchlip redhorse); and from the brassy jumprock by the number of scales around the caudal peduncle (12 in the robust redhorse versus 15–16 in the brassy jumprock) (R. E. Jenkins pers. comm.). Any sucker in the MSRB that exceeds approximately 2.3–2.7 kg is likely to be a robust redhorse.

SIZE The average adult size is approximately 63 cm and 4.1 kg, but some exceed 70 cm and weigh up to 8 kg.

MERISTICS Dorsal rays 12–13, pectoral rays 15–17, pelvic rays usually 10; lateral line scales 40–43, circumbody scales 28–32, caudal peduncle scales 12 (R. E. Jenkins pers. comm.).

CONSERVATION STATUS The robust redhorse was originally described in 1870. Subsequently the name *robustum* was mistakenly applied to the undescribed brassy jumprock, and the true *Moxostoma robustum* was not definitively reidentified until 1992, when it was documented in the Pee Dee, Savannah, and Oconee rivers (R. E. Jenkins pers. comm.). Further study indicated that robust redhorse populations consisted largely of old individuals with little evidence of substantial recruitment (Looney 1998). The robust redhorse is listed as endangered by Georgia and is now the subject of a multiagency recovery effort, which has resulted in the artificial propagation and restocking of juvenile fish in Georgia and South Carolina rivers (DeMeo 2001). Primary threats to this species are habitat loss due to impoundment, siltation, and other types of alteration. Laboratory studies have demonstrated that fine sediment particles that settle in gravel can entrap eggs and larvae and suffocate them (Dilts and Jennings 1999). Predation on the young by introduced flathead and blue catfish may also pose a threat (DeMeo 2001). These problems are compounded by the limited geographical range of this species and its current low numbers.

DISTRIBUTION Historically, the Atlantic slope drainages from the Pee Dee River in North Carolina to the Altamaha River in Georgia (DeMeo 2001). Currently, native populations are known to occur in the Oconee River between Sinclair Dam and Dublin, Georgia; in a short

Moxostoma robustum, robust redhorse; male

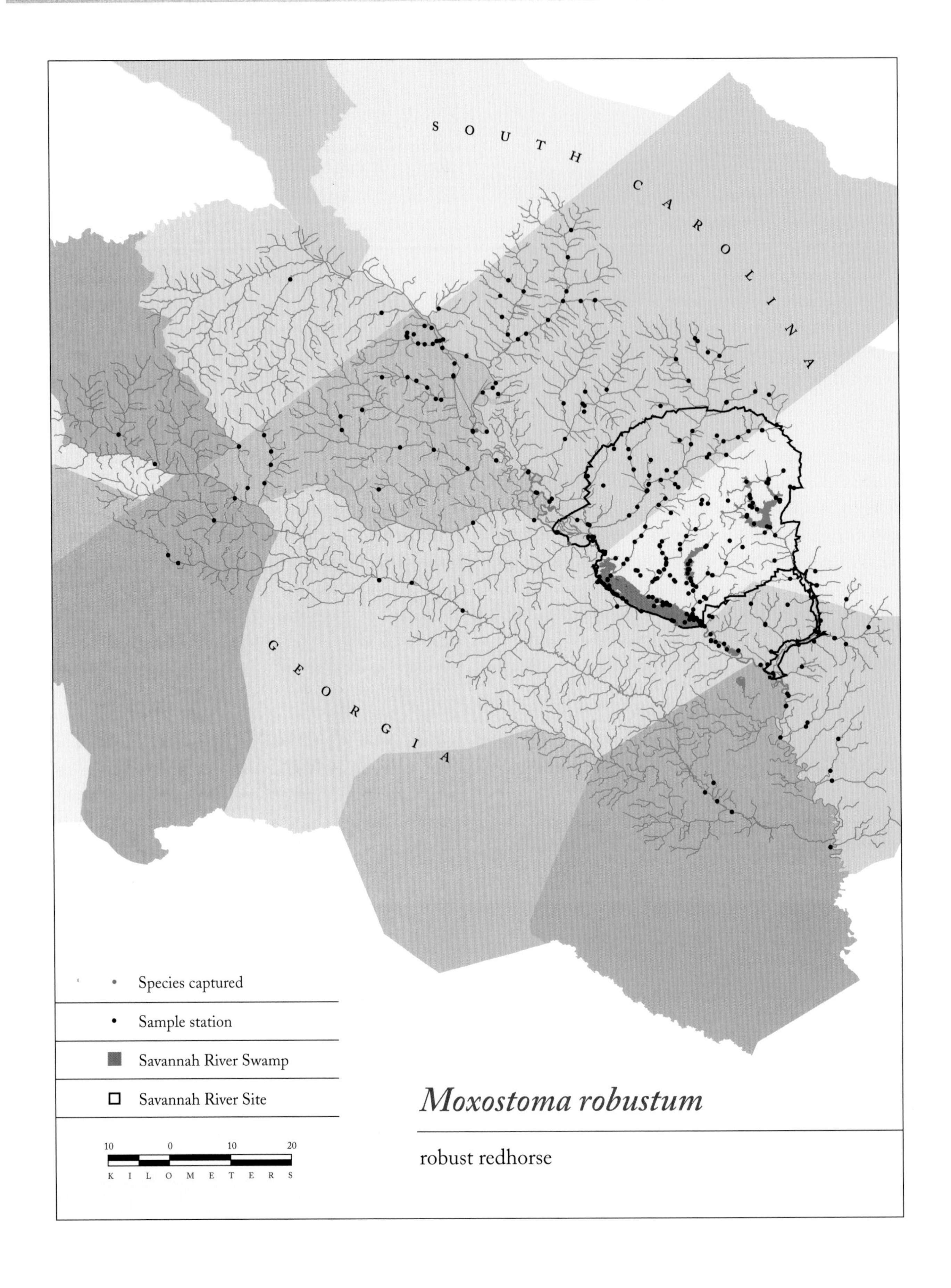

Moxostoma robustum

robust redhorse

Moxostoma robustum, robust redhorse; juvenile

Upper Coastal Plain segment of the Ocmulgee River in Georgia; and in the Savannah River from the Augusta Shoals to far downstream in the Coastal Plain (Barret 2000; DeMeo 2001; R. E. Jenkins pers. comm.). A single specimen was collected from the Pee Dee River in North Carolina (DeMeo 2001). It is possible that small numbers will also be found in other areas as more surveys are conducted that target the habitats preferred by this species.

HABITAT Habitat information is taken largely from the robust redhorse website (www.robustredhorse.com) and from DeMeo 2001. This species occurs primarily in Piedmont and Upper Coastal Plain sections of large rivers. Piedmont reaches are often along the Fall Line and are usually characterized by rocky shoals, outcrops, and pools. Upper Coastal Plain reaches usually have sandy bottoms interspersed with shoals and occasional gravel beds. Nonspawning adults often prefer relatively deep, moderately swift water near outside river bends, often in association with sunken logs, fallen trees, and other woody debris. The presence of juveniles (<40 cm) in Clarks Hill/J. Strom Thurmond Lake on the Savannah River suggests that the robust redhorse is tolerant of lentic habitats for at least a portion of its life cycle. These fish were originally stocked into the Broad River and moved downstream (DeMeo 2001).

BIOLOGY Gravel bars found downstream from the New Savannah Bluff Lock and Dam support substantial numbers of spawning robust redhorse (T. Jones pers. comm.). Spawning was observed by Freeman and Freeman (2001) in this area and in the Oconee River, and the following description refers to those populations. Spawning occurs during late April through late May at temperatures ranging from 17 to 26.7 °C. The preferred spawning habitat includes water depths between approximately 0.3 and 1.1 m, current velocities between 0.26 and 0.67 m/sec, and a substrate dominated by coarse gravel with minimal fine particles and sufficient intragravel flow to aerate the eggs and larvae. Spawning males typically break the surface over the spawning area and occasionally butt each other. Females move into the spawning area from deeper, slower upstream reaches and align with a male. A second male joins the pair, forming a triad with the males flanking the female. The female returns to the upstream pool when spawning is completed.

The robust redhorse is a long-lived fish that can reach an age of 27 years and take 5–6 years to attain sexual maturity (R. E. Jenkins pers. comm.).

The primary food of adults is bivalve molluscs, which are crushed with the heavy molariform pharyngeal teeth. Specimens examined from the Oconee River had consumed the introduced Asian clam, *Corbicula fluminea*. No other species of sucker in the MSRB has similar feeding habits.

SCIENTIFIC NAME *Moxostoma* = "sucking mouth" (Greek); *robustum* = "full-bodied" (Latin).

Scartomyzon sp.
Brassy jumprock

DESCRIPTION Except where otherwise noted, the following description of the poorly known brassy jumprock summarizes information from Jenkins and Burkhead 1993, in which it is described under the common name "smallfin redhorse" (Jenkins and Burkhead 1993, p. 491fn). Body moderate or slender, nearly round in cross section. Head and eyes of moderate size; area between the eyes convex. Mouth subterminal with moderately large, plicate (grooved) lips. Rear margin of the lower lip straight or nearly straight. Swimbladder with three chambers; lateral line complete; outer margin of dorsal fin slightly concave or straight.

Back olive, sides brassy grading to pearly underneath. Median fins are dusky, and paired fins have a pale orange cast.

SIMILAR SPECIES In the MSRB, it is possible to confuse the brassy jumprock with the spotted sucker, notchlip redhorse, and robust redhorse. The complete lateral line and lack of rows of spots on the sides of the body distinguish it from the spotted sucker. It can be distinguished from the notchlip redhorse by its relatively straight rather than V-shaped lower lip and by having 11–13 rather than 14–16 dorsal rays. The robust redhorse is larger (2.3 kg versus 1.4 kg) and has 12 rather than 15–16 scales around the caudal peduncle.

SIZE Adults are 20–36 cm SL.

MERISTICS Dorsal rays 11–13, anal rays 7, pectoral rays 15–18, pelvic rays 9, caudal rays 18; lateral line scales 40–45; gill rakers 27–37 (Jenkins and Burkhead 1993).

DISTRIBUTION From the Cape Fear drainage in North Carolina to the Altamaha River drainage in Georgia. Within the MSRB, it has been collected in only two locations in the Savannah River.

HABITAT Medium-sized streams to large rivers and reservoirs in the Pee Dee drainage. Jenkins and Burkhead (1993) collected brassy jumprock in sometimes turbid, moderate-gradient creeks from 5 to 10 m wide with varied substrates, gravel and rubble being most common in Virginia. Large juveniles and adults inhabit pools and gentle runs, and small juveniles inhabit backwaters.

BIOLOGY Very little is known about the biology. Jenkins and Burkhead (1993) indicated that spawning probably occurs in mid-to-late spring. North Carolina fish were in spawning condition in April and in postspawning condition in May. Gatz (1979 *in* Jenkins and Burkhead 1993) noted the presence of aquatic insects and sand in the guts of small brassy jumprocks.

SCIENTIFIC NAME *Scartomyzon* = "quick or agile sucker" (Greek).

Scartomyzon sp., brassy jumprock, male

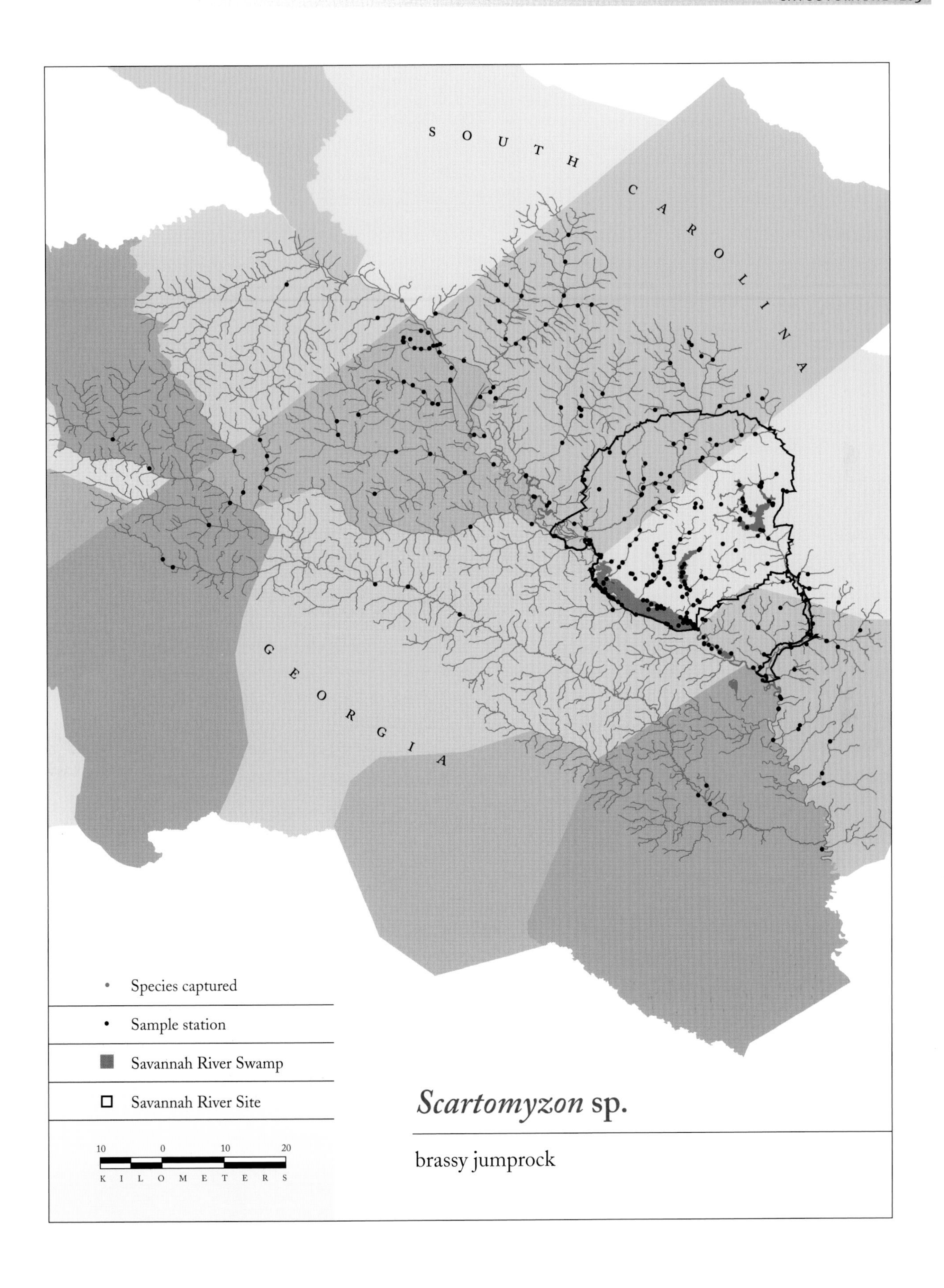

Scartomyzon sp.

brassy jumprock

Ictaluridae (bullheads, catfishes, and madtoms)

All 38–46 species in the seven genera of bullhead catfishes are endemic to North America. The 10 species found in the MSRB include the small madtoms (*Noturus*) as well as species that can reach more than 1.5 m in length such as the channel catfish (*Ictalurus punctatus*). The name "catfish" supposedly comes from the conspicuous facial barbels that vaguely resemble a cat's whiskers. Besides the eight barbels, bullhead catfishes are easily recognized by the lack of scales, the inferior or terminal mouth, the stout spine on the dorsal and pectoral fins, and the presence of an adipose fin that may have a free distal end or may be more or less connected to the caudal fin. In some species, the spines, especially the dorsal ones, produce puncture wounds, with pain caused by a toxin released from glands at the base of the spines.

Ictalurids live in a variety of habitats from large rivers and turbid lakes to swamps and small, clear creeks. Most species are nocturnal, but some forage during the daylight hours. Most ictalurids are opportunistic foragers and consume a variety of organisms. An excellent sense of smell is important for foraging and may also play an important role in the relatively well developed social behavior of some species (Etnier and Starnes 1993). Breeding males of many species develop swollen muscles on the dorsal surface of the head and nape. All species build nests where courtship takes place and the eggs are deposited. The adhesive, demersal eggs are about 2–6 mm in diameter (Jenkins and Burkhead 1993). The eggs are protected and cared for, mostly by the males, but females or both parents are caretakers in some species. The life span varies; smaller species like the madtoms live 4–6 years; the larger species may survive 15 years or more (Jenkins and Burkhead 1993). The catfishes are among the most important families in freshwater fisheries—both recreational and commercial—and in aquaculture, and have been widely introduced outside their native ranges for that reason.

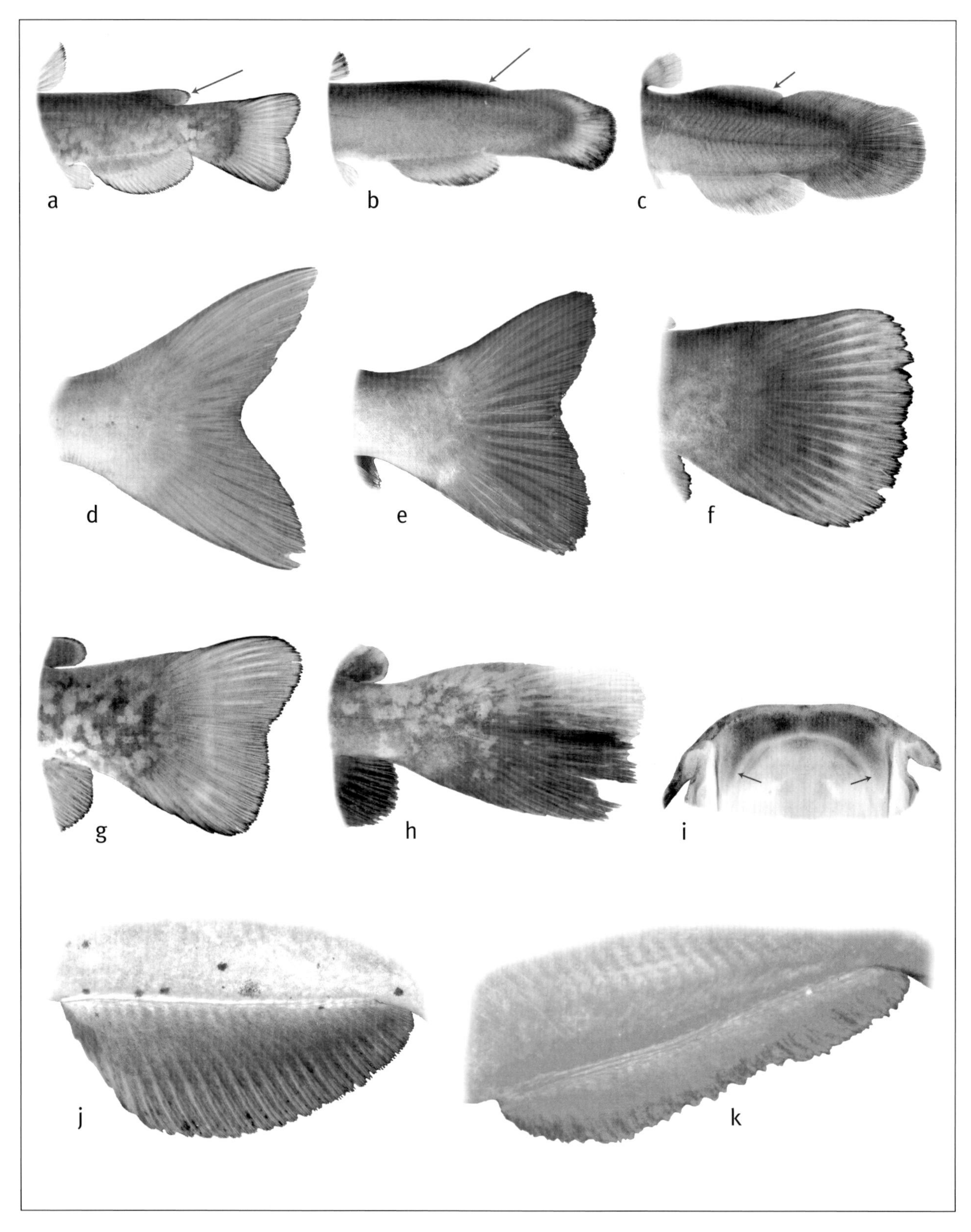

PLATE 13. Morphological characters of Ictaluridae: (a) separate adipose fin, (b) adipose fin joined to caudal fin, (c) adipose fin and caudal fins joined but with a notch, (d) forked caudal fin with nearly equal lobes, (e) forked caudal fin with lower lobe larger, (f) truncate caudal fin, (g) emarginate caudal fin, (h) flathead catfish caudal fin color pattern, (i) palatine teeth of flathead catfish, (j) anal fin with rounded margin, (k) anal fin with straight margin.

PLATE 14. Morphological characters of Ictaluridae: (a) yellow bullhead head showing small eye size, (b) bullhead head showing larger eye size, (c) bullhead dorsal fin lacking a basal dark blotch, (d) bullhead dorsal fin showing a dark basal blotch, (e) flat bullhead premaxillary tooth patch, (f) snail bullhead premaxillary tooth patch, (g) tadpole madtom head showing both jaws equal in length, (h) madtom head showing unequal jaw lengths.

KEY TO THE SPECIES OF ICTALURIDAE

1a. Adipose fin with short base and free posterior margin, not connected to caudal fin (Plate 13 a) **2**
1b. Adipose fin with long base and posterior margin joined to caudal fin (Plate 13 b, c) **9**

[1] **2a.** Caudal fin deeply forked and always symmetrical (Plate 13 d); anal fin rays 24–35. . . . . **3**
2b. Caudal fin emarginate and symmetrical or asymmetrical, or moderately forked and asymmetrical (Plate 13 e–h); anal fin rays (13)16–27(28). **4**

[2] **3a.** Outer margin of anal fin rounded (Plate 13 j); anal fin rays usually 24–29; swimbladder with a single lobe (requires opening the fish) . . . . *Ictalurus punctatus* (channel catfish), p. 227
3b. Outer margin of anal fin straight (Plate 13 k); anal fin rays usually 30–35; swimbladder with 2 lobes (see figure in Pflieger 1975) (requires opening the fish) . . . . *Ictalurus furcatus* (blue catfish), p. 224

[2] **4a.** Lower jaw extending beyond upper jaw; upper tip of caudal fin lighter in color than rest of fin (Plate 13 h); premaxillary tooth patch on upper jaw with lateral backward extensions (Plate 13 i) . . . . *Pylodictis olivaris* (flathead catfish), p. 238
4b. Lower jaw not extending beyond upper jaw; upper tip of caudal fin no lighter in color than rest of fin; premaxillary tooth patch on upper jaw without lateral backward extensions **5**

[4] **5a.** Caudal fin moderately forked and asymmetrical with the lower lobe markedly larger (Plate 13 e); body color plain, unmarked; lower (chin) barbels light colored, white . . . . *Ameiurus catus* (white catfish), p. 212
5b. Caudal fin emarginate and symmetrical or only slightly asymmetrical (Plate 13 f, g); body color mottled or plain; lower (chin) barbels light or dark **6**

[5] **6a.** Small eyes (Plate 14 a); no dark blotch at base of dorsal fin (Plate 14 c) . . . . small-eye bullheads . . **7**
6b. Large eyes (Plate 14 b); blotch at base of dorsal fin (Plate 14 d). . . . . large-eye bullheads . . **8**

[6] **7a.** Lower (chin) barbels light—yellowish or white; olive green body with no or only slight mottling; anal fin rays 24–28 . . . . *Ameiurus natalis* (yellow bullhead), p. 215
7b. Lower (chin) barbels dark—gray or black; body may have distinctive brown mottling; anal fin rays (18)20–24. . . . . *Ameiurus nebulosus* (brown bullhead), p. 218

[6] **8a.** Lateral edges of premaxillary tooth patch clearly rounded on lateral ends and with uniform-sized teeth (Plate 14 e); head flat or slightly convex; anal fin rays (19)22–24(26). . . . . *Ameiurus platycephalus* (flat bullhead), p. 221
8b. Lateral edges of premaxillary tooth patch indented on lateral ends and with two sizes of teeth, anterior teeth and central teeth are larger (Plate 14 f); head laterally compressed; anal fin rays (13)17–20(22) . . . . *Ameiurus brunneus* (snail bullhead), p. 209

[1] **9a.** Terminal mouth (Plate 14 g), jaws equal in length. . . . . *Noturus gyrinus* (tadpole madtom), p. 230
9b. Subterminal mouth (Plate 14 h), upper jaw slightly longer than lower jaw **10**

[9] **10a.** Body and median fins covered with distinctive dark speckling. . . . . *Noturus leptacanthus* (speckled madtom), p. 236
10b. Body color plain, median fins may have dark margin but lack speckling . . . . *Noturus insignis* (margined madtom), p. 233

Note: When identifying ictalurids, count all anal fin rays, both rudimentary rays and principal rays (Hubbs and Lagler 1958). Expand the fin and take care not to miss small rudimentary rays.

Ameiurus brunneus (Jordan, 1877)
Snail bullhead

DESCRIPTION Head relatively flat; snout rounded; anal fin short. Mouth moderately to strongly inferior. Eyes comparatively large for a bullhead. Body shape differs in specimens from mountain and Coastal Plain habitats (Yerger and Relyea 1968). Mountain specimens have a hump-backed appearance, a relatively thick caudal peduncle, shorter fin rays and spines, and a strongly inferior mouth; Coastal Plain specimens are flatter and have longer fin rays.

Dark yellowish green to olive on back, yellow on sides, and white underneath. Body usually mottled or spotted. The dark blotch at the base of the dorsal fin that may vary from vague to prominent is a characteristic feature (Yerger and Relyea 1968).

SIMILAR SPECIES The snail bullhead most closely resembles the flat bullhead but can be distinguished by its lower anal fin ray count ([13]17–20[22] versus [19]22–24[26] for the flat bullhead), premaxillary tooth patch, mouth, and coloration. The lateral ends of the snail bullhead's premaxillary tooth patch are indented, and the anterior and central premaxillary teeth are larger (versus rounded lateral ends and teeth of uniform size in the flat bullhead). The snail bullhead's mouth is slightly subterminal with dark, uniformly colored maxillary barbels; the flat bullhead's mouth is terminal and the maxillary barbels have a light leading edge. We have observed considerable overlap in barbel color, however. The snail bullhead is usually less mottled than the flat bullhead (Yerger and Relyea 1968).

SIZE Adults can reach 290 mm TL (Rohde et al. 1994), but most individuals in the MSRB are considerably smaller.

MERISTICS Dorsal spine 1, dorsal rays 6, anal rays (16)17–20(22), pectoral spine 1, pectoral rays 9, pelvic rays 8; gill rakers (11)12–16(18) (Yerger and Relyea 1968; Jenkins and Burkhead 1993); pectoral spines with numerous small serrations on anterior and posterior edges (Mettee et al. 1996).

DISTRIBUTION Atlantic slope drainages from the Dan River (Roanoke River drainage) of North Carolina and Virginia south to the Altamaha River system of Georgia. On the Gulf Coast, found in the Apalachicola River system of Georgia, Alabama, and Florida. Disjunct populations are present in the middle St. Johns River of Florida

Ameiurus brunneus, snail bullhead

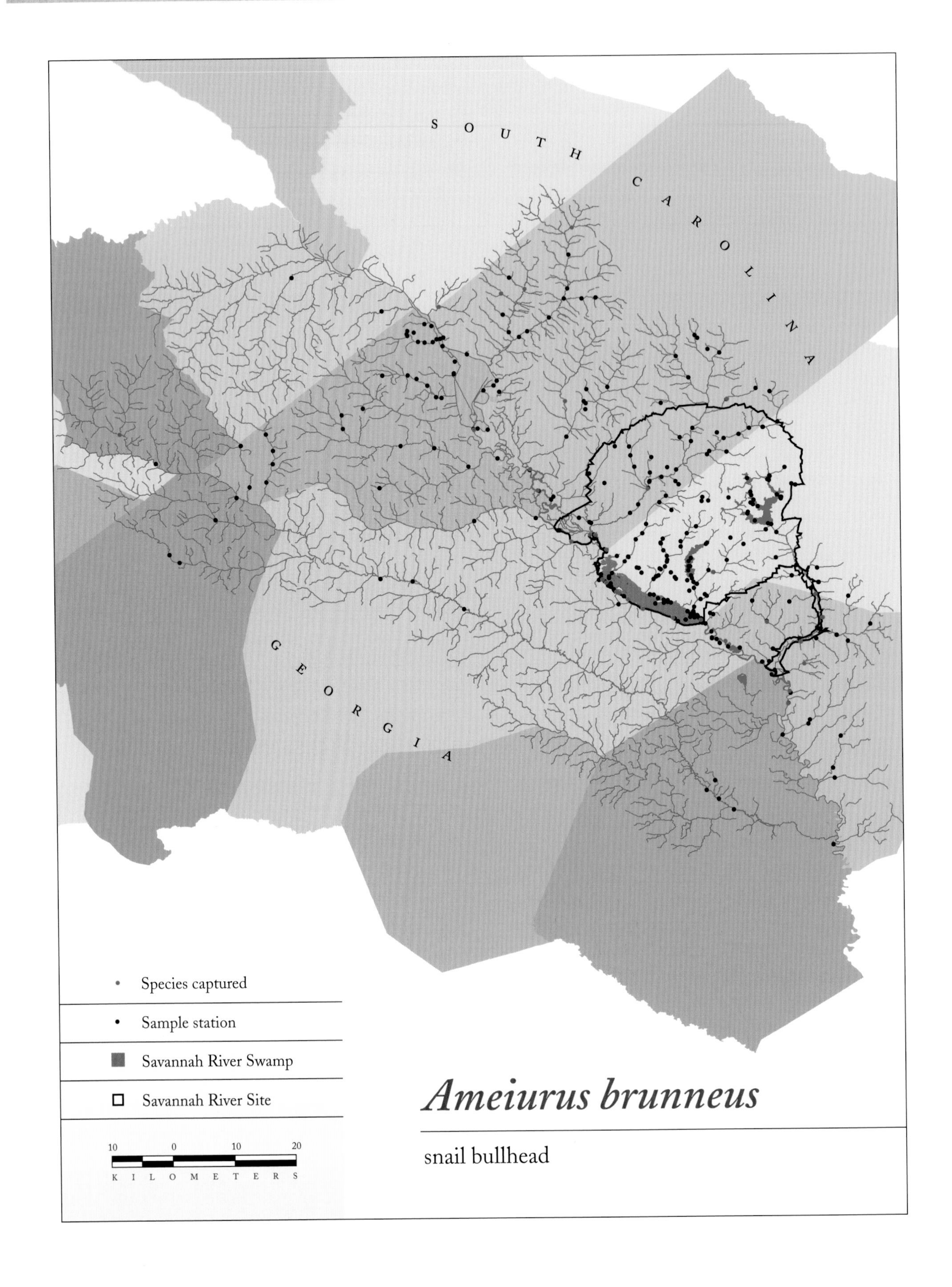

Ameiurus brunneus

snail bullhead

(Burgess et al. 1977; Gilbert and Burgess *in* Lee et al. 1980). In South Carolina, found in mountain, Piedmont, and Upper Coastal Plain habitats (Rohde et al 1994). Collected from various locations in the Savannah River and some of its tributaries, but less abundant than the other bullheads.

HABITAT Streams and rivers with substantial currents and sand, gravel, and rock bottoms, including pools in swift mountain streams (Yerger and Relyea 1968). The flat bullhead, in contrast, is usually found in slow-flowing or standing waters, although the young of both species are sometimes found in relatively small, clear streams (Rohde et al. 1994). In Virginia, collected in slow and moderately flowing runs among rubble and boulders, in river backwaters, and in rivers with soft organic substrates (Jenkins and Burkhead 1993).

BIOLOGY The snail bullhead is the least known bullhead in the MSRB. It may have a prolonged spawning period extending from late winter through midsummer (Yerger and Relyea 1968). Examination of specimens collected from the Dan River system in Virginia suggested that spawning occurs there in May and June (Jenkins and Burkhead 1993).

The snail bullhead is apparently nocturnal and consumes snails, other aquatic invertebrates, small fishes, and some plant material (Yerger and Relyea 1968; Mettee et al. 1996).

SCIENTIFIC NAME *Ameiurus* = "unforked caudal fin" (Greek); *brunneus* = brown (Latin).

Ameiurus catus (Linnaeus, 1758)

White catfish

DESCRIPTION Body heavy; head and mouth wide, especially in older specimens (Stevens 1959b). Caudal fin moderately to weakly and asymmetrically forked; anal fin relatively short and rounded.

Mental barbels white or pale yellow. Back blue-gray to blue-black grading to white on the underside. Pectoral and pelvic fins white or light gray, median fins generally gray. Males in breeding condition often exhibit darker coloration, especially on the head (Jones et al. 1978).

SIMILAR SPECIES White catfish may be confused with channel catfish; however, white catfish lack the spots often found on the channel catfish, and the tail is less deeply forked and somewhat asymmetrical with the lower lobe larger than the upper lobe. The white catfish typically has 22–24 anal rays (Robison and Buchanan 1984), although the number may range from 21 to 26 (Jones et al. 1978); the channel catfish typically has 24–29 anal rays (Jones et al. 1978). The white catfish can be separated from the other bullheads by its forked rather than emarginate tail and from the blue catfish by its shorter anal fin (the blue catfish has 30–36 anal rays) and its wider, flatter body.

Jenkins and Burkhead (1993) summarized the past and present taxonomic status of the white catfish and noted that it is morphologically a typical bullhead (*Ameiurus* spp.) except for its moderately forked tail, which apparently represents an evolutionary reversal.

SIZE Adults are intermediate in size between the larger catfishes and the bullheads, ranging from approximately 210 to 620 mm (Rohde et al. 1994). The South Carolina state record white catfish, caught in Lake Murray, weighed approximately 4.5 kg (SCDNR), but 0.25–1.0 kg fish are more typical (Rohde et al. 1994).

MERISTICS Dorsal spine 1, dorsal rays 5–7, anal rays 21–26, pectoral spine 1, pectoral rays 8–9; gill rakers 18–23 (Stevens 1959b; Jones et al. 1978; Etnier and Starnes 1993).

Ameiurus catus, white catfish

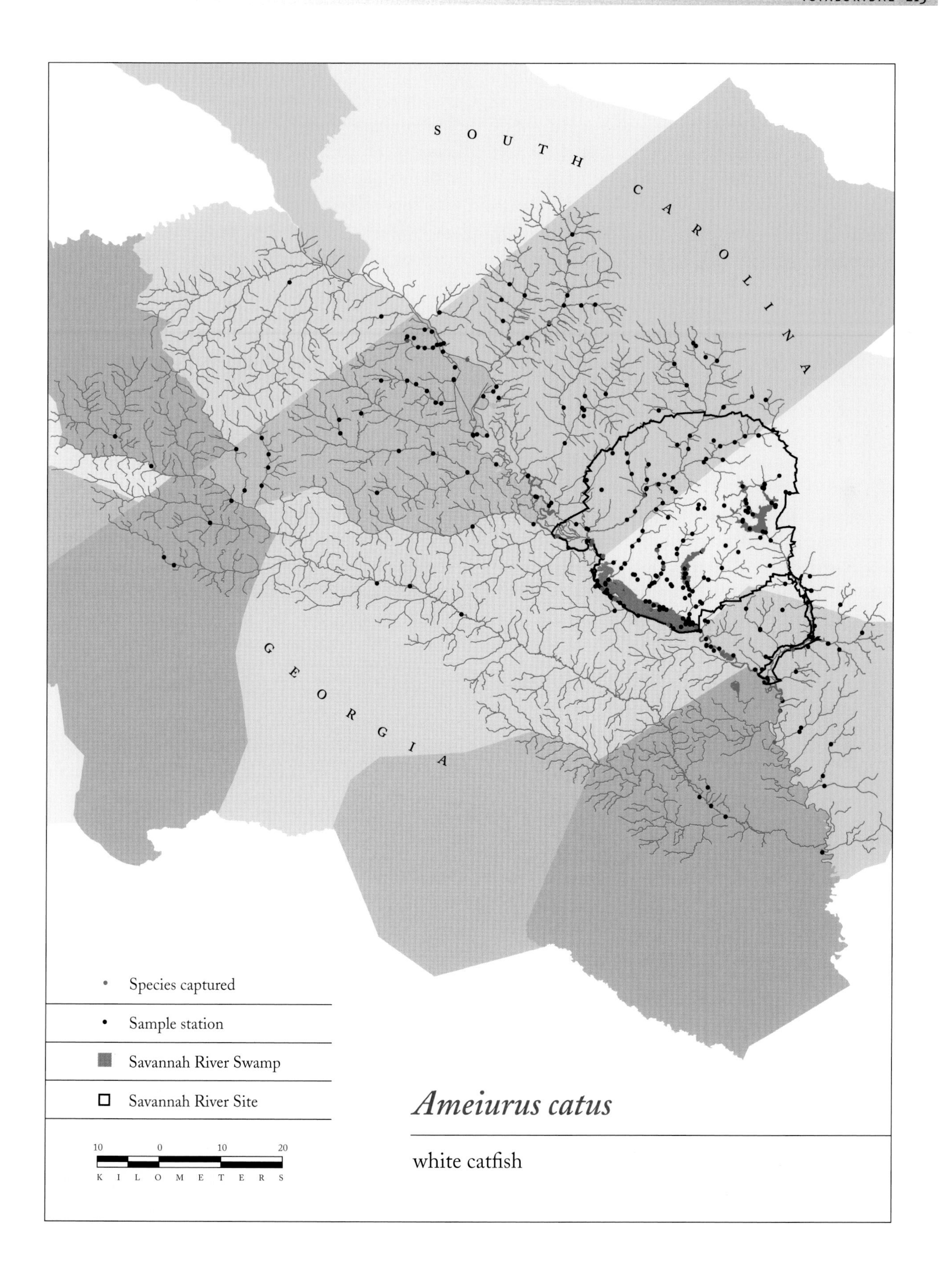

Ameiurus catus

white catfish

DISTRIBUTION Native from southern Florida north on the Atlantic slope to New York and west on the Gulf slope to western Alabama (Glodek *in* Lee et al. 1980). Introduced into Puerto Rico, perhaps as a contaminant in a shipment of channel catfish (Erdman 1984). Found throughout most of South Carolina (Rohde et al. 1994). Within the MSRB, common in the Savannah River and occasionally found in larger Savannah River tributaries. Together with the channel catfish, the white catfish is one of the most abundant large catfishes in the Savannah River, where it sometimes occurs in substantial aggregations in deeper water.

HABITAT Rivers, lakes, reservoirs, and large streams, primarily over silt or sand bottoms. Relatively tolerant of brackish water (Turner and Johnson 1973; Rohde et al. 1994). Less likely than the channel catfish to be found in currents (Etnier and Starnes 1993). In the upper Hudson River estuary, white catfish usually prefer channel border, shoal, and vegetated backwater habitats (Hughes and Carlson 1986). Considered to be pollution tolerant (NCDENR 2001), white catfish declined in abundance as less tolerant species increased in the Ohio River following improvements in water quality and habitat (Pearson and Pearson 1989).

BIOLOGY Spawning probably peaks in June in South Carolina (Stevens 1959b). The minimum spawning temperature in Virginia is about 21 °C (Jenkins and Burkhead 1993). The male and female prepare a relatively large nest (up to about 91 cm across and 46 cm deep) by fanning the bottom near sand, gravel banks, or rock piles (Jones et al. 1978; Hughes and Carlson 1986) and carrying away larger objects in their mouths (Manooch 1984). Nests are usually located under a log or some other type of shelter. The female deposits 1,000–3,500 eggs, which hatch within 6–7 days at 27 °C (Jones et al. 1978; Manooch 1984). The male guards the eggs and the young until they disperse from the nest. Small juveniles often form large schools in shallow water as is typical of juvenile bullheads. They become more solitary as they age.

White catfish typically reach sexual maturity in 3 or 4 years at a minimum size slightly in excess of 200 mm (Etnier and Starnes 1993). Lack of food may result in slow growth and stunting (Schaffter 1997). White catfish may live 11 years or more in South Carolina (Stevens 1959b).

Adults are opportunistic feeders. Approximately 50 different species of organisms were represented in the digestive tracts of white catfish collected from the North Newport River in Georgia, with crustaceans, especially amphipods, dominating (Heard 1975). White catfish collected from Alabama farm ponds had consumed artificial food plus detritus, microcrustaceans, aquatic insect larvae (especially dipterans), and fish (Devaraj 1974). Fish, especially herring and shad, compose most of the diet of white catfish in the Santee-Cooper Reservoir in South Carolina (Stevens 1959b). White catfish collected from the Savannah River had consumed substantial numbers of small Asian clams (*Corbicula fluminea*), sediment, aquatic insect larvae, and small amounts of fish and plant material. White catfish appear to feed more at night than during the day and can be caught with worms, minnows, commercial catfish baits, and shrimp (Rohde et al. 1994).

SCIENTIFIC NAME *Ameiurus* = "unforked caudal fin" (Greek); *catus* = "cat" (Latin).

Ameiurus natalis (Lesueur, 1819)
Yellow bullhead

DESCRIPTION Body stout and heavy. Head broad; wide mouth nearly reaching eye. Eyes comparatively small for a bullhead. Rear margin of the caudal fin straight or slightly rounded, and anal fin base about as long as the head. Well-developed serrations on posterior edges of the pectoral spine can be felt when the spine is grasped and pulled outward.

Body yellow olive to brown or black on top, grading to yellow on the sides and yellow or white underneath. Sides sometimes faintly mottled.

SIMILAR SPECIES The yellow bullhead differs from the brown bullhead in possessing white to yellowish rather than dark chin barbels, and from the flat and snail bullheads in lacking a dark blotch at the base of the dorsal fin. It differs from the white, channel, and blue catfishes in having a square rather than forked tail. Juveniles of the much larger flathead have a relatively lightly pigmented upper caudal fin and backward-projecting lateral tooth patches on the upper jaw.

SIZE Adults are usually 150–460 mm TL (Rohde et al. 1994). The South Carolina state record yellow bullhead (caught in the Edisto River) weighed 2.8 kg (SCDNR).

MERISTICS Dorsal spine 1, dorsal rays 6–7, anal rays 24–28; gill rakers 14–16, branchiostegal rays 8–9 (Jones et al. 1978).

DISTRIBUTION Native to most of the United States east of the Rocky Mountains as far north as North Dakota, the Great Lakes, and New Hampshire; widely introduced outside its native range (Glodek *in* Lee et al. 1980). Widespread throughout South Carolina (Rohde et al. 1994); collected in a number of lotic and lentic habitats in the MSRB.

HABITAT Slow-flowing and standing waters including streams, rivers, ponds, swamps, backwaters, and impoundments (Rohde et al. 1994). Common in small and medium-sized SRS streams (Paller 1994), where it is usually found in slow-flowing pools with soft bottoms of silt or accumulated leaves (Meffe and Sheldon 1988). Often associated with moderate to heavy aquatic vegetation (Jones et al. 1978) or other types of cover such as sunken logs and woody debris. The yellow bullhead prefers clear water (Pflieger 1975) but is comparatively tolerant of pollution (NCDENR 2001). It occurs in inundated caves in Florida where it is apparently capable of successful spawning (Relyea and Sutton 1973).

BIOLOGY In their review, Jones et al. (1978) reported that spawning occurs from mid-May through early June.

Ameiurus natalis, yellow bullhead; breeding male

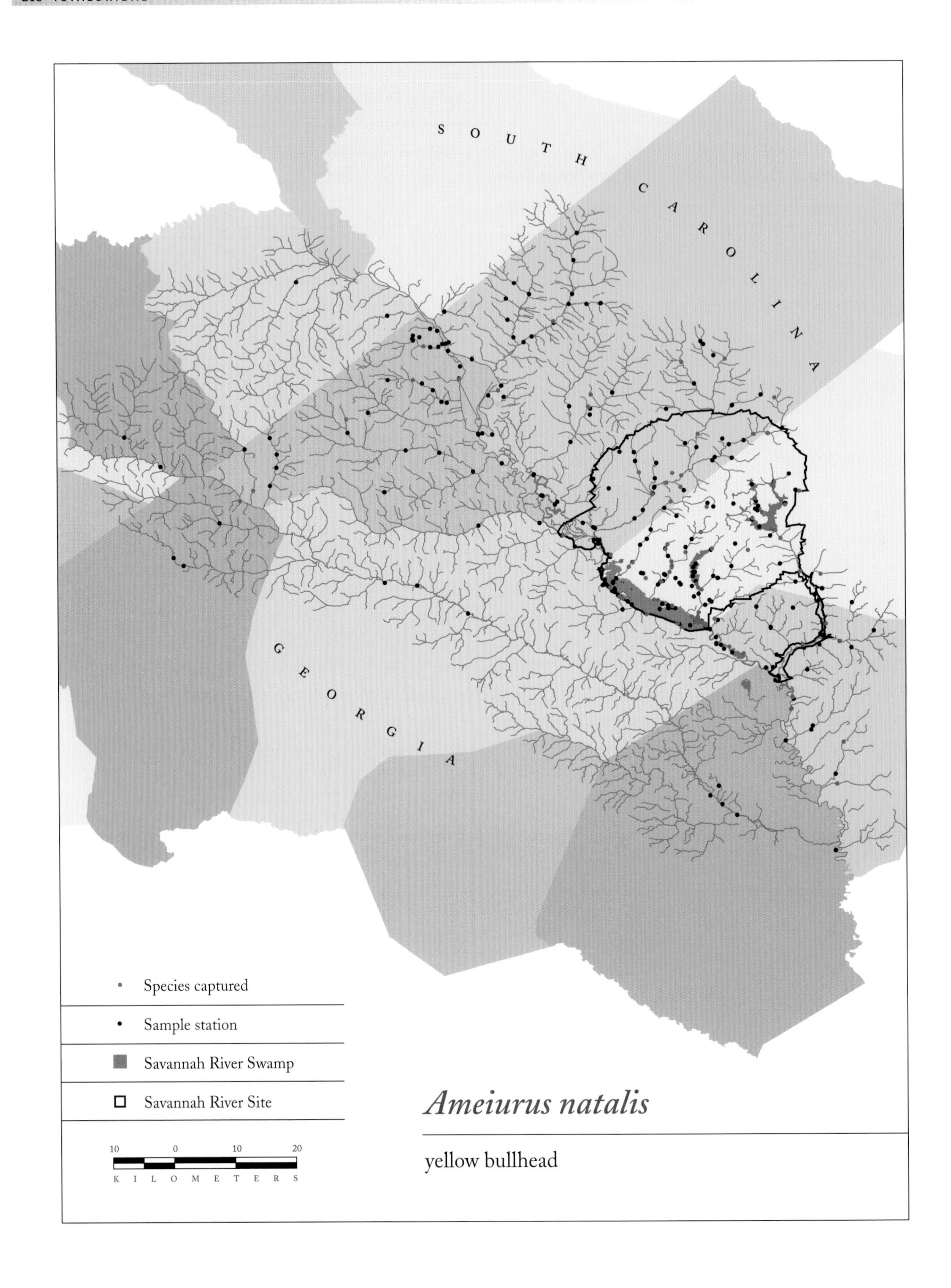
SOUTH CAROLINA
GEORGIA
Species captured
Sample station
Savannah River Swamp
Savannah River Site
10 0 10 20
KILOMETERS
Ameiurus natalis
yellow bullhead

Ameiurus natalis, yellow bullhead; adult of undetermined sex

Spawning probably occurs from late April through June in Virginia (Jenkins and Burkhead 1993), and during April and May in Tennessee (Etnier and Starnes 1993). Both sexes are usually involved in excavating the nest, which ranges from a shallow depression to a burrow; nests are built in shallow water near cover such as submerged logs or tree roots (Etnier and Starnes 1993; Rohde et al. 1994). When spawning, the male and female lie side-by-side facing opposite directions and the male arches ventrally toward the urogenital area of the female (Wallace 1972). The female deposits about 300–700 eggs per clutch. Females may contain several thousand (1,650–7,000) eggs (Jones et al. 1978), suggesting multiple clutches. The eggs hatch in 5–10 days (Jones et al. 1978). Jones et al. (1978) summarized descriptions of yellow bullhead larvae and their distinguishing features. The male guards the eggs and schooling young until they are approximately 50 mm long (Jones et al. 1978). Sexual maturity occurs in 2 or 3 years at a minimum length of 140 mm, and adults can live at least 7 years (Etnier and Starnes 1993).

The yellow bullhead feeds near the bottom, primarily at night (Reynolds and Casterlin 1978), using its keen senses of smell and touch (Pflieger 1975). The diet consists of insects, crustaceans, molluscs, fishes, plant material, and sediment (Rohde et al. 1994). Adults prefer fish and crayfish when these are available (Keast 1985b). Crayfish were the predominant food in the guts of large yellow bullheads collected from SRS streams, while small individuals had eaten crayfish as well as a variety of other invertebrates (Sheldon and Meffe 1993). Yellow bullheads were important in the removal of dead fish from the littoral zone of a small Michigan lake, indicating a role as scavengers (Schneider 1998).

Yellow bullheads are frequently caught by fishing on the bottom with live bait—especially worms—or prepared baits. Although often consumed, they are generally considered less tasty than other catfishes. They are hardy and survive well in aquaria but will eat smaller fish. They engage in relatively complex and sometimes aggressive intraspecific social behaviors mediated by their sense of smell (Todd et al. 1967; McLarney et al. 1974).

SCIENTIFIC NAME *Ameiurus* = "unforked caudal fin" (Greek); *natalis* = "having large buttocks" (Latin), referring to the humps breeding males sometimes develop between the head and the dorsal fin (Mettee et al. 1996).

Ameiurus nebulosus (Lesueur, 1819)
Brown bullhead

DESCRIPTION Body moderately robust; head broad; eyes relatively small; tail square. Mouth terminal and wide, the upper jaw slightly longer than the lower.

Body yellow-brown to gray, usually darkly mottled or spotted. Young specimens are often more uniformly colored and are sometimes black (Robison and Buchanan 1984). Fins gray; anal fin may be slightly mottled.

SIMILAR SPECIES The brown bullhead differs from the yellow bullhead in having darkly pigmented rather than white chin barbels. It differs from the flat bullhead in lacking the dark blotch at the base of the dorsal fin and in having larger eyes and a flatter head.

SIZE Up to 508 mm, although most are 150–450 mm. Reaches 1 kg in Puerto Rico (Erdman 1984).

MERISTICS Dorsal spine 1, dorsal rays 6, anal rays 19–26, pectoral spine 1, pectoral rays 7–9 (Jones et al. 1978). The pectoral spine has 5–8 large serrations on its posterior margin (Rohde et al. 1994); usually 13 or 14 gill rakers (Jones et al. 1978).

DISTRIBUTION Native from Nova Scotia and New Brunswick south to Florida and west through the St. Lawrence and Great Lakes basins into southwestern Ontario; from southeastern Manitoba south through the tier of states on the western banks of the Mississippi to northern Louisiana, but not southern Louisiana or Mississippi or western Alabama (Glodek *in* Lee et al. 1980). Introduced into all of the 48 contiguous U.S. states plus Hawaii except Montana, Wyoming, Utah, and Texas. Stockings in Hawaii failed to become established (Brock 1960), and the species may have been extirpated from New Mexico (Sublette et al. 1990). Also stocked in Puerto Rico (Erdman 1984). Collected at a number of locations on the SRS, although it is usually less abundant than the yellow bullhead and flat bullhead.

HABITAT Standing and slow-flowing waters, including ponds, lakes, reservoirs, swamps, and pools in slow-flowing rivers and streams. Often found over mud or sand bottoms, and seems to have a greater preference for aquatic vegetation than most of its congeners (Robison and Buchanan 1984; Killgore et al. 1989; Rohde et al. 1994). Tolerates poor water quality and is sometimes found in polluted waters with low dissolved oxygen concentrations (Ross 2001). In such circumstances, it may be one of the few species present, and it benefits from re-

Ameiurus nebulosus, brown bullhead

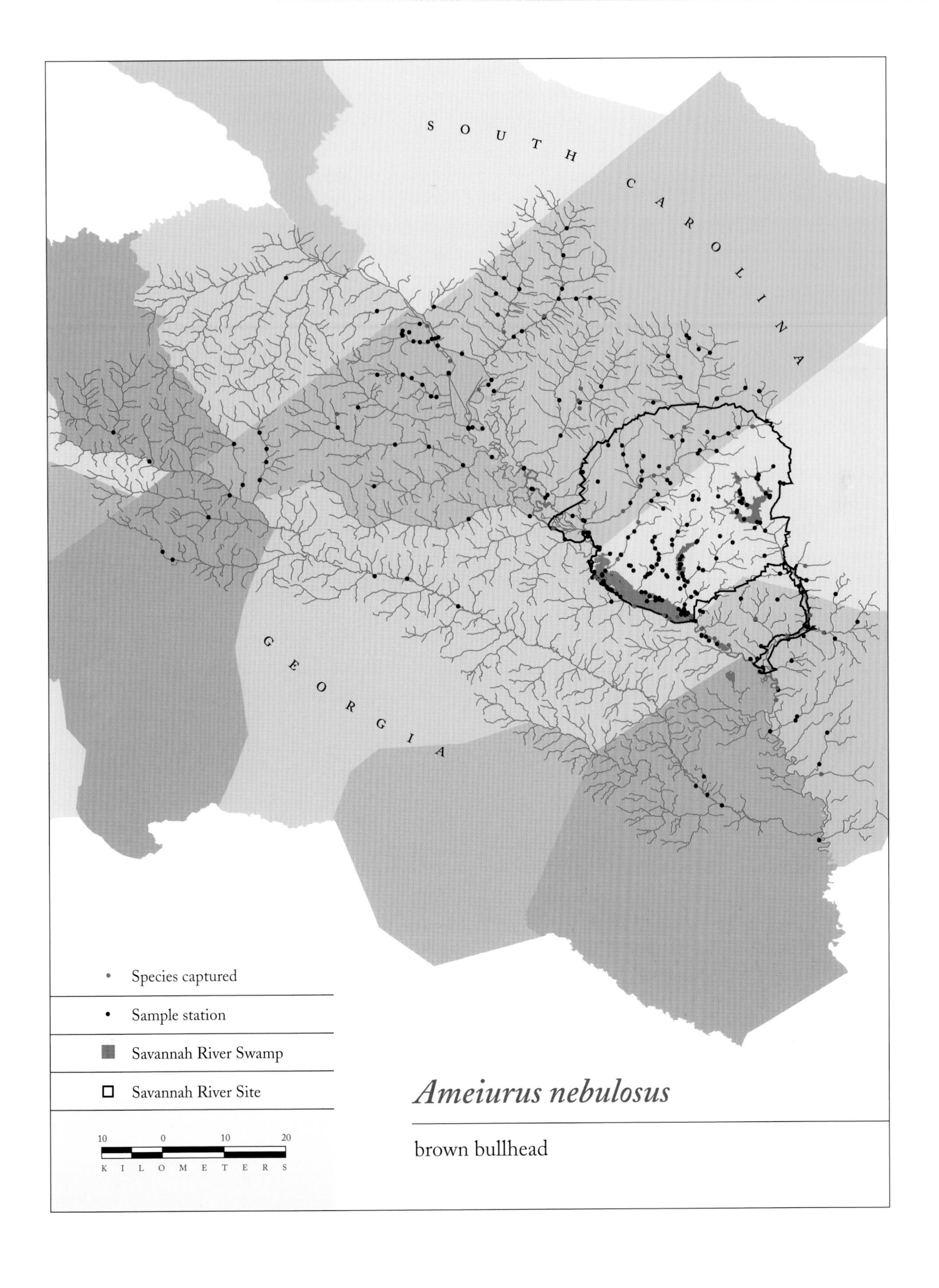

Ameiurus nebulosus

brown bullhead

duced competition and predation (Lesko et al. 1996). While common in the tributary streams of the Suwannee River, it does not enter the Okefenokee Swamp proper (Laerm and Freeman 1986).

BIOLOGY Spawning occurs in the spring and early summer in a nest that is generally in water less than 1 m deep (Robison and Buchanan 1984). Both parents usually participate in constructing the nest, which typically consists of a shallow, circular depression in the shelter of a sunken log or aquatic vegetation (Rohde et al. 1994). The parents guard and care for the eggs, with the male more frequently involved in aerating the eggs by fanning them and manipulating them in his mouth, and the female more frequently involved in chasing away potential predators (Ross 2001). Jones et al. (1978) summarized descriptions of brown bullhead larvae and their distinguishing features. The young form tight schools for the first several weeks, which facilitates guarding by the parents. Parental care may continue for as long as 20 days after egg deposition (Blumer 1985) and is essential for the survival of eggs and larvae (Blumer 1986). Adults mature in 2–3 years (Rohde et al. 1994) and may live 8 or 9 years (Carlander 1969). Territoriality and dominance behavior peak during the reproductive season but also occur at other times (Bryant 1981; Carr et al. 1987). This behavior is triggered by chemical stimuli along with other sensory cues.

The brown bullhead is a benthic feeder that uses its keen senses of taste and smell to find food; it is most active at night, although it may also feed during the day (Robison and Buchanan 1984). The choice of food is determined primarily by availability, and the diet may include insects, worms, molluscs, small fishes, plants, and organic detritus (Rohde et al. 1994). Even sewage may be consumed (Klarberg and Benson 1973). Brown bullheads studied in a New York lake preferred larval Chironomidae (midges) and amphipods, suggesting that feeding is sometimes selective (Kline and Wood 1996).

SCIENTIFIC NAME *Ameiurus* = "unforked caudal fin" (Greek); *nebulosus* = "clouded" (Latin), referring to the mottled sides.

Ameiurus platycephalus (Girard, 1859)

Flat bullhead

DESCRIPTION Head dorsoventrally compressed, as the name implies. Eyes comparatively large for a bullhead. Mouth subterminal to moderately inferior; caudal fin emarginate to slightly forked.

Body brown to gray-brown dorsally, although it may be olive or even yellowish (Yerger and Relyea 1968). Sides mottled or marbled, and underside white or cream. The dark blotch at the base of the dorsal fin is a distinguishing character.

SIMILAR SPECIES Most likely to be confused with the snail bullhead, which has a generally similar shape and color pattern, including the dark blotch at the base of the dorsal fin. The two species can be distinguished by the number of anal fin rays (22–24 in the flat bullhead and 17–20 in the snail bullhead) and the premaxillary tooth patch, which in the flat bullhead has rounded lateral edges and uniformly sized teeth, and in the snail bullhead has indented lateral edges and larger central and anterior teeth. In addition, the maxillary (upper lip) barbel of the flat bullhead often has a pale leading edge while the maxillary barbel of the snail bullhead is uniformly colored (Rohde et al. 1994). However, we noted considerable overlap in barbel color. Other features to note are the mouth, which is in a more inferior position in the snail bullhead than in the flat bullhead; and the color pattern, which is generally more mottled in the flat bullhead (Yerger and Relyea 1968). The brown bullhead is also mottled, but it has smaller eyes and a less flattened head, and lacks a dark blotch at the base of the dorsal fin.

SIZE Seldom exceeds 290 mm (Rohde et al. 1994).

MERISTICS Dorsal spine 1, dorsal rays 6, anal rays (19)22–24(26), pectoral spine 1, pectoral rays 8–9; gill rakers usually 13 or fewer, but up to 17 are possible (Etnier and Starnes 1993; Jenkins and Burkhead 1993). Pectoral spine with numerous small posterior serrations (Etnier and Starnes 1993).

DISTRIBUTION The Piedmont and Coastal Plain from the Roanoke River drainage of Virginia and North Carolina south to the Altamaha River of Georgia; no records for the Tar River drainage of North Carolina (Gilbert and Burgess *in* Lee et al. 1980). Found throughout South Carolina except for the Lower Coastal Plain (Rohde et al. 1994); common in the Savannah River and its tributaries.

Ameiurus platycephalus, flat bullhead; adult

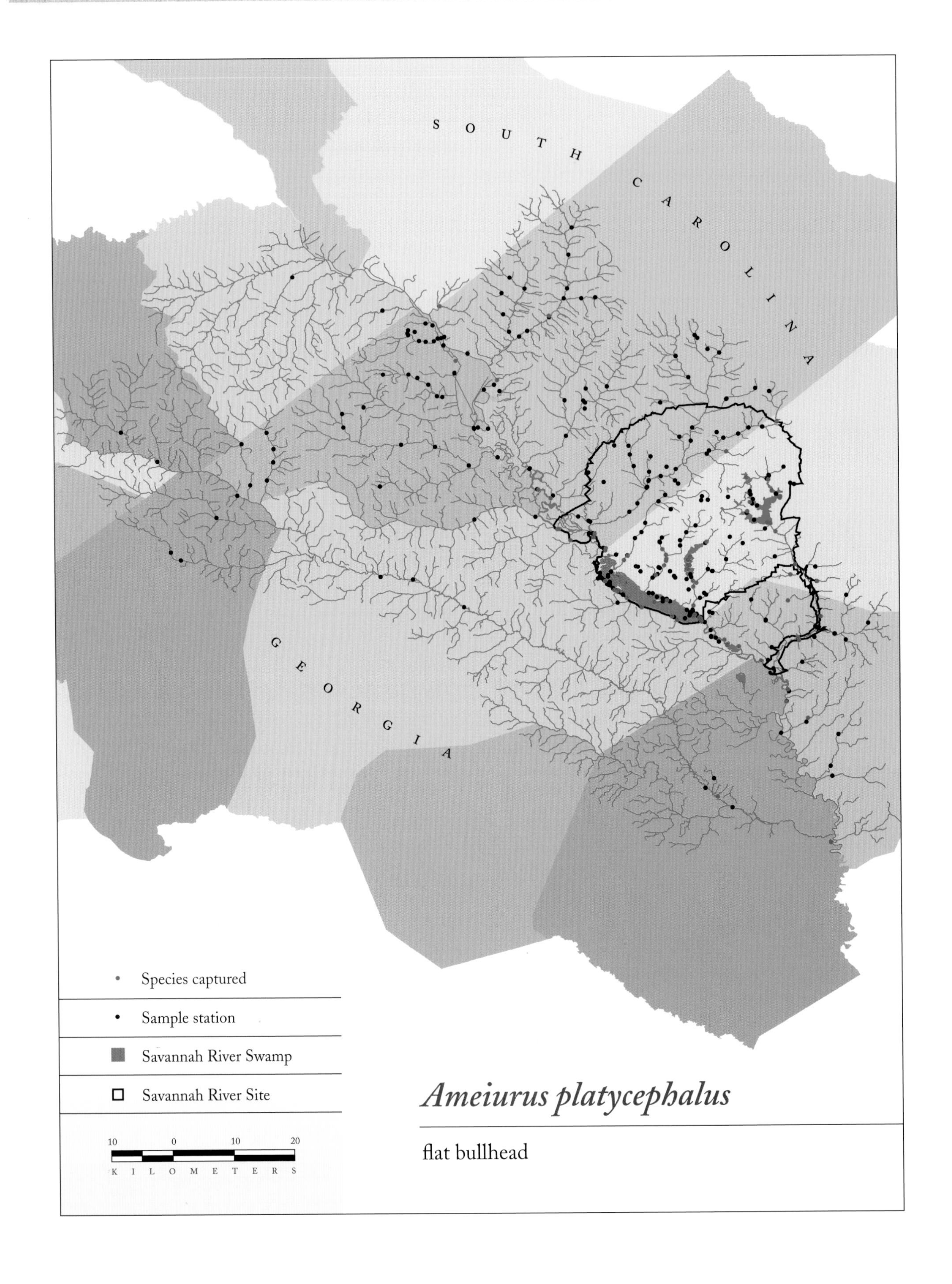

Ameiurus platycephalus

flat bullhead

Ameiurus platycephalus, flat bullhead; juvenile

HABITAT Lakes, reservoirs, ponds, and slow-flowing areas in large rivers, usually over soft substrates of sand or silt (Yerger and Relyea 1968); also found in smaller streams with moderate currents (Rohde et al. 1994). The similar snail bullhead prefers faster-flowing streams with sand, gravel, or rock bottoms (Yerger and Relyea 1968). On the SRS, the flat bullhead is commonest in deeper and wider streams with silty or sandy bottoms (Meffe and Sheldon 1988). Jenkins and Burkhead (1993) noted that it often associates with cover in pools and backwaters, sometimes near swifter water.

BIOLOGY Relatively little is known about the biology of the flat bullhead. In Lake Norman, a North Carolina reservoir on the Catawba River, spawning occurs during June and July at temperatures of 21–24 °C (Olmsted and Cloutman 1979); females reach sexual maturity at 3 years and produce 207–1,742 ova, with larger fish producing more ova; maximum longevity is 7 years; and males grow somewhat faster than females after age 3.

The flat bullhead is an opportunistic bottom feeder that consumes aquatic insects, small fishes, molluscs, bryozoans, other small animals, and plant material (Olmsted and Cloutman 1979; Rohde et al. 1994). The main foods in SRS streams are freshwater shrimp, oligochaetes, and aquatic insects (Sheldon and Meffe 1993). Snails and mussels form a prominent part of the diet in some habitats (Yerger and Relyea 1968).

SCIENTIFIC NAME *Ameiurus* = "unforked caudal fin" (Greek); *platycephalus* = "flat head" (Greek).

Ictalurus furcatus (Lesueur, 1840)
Blue catfish

DESCRIPTION Body large and deep, with back sloping steeply upward from head to dorsal fin. Caudal fin forked; anal fin long, its outer margin straight and tapered like a barber's comb. Lower jaw not protruding beyond the upper jaw.

Back gray to blue, underside white, fins clear or whitish. Without black spots on body (with the exception of Rio Grande specimens [Wilcox 1960]). Young specimens are usually more silver or silver-white than adults (Graham 1999).

SIMILAR SPECIES The blue catfish can be distinguished from both the channel catfish and the white catfish by its greater number of anal rays (30 or more) and the anal fin's straight rather than curved edge. The channel catfish has black spots on the body (except in very large individuals) and a single-chambered swimbladder; the blue catfish has a constriction in the swimbladder that makes it two chambered rather than single chambered (Graham 1999).

SIZE This is the largest catfish in North America, reaching 50.8–165 cm (Rohde et al. 1994). The South Carolina record blue catfish, caught in the Tailrace Canal below Lake Moultrie, weighed 49.6 kg (SCDNR). Historically, large specimens from the Missouri and Mississippi rivers probably exceeded 90 kg or more (Graham 1999).

MERISTICS Dorsal spine 1, dorsal rays 6, anal rays 30–36, pectoral spine 1, pectoral rays 8–9, pelvic rays 8; gill rakers 14–21 (Etnier and Starnes 1993; Jenkins and Burkhead 1993).

CONSERVATION STATUS Introduced; uncertain if it has or will become established.

DISTRIBUTION Native to larger rivers of the Mississippi, Missouri, and Ohio river basins, and Gulf of Mexico drainages from the Coosa-Tallapoosa drainage west; south through Mexico to Guatemala (Glodek *in* Lee et al. 1980). Widely introduced as a food and sport fish and, to a lesser extent, to control populations of the introduced Asian clam, *Corbicula* sp. (Dill and Cordone 1997). Generally more abundant in the southern portion of its range (Graham 1999). There are well-established populations in the Santee-Cooper Reservoir and associated drainages, and the species is found in other South Carolina rivers as well (SCDNR). At the time of this writing, it has been reported from only one site in the MSRB, the Merry Brothers Brickyard ponds. One specimen from these ponds was identified in 2001 by G. P. Friday (pers. comm.). Other reports in newspapers are

Ictalurus furcatus, blue catfish

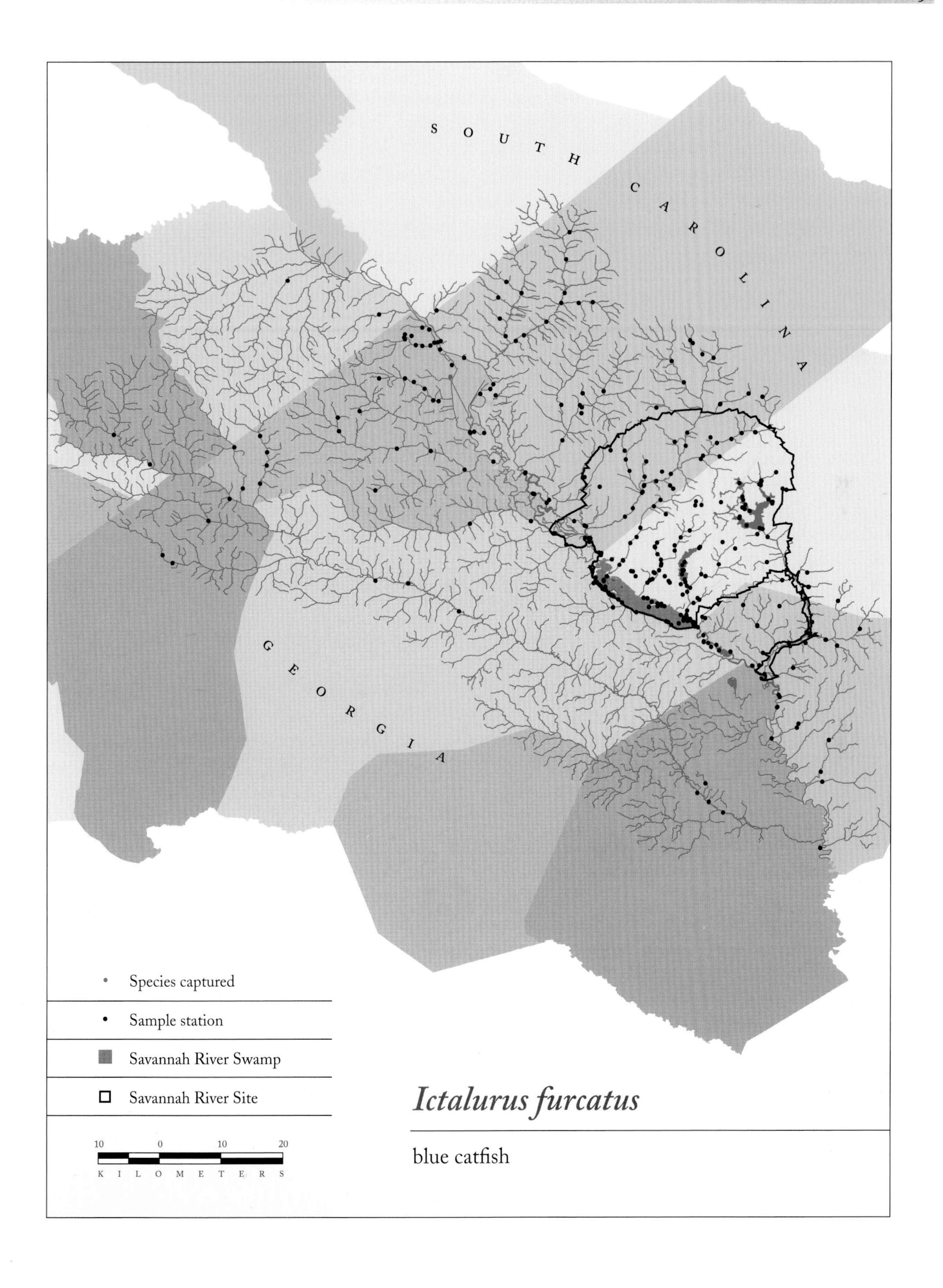

Ictalurus furcatus

blue catfish

not considered here, as channel catfish (*I. punctatus*) without spots are often misidentified as blue catfish.

HABITAT Primarily large, relatively turbid rivers, where this strong swimmer usually prefers deep waters with substantial currents (Graham 1999). Also, often found in river backwaters and large reservoirs, where it occupies deeper water than the channel catfish (Matthews and Gelwick 2002). Catch rates in Alabama rivers were about equal in tailrace, main channel, and tributary habitats (Grussing et al. 1999), and catch rates in the Tennessee-Tombigbee Waterway were likewise generally comparable among channel cut-off, main channel, and tailwater habitats (Jackson 1995). Blue catfish are found over a variety of substrates ranging from gravel to silt (Graham 1999) and are common in swiftly flowing tailwaters below dams (Mettee et al. 1996).

BIOLOGY Spawning occurs during April–June in Louisiana (Perry and Carver 1973). In general, spawning habits are thought to resemble those of the channel catfish (Pflieger 1975). The following account of spawning behavior is summarized from Graham 1999. Blue catfish are cavity nesters that spawn in sheltered areas with minimal currents such as behind rocks, in depressions, under stream banks, and in root wads. The male guards the eggs and fry. The fecundity varies from 900 to 1,350 eggs per kilogram of body weight, and it is not unusual for clutches to contain 40,000–50,000 individuals.

Graham (1999) presented blue catfish age and growth data from a number of reservoirs and rivers in his review of blue catfish biology and management. He concluded that growth is generally rapid throughout life but variable, with intra- and interspecific competition contributing to the variability. Blue catfish mature at lengths ranging from 350 to 662 mm at ages ranging from 4 to 7 years, with maturity occurring earlier in the southern portion of the range. Blue catfish tend to grow faster than channel catfish but slower than flathead catfish in Alabama rivers (Grussing et al. 1999). The maximum age may exceed 20 years (Graham 1999).

Feeding occurs primarily near the bottom and to a lesser extent in mid-water (Pflieger 1975) and is opportunistic. The diet includes insects, crustaceans, clams, mussels, and fish (Graham 1999). The blue catfish is more piscivorous than most catfishes (Pflieger 1975), and an abundant supply of forage fish, such as gizzard and threadfin shads, contributes to rapid growth to large size (W. L. Davis 1980; Graham 1999). Graham (1999) indicated that blue catfish may consume large quantities of the introduced Asian clam (*Corbicula fluminea*) as well as other types of bivalves.

Blue catfish are thought to be more migratory than other North American catfishes, moving upstream in the spring and downstream in the fall in response to fluctuations in water temperature (Graham 1999). These movements can encompass several hundred kilometers and may contribute to the declining abundance of blue catfish in rivers that have been extensively impounded (Etnier and Starnes 1993). Other factors that may reduce blue catfish populations are overharvest, removal of snags, degraded water quality, and channelization (Hesse 1994; Graham 1999). Blue catfish may be more sensitive to low dissolved oxygen levels than channel catfish (Dunham et al. 1993).

Blue catfish have been considered for aquaculture because of their high-quality flesh, relatively high dress-out percentage, resistance to diseases that affect channel catfish, and ease of harvest by seining (Dunham et al. 1993). However, they also have liabilities compared with channel catfish, including slower maturation, lower rates of food conversion, poor spawning success in captivity, and relative intolerance to handling. As a result, they are currently unpopular with the aquaculture industry, although the hybrid between the blue catfish and channel catfish holds some promise (Graham 1999).

SCIENTIFIC NAME *Ictalurus* = "fish cat" (Greek); *furcatus* = "forked" (Latin), referring to the tail.

Ictalurus punctatus (Rafinesque, 1818)

Channel catfish

DESCRIPTION Body comparatively slender except in very large individuals, which become heavy bodied. Caudal fin deeply forked. Upper jaw projecting slightly beyond the lower jaw, premaxillary tooth pad lacking backward-pointing projections; anal fin with convex edge.

Back gray grading to pale gray on the sides, sometimes tinged with olive or yellow; underside white or light yellow. Young and adults have scattered dark spots; spots are often fewer on adults, and on large adults are generally lacking. Breeding males, like other ictalurids, develop an enlarged head, thickened lips, and blue-black coloration on the head, back, and sides (Pflieger 1975).

SIMILAR SPECIES The channel catfish can be confused with the white catfish (*Ameiurus catus*) or blue catfish (*Ictalurus furcatus*). The three species can be separated by the number of anal fin rays: 22–24 in the white catfish (although sometimes as many as 26), 24–29 in the channel catfish, and 30–36 in the blue catfish (Jones et al. 1978). Other useful characteristics are the shape of the anal fin (curved outer edge in the channel catfish, straight in the blue catfish), shape of the caudal fin (asymmetrical and less deeply forked in the white catfish), and coloration (the channel catfish is the only one with spots). The channel catfish can be easily separated from the bullheads by its forked rather than emarginate caudal fin.

SIZE Adults are commonly 45–65 cm but may reach 127 cm (Rohde et al. 1994). The South Carolina state (and world) record channel catfish, caught in Lake Moultrie, weighed 26.3 kg (SCDNR).

MERISTICS Dorsal spine 1, dorsal rays 6, anal rays usually 24–29, pectoral spine 1 with well-developed posterior serrations, pectoral rays 8–10; gill rakers 14–18 (Jones et al. 1978).

DISTRIBUTION Native to most of the area between the Rocky Mountains and the Appalachian Mountains, from the lower St. Lawrence River basin west through the Great Lakes basin into southern Manitoba and southeastern Saskatchewan and Montana, south along the western slope of the Appalachians, east to include most of Georgia and Florida, and west to northeastern Mexico (Glodek *in* Lee et al. 1980). Introduced into all 48 contiguous U.S. states plus Hawaii and apparently established at most introduction sites. Also introduced into Puerto Rico (Erdman 1984). Common throughout South Carolina (Rohde et al. 1994) and often collected in the Savannah River and its larger tributaries. Channel catfish are among the most common of the large catfishes in the Savannah River.

HABITAT Streams, rivers, ponds, lakes, and reservoirs (Etnier and Starnes 1993), but most characteristic of large streams and rivers with slow to moderate current (Jenkins and Burkhead 1993). Adults typically inhabit deep pools during the day, often near cover such as sunken logs and stumps, but may move into shallower water at night to feed. Young channel catfish are often found in shallower water than adults and sometimes occur in streams of moderate size. Jenkins and Burkhead (1993)

Ictalurus punctatus, channel catfish; adult

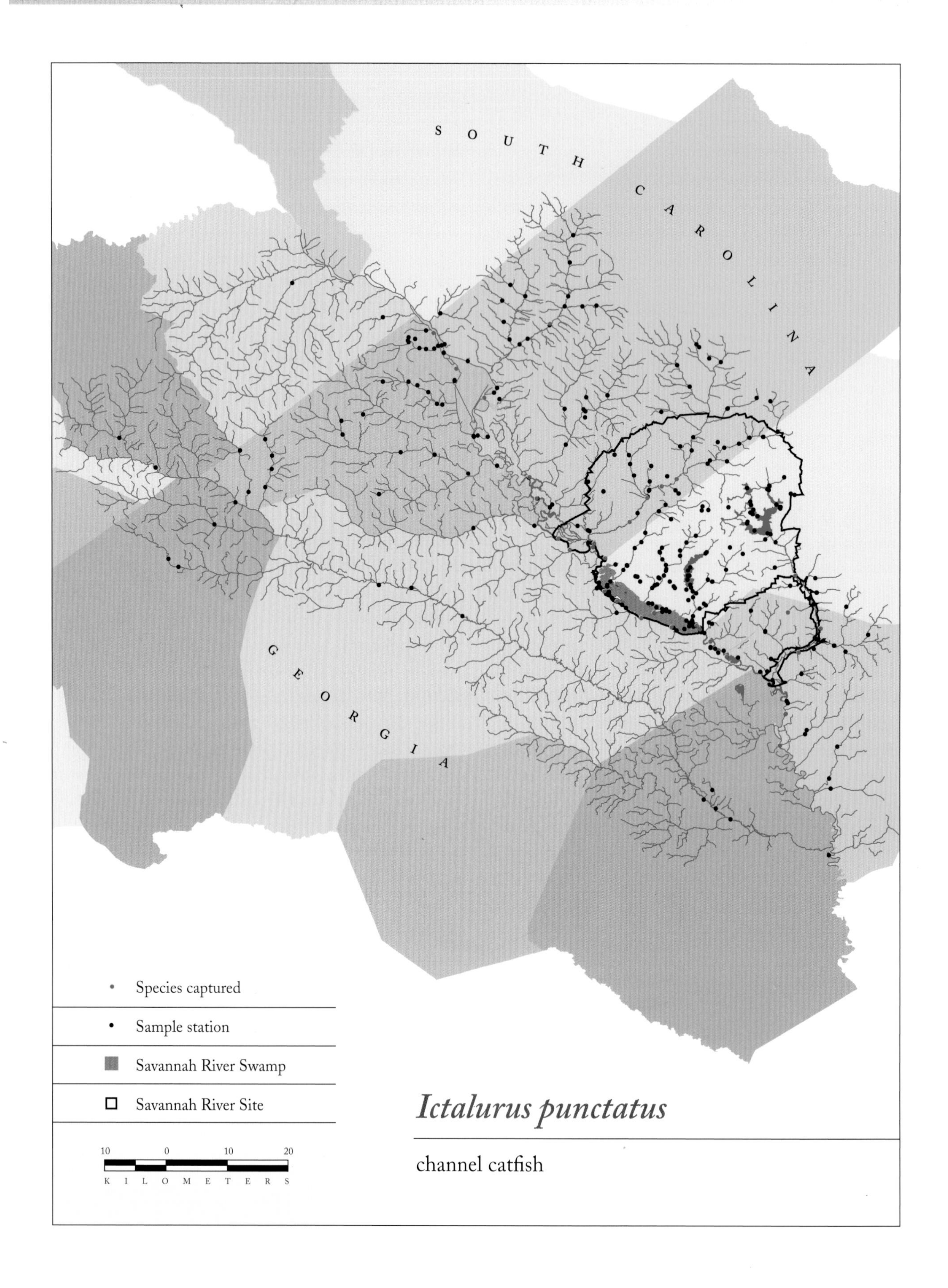

Ictalurus punctatus

channel catfish

Ictalurus punctatus, channel catfish; large adult

noted that the channel catfish is apparently intolerant of highly acidic water in swampy habitats. There are indications that riverine populations occupy rather small home ranges during the summer (Pellet et al. 1998) and typically move 5 km or less (Flotemersch et al. 1997), although they may move between the main river channel, tributary streams, and oxbow lakes depending on the water level (Flotemersch et al. 1997). Channel catfish are more likely to be found in the main river channel during the winter, when they tend to prefer deep water with slow currents (Newcomb 1989).

In Lake Texoma, Oklahoma, channel catfish usually occupy relatively shallow cove habitats while blue catfish are more abundant in deeper offshore waters (Edds et al. 2002). Channel catfish found in shallow water may have been displaced from deeper water by blue catfish since channel catfish are usually more widely dispersed in lakes.

BIOLOGY Spawning occurs during the spring and early summer (Rohde et al. 1994). Spawning dates from late May to early July were reported from the James River in Virginia (Menzel 1945 *in* Jenkins and Burkhead 1993). The male selects and cleans a nest site in a sheltered area such as an undercut bank, under a submerged log, within a debris pile, or in any other secluded cavity, including discarded tires. Females produce a gelatinous mass of 2,000–7,000 eggs (Etnier and Starnes 1993) that usually hatch in about a week (Pflieger 1975). Jones et al. (1978) summarized descriptions of channel catfish larvae and their distinguishing features. The nest is guarded by the male until the young leave. Fingerlings are susceptible to predation and seek shelter by day. In the absence of cover they may form schools near the bottom (Robison and Buchanan 1984). Growth is variable, but sexual maturity is usually attained at 3–6 years and approximately 270 mm (Etnier and Starnes 1993). Channel catfish do not usually exceed 7 years of age, but some may reach 15 years or more (Rohde et al. 1994).

The varied diet includes insects, crustaceans, molluscs, fish, and aquatic plants (Pflieger 1975). Their willingness to take a wide variety of baits—including cheese, chicken parts, soap, and various "stink baits"—suggests that channel catfish will consume almost any type of organic material. Young channel catfish feed predominantly on insect larvae and zooplankton (Walburg 1975), while fish often constitute a significant part of the diet of larger channel catfish (Edds et al. 2002). Channel catfish feed primarily near the bottom at night, using their sensitive senses of touch and smell to find prey.

Channel catfish are highly sought after by anglers and are of considerable recreational and economic significance. They are among the most important aquaculture products in the southern United States, although relatively few are produced in South Carolina. Their importance and availability have made channel catfish the subject of considerable research, and their biology is well known.

SCIENTIFIC NAME *Ictalurus* = "fish cat" (Greek); *punctatus* = "spotted" (Latin).

Noturus gyrinus (Mitchill, 1817)
Tadpole madtom

DESCRIPTION Comparatively small, with adipose fin forming a low, keel-like ridge continuous with upper lobe of caudal fin. The pectoral and dorsal fin spines, which can be locked into an erect position, are unusually sharp and are surrounded by a sheath that contains mild venom (Rohde et al. 1994).

Generally brown to tan above; sides lighter, with a narrow dark line along the mid-side; underside yellow-white or white.

Breeding males may have swollen lips and an enlarged head (Jenkins and Burkhead 1993).

SIMILAR SPECIES Catfish in the genus *Noturus*, commonly known as madtoms, can be distinguished from other catfishes by their comparatively small size and adipose fin. The tadpole madtom can be distinguished from other madtoms in the MSRB by its comparatively stout body, jaws of equal length, and the faint dark line that often runs down the side. It resembles a small bullhead more than other madtoms, but can be separated from the bullheads by the shape of the adipose fin.

SIZE Generally between 29 and 130 mm (Rohde et al. 1994).

MERISTICS Dorsal spine 1, dorsal rays 4–7, anal rays 12–18, pectoral spine 1, pectoral rays 5–10, pelvic rays 5–10, caudal rays 50–66 (Jones et al. 1978).

DISTRIBUTION From the Nueces River of Texas east to Florida, north on the Atlantic slope to New York; north through the Mississippi and Missouri River Valleys to Minnesota, in the Red River in North Dakota and Manitoba east to Quebec. Apparently absent from the Appalachian Highlands from Alabama to New York. Presumed introduced in Massachusetts and New Hampshire (Rohde *in* Lee et al. 1980). Introduced in several states, probably as a contaminant of catfish or bullhead stockings. In South Carolina, widely distributed in the Coastal Plain (Rohde et al. 1994). It has been collected from many sites in the MSRB.

HABITAT Most common in pools within small, slowly flowing streams, but also found in backwaters of larger rivers and shallow margins of lakes (Pflieger 1975; Robison and Buchanan 1984). Tadpole madtoms often occur in association with organic debris or aquatic vegetation over mud or sand bottoms (Robison and Buchanan 1984),

Noturus gyrinus, tadpole madtom, breeding male

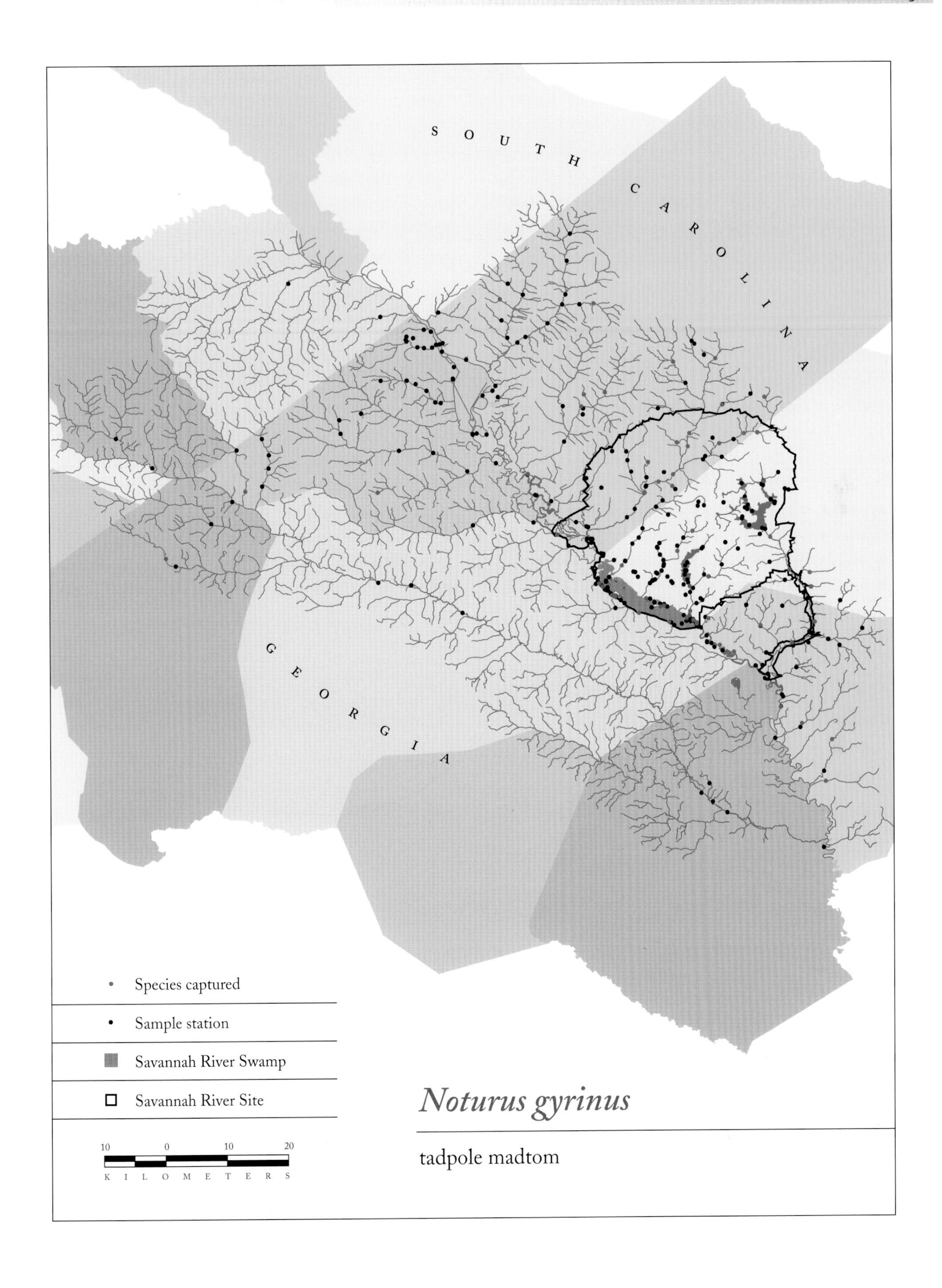

Noturus gyrinus

tadpole madtom

although they disappeared from the littoral zone of Lake Mendota, Wisconsin, following an explosive increase in the Eurasian water milfoil (*Myriophyllum spicatum*) (Lyons 1989). The tadpole madtom prefers slower-flowing waters than other madtoms in the MSRB. On the SRS, it is common in slow, silty areas in smaller streams and is often associated with snags and other woody debris (Meffe and Sheldon 1988; Paller 1994).

BIOLOGY Spawning begins in May and extends into the summer in Illinois (Whiteside and Burr 1986), but may occur earlier (April and May) in southern populations (Jenkins and Burkhead 1993). The small clusters of adhesive eggs (maximum of 117 eggs in a cluster [Jones et al. 1978]) are surrounded by a gelatinous envelope (Scott and Crossman 1973). The eggs are hidden in a sheltered area or cavity, including man-made objects such as tin cans and bottles, and guarded by both parents or by the male alone (Whiteside and Burr 1986). Jones et al. (1978) summarized descriptions of larvae and their distinguishing features. Sexual maturity occurs at approximately 1 year, and few individuals reach 4 years (Whiteside and Burr 1986; Rohde et al. 1994).

These secretive fish usually hide during the day amid submerged leaves, branches, or other heavy cover, or in cavities that they may enlarge by moving bottom material with the mouth (Cochran 1996). They emerge at night to forage on insect larvae, small crustaceans, other types of small invertebrates, and small fishes (Pflieger 1975). In SRS streams, small tadpole madtoms eat primarily chironomid larvae, caddisfly larvae, mayfly nymphs, and oligochaetes. Adults eat primarily crayfish, chironomid larvae, and other aquatic insects (Sheldon and Meffe 1993).

Because of their hardiness, small size, and interesting appearance madtoms are sometimes kept in home aquaria, and they will accept a variety of artificial and live foods. They are shy and usually remain hidden during the day but are active at night. Tadpole madtoms can swallow surprisingly large fish, so they are not the best tankmates for smaller species.

SCIENTIFIC NAME *Noturus* = "back tail" (i.e., tail over the back) (Greek), referring to the adipose fin; *gyrinus* = "tadpole" (Greek).

Noturus insignis (Richardson, 1836)

Margined madtom

DESCRIPTION Body comparatively elongate and slim. Head somewhat flattened; upper jaw projecting slightly beyond lower jaw. Adipose fin forming a low, keel-like ridge contiguous with the upper lobe of the caudal fin; pectoral and dorsal fin spines unusually sharp.

Color ranges from gray to light brown on the back, becoming lighter on the sides and white on the underside. Chin barbels generally white, and other barbels gray or brown.

SIMILAR SPECIES Margined madtoms can be distinguished from other madtoms in the MSRB by the dark outer margins of the median fins. However, Jenkins and Burkhead (1993) noted marked variation in the median fin pigmentation of margined madtoms from different localities.

SIZE The largest madtom of the MSRB, usually between 56 and 150 mm (Rohde et al. 1994).

MERISTICS Dorsal spine 1, dorsal rays 5–7, anal rays 15–21, pectoral spine 1, pectoral rays 7–10, pelvic rays 8–10, caudal rays 54–67 (Jones et al. 1978).

DISTRIBUTION Southeastern Lake Ontario drainages and along the Atlantic slope from New York to Georgia; also in upper Tennessee River tributaries, upper Ohio River tributaries, and the upper Kanawha (New) River (Rohde *in* Lee et al. 1980). Occurs in many mid-Atlantic drainages from southern New York to northern Georgia (Etnier and Starnes 1993). Found throughout most of South Carolina except for the Lower Coastal Plain and the northwest corner of the state (Rohde et al. 1994). It has been sporadically collected from a number of stream sites in the MSRB.

HABITAT Low- to high-gradient creeks and rivers. Commonly in Piedmont streams with moderate to swift currents, clear water, and gravel to rock substrates, where it prefers riffle habitats of moderate depth (Jones et al. 1978; Etnier and Starnes 1993). Also found over softer bottoms in Coastal Plain streams, although it typically avoids swampy areas (Jenkins and Burkhead 1993). On the SRS, found in relatively clear, small to medium-size streams with moderate currents and sand or gravel bottoms (Meffe and Sheldon 1988). Prefers swifter water than the tadpole madtom. Often associated with submerged logs, sticks, planks, and other types of cover.

BIOLOGY Spawning occurs during late June and early July in Pennsylvania (Clugston and Cooper 1960). In Virginia, lowland populations have mature ova in May, and upland populations in June (Jenkins and Burkhead 1993). The adhesive eggs are deposited in compact clusters of 54–200 in shallow excavations under stones or

Noturus insignis, margined madtom, breeding male

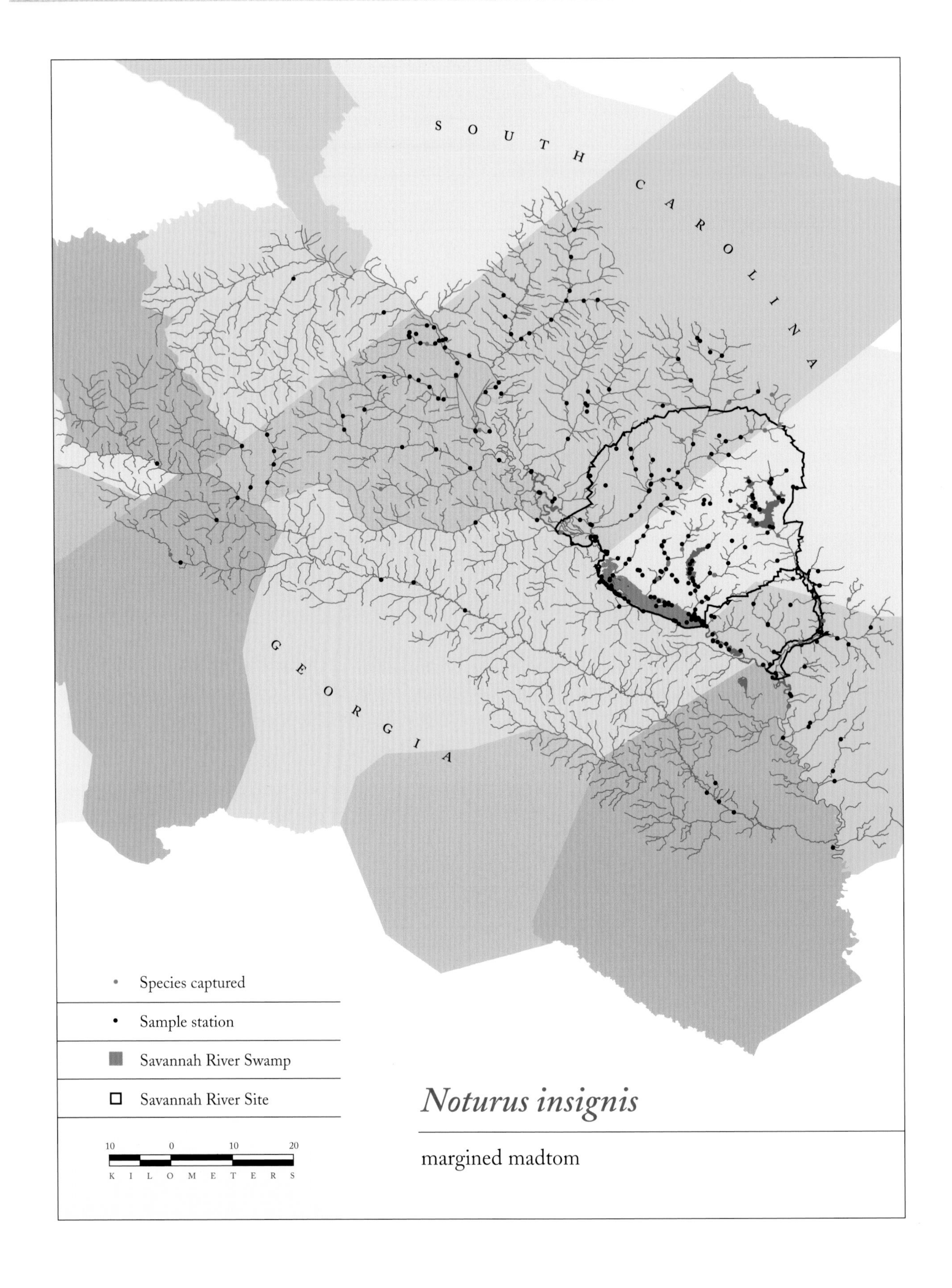
SOUTH CAROLINA
GEORGIA
Species captured
Sample station
Savannah River Swamp
Savannah River Site
10 0 10 20
KILOMETERS
Noturus insignis
margined madtom

other cover in quiet water near riffles (Jones et al. 1978). The nest is guarded by the male (Clugston and Cooper 1960), and hatching occurs in about 7 days at 28 °C (Stoeckel and Neves 2000). Jones et al. (1978) summarized descriptions of margined madtom larvae and their distinguishing features. Growth in Pennsylvania averages about 45 mm per year for the first 2 years and 20 mm per year thereafter (Clugston and Cooper 1960); females reach sexual maturity in their third summer, and males mate in their second or third summer. Longevity is at least 4 years (Clugston and Cooper 1960).

The margined madtom consumes bottom-feeding aquatic insects, other aquatic invertebrates, and small fish (Jenkins and Burkhead 1993). A population studied in the Delaware River fed primarily from dusk to about 0400 with an activity peak at midnight (Gutowski and Stauffer 1993). Individuals fed near the bottom on mayfly nymphs, chironomid larvae, and blackfly larvae; feeding appeared to be gape limited, and the average size of the organisms consumed was smaller than the average size of those in the substrate. On the SRS, small margined madtoms eat chironomid larvae, caddisfly larvae, mayfly larvae, and oligochaete worms; large margined madtoms consume chironomid larvae, crayfish, and a variety of other aquatic insect larvae (Sheldon and Meffe 1993).

SCIENTIFIC NAME *Noturus* = "back tail" (Greek), referring to the adipose fin; *insignis* = "remarkable" (Latin).

Noturus leptacanthus Jordan, 1877
Speckled madtom

DESCRIPTION A small madtom; body comparatively elongate and slender. Adipose fin forming a low, keel-like ridge contiguous with the upper lobe of the caudal fin. Pectoral and dorsal spines unusually sharp and surrounded by a sheath containing mild venom (Rohde et al. 1994). Upper jaw projecting slightly beyond lower jaw. Caudal fin square.

Color generally tan to reddish brown with the underside lighter to white.

SIMILAR SPECIES The speckled madtom can be distinguished from other madtoms by the numerous small black spots scattered over the upper body and fins.

SIZE This is the smallest madtom in the MSRB, generally 37–94 mm (Rohde et al. 1994).

MERISTICS Dorsal spine 1, anal rays 14–18, pectoral spine 1, pectoral rays 7–10, pelvic rays 7–9, caudal rays 47–57; gill rakers 5–8 (Etnier and Starnes 1993).

DISTRIBUTION On the Atlantic and Gulf slopes from the Edisto River of South Carolina to the Amite-Comite River drainage in Louisiana; in peninsular Florida, south into the upper St. Johns River (Rohde *in* Lee et al. 1980). In South Carolina, primarily in the southwestern portion of the state excluding the Lower Coastal Plain (Rohde et al. 1994). Speckled madtoms have been collected, generally in small numbers, from many streams in the MSRB and the Savannah River.

HABITAT Small to large streams with slow to moderate currents (Mettee et al. 1996). On the SRS, the speckled madtom is typically found in larger, faster-flowing streams (Meffe and Sheldon 1988), in this respect differing from the margined madtom, which prefers smaller streams, and the tadpole madtom, which prefers slower waters (Meffe and Sheldon 1988). Often found along stream margins in association with submerged leaves, sunken logs, brush, and other debris, over silt, sand, and gravel bottoms (Mettee et al. 1996).

BIOLOGY Spawning occurs from May through August in Mississippi, often in a discarded can or bottle that has been cleared of debris, presumably by the male (Clark 1978; Etnier and Starnes 1993). Other spawning sites include crevices between sunken logs and stones (Mettee et al. 1996). Females produce relatively few (14–45) large (~5.5 mm) eggs each year; the average clutch size is 17.6. Sexual maturity occurs at approximately 1 year and longevity seldom exceeds 2.5 years (Clark 1978).

On the SRS, small individuals eat primarily caddisfly larvae, chironomid pupae, and mayfly nymphs; large individuals eat primarily caddisfly larvae, chironomid larvae, mayfly nymphs, and other small invertebrates (Sheldon and Meffe 1993). Like other madtoms, this species is a secretive and seldom-seen bottom dweller.

SCIENTIFIC NAME *Noturus* "back tail" (Greek), referring to the adipose fin; *leptacanthus* = "slender spine" (Latin).

Noturus leptacanthus, speckled madtom; (left) breeding male, (right) female

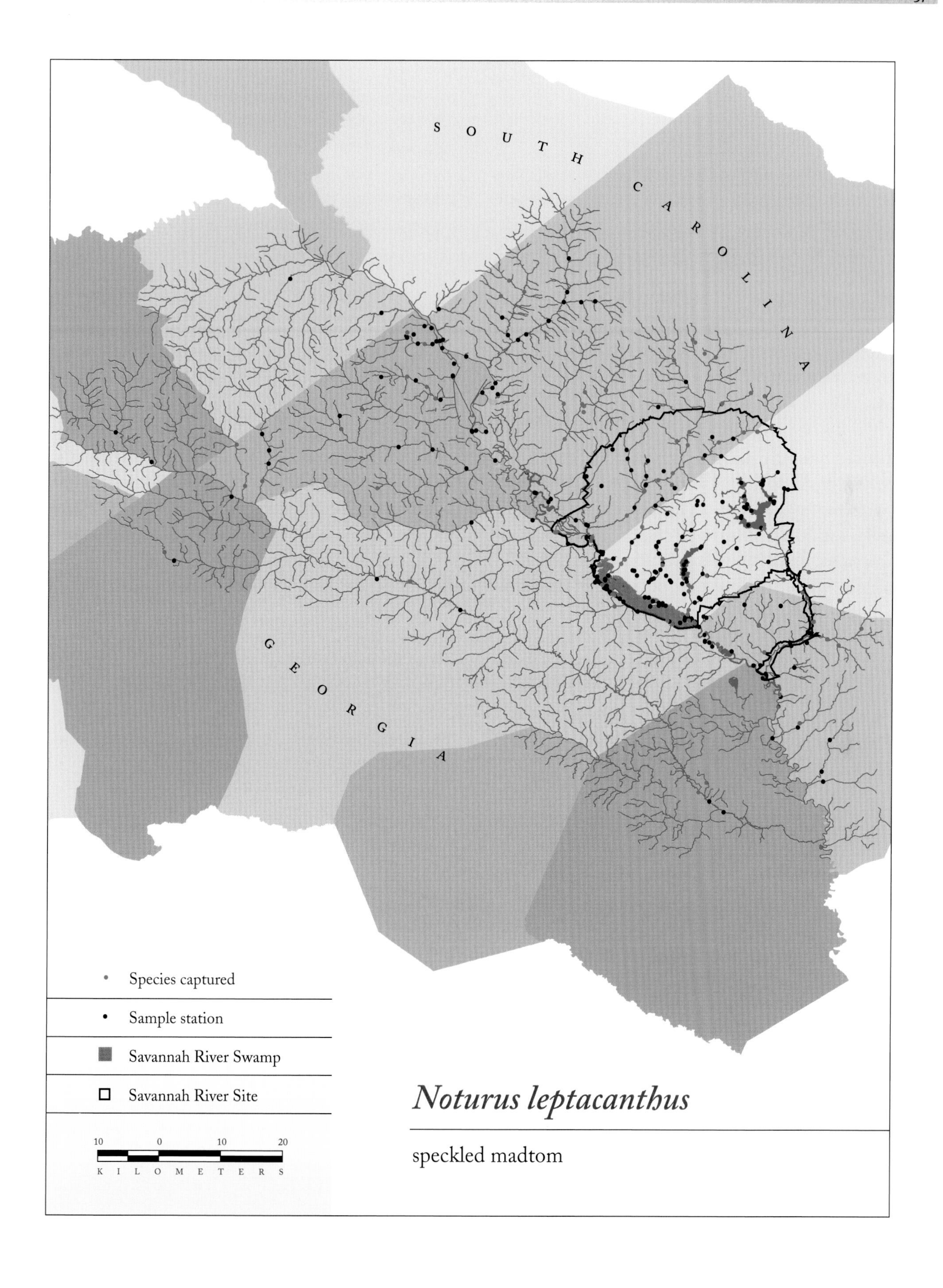

Noturus leptacanthus

speckled madtom

Pylodictis olivaris (Rafinesque, 1818)
Flathead catfish

DESCRIPTION Head broad and flattened; eyes small; lower jaw protruding; caudal fin square or slightly emarginate; anal fin comparatively short (14–17 rays) and rounded.

Mottled yellowish brown on the back and sides grading to creamy white or yellow on the belly; mottling usually stronger on younger fish. Upper lobe of caudal fin usually lighter than the rest of the caudal fin, although this feature may be lost in large adults (Etnier and Starnes 1993). Upper lobe of caudal fin may be orange in juveniles (Rohde et al. 1994).

SIMILAR SPECIES Flathead catfish can be easily distinguished from other large catfishes in the MSRB by the square rather than forked caudal fin. The palatine teeth patches with lateral backward-projecting extensions and protruding lower jaw distinguish it from other local ictalurids. The coloration of the caudal fin of small individuals is also unique.

SIZE It is not unusual for adults to reach 115 cm and 20 kg (Pflieger 1975). The South Carolina record flathead catfish, caught in the Santee-Cooper Diversion Canal, weighed approximately 35.9 kg (SCDNR).

MERISTICS Dorsal spine 1, dorsal rays 6, anal rays 14–17, pectoral spine 1 with serrations on anterior and posterior edges, pectoral rays 10–11, pelvic rays 9 (Etnier and Starnes 1993; Jenkins and Burkhead 1993).

CONSERVATION STATUS Introduced; uncertain whether it has or will become established.

DISTRIBUTION Native to large rivers of the Mississippi, Missouri, and Ohio river basins south and west into northeastern Mexico (Glodek *in* Lee et al. 1980). Introduced into a large number of Atlantic slope drainages from Pennsylvania to Florida and in Gulf Coast drainages in Georgia, Florida, and Alabama; also widely introduced in western states such as California, Idaho, Arizona, New Mexico, Colorado, and Wyoming. At the time of this writing, the flathead catfish is very rare in the MSRB and has only been collected just below the New Savannah Bluff Lock and Dam. It is, however, common in Clarks Hill/J. Strom Thurmond Lake upstream of the MSRB.

HABITAT Large to medium, slow-flowing rivers and reservoirs (Etnier and Starnes 1993). Adults are usually found in deep pools in association with sunken timber or other shelter; juveniles may be found in shallower areas near submerged cover, sometimes in riffle habitats (Etnier and

Pylodictis olivaris, flathead catfish

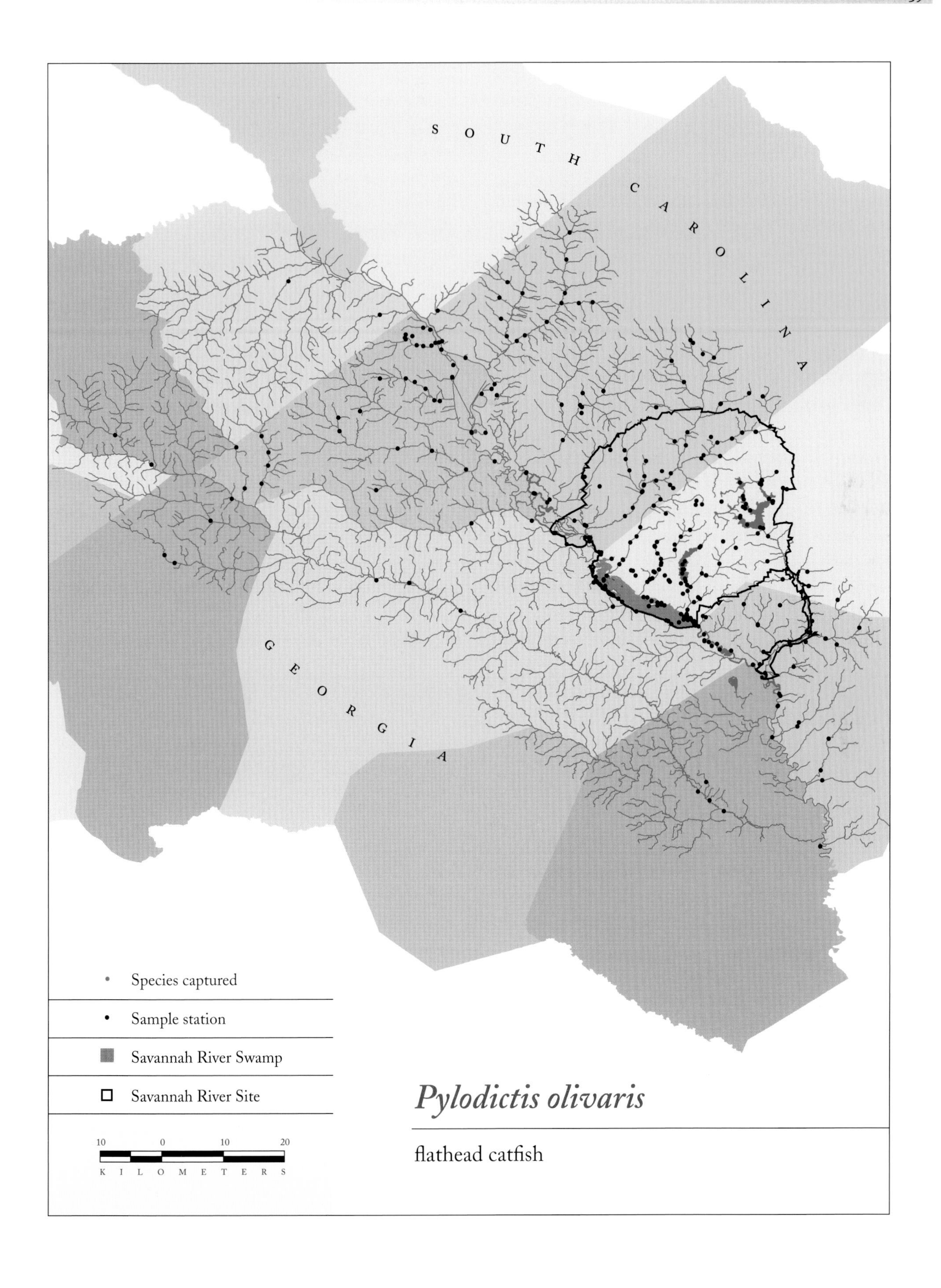

Pylodictis olivaris

flathead catfish

Starnes 1993). Flathead catfish studied in western Texas preferred hard bottoms such as gravel, rocks, and boulders, as well as submerged timber and artificial structures (Weller and Winter 2001).

BIOLOGY Spawning occurs during the summer, mainly June and July (Pflieger 1975; Jenkins and Burkhead 1993), in a nest that has been fanned out by one or both parents in a natural cavity or in the shelter of a submerged structure (Pflieger 1975; Rohde et al. 1994). A single large female can produce a mass of up to 100,000 eggs, which are aerated and swept clean by the fin movements of the male (Etnier and Starnes 1993). The young remain in a compact school near the nest for several days, with the male remaining to guard them, then disperse and begin a solitary existence, which they continue as adults. A single piece of cover, such as a submerged log, will generally harbor only one individual (Pflieger 1975). Sexual maturity is reached in 4 or 5 years (Rohde et al. 1994) at a minimum size of approximately 390 mm (Munger et al. 1994). Individuals may live as long as 26 years (Rohde et al. 1994).

The flathead catfish is primarily a nocturnal feeder and may move from the deeper water occupied during the day into shallow water to feed (Pflieger 1975). Although some catfishes are considered to be opportunistic scavengers, adult flathead catfish are highly piscivorous, consuming shad, bullheads, sunfishes, and other species (Turner and Summerfelt 1970; Layher and Boles 1980). The young rely largely on aquatic insects and crayfish until they become large enough to eat fish (Layher and Boles 1980).

In areas where they have been introduced, flathead catfish often become the dominant predators and may adversely affect prey populations (Barr and Ney 1993; Weller and Robbins 1999). In the Edisto River system in South Carolina, for example, flathead catfish are thought to be responsible for reductions in native bullhead and redbreast sunfish populations. Studies in Illinois suggest that they may prefer to consume gizzard shad when available rather than sunfishes or channel catfish (Hass et al. 2001).

SCIENTIFIC NAME *Pylodictis* = "mud fish" (Greek); *olivaris* = "olive colored" (Latin).

Esocidae (pikes and pickerels)

Pikes and pickerels are freshwater fishes that live in Arctic and temperate waters of northern Europe and Asia and northern and eastern North America. The single genus has five species. *Esox lucius*, the northern pike, occurs in both Eurasia and North America; *E. reicherti* is restricted to the Amur River of Asia; and the other three species are native to North America (Crossman 1996). Of these, the two pickerels (redfin pickerel, *E. americanus*, and chain pickerel, *E. niger*) are found in the MSRB and are native to Georgia and South Carolina. The muskellunge (*E. masquinongy*) originally had a more northern distribution but has been introduced in the southeastern United States. An artificial hybrid between the muskellunge and the northern pike, the tiger musky, has also been introduced in the southeastern United States. The fossil record indicates that the geographical range of this family has been shrinking. The closest living relatives of the esocids are the mudminnows (family Umbridae). Traditionally these two families have been considered together as a suborder of the order Salmoniformes, although some recent classifications place pikes and mudminnows in the order Esociformes.

Because esocids are ambush predators, they generally prefer still or quietly flowing water with rooted aquatic plants or other cover. They rarely enter brackish water.

Esocids vary in size from the 1 m muskellunge to the 30 cm redfin pickerel. Pikes and pickerels are elongate fish with dorsal and anal fins set far back on the body and a complete lateral line. The elongated jaws roughly resemble a duck's beak and have strongly toothed jaws, palate, vomer, and tongue. The lower jaw protrudes slightly. The gill rakers are reduced to patches of sharp denticles. The scales are cycloid, and there are no spines in any of the fins. All members of this family are early spring to early summer spawners. They typically spawn in groups of one female and two or more males, scattering eggs across the substrate.

In many areas, pikes and pickerels are valued as food, sport, and trophy fish. Because of their efficiency as predators, they may be keystone species in some ecosystems. *Pike—Biology and Exploitation* (Craig 1996) provides detailed information on the biology of this family.

KEY TO THE SPECIES OF ESOCIDAE

1a. Long snout. Distance from tip of snout to center of eye greater than distance from center of eye to posterior edge of gill cover, the ratio greater than 1.1 (Plate 15 b); body of adults—but not juveniles—with chainlike pattern; highly reflective patches in center of all scales on sides of body forming distinct horizontal rows (Plate 15 d); dorsal rays 19–21; anal rays 17–19; branchiostegal rays (13)14–17; lateral line scales 110–138; subocular bar generally vertical or slanted forward from the eye (Plate 15 b); fins not heavily colored . . . . *Esox niger* (chain pickerel), p. 246

1b. Shorter snout. Distance from tip of snout to center of eye generally less than distance from center of eye to posterior edge of gill cover, the ratio less than 1.1 (Plate 15 a); body never with chain pattern, but generally with olive bars; fewer scales with highly reflective patches on sides of body (Plate 15 c); dorsal rays (15)16–17(18); anal rays (14)15–16(17); branchiostegal rays (10)12–14(15); lateral line scales 92–118; subocular bar angled posterior or vertical from the eye (Plate 15 a); fins of adults red . . . . *Esox americanus* (redfin pickerel), p. 243

Note: In Esocidae all fin rays should be included in count, both rudimentary and principal rays (Hubbs and Lagler 1958).

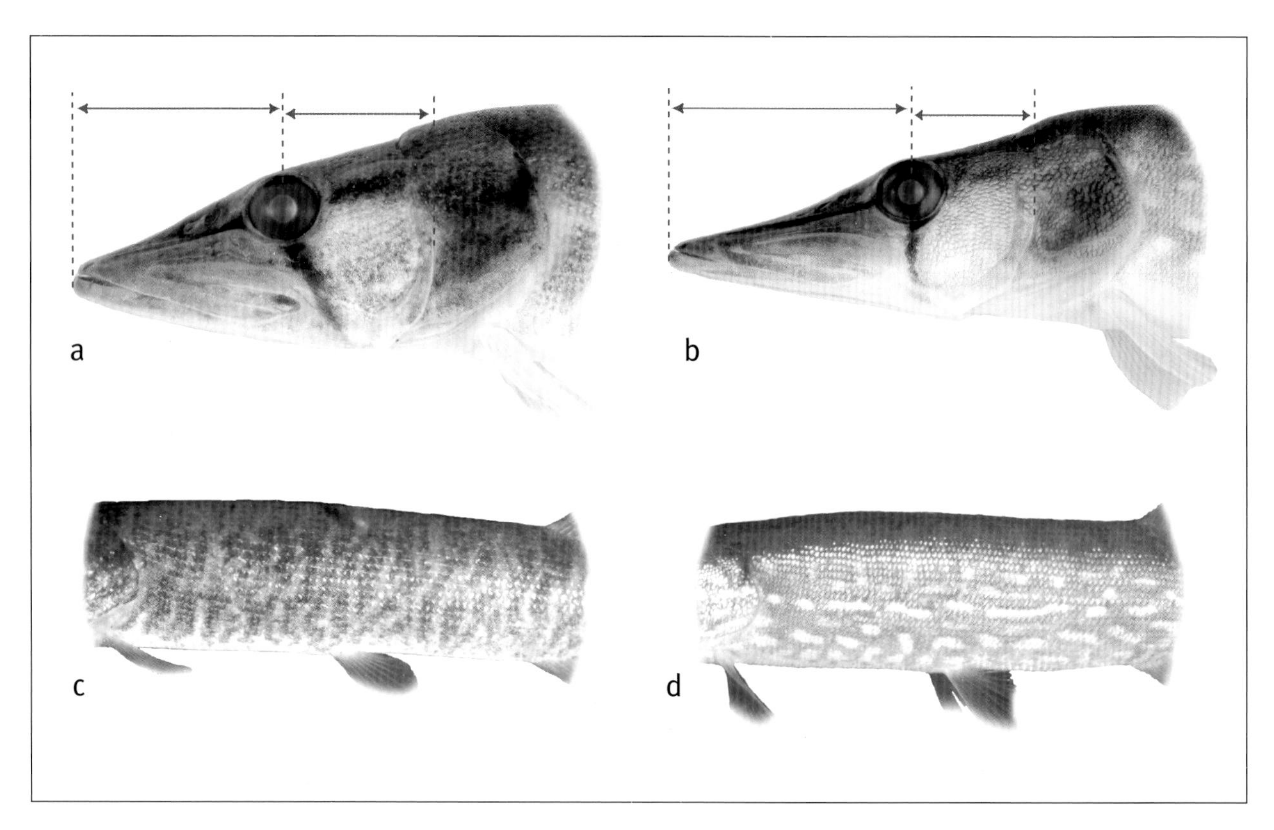

PLATE 15. Morphological characters of Esocidae: (a) redfin pickerel head showing snout shape and length, and suborbital bar, (b) chain pickerel head showing snout shape and length, and suborbital bar, (c) redfin pickerel lateral color pattern, (d) chain pickerel lateral color pattern.

Esox americanus Gmelin, 1788
Redfin pickerel

DESCRIPTION Body robust, cylindrical. Dorsal and anal fins located above each other and relatively close to caudal fin (on posterior one-third of body). Mouth large and terminal, snout resembling a duck's bill. Undersurface of lower jaw with four sensory pores on each side. Cheek and operculum fully scaled, while top of head lacks scales. Lateral line complete.

Fins generally bright red in larger adults, but orange or dusky in smaller individuals. Adults have a dark olive green dorsum, olive to black wavy vertical bars on sides, and white ventrum. Suborbital bar (under eye) vertical or slanted backward. Small juveniles have an olive mottled dorsum; dark sides with a thick, iridescent, light yellow green midlateral stripe (light in preservative); and transparent or dusky fins, but dorsal and anal fins may have orange tinge.

SIMILAR SPECIES The young of the chain pickerel (*E. niger*) and the redfin pickerel are easy to confuse. The snout of the redfin pickerel is shorter and somewhat wider than that of the chain pickerel. The iridescent green midlateral stripe (light in preservative) on small juveniles is generally thinner and more wavy in the chain pickerel. Coloration and meristic features will aid in distinguishing between the species. The redfin pickerel generally has fewer branchiostegal rays, dorsal rays, and anal rays than the chain pickerel. The redfin pickerel hybridizes with the chain pickerel (Crossman and Buss 1965), which can complicate identification.

SIZE Adults may reach 350 mm and 288 g in the SRS (Bennett and McFarlane 1983). In the MSRB, most are less than 200 mm SL, and individuals over 250 mm SL are rare.

MERISTICS Dorsal rays (15)16–17(18), anal rays (14)15–16(17) (Crossman 1960), pectoral rays (13)14–15(17), pelvic rays (8)9–10 (Jenkins and Burkhead 1993); branchiostegal rays (10)12–14(15), lateral line scales 92–118 (Crossman 1960).

DISTRIBUTION The St. Lawrence River of Quebec south along the Atlantic Coast east of the Appalachian Mountains well into Florida, west along the Gulf Coast to eastern Texas and Oklahoma, and from there north through the Mississippi Valley into the Great Lakes drainages (Crossman *in* Lee et al. 1980). West of the Appalachians and throughout the Mississippi River and Missouri River drainages, represented by the subspecies *E. a. vermiculatus*; east of the Appalachians, represented by *E. a. americanus*. Populations in Florida, western Georgia, Alabama, and eastern Mississippi are intergrades between subspecies (Crossman *in* Lee et al. 1980). Abundant in small streams of the MSRB year-round.

HABITAT Vegetated shallow waters (Meffee and Sheldon 1988), especially in sluggish streams and backwaters of faster-flowing streams; can be abundant in darkly stained

Esox americanus, redfin pickerel; adult or older juvenile

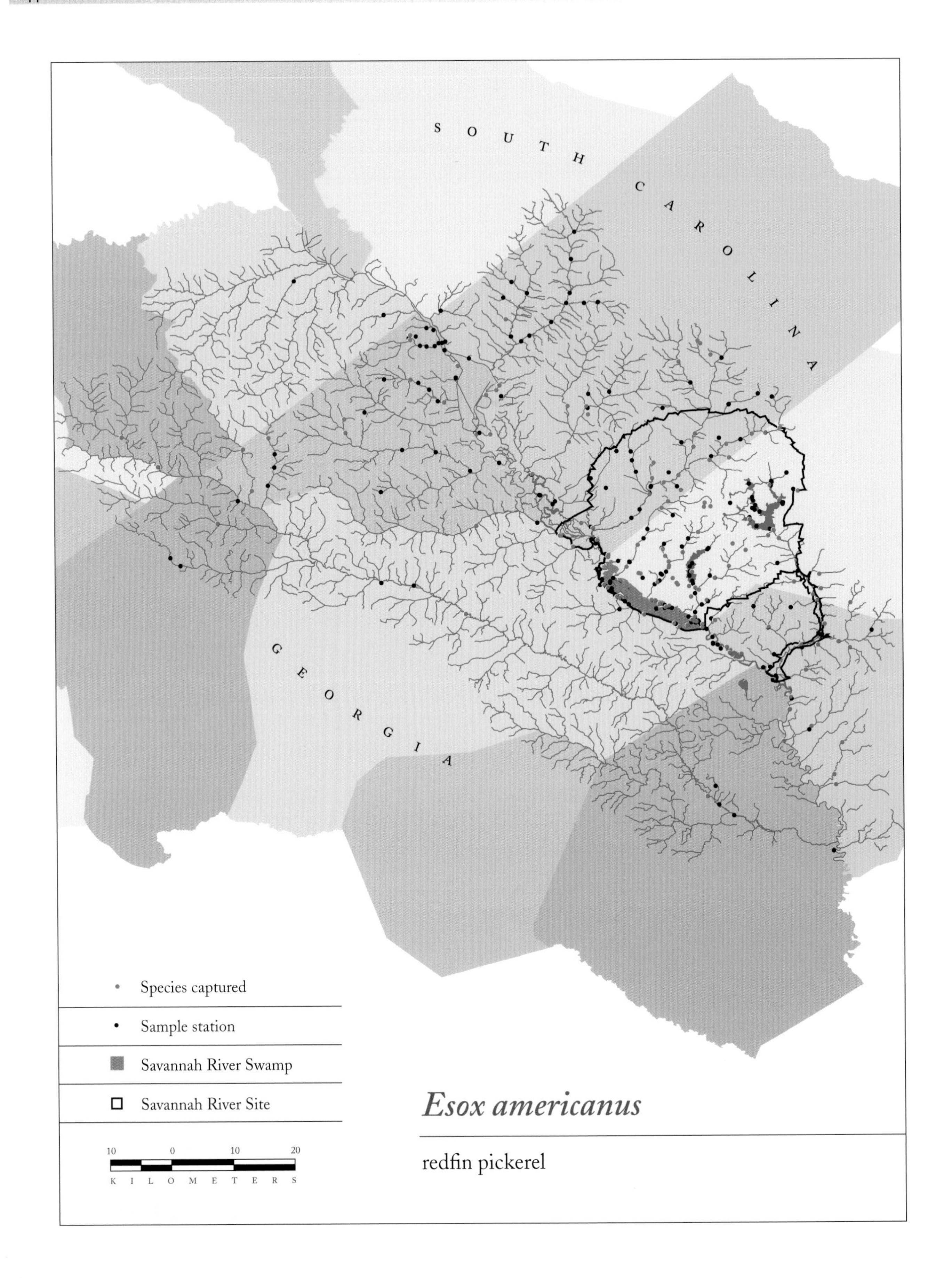

Esox americanus

redfin pickerel

Esox americanus, redfin pickerel; juvenile

waters with a low pH. The redfin pickerel seems to avoid lakes, ponds, and areas lacking vegetation. The closely related chain pickerel is found in the same general habitats, but seems to prefer clearer and more heavily vegetated waters (Etnier and Starnes 1993).

BIOLOGY Spawning occurs over a period of 2–4 weeks in late winter and early spring when the water temperature is between 4 and 10 °C as reported for North Carolina by Crossman (1962). There is some evidence that spawning can take place over a longer period. Based on back-calculated hatch dates, Ballek (1994) reported hatching from November through March in South Carolina; Brim (1991) found larvae in October in the same area, possibly in response to floods. Malloy and Martin (1982) found gravid females at 3.5 °C and got apparently normal development of eggs in the range of 1–6 °C. Larval development appeared normal at temperatures up to 12 °C, but the authors reported significant egg and larval mortality at 15 °C. The small, adhesive eggs are scattered in small numbers among underwater plants in floodplains, grassy banks, and other areas with heavy vegetation (Crossman 1962; Wang and Kernehan 1979). See Jones et al. 1978 for a description of the larvae.

Based on information from North Carolina, sexual maturity is probably reached at the age of 2 in the MSRB (Crossman 1962), and longevity is 3 or 4 years. Oklahoma populations (subspecies *E. a. vermiculatus*) seldom live more than 3 years (Ming 1968), while northern populations may live 7 or 8 years (possibly also the subspecies *E. a. vermiculatus*) (Crossman 1960).

The redfin pickerel is a voracious predator. Larvae and smaller juveniles eat zooplankton and insects. Individuals as small as 31 mm SL will feed opportunistically on fish larvae, and young-of-the-year may switch to piscivory at 50 mm TL (Mittelbach and Persson 1998). Adults eat fishes predominantly but often include crayfish and other invertebrates in the diet. Prey items may be up to 50% of the pickerel's own length (Ballek 1994).

SCIENTIFIC NAME *Esox* = "pike" (Latin); *americanus* = "from America."

Esox niger Lesueur, 1818

Chain pickerel

DESCRIPTION Body slender, elongate, and somewhat compressed. Mouth large and terminal. Lateral line complete. Snout resembles a duck's bill. Dorsal and anal fins located above each other and relatively close to the caudal fin (on posterior one-third of the body). Undersurface of lower jaw with four sensory pores on each side. Cheek and opercle fully scaled, while top of head lacks scales.

Dorsum of adults dark green to olive, but may be mottled; the sides have a light yellow-green ground color with darker olive green chainlike markings. The belly is generally white but may be yellowish. Scales on sides have highly reflective iridescent patches that form distinct horizontal rows. Subocular bar (under the eye) generally vertical or slightly slanted forward. Dorsal and anal fins generally olive, paired fins pale to orange. No fins are ever bright red as in adult redfin pickerel. Juveniles lack the chainlike markings on the sides and are often darkly colored. Small juveniles have an iridescent green midlateral stripe (light in preservative) similar to the redfin pickerel's but generally thinner and more wavy.

SIMILAR SPECIES The snout of the chain pickerel is longer and somewhat narrower than that of the redfin pickerel (*E. americanus*). The young of the chain pickerel and the redfin pickerel are easy to confuse (see redfin pickerel species account for distinguishing characters). The chain pickerel hybridizes with the redfin pickerel (Crossman and Buss 1965) and the northern pike (*E. lucius*) (Crossman *in* Lee et al. 1980), which can complicate species identification.

SIZE Adults may reach 99 cm TL (Rohde et al. 1994) but are generally no longer than 60 cm in the MSRB.

MERISTICS Dorsal rays 19–21, anal rays 17–19 (Crossman 1960), pectoral rays 12–15, pelvic rays 9–10 (Jenkins and Burkhead 1993); lateral line scales 110–138; branchiostegal rays (13)14–17 (Crossman 1960).

DISTRIBUTION Native to the Atlantic slope from New England to Florida, the Gulf slope, and the lower Mississippi Valley (Crossman 1978). Populations along the Texas-Oklahoma border may be relicts or introduced. Introduced and/or dispersed into Quebec, Nova Scotia, New Brunswick, the Lake Erie drainages of New York, and parts of Ohio, Kentucky, Iowa, Minnesota, and Colorado (Crossman *in* Lee et al. 1980). East of the Appalachian Mountains, more abundant on the Coastal Plain, but also found above the Fall Line (Crossman *in* Lee et al. 1980). Introduced and established on the New York side of Lake Ontario (Cudmore-Vokey and Crossman 2000). Chain pickerel are common in the MSRB year-round.

HABITAT Clear, well-vegetated water bodies; they seem to prefer sluggish creeks, rivers, ditches, and natural and artificial ponds and lakes over substrates of silt or mud.

Esox niger, chain pickerel; adult or older juvenile

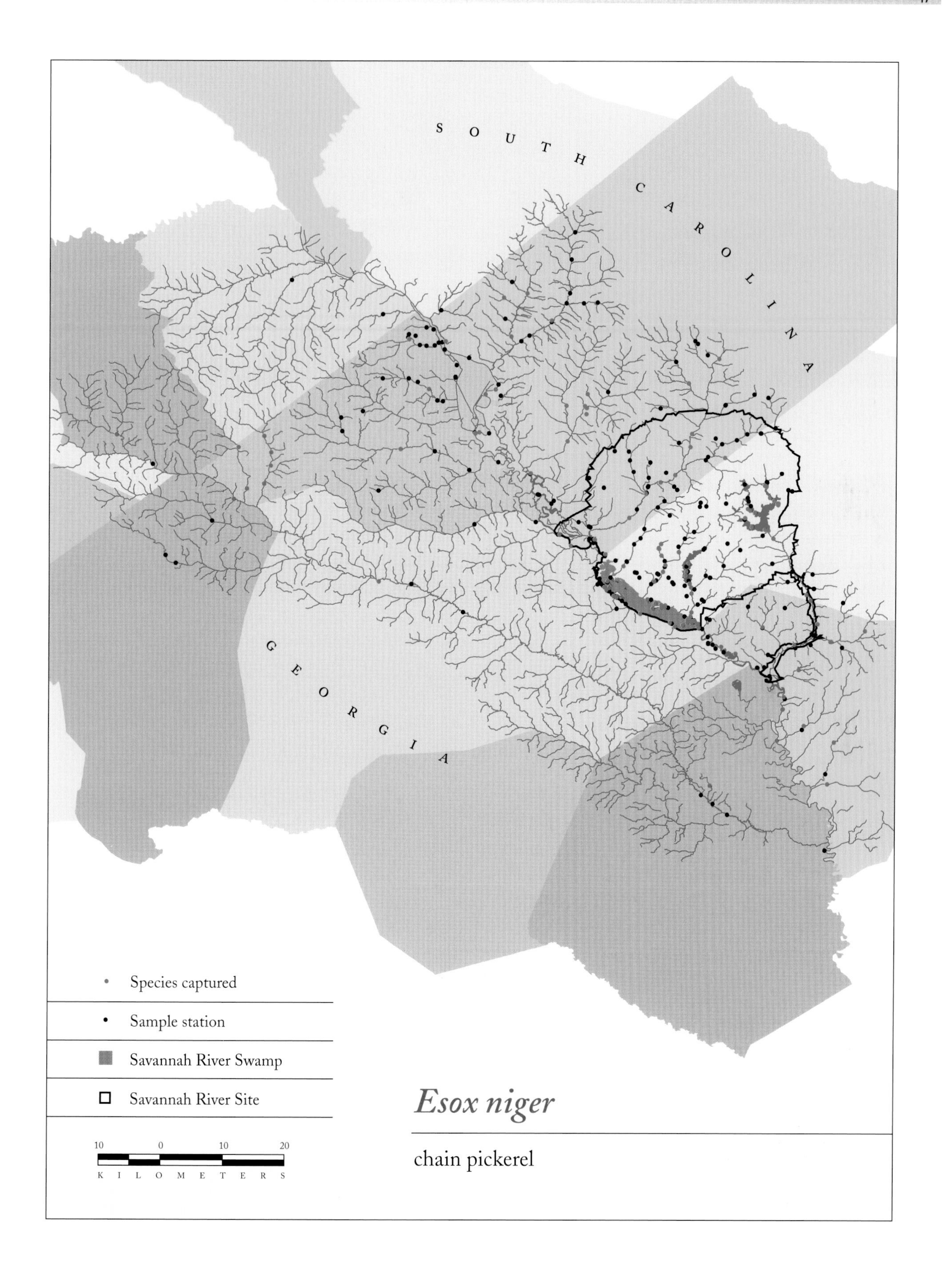

Esox niger

chain pickerel

Esox niger, chain pickerel; juvenile

In Arkansas, most abundant in lakes below the Fall Line (Robison and Buchanan 1984). Based on catch rates, clearly more common in oxbow lakes than in rivers in Florida (Beecher et al. 1977). Adults appear to be more abundant in deeper water, moving to shallow areas at night, while the juveniles are mostly found in shallow waters. The chain pickerel is probably more abundant in lakes and less likely to penetrate smaller creeks than the closely related redfin pickerel (Laerm and Freeman 1986), which seems to prefer waters less vegetated and less clear (Etnier and Starnes 1993).

BIOLOGY The chain pickerel is a predaceous, solitary fish. Spawning occurs in shallow, well-vegetated waters in winter and spring, earlier in the south (December–February in Florida) than in the north (March–May in Delaware) (Underhill 1949; Jenkins and Burkhead 1993). See Mansueti and Hardy 1967; or Jones et al. 1978 for descriptions of larvae. Sexual maturity can be reached at age 1 in the south and at age 2 or 3 in the northern part of the distribution (Etnier and Starnes 1993). The maximum life span is 8 or 9 years in the north, possibly less in the southern range (Etnier and Starnes 1993). Considered a game fish on the east coast of the United States (Etnier and Starnes 1993).

Larvae and juveniles eat zooplankton and insects; the adults are piscivores, but the diet continues to include crustaceans, insects, and other food items (Meyers and Muncy 1962; Germann and Swanson 1978). In the MSRB, chain pickerels ranging from 17 to 47 cm TL eat primarily fish, along with some aquatic insects (Wiltz 1993). Young-of-the-year may become piscivorous at 100–150 mm TL (Mittelbach and Persson 1998); in Lake Conway, Florida, fish enter the diet of juveniles between 51 and 75 mm SL (Guillory 1979a). The most frequent food item may be sunfishes and other game fishes (McIlwain 1970). Larger *E. niger* eat more fish species than smaller ones do, and prey size rather than prey abundance may determine prey selection (Guillory 1979a).

SCIENTIFIC NAME *Esox* = "pike" (Latin); *niger* = "dark" or "black" (Latin) (coloration of juveniles).

Umbridae (mudminnows)

The family Umbridae contains three genera and six species. The genus *Novumbra*, with one species (Olympic mudminnow, *N. hubbsi*), is restricted to the Olympic peninsula of Washington State. *Dallia* has two species, one restricted to Siberia and the other found in Alaska and Siberia (Alaska blackfish, *Dallia pectoralis*). The other genus, *Umbra*, has three species, one found in central North America (central mudminnow, *U. limi*), one found in eastern North America (the eastern mudminnow, *U. pygmaea*; the only representative of the family in the MSRB), and a third species native to the Danube basin of Europe. Systematically, umbrids are most closely related to the pikes and pickerels (family Esocidae). During the flexion of the notochord in the larval development of both Esocidae and Umbridae, the tip of the notochord may extend well beyond the developing hypural plates to form a transient separate lobe to the developing caudal fin. The only other taxa in the MSRB that exhibit this larval phenomenon are the gars, in which it is much more pronounced.

Mudminnows are small freshwater fishes that live in waters that are frequently low in dissolved oxygen. They have an elongate and moderately compressed body with a short snout and cycloid scales; they have no lateral line, and there are no spines in any of the fins. The caudal fins are rounded and the dorsal and pelvic fins are located posteriorly. The physostomous swimbladder can be used for breathing atmospheric oxygen. Where they are abundant, mudminnows are collected for bait.

Umbra pygmaea (DeKay, 1842)

Eastern mudminnow

DESCRIPTION Body relatively plump. Head slightly flattened; snout blunt. Caudal fin rounded. Distinctive brown color with black bar or elongated blotch located just before base of caudal fin. Ten to 14 thin, dark stripes may be present along top and sides of body.

SIMILAR SPECIES The eastern mudminnow is a small, soft-rayed, minnowlike fish that can be distinguished from the true minnows by its lack of a lateral line and by the position of its dorsal and anal fins, which are located far back on the body. The short, rounded dorsal fin distinguishes this species from juvenile bowfin.

SIZE Adults are typically 50–100 mm TL (Rhode et al. 1994), but some reach 110 mm.

MERISTICS Dorsal rays 13–17, anal rays 9–12, pelvic rays 6, pectoral rays 12–16; lateral series scales 30–34 (Jones et al. 1978; Jenkins and Burkhead 1993).

DISTRIBUTION From southeastern New York, including Long Island, south, mostly on the Coastal Plain, to the St. Johns River of Florida and west to the Aucilla River drainage of Georgia and Florida (Gilbert *in* Lee et al. 1980). Introduced in the Netherlands (Dederen et al. 1986) and elsewhere. Sparse to common throughout the South Carolina Coastal Plain (Rohde et al. 1994). The eastern mudminnow is occasionally collected from stream and swamp habitats in the MSRB, but it is seldom abundant.

HABITAT Small, sluggish streams, ponds, swamps, sloughs, and beaver impoundments (Etnier and Starnes 1993; Jenkins and Burkhead 1993) in areas where the bottom is soft and silty, and there is considerable organic debris or vegetation. The eastern mudminnow is tolerant of low pH. In introduced populations studied in the Netherlands, abundance was inversely related to pH and the presence of predatory fishes, and the eastern mudminnow was one of the few species found at a pH below 4.0 (Dederen et al. 1986).

BIOLOGY Spawning occurs during March through at least April at water temperatures of 14–15 °C (see summary in Jones et al. 1978). However, Malloy and Martin (1982) in Maryland found gravid females at 3.5 °C, apparently normal development of eggs in the range of 1–6 °C, normal larval development at temperatures up to

Umbra pygmaea, eastern mudminnow

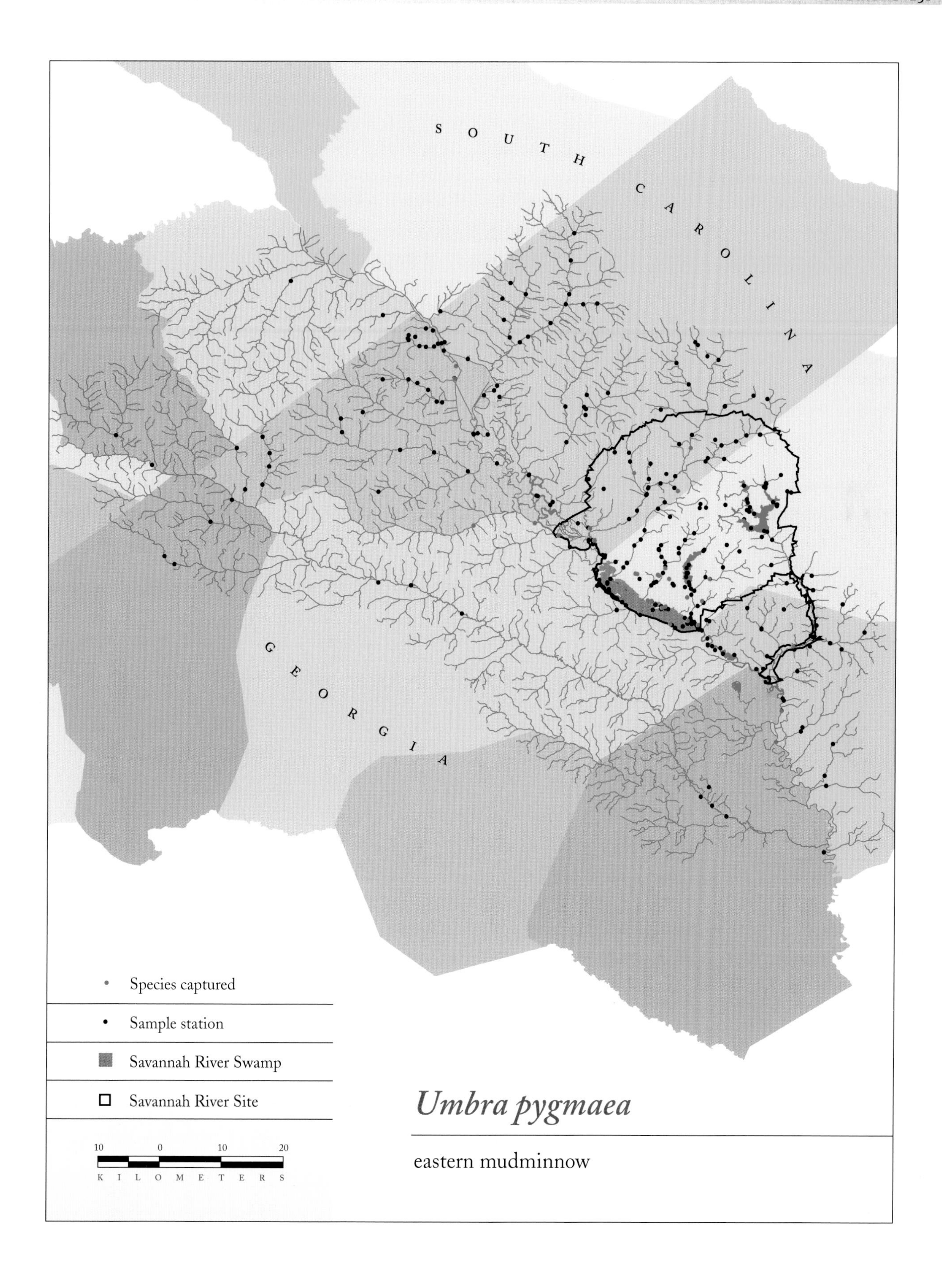

Umbra pygmaea

eastern mudminnow

12 °C, but significant egg and larval mortality at 15 °C. Spawning occurs in shallow, heavily vegetated waters under debris or in other sheltered areas following a spawning display by the male (Breder and Rosen 1966). The adhesive eggs are guarded and fanned by the female (Rohde et al. 1994). Fecundity is 31–2,566, with an average of 342 eggs per clutch (Jones et al. 1978). Sexual maturity is reached in 1 or 2 years, and longevity seldom exceeds 4 years.

Mudminnows consume a variety of small invertebrates, including aquatic insect larvae, small insects, amphipods, isopods, ostracods, small crayfish, and snails (see summary in Jenkins and Burkhead 1993). They may also eat small amounts of plant matter, protozoans, and small fishes. They may stalk or ambush their prey (Etnier and Starnes 1993).

Eastern mudminnows are physostomous; that is, a duct connects the pharynx to a vascularized swimbladder, allowing them to obtain oxygen by gulping air at the water surface and enabling them to inhabit sluggish waters with little dissolved oxygen (Etnier and Starnes 1993). They may burrow into the mud to avoid predators and to survive adverse environmental conditions (Robison and Buchanan 1984; Etnier and Starnes 1993; Rohde et al. 1994).

Because of their tolerance for acidity and low dissolved oxygen concentrations, eastern mudminnows are sometimes found in habitats that support few other fishes. However, they were largely extirpated from an Ohio stream as a result of habitat destruction caused by channelization (Trautman and Gartman 1974), indicating they are not necessarily tolerant of anthropogenic disturbance.

SCIENTIFIC NAME *Umbra* = "a shadow" (Latin); *pygmaea* = "pygmy" (Latin).

Salmonidae (salmons, trouts, and chars)

The 65 or so species of salmons, trouts, and chars live in cold waters of North America and Eurasia. Some species reside in fresh water year-round, others are anadromous. Worldwide they are prized by commercial and recreational fishermen. Many species are extensively cultured for food and recreational purposes. Anglers' enthusiasm for trout has led to the introduction of many species in areas outside their native range, even in locations where reproduction, or even long-term survival, is unlikely. Some trout species have been thoroughly domesticated and are genetically quite distinct from the original stocks brought into hatchery programs more than a century ago. The rainbow trout, *Oncorhynchus mykiss*, and more recently the brown trout, *Salmo trutta*, have been reported from the MSRB. Both species have been stocked into the section of the Savannah River between Clarks Hill/J. Strom Thurmond Dam and New Savannah River Bluff Lock and Dam (Bettross 2000). This reach of the Savannah River was unable to support a trout fishery because of high water temperatures and low dissolved oxygen concentrations. Since there is no reliable indication that the introduced fish survived for a significant length of time, current capture is unlikely unless more are introduced. The rainbow trout produced in hatcheries is probably the result of hybridization of two, three, or more trout species and contains genetic material from several different natural populations. The brook trout (*Salvelinus fontinalis*) occurs in headwater streams of the Savannah River and is native to mountain streams in western South Carolina. It does not occur naturally in the Piedmont or Coastal Plain of South Carolina or Georgia, and has not been recorded from the MSRB.

Salmo trutta, brown trout. This species was introduced in the MSRB, but we know of no subsequent recaptures.

KEY TO THE SPECIES OF SALMONIDAE

1a. Caudal fin profusely covered with small black spots; in life, side marked with a pink or red longitudinal stripe and many small black spots . . . . *Oncorhynchus mykiss* (rainbow trout), p. 255

1b. Spots on caudal fin absent or faint; in life, side not marked with a pink longitudinal stripe, but with large reddish spots . . . . *Salmo trutta* (brown trout)—no species account

Oncorhynchus mykiss (Walbaum, 1792)

Rainbow trout

The rainbow trout, also called the steelhead trout, was previously known as *Salmo gairdneri* but was subsequently merged with *Salmo mykiss* when the two were determined to be conspecific. Studies also determined western trout to be more closely related to *Oncorhynchus*, the Pacific salmons, than to *Salmo*. Consequently, salmon and trout taxonomy was adjusted, leaving only the Atlantic salmon, brown trout, and their Eurasian relatives in the genus *Salmo* (Smith and Stearley 1989).

DESCRIPTION Body moderately compressed. Mouth large, terminal, and slightly oblique. Posterior edge of maxillary usually extending beyond eye. Males develop a hook in the lower jaw during spawning. Fins without spines; fleshy adipose fin present between the dorsal and caudal fins; axillary process present at base of the pelvic fin. Caudal fin strongly emarginate or slightly forked. Scales generally very tiny and cycloid, but variable in different stocks (Vernon and McMynn 1957).

The broad pink to red lateral stripe along the sides that gives this fish its common name is accentuated in breeding adults; the skin is green dorsally and white to yellowish ventrally (Rohde et al. 1994). Many small dark spots are scattered across the back, sides, head, and caudal fin. Pectoral, pelvic, and anal fins are generally pale. Large, anadromous adults (steelhead form) are lustrous blue or silvery with few or no spots (Becker 1983). No nuptial tubercles are present, but minor changes to the head, mouth, and color occur in spawning males (Mettee et al. 1996). Juveniles have 5–10 distinctive, dark oval blotches on the sides called parr marks.

SIMILAR SPECIES The brown trout, also at times introduced into the Savannah River, lacks the pink or reddish longitudinal stripe, has large reddish spots on its side, and lacks distinct spots on the caudal fin. Rainbow trout also have dark spots on the snout anterior to the nares that are absent in brown trout.

SIZE Adults can reach 737 mm (Mettee et al. 1996) in the United States and up to 900 mm in lakes in British Columbia (Larkin et al. 1957).

MERISTICS Dorsal rays 11–17, pectoral rays 11–17, pelvic rays 9–10, caudal fin rays 19; branchiostegal rays 9–13; lateral line scales 100–150 (Scott and Crossman 1973); anal rays (12)14–16 (Jenkins and Burkhead 1993); pyloric caeca 27–80 (varies with size) (Northcote and Paterson 1960).

CONSERVATION STATUS At times, seasonally introduced into the Savannah River.

DISTRIBUTION Native in Asia from the Amur River and the Kamchatka peninsula plus the Commander Islands, and in North America from the Kuskokwim River of Alaska along the coast to Baja California Norte and the

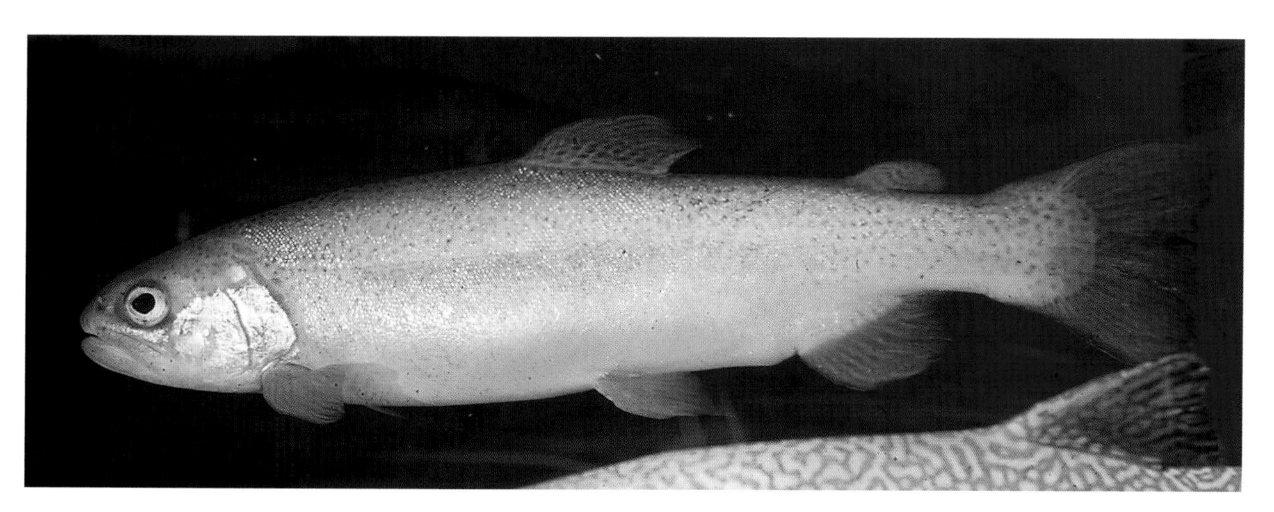

Oncorhynchus mykiss, rainbow trout

Río Presidio, Durango, Mexico; and inland in the headwaters of the Fraser River, British Columbia, Canada, the Columbia River and Snake River basins, plus the headwaters of the Mackenzie River basin, Canada; introduced and established in North America throughout the mountainous areas of the western United States and Canada, parts of North Dakota, Manitoba, and Ontario; the Great Lakes and St. Lawrence River basins; the Appalachian Mountains south into Georgia and Alabama; the Ozark and Ouachita highlands of Arkansas, Missouri, and Oklahoma; the Sierra Madre Occidental of southern central Mexico; and areas of South Dakota and Nebraska (Behnke *in* Lee et al. 1980). Introduced and reproducing in streams above 800 m elevation in Hawaii (Brock 1960; Maciolek 1984).

HABITAT Cold-water streams (12–19 °C) in mesohabitats ranging from pools to calm eddies in riffles (Mettee et al. 1996). Widely adaptable and widely stocked in creeks, rivers, lakes, and reservoirs. Reproducing populations require a seasonal drop in temperature below 13 °C (Mettee et al. 1996). The original native rainbow trout populations were extremely variable over the wide distribution, and that genetic diversity has been utilized by fish culturists to produce strains that tolerate warm water, grow rapidly, are disease resistant, spawn at different times of the year, and have different migratory patterns (Etnier and Starns 1993). While cooler temperatures are preferred, temperatures up to 28 °C may be tolerated (Becker 1983).

BIOLOGY Rainbow trout spawn in smaller tributaries of rivers or in inlets or outlet streams of lakes from March to August (Lindsey et al. 1959). In the mid-Atlantic area, spawning generally occurs from mid-April to late June (Rohde et al. 1994) at water temperatures of 10–15 °C (Scott and Crossman 1973). Fecundity of females averages 2,500–4,500 (Carlendar 1969). Nests (or redds) are built in gravel or sandy substrate at all times of day. The female sweeps a depression in the substrate and deposits her eggs (800–1,000 in each redd), which are fertilized sequentially by nearby males. The female then covers the fertilized eggs with gravel (Scott and Crossman 1973). No additional parental care is provided.

Etnier and Starnes (1993) reported growth in smaller streams in the Blue Ridge area of Tennessee as follows: age 1, 80–120 mm; age 2, 160–200 mm; age 3, 200–240 mm. Wales (1941) described the early development of eggs and alevins (newly emerged young). Brayton (1981) set longevity at 3–4 years in Virginia, and Carlander (1969) put it at 7 years.

Rainbow trout feed close to the bottom on aquatic and terrestrial insects, molluscs, crustaceans, fish eggs, and small fishes, including other trout. In the Great Lakes, adults feed mostly on alewives (Becker 1983). The young feed predominantly on zooplankton (Scott and Crossman 1973). Rainbow trout raised in hatcheries are a domesticated form derived primarily from a mixture of coastal steelhead populations (Behnke 1992); this is probably the form introduced into the Savannah River.

SCIENTIFIC NAME *Oncorhynchus* = "hook snout" (Greek), referring to the hooked jaw of breeding males; *mykiss* (Latin) = derivation of *mikizha* or *mykyz*, the Kamchatkan word for trout.

Aphredoderidae (pirate perch)

Aphredoderidae is a monotypic family, and the pirate perch, *Aphredoderus sayanus*, is the only living species. While the living pirate perch is restricted to the eastern United States, fossil forms have been found in the western United States as well; all fossil forms appear to have been restricted to North American fresh water.

The location of the adult's vent in the throat region anterior to the pelvic fins has long intrigued biologists. In very small juveniles, the vent is in a more typical posterior position just ahead of the anal fin, but as the fish grows, the vent migrates forward to a position behind a knoblike mass of thoracic muscle just posterior to the gill membranes (Jordan 1878; Mansueti 1963). Advantages of this placement have only recently been discovered (see details in species account). The pirate perch is generally thought to be most closely related to the cavefishes (family Amblyopsidae), which likewise have an anteriorly placed vent.

Aphredoderus sayanus (Gilliams, 1824)

Pirate perch

DESCRIPTION Body stout; head large. Mouth terminal and large, the lower jaw projecting slightly. Anus positioned in the throat region ahead of the pelvic fins in adults. Scales ctenoid. Lateral line incomplete or occasionally complete, but generally best developed as an arch extending posteriorly from behind the gill cover to above the anal fin origin. Caudal fin margin square or slightly emarginate. Head with an extensive array of sensory pores (Moore and Burris 1956). Operculum with a sharp spine, and rear margin of preoperculum strongly serrate.

The body may be slate gray to black with a pink or purple cast but is usually grayish chocolate brown, often with a purplish sheen on the lower side of the head; lighter colored ventrally. A dark vertical bar lies near the base of the caudal fin. Males may be nearly black during breeding. A dark vertical bar is present below the eye (suborbital bar), and usually a horizontal bar behind the eye (postorbital bar). The caudal fin may have white margins at the upper and lower posterior corners. Posterior margins of the dorsal and anal fins also generally have a light margin.

SIZE To about 144 mm.

MERISTICS Dorsal spines 3–4, dorsal rays (9)10–12, anal spines (2)3, anal rays 6–7, pectoral rays 12–13, pelvic spine 1, pelvic rays 6–7, principal caudal rays 17–19; lateral series scales 47–60; branchiostegals 6, gill rakers on first arch 10–12, rakers are blunt.

Note: The first of the three anal spines is very small compared with the others, but it is almost always present. Care must be exercised in detecting the first spine.

DISTRIBUTION The Atlantic Coast from New York to Florida (not in southern Florida [Loftus and Kushlan 1987]), west to Texas, and up the Mississippi Valley to the Lower Peninsula of Michigan with disjunct populations in Lake Erie and Lake Ontario drainages of New York (Lee *in* Lee et al. 1980). Pirate perch are common and widely distributed in the MSRB, especially in tributary streams, where they are among the few fishes that may be common from the smallest headwater to the largest tributary.

HABITAT Small to large tributaries, swamps, ponds, and ditches; frequently associated with undercut banks, dense vegetation, complex woody debris, and root masses (Abbott 1868; Becker 1923; Monzyk et al. 1997).

Aphredoderus sayanus, pirate perch

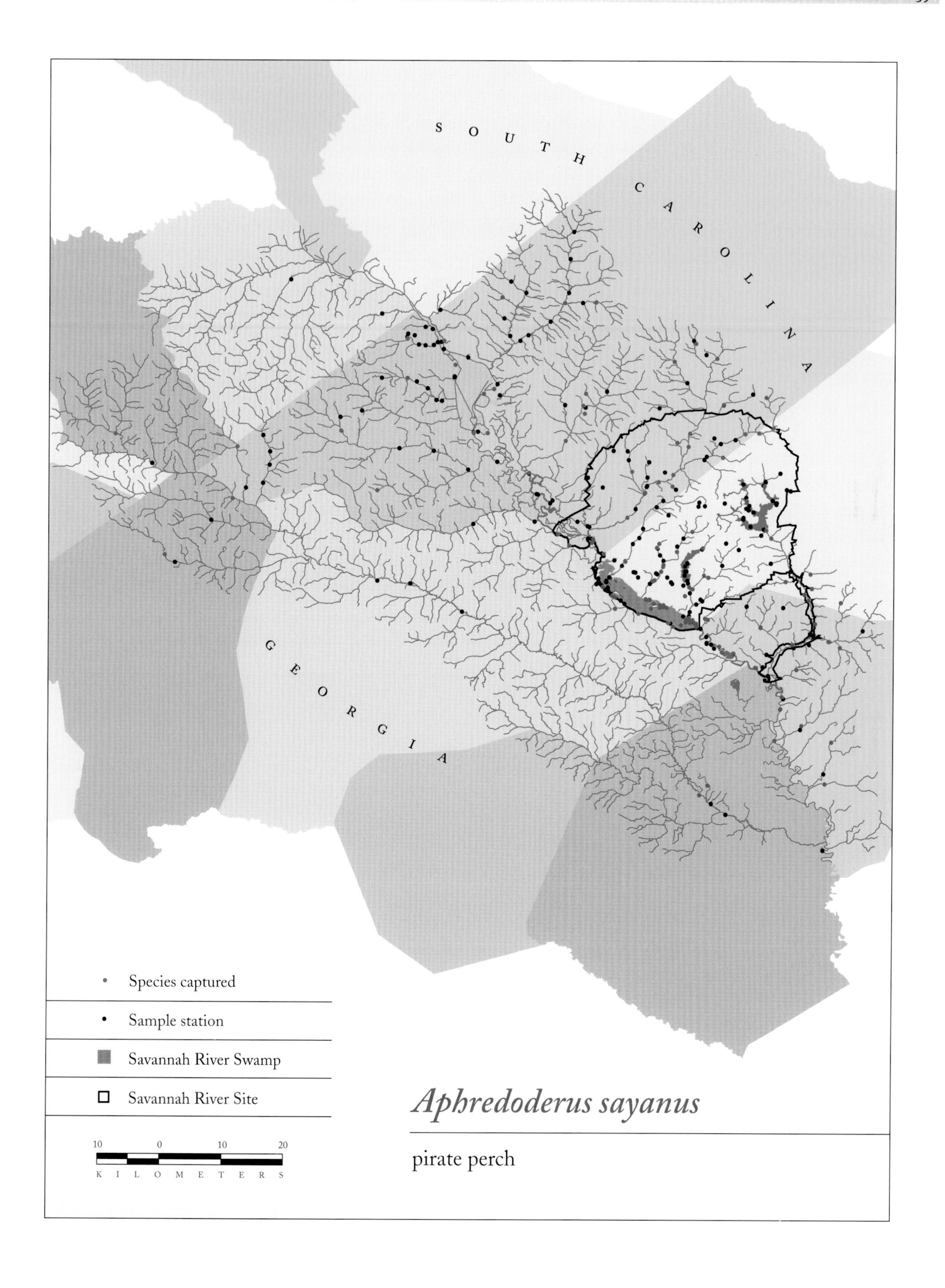

Aphredoderus sayanus

pirate perch

BIOLOGY Breeding begins in late winter or early spring. In the MSRB, males may begin releasing milt (when the abdomen is lightly pressed) in December, but spawning generally occurs from January through April, peaking in the middle of this range (DEF unpub. data). Spawning is reported to occur from January through March in North Carolina (Shepherd and Huish 1978), as early as February in east Texas (Martin and Hubbs 1973), and in January in northern Florida and southern Georgia (Swift et al. 1977).

The anterior (jugular) location of the urogenital pore in the pirate perch has led to controversy concerning its spawning habits. Some authors suggested oral incubation of eggs (Martin and Hubbs 1973; Pflieger 1975), based primarily on three observations: (1) the eggs artificially stripped from a female seemed to move forward along a groove from the vent to the branchial chamber (Martin and Hubbs 1973); (2) at least one species of cavefish (Amblyopsidae, *Amblyopsis spelea*), which also has an anteriorly located vent, has been reported to be a gill brooder (Breder and Rosen 1966); and (3) three eggs were found in the branchial cavity of a preserved pirate perch specimen (Boltz and Stauffer 1986). Other authors, based on aquarium (Brill 1977; Katula 1987, 1992; Fontentot and Rutherford 1999) or field (Abbott 1861, 1868, 1870) observations, suggested that the eggs are broadcast across the bottom substrate, or that parents construct a nest for the eggs, or that old sunfish nests are used for spawning. Fletcher et al. (2004) assessed actual spawning behavior, and their field observations, underwater filming, and genetic analyses of parentage revealed that pirate perch hid their eggs in cavities within dense root masses of woody riparian plants, or occasionally within the roots of aquatic macrophytes. The canals used for spawning were generally about 1 cm in diameter and up to several centimeters deep, and eggs were deposited in loose clusters. While the adults appeared to construct some of the canals, others had been made by burrowing aquatic insects and salamanders. Spawning was sometimes sequential, with the female first entering the narrow canal and depositing eggs while a male awaited his turn to enter and fertilize them. Since both sexes swim forward into the narrow canals and then back out, the anteriorly located urogenital pore facilitates these behaviors. At times, multiple canals led to the same egg cluster, and it may be that multiple males spawn with a single female. Although the male defended the oviposition site, sometimes by plugging the entrance with his body, no evidence of extended parental care was observed. The roots protected the nests from the swift-flowing water—up to 39 cm/sec at the nest site. Some nests were adjacent to water flowing 45–60 cm/sec or more, including a nest adjacent to a cobble riffle.

Most pirate perch live only 2–3 years (Hall and Jenkins 1954; Shepherd and Huish 1978), and few live more than 4 years. Pirate perch are primarily nocturnal, with activity peaking at dawn and dusk (Abbott 1861; Parker and Simco 1975). They are reported to be rather sedentary, with the greatest movements in the winter (Whitehurst 1981).

Pirate perch feed primarily on aquatic insects, especially chironomids, and crustaceans such as shrimp, amphipods, isopods, cladocerans, and copepods, but occasionally eat terrestrial insects and small fishes (Gunning and Lewis 1955; Shepherd and Huish 1978; Sheldon and Meffe 1993).

SCIENTIFIC NAME *Aphredoderus* = excrement throat (Greek), referring to the position of the urogenital opening near the gills; *sayanus* = named in honor of the naturalist Thomas Say.

Amblyopsidae (cavefishes)

The secretive cavefishes are confined to fresh water in caves, springs, and swamps. All are small fish and either have small rudimentary eyes or totally lack obvious eyes. The scales are small, cycloid, and embedded. All species except the northern cavefish, *Amblyopsis spelaea*, lack pelvic fins. The anal and urogenital openings are located on the throat near the gill openings and the large mouth. Most species incubate the developing eggs in their large oral cavity. Distribution of this family is restricted to eastern North America south of the areas formerly covered by Pleistocene glaciation. There are six species in either four or five genera (depending on the authority). All species living in caves are rare and potentially endangered. In some karst areas, cavefish are occasionally found in well water and are considered by the locals to be a sign that the water is safe to drink (Warren et al. 2000). The swampfish (*Chologaster cornuta*) is the only member of the family not restricted to limestone areas. It is not rare and may be much more abundant than is commonly thought because of its presumed nocturnal behavior and inaccessible habitat.

Chologaster cornuta Agassiz, 1853

Swampfish

DESCRIPTION Body spindle shaped with a deep and slightly compressed caudal peduncle. Head flattened for bottom dwelling. Eyes dorsolaterally placed and small, but functional. Pelvic fins absent; pectoral fins laterally placed, rather large, and rounded; caudal fin sharply rounded. Anus in the more normal position ahead of anal fin in juveniles, but migrates forward to the jugular position in adults. Scales tiny, embedded, and cycloid. Males develop a fleshy Y-shaped protuberance on the upper jaw (see photo below); females may have small bumps there.

Dorsum and dorsolateral sides are a distinctive dark brown with a reddish orange hue—contrasting sharply with the lighter ventral sides with an oranger cast—anteriorly fading to pearly white on the flanks. With two or three darker longitudinal stripes on each side: a dark midlateral stripe extending from snout through eye to caudal peduncle and separating the dark dorsal and light ventral sides; a dark dorsolateral stripe above it that may fade above the pectoral fins or near the caudal base; and sometimes a ventrolateral stripe that is highly variable when present and may consist of a thin, fine line or may become wide and diffuse anteriorly between origins of anal and pectoral fins. Pectoral and anal fins whitish gray. Dorsal fin often with dark mottling. Caudal fin with dark vertical blotch at base. Central fin rays and membranes dark; shorter dorsal and ventral rays light. During the spawning season, the gonads of both sexes are visible through the ventral body wall (DEF pers. obs.).

SIZE To about 57 mm SL (Cooper and Rohde *in* Lee et al. 1980); adults in the MSRB are more commonly 29–40 mm SL.

MERISTICS Dorsal rays (9)11(12); anal rays 9–10; pectoral rays 10(11) (Woods and Inger 1957) or (9)10(11) (Jenkins and Burkhead 1993), branched caudal rays 9–11 (Woods and Inger 1957); lateral line scales 60–62 (Sterba and Tucker 1973) or about 70 (Smith 1907).

DISTRIBUTION The Coastal Plain from southeast Virginia to east-central Georgia (Cooper and Rohde *in* Lee et al. 1980). Most of our records for the MSRB are from the Savannah River Swamp. This species is often thought

Chologaster cornuta, swampfish; adult

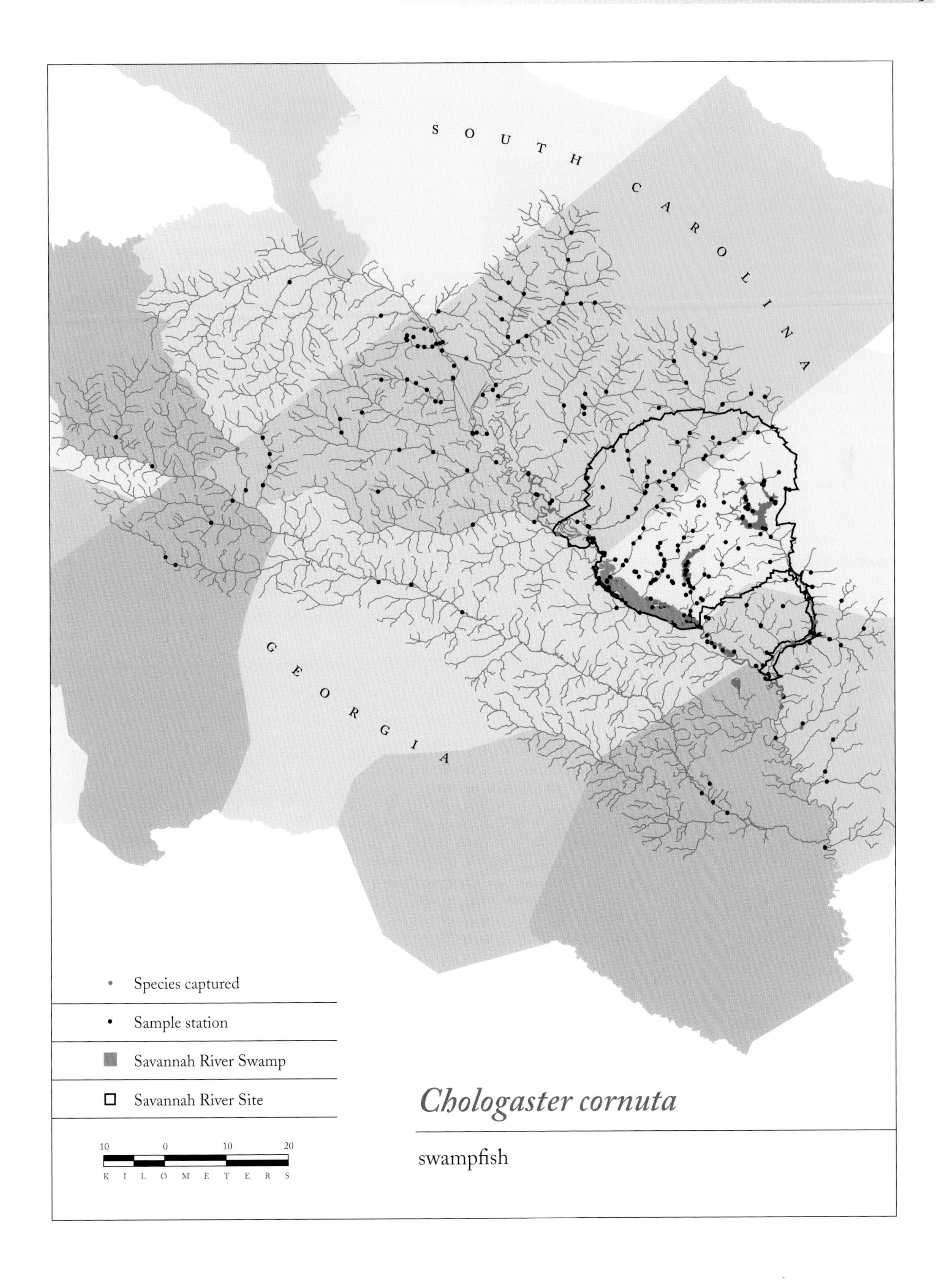
SOUTH CAROLINA
GEORGIA
Species captured
Sample station
Savannah River Swamp
Savannah River Site
10 0 10 20
KILOMETERS
Chologaster cornuta
swampfish

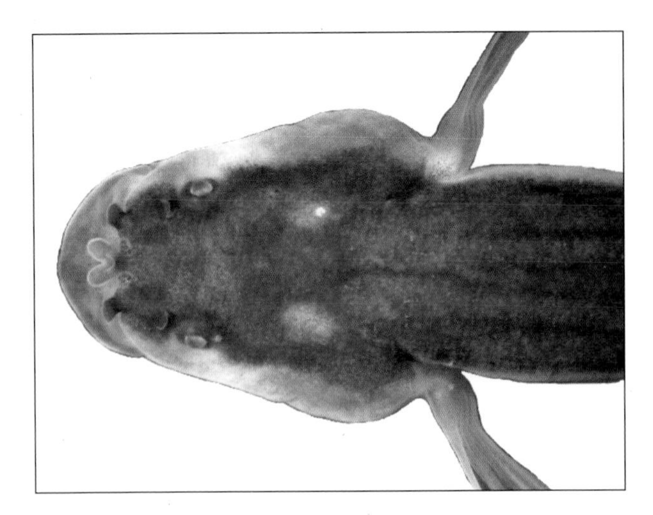

Chologaster cornuta, swampfish; snout of male

to be rare, but that may be because populations are localized and nocturnal (Poulson 1963). The habitat is also frequently difficult to sample efficiently.

HABITAT More frequently in acidic water heavily stained with tannins and humic acids, but also found in clearer streams (e.g., Fourmile Branch and Steel Creek of the MSRB). As their name suggests, swampfish are primarily found in calm waters such as swamps, ditches, ponds, and sluggish creeks, often in dense aquatic vegetation, especially around sphagnum moss (Rohde et al. 1994). They prefer open, well-shaded streams where the temperature remains below 23 °C and there are tangles of roots and debris (Poulson 1963).

BIOLOGY Spawning occurs primarily in March and April, but gravid females have been found in the MSRB as late as the first week of May (DEF pers. obs.). Although incubation in the buccal cavity has been reported in other members of the Amblyopsidae, such as *Amblyopsis spelaea* (summarized in Breder and Rosen 1966), and Poulson and White (1969) reported buccal incubation in cavefish, this is probably not the case for swampfish, as the volume of the eggs is too large to fit (Rohde et al. 1994). In light of the observations by Fletcher et al. (2004) on the spawning behavior of pirate perch, which share the character of a jugular placed anus, it seems probable that the swampfish likewise deposits its eggs inside root masses. Longevity is about 2 years.

Swampfish feed nocturnally, primarily on small crustaceans and insects. Feeding appears to be primarily by stalking or ambush (Poulson 1963). Their directional sense of smell seems to be well developed; the eyes "probably form images" (Poulson 1963).

SCIENTIFIC NAME *Chologaster*, from *cholos* = "maimed" (Greek), and *gaster* = "belly" (Greek), referring to the poorly developed pelvic fins; *cornuta* = "horned" (Latin), referring to the flaps of the nostril.

Belonidae (needlefishes)

Needlefishes are primarily coastal marine fishes that enter fresh water, although a few species are confined to fresh water. Some species live in the open ocean far from the continental shelf. Needlefishes are recognized by their elongate shape, and most species have greatly elongated upper and lower jaws of equal length with numerous teeth. All have dorsal fins located far back on the body, emarginate to forked caudal fins, and pectoral fins set high on the body just behind the gill openings. The back of most species is greenish or bluish with a reflective, silvery horizontal stripe much like that on silversides. The body of larger marine species may be almost totally silver with a darker back. Smaller species such as the Atlantic needlefish tend to have translucent musculature that probably reduces their visibility to predators. Larval needlefish have short, equal-length jaws. However, during development the lower jaw begins to elongate first, and at that stage the fish resemble members of the closely related family Hemiramphidae (halfbeaks) (Boughton et al. 1991). Needlefish eggs have varying numbers of chorionic filaments, some or all of which may be elongate. These filaments become tangled together so that the eggs tend to occur in clusters. Some species have internal fertilization and bear live young. All needlefishes are predatory and largely piscivorous. They usually do not attack prey items that are straight ahead, but instead snap sideways. Some of the larger species are considered to be high-quality food, but the smaller ones are usually either ignored by fishermen or used as bait. The Atlantic needlefish, *Strongylura marina*, is the only species described from the MSRB.

Strongylura marina (Walbaum, 1792)
Atlantic needlefish

DESCRIPTION Body slender, elongate, nearly cylindrical. Jaws long and slender with sharply pointed teeth. Dorsal fin and anal fins set far back on body, dorsal fin origin behind origin of anal fin; distal edge of both fins falcate. Pelvic fins slightly closer to caudal fins than to pectoral fins. Scales tiny and cycloid.

Body translucent yellowish to greenish above, silvery on the sides, and whitish pale below. A dark green middorsal stripe is present when viewed from above. Fins translucent to white.

SIMILAR SPECIES Gars (*Lepisosteus* spp.) have a generally similar body shape but are easily distinguished from the Atlantic needlefish by their larger scales and rounded anal fin.

SIZE Up to 122 cm TL but generally much smaller (32.5 cm) on the SRS (Bennett and McFarlane 1983).

MERISTICS Dorsal rays 14–17, anal rays 16–20, pectoral rays (10)11(12) (Berry and Rivas 1962), gill rakers absent or vestigial.

DISTRIBUTION From Casco Bay, Maine, to Rio de Janeiro, Brazil, ascending upstream in coastal rivers to the Fall Line (Burgess *in* Lee et al. 1980). This species appears to have entered the Tennessee River through the Tennessee-Tombigbee Waterway and has been taken in the Kentucky Reservoir, Tennessee (Etnier and Starnes 1993). It is uncommon in the MSRB.

HABITAT Generally in shallow marine waters, but may ascend several hundred kilometers up rivers. Found in larger stream systems, usually near the water surface and around structures like trees and docks. This behavior and the fast swimming speed complicate efforts to collect needlefishes. In the MSRB, they are reported only from the Savannah River and appear to be most common in the summer months.

BIOLOGY Spawning takes place in estuaries and bays, and perhaps also in fresh water (Virginia: Massman 1954; Florida: Hellier 1967). The spawning season is May–August (Breder 1962; Foster 1974; Able and Fahay 1998). The eggs are around 3 mm in diameter, demersal, and are attached via their chorionic filaments in dense masses to algae, underwater vegetation, or other structures (Foster 1974; Able and Fahay 1998). Eggs and larvae were described by Foster (1974) and Hardy (1978). The larvae are darkly pigmented, about 14 mm at the time of hatching, and are elongate with elongate jaws (Able and Fahay 1998). During larval and juvenile development the lower jaw grows faster than the upper jaw, and the fish resemble halfbeaks (Able and Fahay 1998). When the juveniles are about 100 mm, both jaws are again about the same length. Growth is rapid, and juveniles reach 19–40 cm by

Strongylura marina, Atlantic needlefish

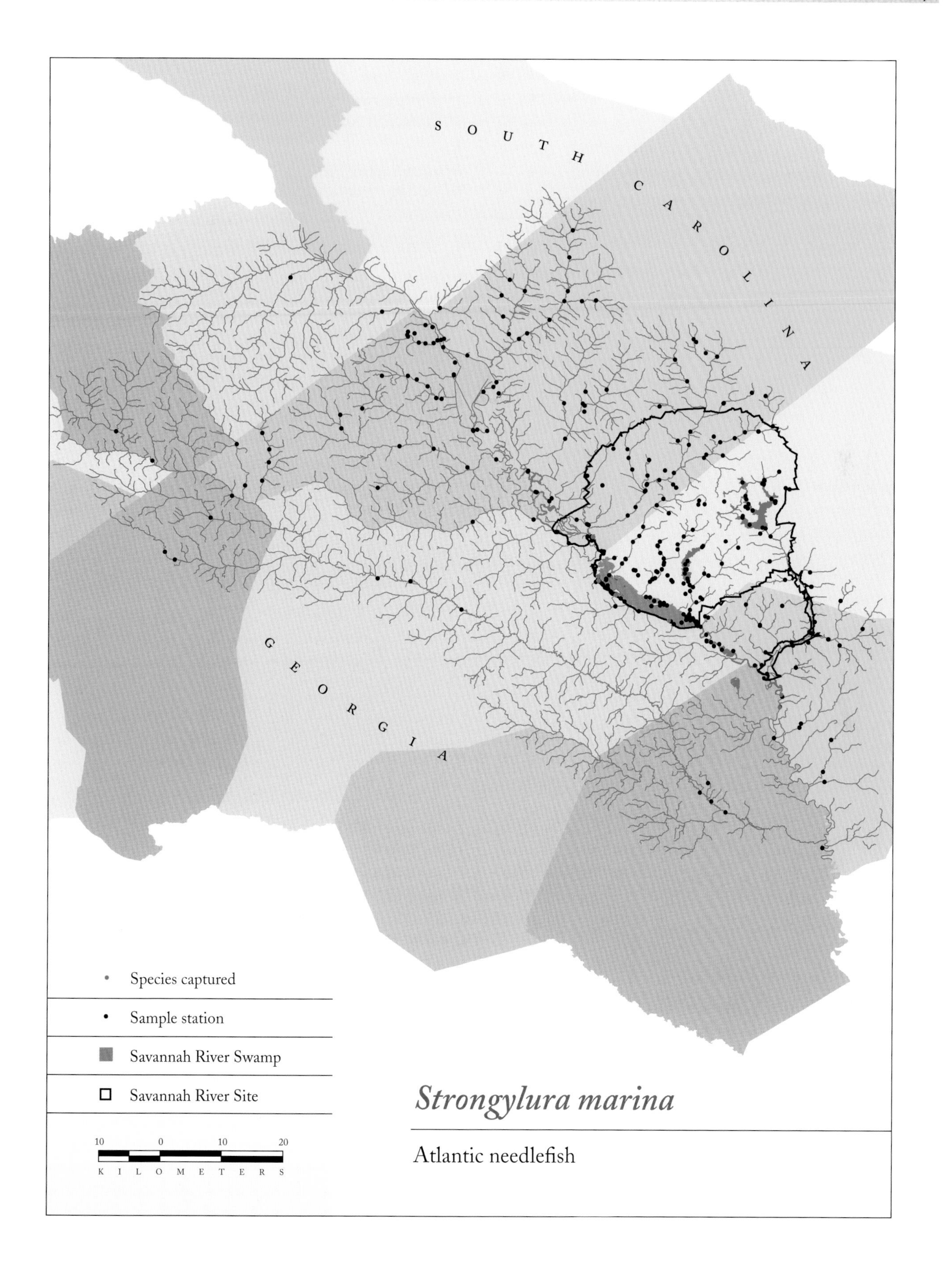

Strongylura marina

Atlantic needlefish

the fall of their first year (Able and Fahay 1998). Atlantic needlefish can reach maturity in their second year at lengths as short as 205 mm (Tracy 1910 *in* Hardy 1978). There is little information on the life span.

Needlefishes are collected in rivers and estuaries primarily during the warmer spring and summer months; in the colder months they migrate out to sea (Ogburn et al. 1988; Able and Fahay 1998).

The Atlantic needlefish is a piscivore that feeds mostly on shiners, shads, brook silverside (Alabama: Mettee et al. 1996), and Atlantic silverside (Maryland: Foster 1974).

SCIENTIFIC NAME *Strongylura* = "rounded" (Greek), referring to the rounded body in cross section; *marina* = "of the sea" (Latin).

Fundulidae (topminnows)

Topminnows are freshwater, marine, or estuarine fishes, and many species have a wide tolerance for salinity variations. Martin (1968) reported the Gulf killifish (*Fundulus grandis*), an estuarine species, in waters with salinity ranging from 0.1 to 97.3 ppt (surface seawater salinity normally lies in the range 35–37 ppt). The topminnows are restricted to North America from Yucatán to southeastern Canada, plus Bermuda and Cuba. There are five genera and about 50 species, mostly in the genus *Fundulus*, including the 2 species present in the MSRB: the golden topminnow (*F. chrysotus*) and the lined topminnow (*F. lineolatus*).

The dorsal surface of the head, and sometimes the nape, is generally flattened, and the mouth is generally superior with a small gape (a few species, such as the mummichog, *F. heteroclitus*, have a rather rounded head and nearly terminal mouth). Scales are cycloid. The origin of the pelvic fins is nearer to the tip of the snout than to the base of the caudal fin. The topminnows have a deep groove separating the upper jaw from the snout. All of these features are shared with the New World livebearers of the family Poeciliidae. The following three characters can be used to distinguish the two families: (1) all New World male livebearers have an anal fin that is modified into a gonopodium (anal fin modified as an intromittent organ for internal fertilization), but male topminnows do not have a gonopodium; (2) the third ray of the anal fin is branched in the topminnows, but not in the New World livebearers; and (3) topminnows have more lateral series scales (25 to about 70, usually more than 30) than the livebearers (usually fewer than 32). The color pattern of adult male topminnows may be distinctly different from that of females and juveniles, especially during breeding season. Males also often have longer and more pointed anal fins than females.

Some *Fundulus* species are capable of breathing atmospheric oxygen (Halpin and Martin 1999), and individual Gulf killifish have been observed voluntarily stranding themselves to avoid capture by seine nets (FDM unpub. data). Mummichogs strand themselves to escape deoxygenated water (Halpin and Martin 1999). During the often prolonged breeding season, males of some species have spinelike structures on the fin rays and scales called "contact organs." Eggs of all species have chorionic filaments; in some species these may be quite long. The eggs of many species can tolerate stranding at low tide, and there is some evidence that they hatch sooner after experiencing stranding (FDM unpub. data). Topminnows are small, mostly less than 10 cm in length. The shape of the head and mouth allows them to take in water from oxygen-rich microlayers of surface water so that they can live in waters with low oxygen levels where few other species can survive. Topminnows are often kept as aquarium fishes. They prey on mosquito larvae and other aquatic invertebrates and terrestrial insects and may be important in controlling these insects' populations.

KEY TO THE SPECIES OF FUNDULIDAE

1a. Black suborbital bar usually present; side of body marked with about 6–8 thin, dark horizontal lines, lines may be faded and replaced by dark vertical bars in males; no iridescent gold flecks on side . . . . *Fundulus lineolatus* (lined topminnow), p. 273

1b. Black suborbital bar always absent; sides of body never marked with horizontal lines; males sometimes have vertical bars or blotches that may be similar in shape and size to those of *F. lineolatus* but are generally not as dark; iridescent gold flecks usually present on sides in life . . . . *Fundulus chrysotus* (golden topminnow), p. 270

Fundulus chrysotus (Günther, 1866)

Golden topminnow

DESCRIPTION Body robust. Head short; mouth small and terminal. Eyes moderate in size, in diameter about same as snout length. Lateral line absent.

Dorsum olive, sides pale olive green to yellowish with conspicuous small iridescent gold or greenish and pearl white or bluish flecks on the sides, pale yellow or white ventrally. A short golden stripe is present along dorsal ridge anterior to the dorsal fin, and a small golden spot near each nostril. No distinct dark spots or bars present below the eyes. Fins of males with reddish or yellowish spots. Adult males may have 8–11 vertical, generally faint, greenish bars on each side; these bars are variable and frequently may be elliptical blotches or wavy lines, sometimes irregularly spaced; bars are never present on females. Females and immature males have a scattering of pearl-blue spots on the sides in life. Melanistic individuals of both sexes occur (Loftus and Kushlan 1987).

Breeding males develop reddish orange flecks on the rear half of the body, generally darker posteriorly, and on the median fins. Flecks on the body may form an irregular pattern or broken rows. Membranes of pelvic and anal fins develop bluish green iridescence. Breeding males have tubercles on the anal fin (Etnier and Starnes 1993). Breeding females are yellowish with black eyes but no red flecks.

SIZE The size range is 35–84 mm, usually less than 64 mm (Robison and Buchanan 1984) or 70 mm TL (Etnier and Starnes 1993).

MERISTICS Dorsal fin rays (7)9–10, anal fin rays (9)11(12), pectoral fin rays 14–15(16), pelvic fin rays 6, principal caudal fin rays 13–18; scales in lateral series 30–35; branchiostegals 5; total gill rakers 6–7.

CONSERVATION STATUS This species is listed as "rare or uncommon in state" (S3) and as "demonstrably secure globally" (G5) by the Georgia Department of Natural Resources.

Fundulus chrysotus, golden topminnow; (left) male, (right) female

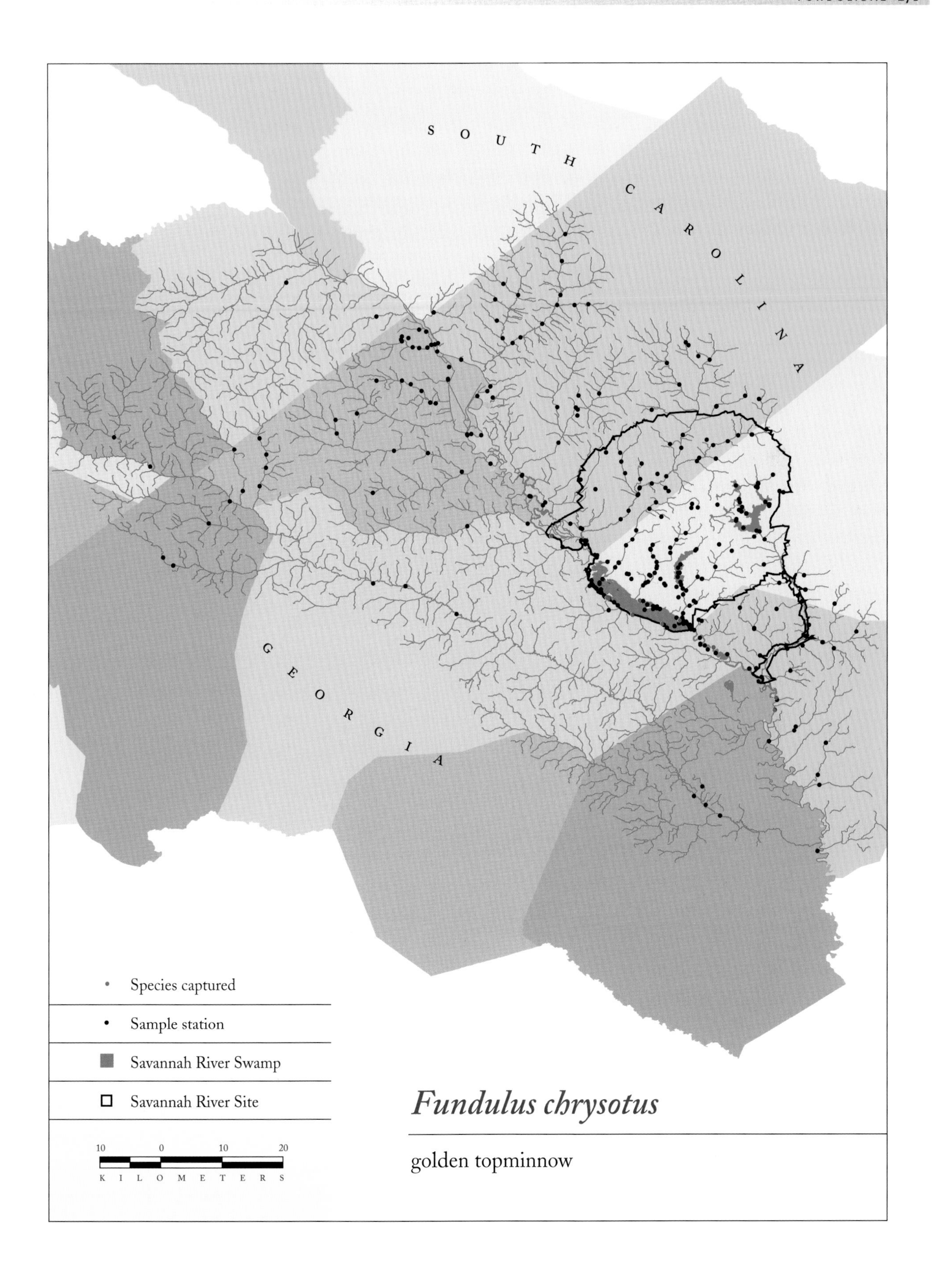

Fundulus chrysotus

golden topminnow

DISTRIBUTION The Coastal Plain from South Carolina, Georgia, and Florida west to eastern Texas and up the Mississippi Valley to southeastern Missouri (Shute *in* Lee et al. 1980). The isolated population in the Guadalupe River drainage of central Texas may or may not be natural (Whiteside and Berkhouse 1992); the Texas Natural History Collection has specimens from even farther west in Aransas County. Reports for Oklahoma may represent range extensions or previously overlooked populations (Cashner and Matthews 1988).

HABITAT Quiet, shallow water with dense vegetation, often in black water. Most frequently encountered in marshes and swamps, along lake and pond margins, and in sloughs, ditches, and creek backwaters. They also frequent floodplain pools (Hoover and Killgore 2002) and may enter brackish water. Most records in the MSRB are from the Savannah River Swamp and from backwaters of the main river downstream from the Savannah River Swamp.

BIOLOGY Spawning occurs in summer, and eggs are laid singly on submerged vegetation. In a population studied in Louisiana, courtship occurred in open water near submerged vegetation with the male leading the female in among the plants, where egg deposition occurred. These spawning events took place less than 5 cm below the surface in water more than a meter deep (FDM unpub. data). There is indirect evidence that golden topminnows lay "resting" eggs that can survive exposure during drought conditions in the Everglades (Loftus and Kushlan 1987). Growth is rapid, and individuals hatched early in the season may be reproductive in late summer in South Carolina (Rohde et al. 1994). Some individuals may live into their third summer.

Golden topminnows live singly or in small groups in the upper part of the water column (Loftus and Kushlan 1987). There is some evidence that they are omnivores feeding on insects, mites, small crustaceans, small snails, and plants such as watermeal (*Wolffia* sp.) (Rohde et al. 1994). Most food items are taken at or near the surface (Hunt 1953).

SCIENTIFIC NAME *Fundulus* = "bottom" (Latin); *chrysotus* = "gilded" or "golden" (Greek).

Fundulus lineolatus (Agassiz, 1854)

Lined topminnow

DESCRIPTION Head elongate and flattened; mouth terminal and slightly oblique. Lateral line absent. A large gold spot is present on the dorsal surface of the head, and usually a conspicuous black bar, more distinct in preservative, below the eye; the bar is occasionally faded in MSRB specimens. The bar is generally diffuse, wider than high, and seldom extends to the preopercular canal (Rivas 1966). A triangular pigment patch is present on the upper half of the pectoral fin base (Wiley 1977).

Dorsum olive, sides lighter olive yellow, whitish yellow ventrally. Females with six to eight dark lengthwise stripes on sides; males with 11–15 narrow, blackish vertical bars. Smaller females may have distinct vertical bars similar to those of males (Wiley 1977). Dorsal and anal fins of males may have three to five wavy lines or may be variegated with darker pigment. Adults may have orange-red coloring around the mouth and cheeks in life (Carr and Goin 1959).

SIZE Range, 33–86 mm.

MERISTICS Dorsal rays 7–8, mode 7; pectoral fin rays 11–14, mode 13 (Wiley 1977); anal fin rays 8–10, mode 9 (Brown 1958, anal count adjusted by Wiley 1977 to match criteria of Hubbs and Lagler 1958); lateral series scales (62)64–70(72) (Brown 1958) (Wiley 1977 gives lateral line scales as 32–36); scales around caudal peduncle 16–19, mode 16 (Wiley 1977).

DISTRIBUTION The Atlantic Coastal Plain from southeastern Virginia to Dade County, Florida, west to the Ochlockonee River drainage of Florida and Georgia (Brown 1958; Rivas 1966; Wiley *in* Lee et al. 1980). Widespread and common in the MSRB.

HABITAT Sloughs, ditches, ponds, along lake margins, and in stream and river backwaters. Usually found near aquatic vegetation in blackwater areas where the water is soft, acidic, clear, and stained by tannins and humic acids.

BIOLOGY Spawning occurs in spring and summer. Males defend territories in shallow water (less than 0.6 m) along shorelines. The eggs are approximately 1.6 mm in diameter (Rohde et al. 1994).

Lined topminnows feed on a wide variety of small invertebrates and plants (Rohde et al. 1994).

SCIENTIFIC NAME *Fundulus* = "bottom" (Latin); *lineolatus* = "having lines" (Latin).

Fundulus lineolatus, lined topminnow; (left) male, (right) female

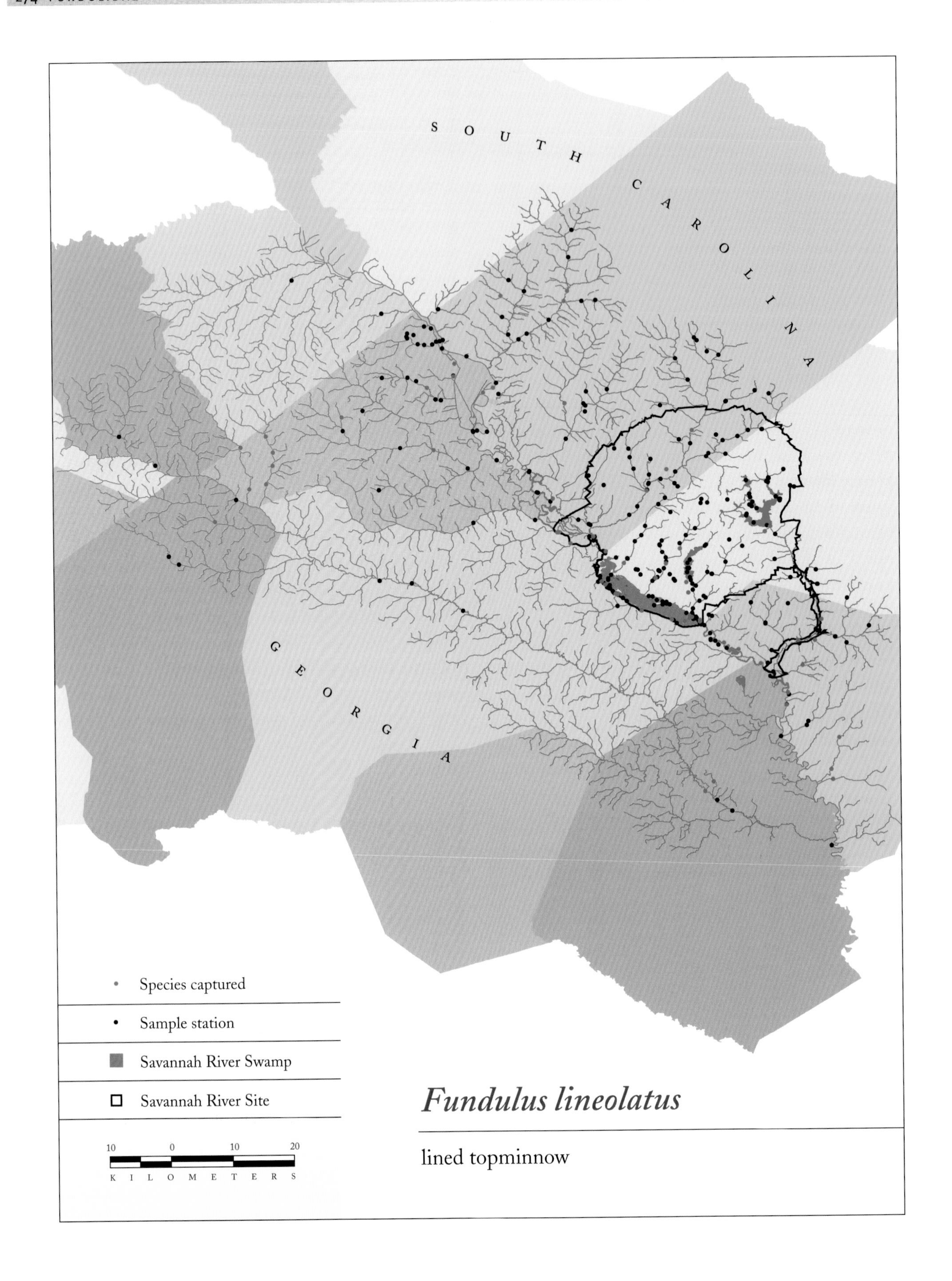

Fundulus lineolatus

lined topminnow

Poeciliidae (livebearers)

Livebearers are small freshwater fishes; a few species live in estuaries or inshore marine habitats (mostly around mangroves). They are anatomically, behaviorally, and ecologically very similar to the topminnows (family Fundulidae). The introduction to Fundulidae describes a suite of characters useful for separating the two families. The livebearers are found in South, Central, and North America; on a number of Caribbean islands; and in Africa. There are 24 genera and about 290 species.

The common name, "livebearers," suggests that all species exhibit internal fertilization and bear live young, but that is not the case. Breeding males of the subfamily Poeciliinae do in fact possess a gonopodium; the females may store live sperm for a number of months; and nearly all species give birth to precocial neonates (newborns with juvenile rather than larval characters). In contrast, the African members of the family, all in the subfamily Aplocheilichthyinae, lay externally fertilized eggs and are more similar to the Fundulidae or other Cyprinodontiformes in their reproduction. Certain livebearers, particularly the guppy (*Poecilia reticulata*) and the mosquitofishes (genus *Gambusia*), have been widely introduced to control mosquitoes that serve as vectors for diseases, especially malaria. While livebearers do eat mosquito larvae, topminnows and mudminnows are probably more effective in controlling mosquito populations. Introductions of mosquitofishes have endangered native fish and amphibians, especially in the desert Southwest of the United States. Many of the most popular aquarium fishes are in this family (e.g., guppies, mollies, swordtails, and platys). The eastern mosquitofish, *Gambusia holbrooki*, is the only common livebearer species in the MSRB. However, the least killifish, *Heterandria formosa*, occurs in the Savannah River drainage on the Lower Coastal Plain, of which the western edge just barely lies within the MSRB. Also, the sailfin molly, *Poecilia latipinna*, is found in the backwaters of the estuarine portion of the Savannah River.

Heterandria formosa, least killifish

KEY TO THE SPECIES OF POECILIIDAE

1a. Conspicuous black spot on anterior dorsal fin base of both sexes and on anal fin base of female; prominent dark stripe on side from chin to a dark spot near base of caudal fin (anterior portion of stripe may be faded); 6–9 sometimes vague dark bars cross the stripe .... *Heterandria formosa* (least killifish)—no species account

1b. Only conspicuous dark pigment always present is the black suborbital bar; sides of body never marked with horizontal stripe or vertical bars, bases of dorsal and anal fins not marked with conspicuous dark spot; males sometimes have irregular black spots or blotches and gravid females have a dark blotch just anterior to vent .... *Gambusia holbrooki* (eastern mosquitofish), p. 277

Gambusia holbrooki Girard, 1859

Eastern mosquitofish

DESCRIPTION Body small and somewhat drab; males more slender than the deeper-bodied females. Head flattened on top; mouth opening even with or above level of pupil of the eye. Caudal fin rounded. Dorsal fin origin behind origin of anal fin. Anal fin of males modified into a gonopodium (a copulatory organ). Males, juveniles, and smaller females not in advanced stages of pregnancy are slender; larger females are proportionally deeper bodied.

Body color more or less uniform gray, tan, or greenish olive, with the belly lighter than the back. Speckling or black spots may be present, but usually the most obvious dark marking is a dark vertical bar below the eye. Infrequently, males are melanistic. Mature females usually have a dark spot near the urogenital opening, which is more obvious late in pregnancy.

SIMILAR SPECIES Topminnows and mosquitofish share several characters (see Fundulidae family account for details). The eastern and western mosquitofish are very similar but can be distinguished by the morphology of the gonopodium and by the larger number of dorsal (8 versus 7) and anal (10 versus 9) fin rays in the eastern mosquitofish (see Etnier and Starnes 1993).

SIZE Range, 20–65 mm TL; males are usually less than 26 mm TL, females may reach 65 mm TL.

MERISTICS Dorsal rays (6)7–8 (Walters and Freeman 2000), pectoral rays 13–14, pelvic rays 6 (Sterba and Tucker 1973), anal rays 8–11(12) in females (Walters and Freeman 2000), male with anal fin modified into copulatory organ; lateral series scales 30–32 (Sterba and Tucker 1973).

DISTRIBUTION Native from New Jersey south along the Atlantic Coast to southern Florida and west to the Mobile Bay drainages of Alabama and Georgia. Fin ray counts and gonopodial features indicate that eastern mosquitofish intergrade with western mosquitofish (*G. affinis*) in part of the Chattahoochee River drainage of Georgia and Alabama and in the Mobile Bay area (Angus and Howell 1996). Populations in the Conasauga River of western Georgia are probably the result of human introduction and appear to be displacing the native western mosquitofish populations (Walters and Freeman 2000). There are populations of western mosquitofish in the headwaters of the Savannah and Chattahoochee rivers, while the lower reaches are populated by eastern mosquitofish or hybrids, introducing the possibility that the eastern mosquitofish has only fairly recently replaced western mosquitofish in the MSRB (Smith et al. 1989; Lydeard et al. 1991). Because of chromosomal incompatibilities, hybrids with western mosquitofish females are not viable, while those with eastern mosquitofish females as parents are fully viable and fertile (Black and Howell 1979), providing part of the mechanism for species replacement when eastern mosquitofish invade western mosquitofish territory. Except for the Savannah River drainage, eastern mosquitofish in Georgia, Florida, and Alabama are genetically distinct from other populations (Wooten et al. 1988). Eastern mosquitofish have been introduced into Australia (Congdon 1995). North American *Gambusia*, probably eastern mosquitofish, have been

Gambusia holbrooki, eastern mosquitofish; (left) female, (right) male

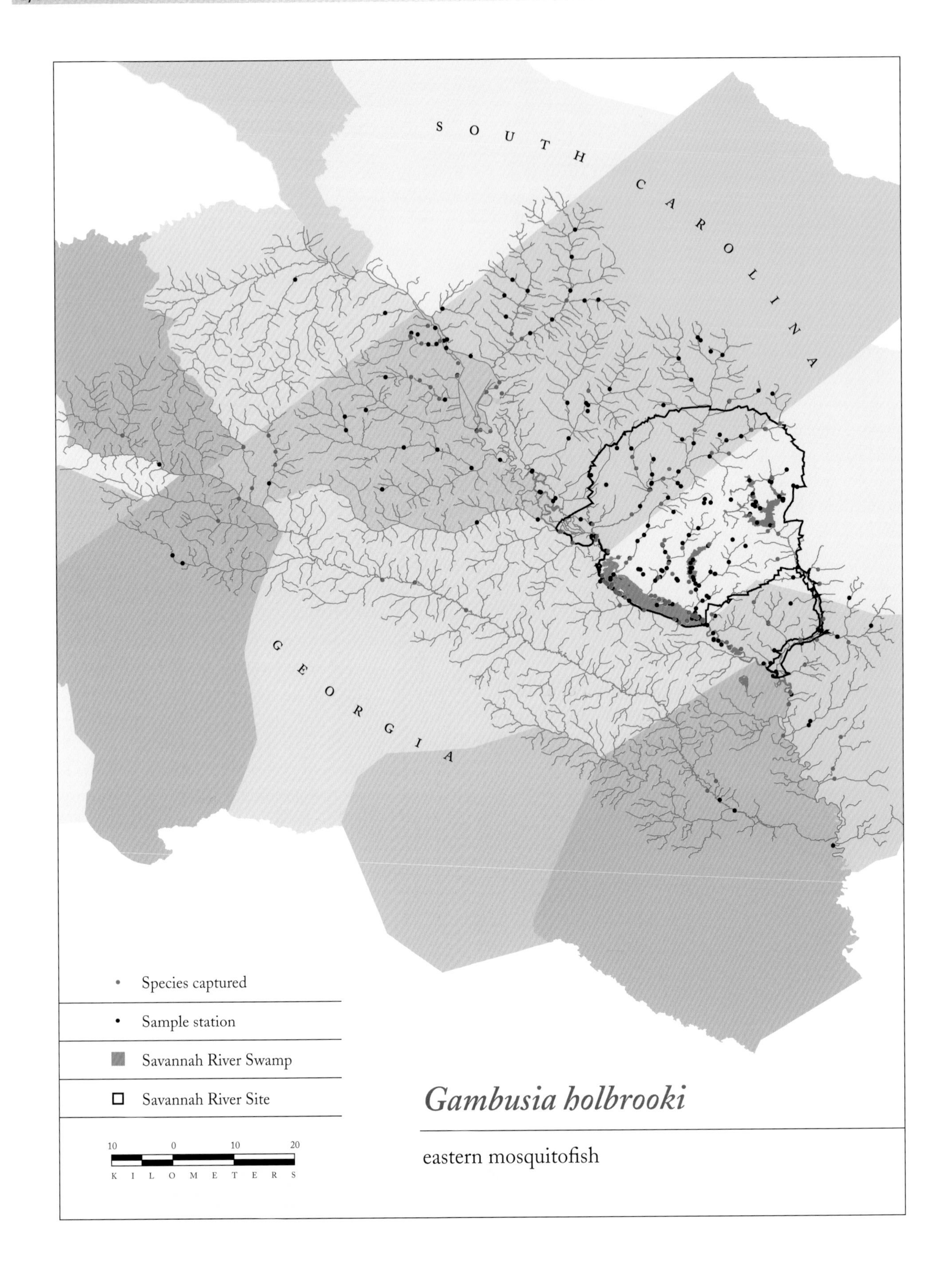
SOUTH CAROLINA
GEORGIA
Species captured
Sample station
Savannah River Swamp
Savannah River Site
10 0 10 20
KILOMETERS
Gambusia holbrooki
eastern mosquitofish

widely introduced in southern Europe, Asia, Africa, and the islands of the Pacific and Caribbean. Widely distributed and locally abundant within the MSRB.

HABITAT Found mostly in vegetated areas of lakes, ponds, streams, ditches, sloughs, swamps, marshes, and river margins. Eastern mosquitofish are very tolerant of low oxygen conditions and warm water and may be the only living fish present in some bodies of water. They live primarily at the surface and are more abundant where vegetation is not extremely dense. They occur in fresh water to salinities nearing 80 ppt (Loftus and Kushlan 1987).

BIOLOGY This livebearer breeds throughout the year in areas where the water temperature remains high (e.g., power plant cooling basins, extreme southern Florida), with males courting and fertilizing females whenever water temperatures are high enough for them to be active. In most places, courtship is rare between October and March. The gestation period is 21–28 days, and three or four broods may be produced in a season. Brood size is related to female size and can vary from 1 to more than 300 young, although broods of more than 100 are rare. Young born in the spring may be sexually mature and breeding by the fall.

Eastern mosquitofish feed mostly on insects and insect larvae, crustaceans, algae, and very small fish that they encounter at or near the water surface. Cannibalism occurs but is rare; Nesbit and Meffe (1993) found only nine cases in 1,883 individuals examined. Western mosquitofish also feed on amphibian larvae and may be major predators on some endangered amphibians in California where they have been introduced (Kittl 1999). Eastern mosquitofish may have similar appetites. They have been observed picking at the skin of alligators, presumably feeding on dead skin or parasites (Loftus and Kushlan 1987). This species has been widely introduced for mosquito control; where there is much vegetation it does not adequately control mosquito larvae, but in artificial habitats such as abandoned swimming pools, ornamental pools, and mine pits it is often effective (Duryea et al. 1996).

SCIENTIFIC NAME *Gambusia*, from *gambusino* = "small," "worthless," or "nothing" (Cuban Spanish); *holbrooki* = refers to J. E. Holbrook, a nineteenth-century American naturalist.

Atherinopsidae (New World silversides)

Most New World silversides are freshwater species from Central and South America, although there is a sizable number of marine and estuarine species. Three genera (*Atherinella*, *Chirostoma*, and *Odontesthes*) account for most of the freshwater species. Marine forms occur from Oregon to Chile in the eastern Pacific, and from Newfoundland to Argentina and the Falkland Islands in the western Atlantic. Freshwater forms are most abundant in Mexico and South America. Twelve of the 100 or so species in this family occur in the United States and Canada.

Silversides can be recognized by a suite of characters: two dorsal fins that are widely separated; a terminal or slightly superior mouth with slightly to moderately elongated jaws; cycloid scales; flattened dorsum; pectoral fins high on the sides near the top of the gill opening; forked caudal and anal fins with numerous rays; body silvery white or, more commonly, yellowish to straw and translucent, with the swimbladder and other internal organs easily visible under the proper light; a horizontal stripe down the side of the body that is reflective and silvery in life and may be on a wider opaque band of white or off-white.

Silversides are schooling fishes, and most feed at or near the surface. All species have eggs with one or more long chorionic filaments. On the Atlantic Coast, the Atlantic (*Menidia menidia*) and tidewater (*Menidia peninsulae*) silversides mate in mass spawnings that are timed to certain phases of the moon and tidal cycles during the spring and summer months (Middaugh 1981; Middaugh et al. 1981, 1984, 1986; Conover and Kynard 1984; Middaugh and Hemmer 1987). The chorionic filaments tangle these eggs together and attach them to plants and debris in the water, producing large mats of eggs and debris. Although these mats are stranded at low tide, the eggs resist desiccation to have high survival. Some of the larger marine species such as the grunions are fished commercially, but the silversides' major importance to fisheries is as food for larger sport and commercial fishes. In the MSRB, the brook silverside (*Labidesthes sicculus*) is the only representative of the Atherinopsidae.

Labidesthes sicculus (Cope, 1865)
Brook silverside

Brook silversides in Florida, Georgia, and South Carolina are in the subspecies *L. s. vanhyningi*. This subspecies will become a species name in the near future. It will be *L. vanhyningi* and the common name will be the southern brook silverside.

DESCRIPTION Body elongate and slender. Mouth terminal; jaws rather elongated and beaklike. Eyes large. Two widely separated dorsal fins located on the posterior half of the body (excluding the caudal fin). Anal fin long and falcate, its origin nearly even with origin of first dorsal fin. Second dorsal fin falcate; caudal fin moderately forked; pectoral fins laterally placed. Lateral line incomplete.

Body translucent to silver with a visible peritoneum. Sides silvery with aqua bluish, with an iridescent midlateral stripe that becomes dark in preservative. The iridescence may extend onto the chin. Dorsum pale straw to light olive; fins transparent. Breeding males may have red on the snout.

SIZE The maximum size is about 90 mm SL (Jenkins and Burkhead 1993), but SRS specimens are usually somewhat smaller (84 mm) (Bennett and McFarlane 1983), and those in tributary streams are mostly less than 70 mm SL.

MERISTICS Adult counts for all fins are attained between 15.5 and 20.1 mm TL (Rasmussen 1980). First dorsal spines 3–6, second dorsal spine 1, dorsal rays 9–13, anal spine 1, anal rays 20–27, pectoral rays 12–13, pelvic spine 1, pelvic rays 5; midlateral series scales 74–95 (Rasmussen 1980; Jenkins and Burkhead 1993).

DISTRIBUTION The St. Lawrence River, Mississippi River, and Great Lakes drainages from Canada, Pennsylvania, and Minnesota south through the Great Plains and Louisiana and east along the Coastal Plain to South Carolina (Lee *in* Lee et al. 1980). Brook silversides are widespread and very common in the MSRB.

HABITAT The brook silverside occurs in a wide variety of habitats ranging from lakes to creeks. Most authors suggest that it prefers clear water (Jenkins and Burkhead 1993). Catch per effort studies indicate a preference for oxbow lakes over river habitats in Florida (Beecher et al. 1977). Adults occur in schools and are mostly seen near the water surface in shallow areas (Hubbs 1921).

Labidesthes sicculus, brook silverside

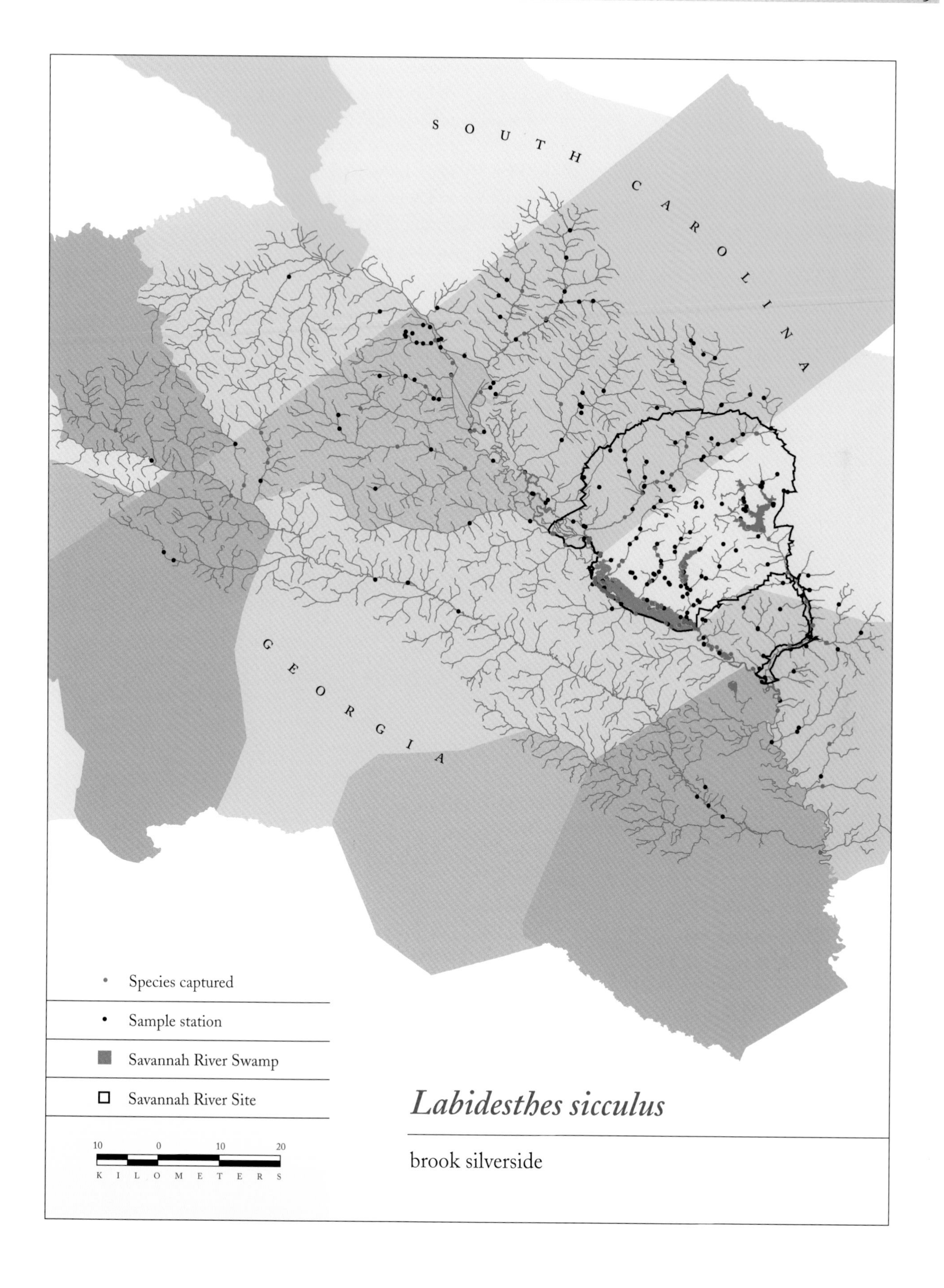

Labidesthes sicculus

brook silverside

BIOLOGY Spawning in the northern states occurs from May through August at 17–23 °C (Hubbs 1921; Cahn 1927; Nelson 1968). Eggs are 0.8–1.4 mm in diameter (0.8–1.2 mm in Indiana, 1.2 mm in Florida) and are attached or attach themselves by one (Indiana) or two (Florida) long, very adhesive chorionic filaments to aquatic plants or underwater structures (Hubbs 1921; Nelson 1968; Rasmussen 1980). No nest is prepared, and no care is provided for the deposited eggs or embryos. The eggs and larvae were described by Rasmussen (1980). Larvae are 4.0–5.6 mm TL at hatching (Nelson 1968; Rasmussen 1980) and grow very quickly. Grier et al. (1990) reported internal fertilization in Florida. The brook silverside is a short-lived species with a maximum life span of 21–23 months; most individuals do not live more than 12 months (Nelson 1968).

Brook silversides in the MSRB feed predominantly on Cladocera, followed by aquatic insects, especially chironomids (Wiltz 1993); this is consistent with the diet reported elsewhere (Hubbs 1921; Cahn 1927; Mullan et al. 1968). Larvae and early juveniles stay near the surface day and night. Adults swim very actively during the day in pursuit of prey and occasionally leap from the water to escape predation; at night they are quiescent and mostly near the surface (Hubbs 1921).

SCIENTIFIC NAME *Labidesthes* = "pair of forceps" (Greek), in reference to the elongated jaws (Jenkins and Burkhead 1993); *sicculus* = "dried" (Latin), in reference to its capture in drying river edge pools.

Moronidae (temperate basses)

Temperate basses are predators that live in freshwater, estuarine, and coastal marine habitats. They are characterized by a strongly compressed body, ctenoid scales, a complete lateral line that reaches well onto the middle caudal fin rays, nine stout spines in the first dorsal fin, one spine and 10–14 rays in the second dorsal fin, an emarginate or forked caudal fin, pseudobranchae on the inside of the opercle, a serrate preopercular margin, and a spine on the opercle. All North American species have stripes, although the stripes may be very indistinct or even absent in white perch, especially larger individuals. According to some systematists the family Moronidae consists of only two genera: *Morone*, with four species that are all native to North America, and *Dicentrarchus*, with two species that are native to Europe and Africa (Nelson 1994). Other authors combine *Morone* and *Dicentrarchus* into *Morone*. Secor (2002), in part based on McCulley 1962, includes the genera *Lateolabrax* (with two species) and *Siniperca* (with four species), and two Asian Pacific forms in the Moronidae. We refer only to the North American forms when we speak of the genus *Morone*. Older literature will have the temperate basses combined with either the Percichthyidae or the Serranidae, and some accounts will refer to the genus as *Roccus* instead of *Morone*.

All species of *Morone* spawn in fresh or slightly brackish water; the *Dicentrarchus* species spawn in marine water. Some *Morone* species run up rivers and streams for some distance in order to spawn. Others, especially in the southern United States, spawn just above the estuarine zone. All species are egg scatterers, and no nests are built. The temperate basses are highly valued as food and sport fishes, and striped bass (*Morone saxatilis*) have been widely introduced well outside the species' natural range. Various crosses of striped bass and white bass (*M. chrysops*) have been artificially created for sportfishing and aquaculture purposes, and some can be found in the MSRB.

KEY TO THE SPECIES OF MORONIDAE

1a. No median tooth patches on back of tongue, may have teeth in front; dark parallel horizontal lines on dorsolateral side above lateral line absent or indistinct and heavily interrupted. . . . . *Morone americana* (white perch), p. 286

1b. One or 2 median tooth patches present on back of tongue; dark parallel horizontal lines on dorsolateral side above lateral line present, may be faint or bold, but nearly always continuous 2

2a. Narrow bodied with head length usually greater than body depth; dorsal profile not strongly arched behind head; lines on side of adult body generally continuous and bold; 2 distinct parallel tooth patches on back of tongue; usually 12 or fewer dorsal rays. . . . . *Morone saxatilis* (striped bass), p. 292

[1] **2b.** Relatively deep bodied with head length usually less than body depth; dorsal profile strongly arched behind head; lines on side of body generally faint and lines below the lateral line usually discontinuous; single median tooth patch on back of tongue (may be weakly divided); usually 13 or more dorsal rays . . . . *Morone chrysops* (white bass), p. 289

Note: The teeth on the tongue can be easily detected by using a blunt probe, finger, or pen to investigate the tongue, especially at the back. *Morone* hybrids, commonly released in the Savannah River drainage, may have characters intermediate to the parent species (see hybrid striped bass account below).

Morone americana (Gmelin, 1789)

White perch

DESCRIPTION Body oblong, ovate, and compressed, with a moderately elevated back anteriorly. Dorsal profile of head above eyes slightly concave. Mouth oblique and terminal, with lower jaw only slightly protracted and nearly even with upper jaw (Woolcott 1962; Hardy 1978; Jenkins and Burkhead 1993). Dorsal fins nearly separated, slightly joined by membrane, and spinous and soft portions have about the same base lengths. Caudal fin forked. Second and third anal spines relatively long and nearly equal in length. Gill rakers long; median tooth patches absent from base of tongue.

Dorsum dark, sometimes brassy, and sides pale silvery green fading to pale silvery white ventrally (Hardy 1978). Young may have generally faint and broken horizontal stripes on the sides, especially below the lateral line, and may also have dusky vertical parr bars. Both stripes and bars disappear after 1 year (Lippson and Moran 1974; Page and Burr 1991). Adults usually lack dark horizontal stripes.

SIZE Adults can reach 495 mm (Hardy 1978).

MERISTICS First dorsal spines 7–11, second dorsal spine 1, dorsal rays 11–13, anal spines 3, anal rays 9–10; lateral line scales 44–55; gill rakers 19–22, branchiostegal rays 7 (Scott and Crossman 1973); pectoral rays (10)14–17(18) (Jenkins and Burkhead 1993).

Morone americana, white perch

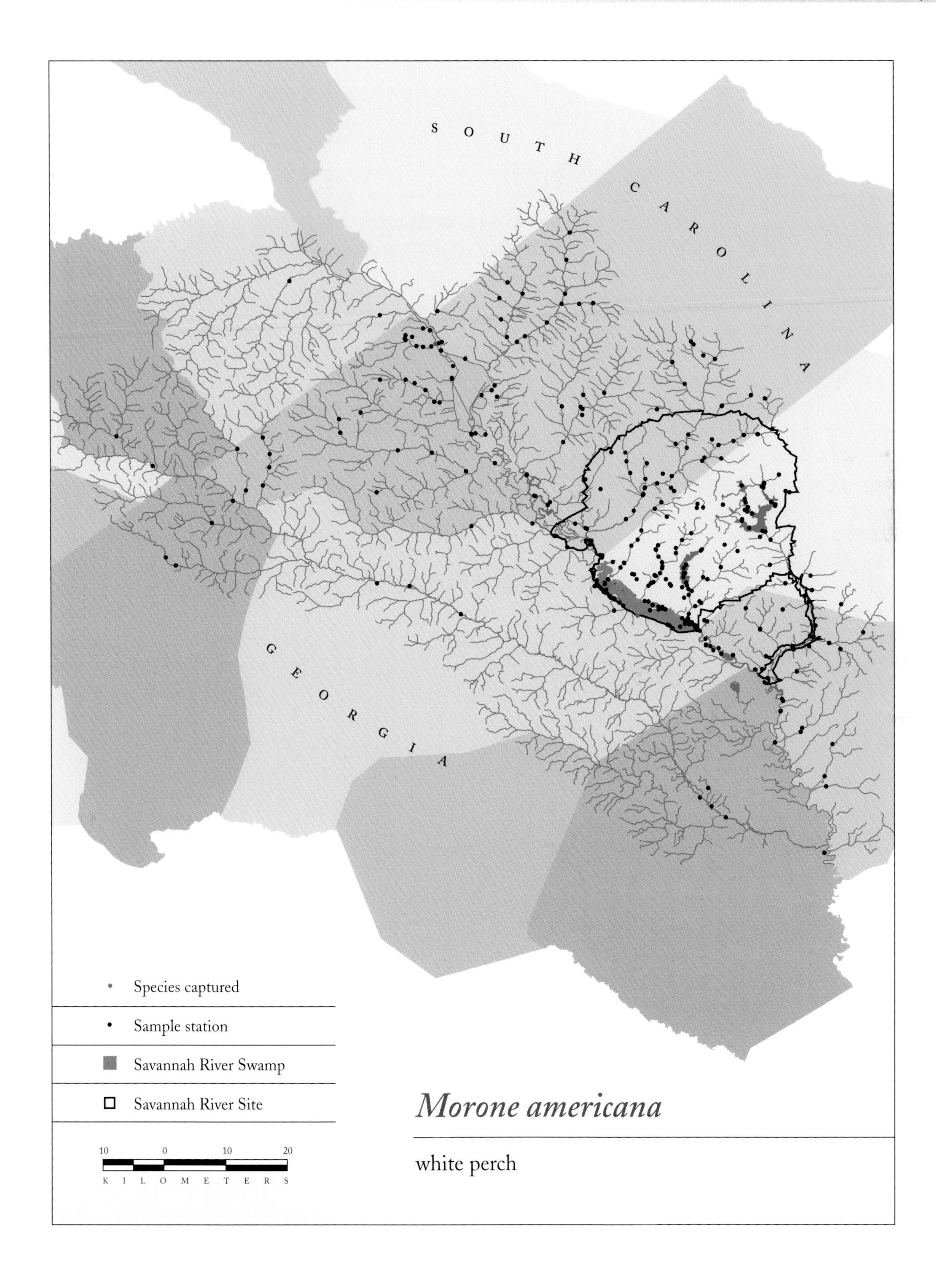

Morone americana

white perch

CONSERVATION STATUS Probably introduced in the Savannah River.

DISTRIBUTION The native range is from Cape Breton Island, Nova Scotia, to South Carolina, with introduced populations in New Brunswick (Burgess *in* Lee et al. 1980); most abundant resident species in the Connecticut River (Marcy and Richards 1974). The native range is near the winter thermal minimum for the species (Johnson and Evans 1990). White perch have been introduced into Nebraska and are now spreading through the Kansas, Platte, and Missouri river systems (Hergenrader 1980). Rohde et al. (1994) reported the species present only in northern South Carolina. White perch are not often collected in the MSRB. We speculate that the population in the Savannah River came from individuals stocked in Clarks Hill/J. Strom Thurmond Lake.

HABITAT White perch are schooling, open-water fish (Rohde et al. 1994) that are generally semianadromous, as the adults migrate to tidal fresh and slightly brackish waters each spring to spawn (Stanley and Danie 1983). They seem to prefer small ponds for spawning (Werner 1980). Populations have become landlocked in many freshwater lakes and ponds. They are usually found in salinities ranging from 5 to 19 ppt, with 30 ppt the upper limit reported (Ryder 1883). Inshore zones of creeks and estuaries are important nursery areas for young-of-the-year (Wang and Kernehan 1979).

BIOLOGY Spawning, which is reported to occur from April through June over much of the distribution (Collette and Klein-MacPhee 2002), takes place in fresh to low-salinity waters in large rivers, generally over fine gravel or sand. Marine and estuarine populations move shoreward and generally upstream in the spring, often in large schools. Lake-dwelling populations may make no spawning migration (Stanley and Danie 1983). The eggs are pelagic in flowing streams and tidal waters, but in calmer water they generally sink to the bottom, where they attach to structures and each other (Collette and Klein-MacPhee 2002). Eggs usually hatch in 44–50 hours, and the yolk-sac larvae settle to the bottom and assume a head-up position (Mansueti 1964; Hardy 1978). See Lippson and Moran 1974; Hardy 1978; and Wang and Kernehan 1979 for descriptions of larvae and early juveniles. Females produce 50,000–150,000 eggs in a season, spawned episodically over a period of 10–21 days.

Juveniles use inshore areas of rivers and estuaries downstream of the spawning area during their first summer and fall. Growth is rapid during the first few years, and almost 80% of the total growth is attained in the first 3 years of life (Collette and Klein-MacPhee 2002). Males mature by age 2 and females at age 3 or 4 (Mansueti 1961; Rohde et al. 1994). White perch generally live about 6–7 years, but some individuals may survive through age 17 (Cooper 1941; Marcy and Richards 1974; Rohde et al. 1994; Collette and Klein-MacPhee 2002).

The diet of these predaceous bottom dwellers consists of crabs, shrimp, and small fishes. Younger fish feed on copepods, insects (primarily chironomids), and cladocerans in the Connecticut River (Marcy and Richards 1974).

When this species can interact with other *Morone* species, either through introduction or through invasion, hybridization often occurs (Irons et al. 2002).

SCIENTIFIC NAME *Morone*, origin unknown; *americana* = "of or from America" (Latin).

Morone chrysops (Rafinesque, 1820)

White bass

DESCRIPTION Body relatively deep and compressed; head length usually less than body depth. Dorsal profile strongly arched behind head. Caudal fin moderately forked. First (spiny) and second (soft) dorsal fins separated, but near each other. Third anal spine distinctly longer than second one. Posterior margin of each operculum with two spines, and margin of preopercle serrate. Tongue with elongate, sometimes weakly separated, single median tooth patch.

Dorsum dark green to gray, sides silvery gray to pale green with dark horizontal stripes that are often interrupted, belly silvery white.

SIZE White bass can reach 450 mm but are usually much smaller; females grow faster and reach larger sizes than males (Webb and Moss 1968; Guy et al. 2002b).

MERISTICS First dorsal spines 9, second dorsal spine 1, dorsal rays (12)13(14), anal spines 3, anal rays (11)12(13), pectoral rays (15)16(17); lateral line scales (52)54–58(60) (Jenkins and Burkhead 1993).

CONSERVATION STATUS White bass are introduced in the MSRB.

DISTRIBUTION Native to the St. Lawrence River through the Great Lakes, north and west to Lake Winnipeg (a recent extension via the Red River), south through the Ohio and Mississippi River Valleys to Louisiana, and west in Gulf of Mexico drainages to the Rio Grande; introduced in various eastern Gulf of Mexico drainages and Atlantic drainages as far north as Pennsylvania; also introduced in western Kansas, Colorado, Utah, Arizona, California, and Washington (Burgess *in* Lee et al. 1980). White bass are not very common in the MSRB and are almost exclusively found in the Savannah River.

HABITAT Slow-flowing rivers, reservoirs, and lakes, generally in the deeper pools of streams around vegetation or other cover. White bass seem to prefer clear water and have become more common in the Missouri River since the construction of larger reservoirs has reduced turbidity (Pflieger 1975).

BIOLOGY There is an enormous body of literature on various aspects of the biology of this popular sport fish (e.g., see *Journal of Fisheries Management* 22[2], 2002). White bass are most commonly found in schools. They can be

Morone chrysops, white bass

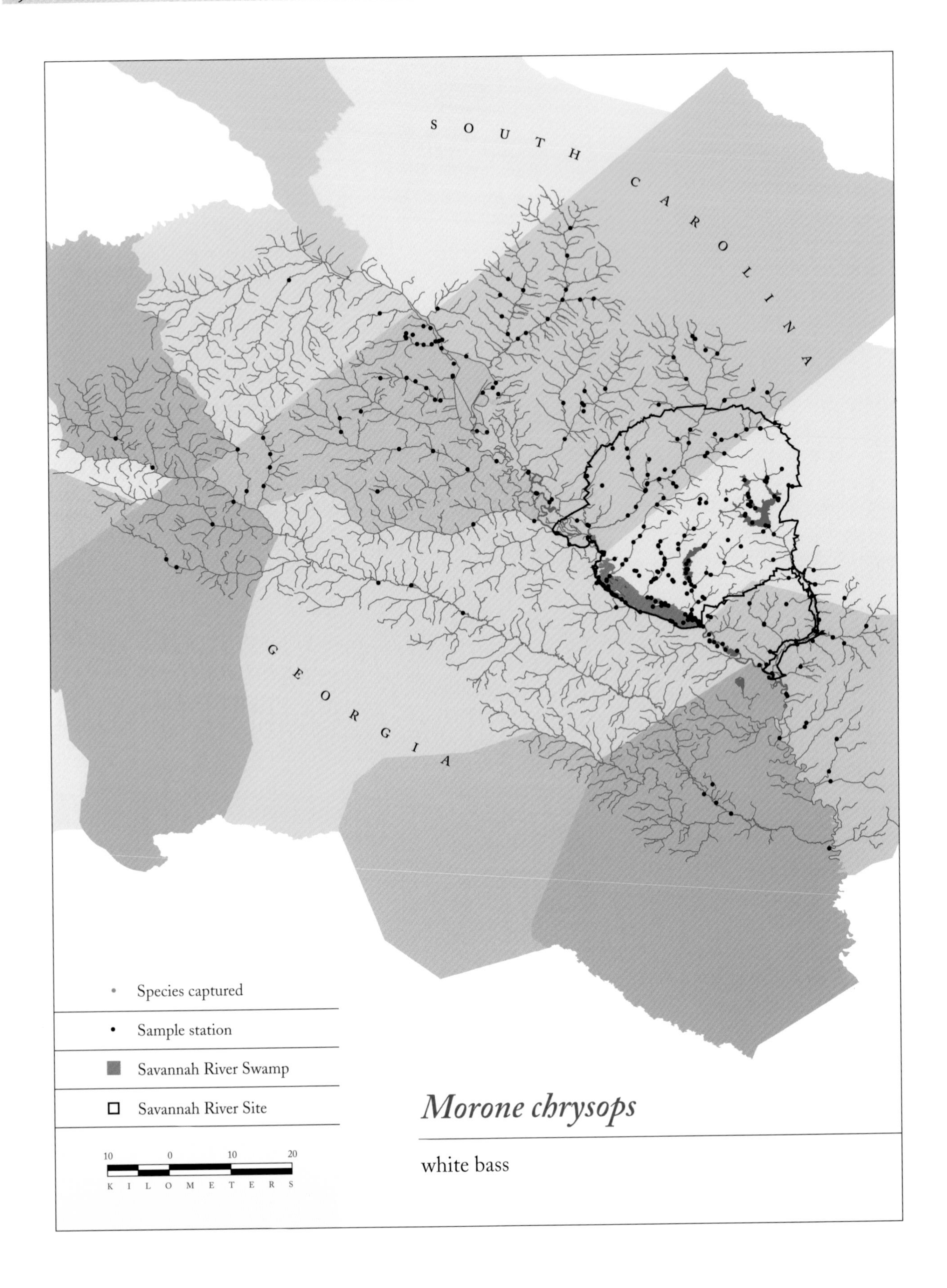
SOUTH CAROLINA
GEORGIA
Species captured
Sample station
Savannah River Swamp
Savannah River Site
10 0 10 20
KILOMETERS
Morone chrysops
white bass

sexually mature at the age of 2, and spawn as early as March in the southern regions and in May farther north (Guy et al. 2002a, 2002b). They are potandromous, migrating up rivers to shoals and smaller streams to spawn. Spawning occurs near the surface while the fish are aggregating in shallow waters (Webb and Moss 1968; Bennett and McFarlane 1983). Females deposit between 125,000 and more than 1 million eggs (Madenjian et al. 2000), which sink to the bottom and adhere to the substrate (Jenkins and Burkhead 1993). White bass can live 7 years or more, but most do not survive past 5 years (Lovell and Maceina 2002).

White bass are visual predators that seem most active at dawn and dusk (Webb and Moss 1968; Willis et al. 2002). The young prey on aquatic insects and crustaceans; the adults predominantly eat fish. Gizzard shad make up a large part of the diet in Missouri reservoirs (Pflieger 1975). However, northern populations tend to feed heavily on invertebrates; southern populations tend to feed more on fish (Madenjian et al. 2000).

Natural hybridization of *M. americana* and *M. chrysops* can occur (Todd 1986).

SCIENTIFIC NAME *Morone*, unknown; *chrysops* = "golden eye" (Greek).

Morone saxatilis (Walbaum, 1792)
Striped bass

DESCRIPTION Body elongate, compressed. Head with acute snout; mouth large, with lower jaw projecting ahead of upper jaw. Maximum body depth three or more times into standard length. Margin of operculum with two sharp spines; margin of preopercle serrate. Lateral line complete and almost straight. Two elongate tooth patches on back of tongue. First and second dorsal fins completely separate. Caudal fin forked.

Back dark gray to green. Sides green to silver, shading to white or cream on the belly. Seven or eight continuous black stripes on side: three or four stripes above the lateral line, one containing the lateral line, and three below. Juveniles smaller than 12.7 cm may have vertical bars on the sides.

SIZE Males to 116 cm FL, females possibly to 183 cm. Landlocked striped bass reach 27.5 kg; marine fish may exceed 45 kg. Maximum size in SRS is about 10 kg (Bennett and McFarlane 1983).

MERISTICS First dorsal spines 8–9, second dorsal spine 1, dorsal rays (9)12(14), anal spines 3, anal rays (7)9–12(13); lateral line scales 50–72, Gulf Coast populations 63–72, Atlantic Coast populations 50–67 (Hardy 1978); gill rakers on lower arch 12–15, on upper arch 6–12.

DISTRIBUTION On the Atlantic Coast, from the St. Lawrence River of Canada and Nova Scotia to the St. Johns River of Florida; in the Gulf of Mexico, from western Florida through the Mobile Bay drainage of Alabama west to Lake Ponchartrain, Louisiana (Burgess *in* Lee et al. 1980). Striped bass (uncertain whether native or introduced) have been taken in marine water in Texas (Hoese and Moore 1977), and may be native in east Texas (Pam Fuller pers. comm.). They were introduced in California and now occur from Ensenada, Mexico, to the Columbia River of Washington (Bain and Bain 1982). Also introduced into the former Soviet Union and South Africa. In the MSRB, most frequently found in the Savannah River, but relatively large numbers ascend larger tributaries such as Upper Three Runs in the summer to avoid warmer river water.

HABITAT Free-flowing rivers, reservoirs, and coastal marine waters with at least some current, over bottoms with gravel, rocks, and vegetation. In marine waters, they are seldom more than 6–8 km from shore (Bigelow and Schroeder 1953). Spawning habitat appears to be fresh water well above the salt wedge in the estuary. Enough current to keep the semipelagic eggs in the water column

Morone saxatilis, striped bass; adult

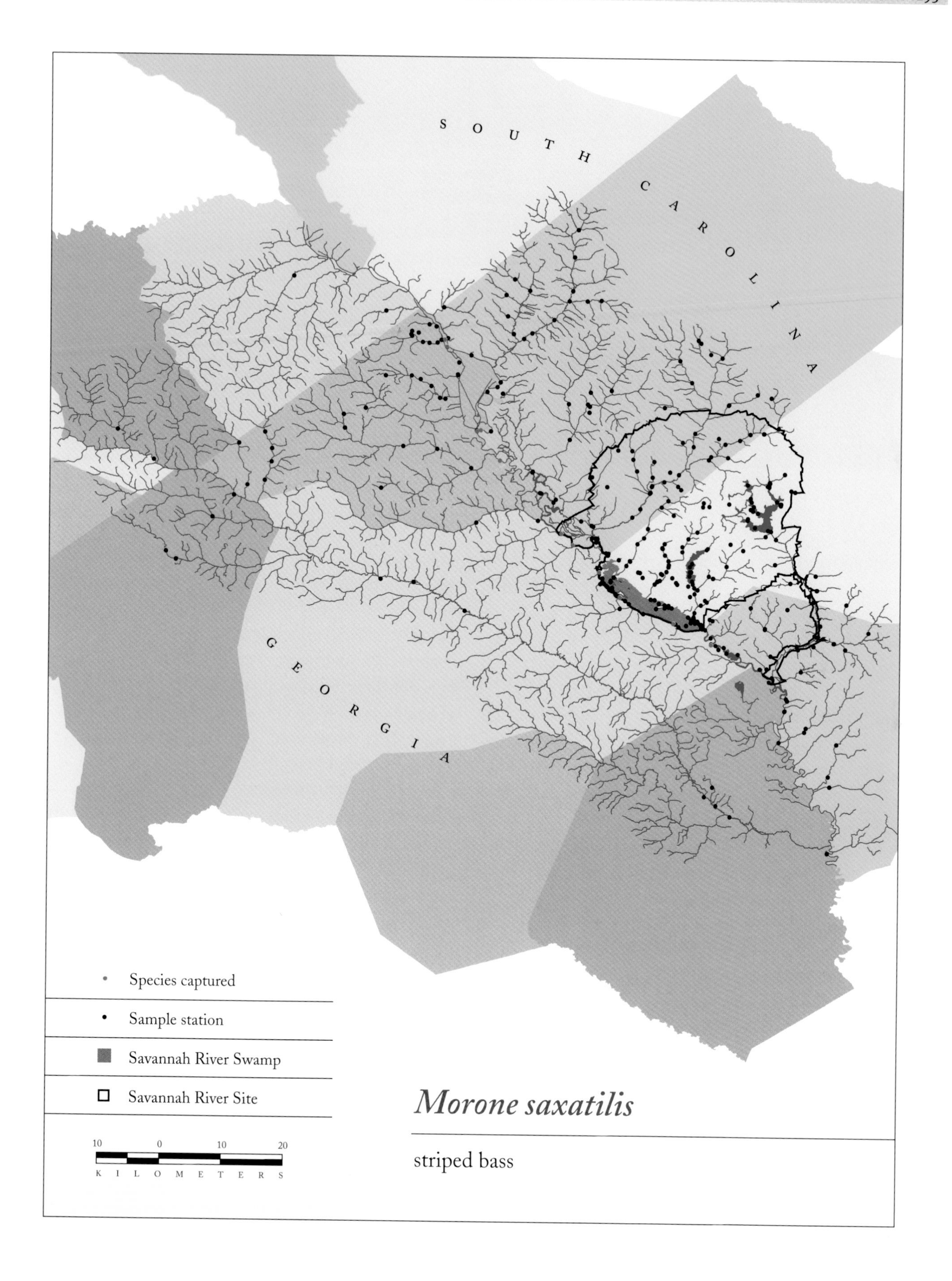

Morone saxatilis

striped bass

Morone saxatilis, striped bass; juvenile

is required as well as enough distance above the salt wedge or still water to allow the eggs to hatch before sinking to the bottom. Striped bass seem to avoid temperatures above 21 °C, and larger adults are generally found in deeper water.

BIOLOGY There is an enormous body of literature on the biology of this species, not least because of its economic importance as a sport fish (e.g., see *Transactions of the American Fisheries Society*, especially volumes 110[1], 1981, and 114, 1985; and *Fisheries Management and Ecology*, volume 10[5], 2003). Although the striped bass is generally considered anadromous, some reproducing landlocked and largely riverine populations exist. Striped bass migrate upriver and into tributaries for spring spawning in March, April, and May; larger fish migrate longer distances than smaller fish. Spawning occurs in strong currents of large rivers when the water temperature is above 14.4 °C. Hardy 1978 provides a detailed summary of larval development. Adults usually return to estuaries or coastal marine waters after spawning, but in some populations they may move farther upriver (Dudley et al. 1977); this tendency is more pronounced in the southern part of the range (Crance 1984). The young of nonlandlocked populations spend the first year in coastal estuaries. Where not prevented by dams, they overwinter in estuaries and in coastal marine waters. Populations from New England northward and from Cape Hatteras southward do not move great distances from the mouth of the natal stream; those in the Mid-Atlantic Bight migrate long distances and mingle breeding stocks (Hardy 1978). There is evidence of stream fidelity (Mulligan et al. 1987). Young-of-the-year grow significantly larger in low-salinity estuarine environments than in fresh water (Secor et al. 2000). Males reach sexual maturity at age 2 or 3, females at age 4 or 5. Males may be sexually mature at 18 cm; females can be mature at 36 cm.

Striped bass occur in large, continuously moving schools and feed on schooling fishes such as shads (Stevens 1958; Etnier and Starnes 1993). The larvae feed primarily on crustacean nauplii, switching to larger zooplankton and macroinvertebrates as they grow (Humphries and Cumming 1973). High concentrations of small zooplankton appear necessary for the survival of larvae during the first few days after hatching (Setzler et al. 1980; Cooper and Polgar 1981; Eldridge et al. 1981; Martin et al. 1985; Martin and Wright 1987; and Setzler-Hamilton et al. 1987). There is high mortality during the first winter, and the nutritional condition of the fish determines success in overwintering (Hurst et al. 2000).

SCIENTIFIC NAME *Morone*, meaning unknown; *saxatilis* = "rock dwelling" (Latin).

Morone saxatilis × *Morone chrysops*
Hybrid striped bass

DESCRIPTION Body slightly compressed with lateral stripes generally broken behind pectoral fin and below lateral line. Color silver to black dorsally and white on belly, with shading that depends on water color (Hodson 1989).

SIZE Hybrid striped bass can exceed 6.8 kg, but anglers generally catch fish between 0.9 and 2.2 kg (Hodson 1989).

MERISTICS First dorsal spines 8–9, second dorsal spine 1, second dorsal rays 13–14, anal spines 3, anal rays 9–13 (Hodson 1989). See Jenkins and Burkhead 1993 (table 9) for an overview of the characters of *M. chrysops*, *M. saxatilis*, and hybrids.

DISTRIBUTION Widely introduced in lakes and reservoirs throughout the United States. In the MSRB, Avise and Van Den Avyle (1984) found F_1 hybrids much more common near Augusta than white bass or striped bass. They are popular sport fish in Clarks Hill/J. Strom Thurmond Lake. No distribution map was made for this introduced form. Some of the historical white bass records may be based on hybrids.

HABITAT Most often found in open-water areas of slow-moving streams, reservoirs, lakes, and ponds; they seem to avoid shallow waters with dense vegetation (Hodson 1989). Hybrid striped bass tend to concentrate in deep areas near inflowing streams in late winter and spring. Found at temperatures from 4 to 33 °C, but 25–27 °C seems to be the range for optimum growth (Hodson 1989). They do well in salinities up to 25 ppt, and tolerate full-strength seawater (Myers and Kohler 2000).

BIOLOGY The "palmetto" bass is the hybrid offspring of a female striped bass (*Morone saxatilis*) and a male white bass (*Morone chrysops*). It was originally produced by Bob Stevens in 1965 (Van Olst and Carlberg 1990; Kohler 2000). The shape and color are intermediate between those of striped bass and white bass. The "sunshine bass" (a hybrid between male striped bass and female white bass) is also a popular *Morone* hybrid and is also intermediate in appearance to the parental species. Hybrid striped bass are produced as food or sport fish and to control gizzard shad in some reservoirs (Kerby and Harrell 1990; Myers and Kohler 2000). An extensive litera-

Morone saxatilis × *Morone chrysops*, hybrid striped bass

ture exists on various aspects of the selection, cultivation, and biology of *Morone* hybrids (e.g., Kerby and Joseph 1979; Hodson 1989; Harrel 1997; Rudacille and Kohler 2000). Much of the following life history information is taken from Hodson 1989.

Fish are stocked in reservoirs as fry or juveniles. Most are pond cultured, but hybrids are fertile, and successful spawning has been reported in a few reservoirs. Some populations may make spawning migrations into upstream areas (Hodson 1989). Natural spawning takes place in the spring at temperatures between 15 and 20 °C. Females produce about 160,000 eggs per kilogram of body weight and spawn once; males can fertilize eggs multiple times (Hodson 1989).

Males start to mature at age 1 (about 250 mm and 500 g), and all are sexually mature at age 2; females mature at age 2 (a few individuals) or 3 (all) (Hodson 1989). Significant growth takes place above 15 °C, but optimum growth occurs between 25 and 27 °C. Hybrids grow fast during the first 2 years, to 225–350 grams in year 1 and 450–550 grams in year 2 (Hodson 1989). A typical hybrid lives between 5 and 6 years, a life span more like that of white bass than striped bass, which can live 30–40 years (Hodson 1989). They are most active at dusk and dawn (Hodson 1989).

Small juveniles eat crustaceans, zooplankton, and insects, and zooplankton continues to be part of the diet until they are at least 125 mm long. Juveniles switch to piscivory at a young age, with shad being an important prey item.

SCIENTIFIC NAME *Morone*, meaning unknown; palmetto bass = after the nickname for South Carolina (the Palmetto State), where the first female striped bass × male white bass hybrid was produced; sunshine bass = after the nickname for Florida (the Sunshine State), where the male striped bass × female white bass hybrid was developed.

Elassomatidae (pygmy sunfishes)

The six known species of the family Elassomatidae are all from one genus, *Elassoma*. All pygmy sunfishes are small—less than 5 cm in length. Three species have been recorded from the MSRB: the banded pygmy sunfish (*E. zonatum*), the Everglades pygmy sunfish (*E. evergladei*), and the relatively recently described bluebarred pygmy sunfish (*E. okatie*). The remaining species—the Carolina pygmy sunfish (*E. boehlkei*), the Okefenokee pygmy sunfish (*E. okefenokee*), and the spring pygmy sunfish (*E. alabamae*)—have not been confirmed from the MSRB. The one report of *E. boehlkei* within the MSRB is probably a misidentification. Elassomatids were long considered aberrant sunfishes (family Centrarchidae), but they are now recognized as morphologically, biochemically, and genetically distinct (Rohde et al. 1994).

Most pygmy sunfishes are found in the Coastal Plain. All species lack a lateral line and have a rounded caudal fin, cycloid scales, four spines in the dorsal fin, and three spines in the anal fin. They occur in calm, shallow, often tannin-stained waters and are often found motionless among heavy vegetation. Their secretive lifestyle makes collecting these fish somewhat problematic. During the early to mid-spring breeding season the males become colorful, and spawning and courtship occur in submerged vegetation (Rohde et al. 1994). The females deposit a relatively small number of eggs on underwater plants. No nests are built and the eggs are not cared for. The larvae grow rapidly and the life span rarely exceeds 18 months (Rohde et al. 1994), although banded pygmy sunfish (*E. zonatum*) may live 2 years or more (Barney and Anson 1920; Walsh and Burr 1984). The diet consists of small aquatic invertebrates. The three species that occur in the MSRB can be confused with juveniles of the larger sunfishes (family Centrarchidae), which have six or more dorsal spines; and pirate perch, which have an opercular spine and ctenoid scales. Adult pirate perch also differ in having the anus in the jugular (throat) position. Pygmy sunfish breed well in aquaria if provided with clear, slightly acidic water and live prey (Rohde et al. 1994).

FIGURE 22. Bluebarred pygmy sunfish, *Elassoma okatie*. A large portion of the known range of this threatened species lies within the Savannah River drainage.

KEY TO THE SPECIES OF ELASSOMATIDAE

1a. Scales present on top of head; side of body between the pectoral fin and caudal peduncle marked with streaks or mottling (Plate 16 a) . . . . *Elassoma evergladei* (Everglades pygmy sunfish), p. 300

1b. Scales absent from top of head; side of body between the pectoral fin and caudal fin marked with 7–16 dark vertical bars (Plate 16 b–d) **2**

[1] **2a.** Postocular stripe present; 1 or 2 dark dorsolateral blotches usually present on side of body between pectoral fin origin and anterior portion of dorsal fin (Plate 16 b) . . . . *Elassoma zonatum* (banded pygmy sunfish), p. 304

2b. Postocular stripe absent; no dark dorsolateral blotches present on side of body between pectoral fin origin and anterior portion of dorsal fin (Plate 16 c, d) **3**

[2] **3a.** Wide, dark bars present on body between pectoral fin and caudal peduncle, bars usually 3 times the width of the narrow, light-colored interbar spaces, bars number (8)10–12(14) (Plate 16 c) . . . . *Elassoma okatie* (bluebarred pygmy sunfish), p. 302

3b. Narrow dark bars on body between pectoral fin and caudal peduncle, bars nearly equal in width to the light-colored interbar spaces, bars number (10)12–14(16) (Plate 16 d) . . . . *Elassoma boehlkei* (Carolina pygmy sunfish)—no species account

Note: This key is largely derived from Rohde and Arndt 1987.

PLATE 16. Morphological characters of Elassomatidae: (a) Everglades pygmy sunfish showing mottled color pattern, (b) banded pygmy sunfish showing lateral color pattern, (c) bluebarred pygmy sunfish showing lateral color pattern, (d) Carolina pygmy sunfish showing lateral color pattern.

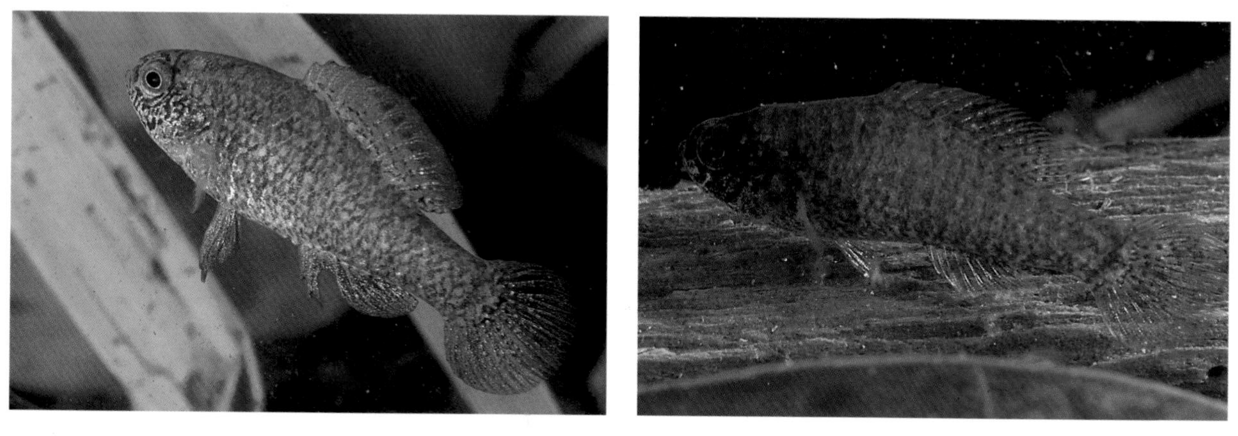

Elassoma boehlkei, Carolina pygmy sunfish; (left) male, (right) female

Elassoma evergladei Jordan, 1884
Everglades pygmy sunfish

DESCRIPTION Mouth small and oblique with the maxillary barely reaching the front margin of the pupil (Carr and Goin 1959). Embedded scales present on top of head. Scales on body cycloid. Lateral line absent.

No dark shoulder blotches or postocular stripe. Sides of nonbreeding fish marked with light streaks, mottling, or blotches, but lack dark vertical bars. Lips dark. Females generally brown on back and mottled brown and cream to white on lower sides. A crescent-shaped area under and behind the eye may appear gold or iridescent blue in life. Breeding males black with iridescent blue patches in life. In MSRB, similar in appearance to the banded pygmy sunfish (*E. zonatum*) and the bluebarred pygmy sunfish (*E. okatie*).

SIZE Up to 32 mm.

MERISTICS Dorsal spines (3)4(5), dorsal rays (8)9–10, anal rays 4–5, pectoral rays 13–15 (Böhlke 1956); lateral series scales 23–32 (Mettee et al. 1996).

DISTRIBUTION Coastal streams below the Fall Line from Cape Fear, North Carolina, south into southern Florida and west to the Mobile Bay basin of Alabama (Böhlke and Rohde *in* Lee et al. 1980).

HABITAT Vegetated areas along margins of low-gradient streams, sloughs, canals, overflow pools, and swamps where the current is weak, water levels undergo radical changes (Rubenstein 1981a), and the bottom consists of mud, silt, or detritus. They may also occur in sphagnum bogs (Laerm and Freeman 1986). They are more abundant in natural than constructed marshes in central Florida (Streever and Crisman 1993), and are more restricted to stained (black) water than other species of *Elassoma* (Swift et al. 1977).

BIOLOGY Everglades pygmy sunfish are generally solitary (Rubenstein 1981a). Spawning occurs from March through November in the Okefenokee Swamp of Georgia (Freeman and Freeman 1985), and throughout the year in Florida (Laerm and Freeman 1986). The demersal eggs are scattered in aquatic vegetation (Mettee et al., 1996).

Population densities fluctuate widely in response to variations in environmental conditions (Rubenstein 1981b). Ovaries of females held in captivity at varying densities contained an average of 115–500 eggs, and the number of eggs increased with female size (Rubenstein 1981a).

Everglades pygmy sunfish are invertivores that feed primarily on cladocerans and dipteran larvae in the Okefenokee Swamp (Freeman and Freeman 1985), but also on annelids, chironomids, copepods, and ostracods. In captivity, dominant males established territories in feeding areas (Rubenstein 1981b).

SCIENTIFIC NAME *Elassoma* = "small" (Greek); *evergladei* = "of the Everglades," where the type specimens were captured.

Elassoma evergladei, Everglades pygmy sunfish; (left) male, (right) female

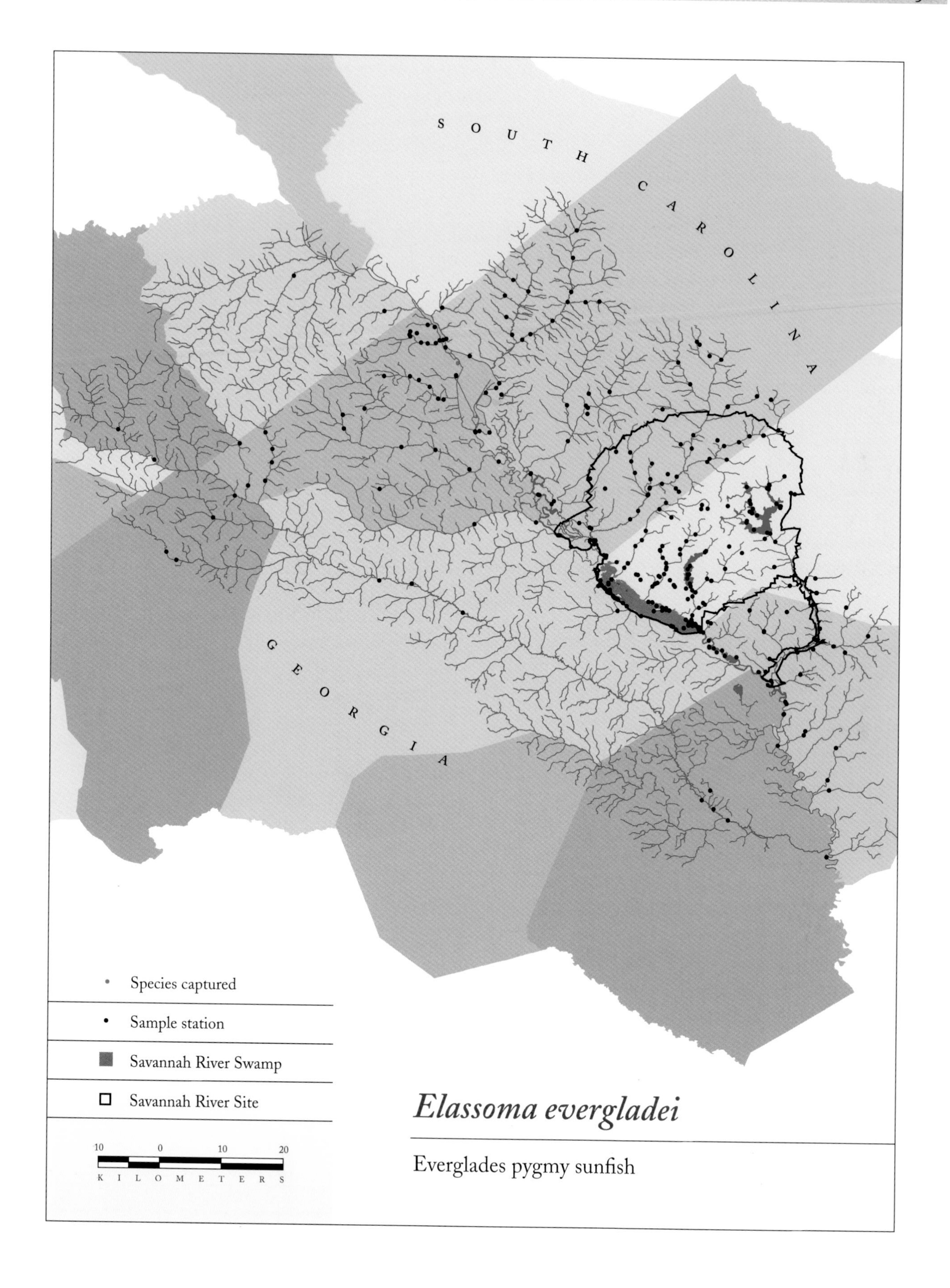

Elassoma evergladei

Everglades pygmy sunfish

Elassoma okatie Rohde and Arndt, 1987
Bluebarred pygmy sunfish

DESCRIPTION Body compressed, with a deep caudal peduncle. Mouth small and terminal to slightly superior. Scales absent on top of head. Pelvic fin long and pointed, reaching anal fin.

Dark shoulder blotches and postocular stripe absent. Nine to 12 dark vertical bars present on sides, rarely 8 or 14. Bars are relatively wide, about three times wider than lighter-colored interspaces, and average 1.1 mm in width in males and 1.0 mm in females. Breeding males are typically black with blue-green markings in life and a conspicuous brilliant spot at the anterior edge of the eye. Females are lighter in color than males but may have blue, green, or yellow flecks on thorax and mid-trunk.

SIMILAR SPECIES Everglades pygmy sunfish (*E. evergladei*) and banded pygmy sunfish (*E. zonatum*); see species accounts for distinguishing characters.

SIZE To 28.7 mm SL; adult males and females average 21 mm SL (Rohde and Arndt 1987).

MERISTICS Dorsal spines (3)4–5(6), dorsal rays (8)9–11(12), anal rays (4)5–7(8), pectoral rays (13)14–17; lateral series scales (24)25–30; lower gill rakers 6, rarely 5 or 7 small fleshy knobs; branchiostegals 5 (Rohde and Arndt 1987).

CONSERVATION STATUS This species is considered imperiled or potentially so (G2/G3) because of its rarity and restricted range. In South Carolina, it is listed as a species of special concern. The state of Georgia lists it as S1, critically imperiled, because of its extreme rarity. The Savannah River and Edisto River populations are sufficiently genetically distinct that they should be managed separately (Quattro et al. 2001a).

DISTRIBUTION Known only from the New River, Edisto River, and Savannah River drainages in South Carolina and Georgia (Quattro et al. 2001a). Within the Savannah River drainage, found as far west as Boggy Gut Creek in Richmond County, Georgia (Hoover et al. 1998).

HABITAT The primary habitat is roadside ditches and backwaters of creeks or rivers with brown-stained water and abundant vegetation including bladderwort, duckweed, alligatorweed, pondweed, spatterdock, rushes, and grasses.

BIOLOGY Virtually nothing is known about the biology of the bluebarred pygmy sunfish. The genetic evidence available fails to rule out the possibility that *E. okatie* and *E. boehlkei* are the same species, although they probably are separate (Quattro et al. 2001b).

SCIENTIFIC NAME *Elassoma* = "small" (Greek); *okatie*, from Native American (Muskhogean) *oka* = "water" and *ateeh* = "coming from," and by derivation, "aquatic."

Elassoma okatie, bluebarred pygmy sunfish; (left) male, (right) female

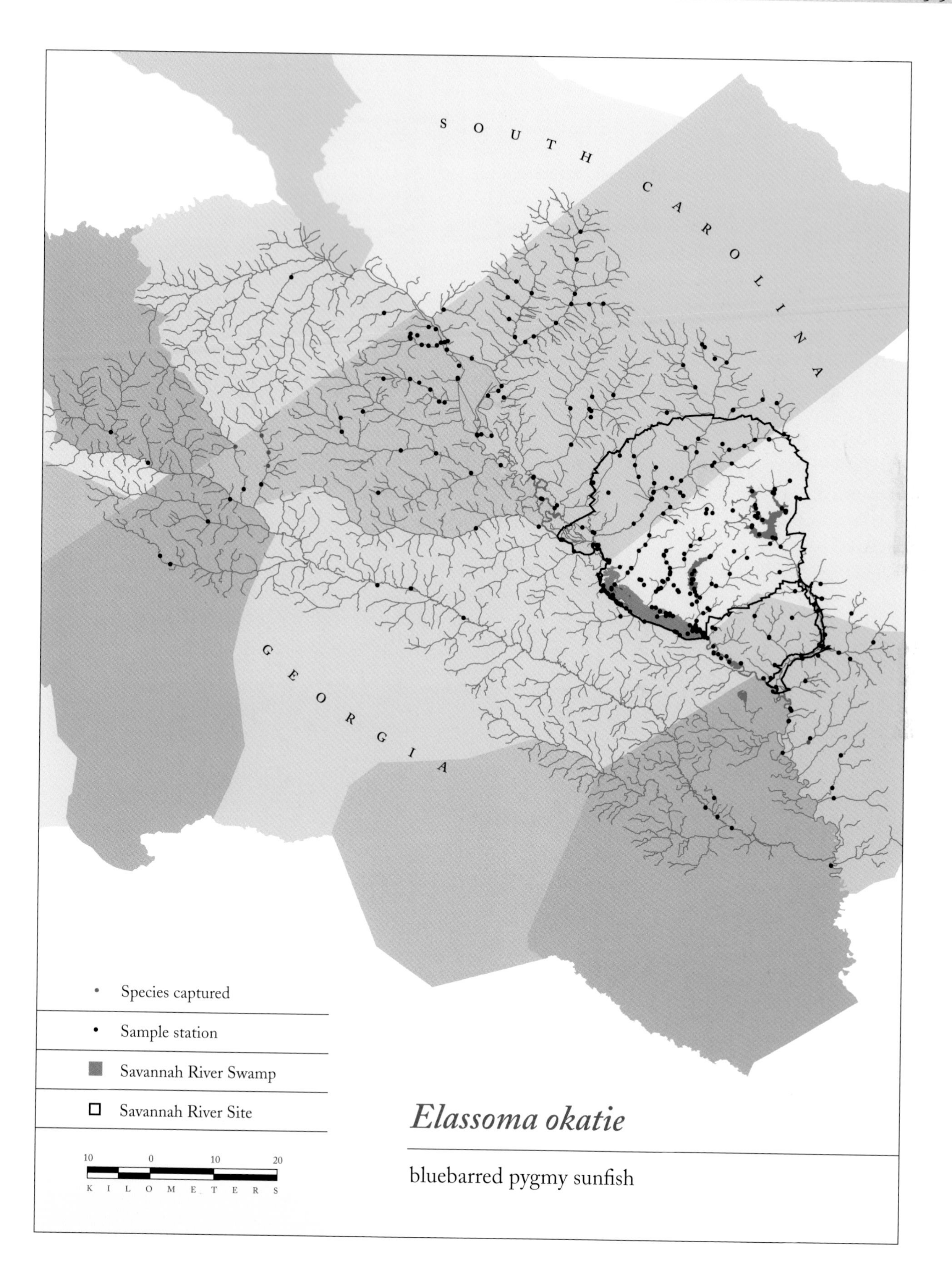

Elassoma okatie

bluebarred pygmy sunfish

Elassoma zonatum Jordan, 1877

Banded pygmy sunfish

DESCRIPTION Body compressed and moderately deep. Mouth terminal. No lateral line. Caudal fin rounded; pectoral fins short and rounded. Scales cycloid, those on cheeks and opercles mostly embedded; no scales present on top of head. Spinous and soft portions of dorsal fin broadly joined with no notch.

Body olive green to brown and heavily stippled with black in life. Sides with series of 8–12 blackish vertical bars, slightly wider than lighter interbar spaces. Postocular stripe dark. Large black shoulder blotch present below front of dorsal fin, may be broken into two or three smaller blotches. Caudal and soft portions of anal and dorsal fins of males with black banding. Fins of females may be clear with rows of small brown spots or faint mottling.

Breeding males may be almost solid black (sometimes obscuring vertical bars and shoulder blotch) with iridescent blue and green flecks and whitish vertical bars. Breeding females may be lighter in color than during the rest of the year and tend to have clear pectoral and anal fins (Barney and Anson 1920).

SIMILAR SPECIES Everglades pygmy sunfish (*E. evergladei*) and bluebarred pygmy sunfish (*E. okatie*).

SIZE Relatively large for *Elassoma*, reaching about 45 mm SL (Jones and Quattro 1999). Breeding males average 24 mm TL; breeding females average 25 mm TL (Barney and Anson 1920).

MERISTICS Dorsal spines 4–5(6), dorsal rays (7)8–11(12), anal spines 3, anal rays (3)4–6(7) (Rohde and Arndt 1987), pectoral rays (13)14–16(17); lateral series scales 38–45 (Robison and Buchanan 1984), 31–36 (Etnier and Starnes 1993; Mettee et al. 1996), or (28)30–34(36) (Rohde and Arndt 1987).

DISTRIBUTION The Coastal Plain from North Carolina to northern Florida, west along the Gulf Coast to eastern Texas, and north in the Mississippi Valley to south-

Elassoma zonatum, banded pygmy sunfish; male

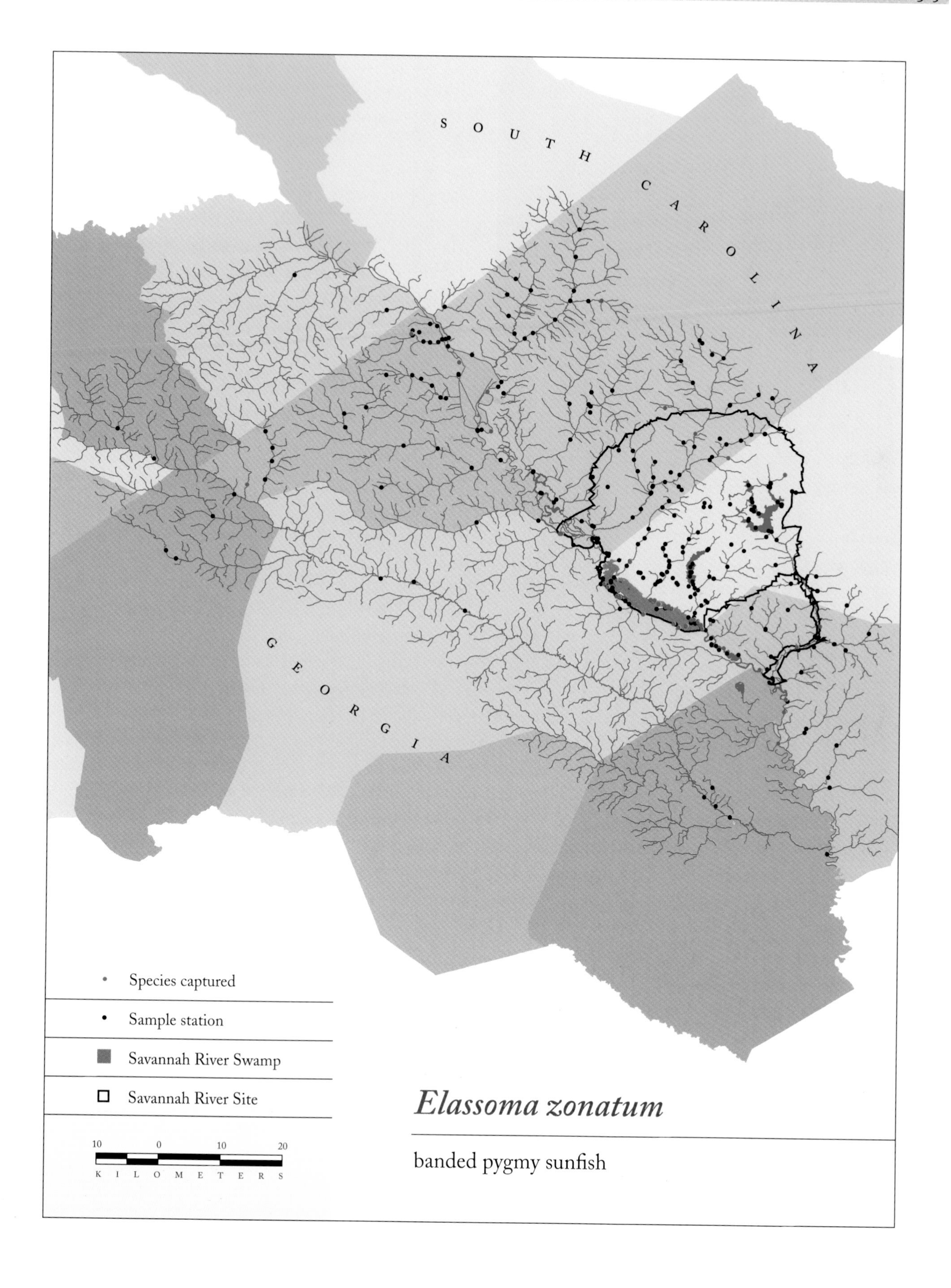

Elassoma zonatum

banded pygmy sunfish

Elassoma zonatum, banded pygmy sunfish; female

ern Illinois; rarely above the Fall Line (Böhlke and Rohde *in* Lee et al. 1980). In the MSRB, relatively widespread and the most common of the pygmy sunfishes.

HABITAT Well-vegetated, sluggish or still waters such as swamps, backwaters, bayous, sloughs, and ponds; usually over bottoms of soft mud or detritus. In Illinois, common around mosquito fern (*Azolla*), duckweed (*Lemna*), and hornwort (*Ceratophyllum*) (Gunning and Lewis 1955).

BIOLOGY This small fish is secretive and solitary except for congregations at spawning sites (Mettee et al. 1996). It is an early spring spawner. Ovaries begin to significantly enlarge in February, and spawning occurs in mid-March, or in April in southern Illinois (12–16 °C) and northern Louisiana (16.7–20 °C); they will spawn at warmer temperatures in aquaria (Barney and Anson 1920; Walsh and Burr 1984). Although no nest is prepared, dominant males defend territories and use distinctive displays to court females. Spawning occurs over or among submergent vegetation, and the adhesive, demersal eggs are scattered over the leaves and stems. Females spawn repeatedly with the same or different males at intervals of several hours to a week (Barney and Anson 1920; Mettee 1974 *in* Walsh and Burr 1984). Males guard the site for up to 2 days after spawning. Two males have been observed spawning as a trio with a single female, and on one occasion two males guarded the same clutch for several hours after such a spawning event (Poyser 1919 *in* Walsh and Burr 1984; Walsh and Burr 1984). Gravid females regularly contain more than 300 eggs, and some have up to 970; egg numbers increase with female size (Barney and Anson 1920). Eggs are spawned in batches of 40–60 eggs (Barney and Anson 1920) or 6–76 with an average of 38 (as reported by Walsh and Burr 1984). Eggs hatch in 4–5 days at 21 °C (Walsh and Burr 1984) and in about 7 days at 18.3 °C (Barney and Anson 1920). Larvae are described in Walsh and Burr 1984. Males and females mature at 1 year, at sizes as small as 18 mm TL in Louisiana and 19 mm SL in Illinois (Barney and Anson 1920; Walsh and Burr 1984). In Kentucky, they live 1–2 years, growing more slowly in the second year. They may live 3 years at southern latitudes (Walsh and Burr 1984).

Banded pygmy sunfish are visual invertivores that feed on crustaceans, molluscs, and aquatic insects, especially chironomids, during daylight hours (Walsh and Burr 1984). Adults in densely vegetated habitat may have few vertebrate predators but may be preyed on by predaceous water beetles, giant water bugs, and odonate nymphs (Barney and Anson 1920). Predation by redfin and chain pickerels (*Esox americanus* and *E. niger*) and warmouth (*Lepomis gulosus*) has also been reported (Gunning and Lewis 1955; Walsh and Burr 1984; Fletcher and Burr 1992).

SCIENTIFIC NAME *Elassoma* = “small” (Greek); *zonatum* = “banded” (Latin).

Centrarchidae (sunfishes)

The eight genera of Centrarchidae are indigenous to North America and are represented by 30 living species, all but 1 of which have a native range east of the Rocky Mountains. Only the Sacramento perch (*Archoplites interruptus*) resides in the Pacific Coast drainages of California. With the exception of the bullhead catfishes (family Ictaluridae), the sunfishes are the largest indigenous fish family in North America. Seventeen species are found in the MSRB.

The body of sunfishes is somewhat to very laterally compressed and ranges from moderately elongate to deep. The head is medium to large, and the mouth is small to large. The lateral line is usually complete, and the body is fully scaled with ctenoid scales. Only the mud sunfish (*Acantharchus pomotis*) has scales lacking ctenii. A spinous dorsal fin is immediately followed by a soft (rayed) dorsal fin. These fins may be broadly joined, as in *Lepomis*, or narrowly connected, as in largemouth bass (*Micropterus salmoides*). The anal fin and pelvic fins have spines anteriorly. The pelvic fin usually has one spine and five soft rays. The coloration of sunfishes varies greatly, but many are spectacularly colored. Many species exhibit striking sexual dimorphism in both color patterns and morphological characters, especially during the breeding season; in other species the sexes are similar. Sunfishes are visual predators, and their diet includes crustaceans, insects, molluscs, and fish.

The reproductive behavior, which has been well studied, is quite variable. Depending on the taxon, spawning in the MSRB is generally in early to late spring, or late spring to midsummer. All species build nests, but the structure and location of the nests vary greatly among species. Most nests are circular and are swept in the sediment, or among vegetation as in *Enneacanthus*, or on a stump or log as in *Micropterus* (see species accounts for details). Eggs are fertilized and deposited in the nest while the male and female move over the nest. Most species guard the nest aggressively until the young leave it, but care may be extended to the free-swimming young as well. Offspring may be guarded by only the male (e.g., *Lepomis*) or by both parents (e.g., largemouth bass). In many species, multiple females may spawn in a single nest, but some species are monogamous. Larger males spawn by attracting females to a nest; some smaller males exhibit "parasitic" spawning behaviors: male "sneakers," for example, may join a spawning pair by briefly darting into the nest to fertilize eggs, and "female mimics" may exhibit female behavior to gain access and fertilize eggs.

Hybridization is common, especially among the *Lepomis* species (often called "bream" by anglers), and can complicate species identification. The identification of small individuals (<2.5 cm), especially of *Lepomis* species, can also be difficult because they lack many morphological characters that distinguish adults. Color patterns also may change with age. Sunfishes have been introduced all over the world, and many species are important to recreational fisheries. Largemouth bass are coveted sport fish, and many a child's first fishing trip involves a stringer of bluegills or other bream.

KEY TO THE SPECIES OF CENTRARCHIDAE

1a. Posterior margin of caudal fin straight or rounded (Plate 3 c) **2**
1b. Caudal fin emarginate or slightly forked (Plate 3 e) **5**

[1] **2a.** Anal spines (4)5(6); mouth large with upper jaw (posterior edge of maxilla) extending behind middle of eye; 3 or 4 parallel dark stripes across forehead . . . . *Acantharchus pomotis* (mud sunfish), p. 312
2b. Anal spines (2)3(4); mouth small with upper jaw (posterior edge of maxilla) not extending behind middle of eye; no parallel dark stripes across forehead **3**

[2] **3a.** Distinctive black blotch on front of spinous dorsal fin (Plate 17 a); side marked with 6 dark vertical bars on a "light" (whitish) background; deep notch between spinous and soft (rayed) dorsal fins, spines decreasing markedly in length before and after the fourth spine, best seen by pulling spine erect to fully expand the fin (Plate 17 a) . . . . *Enneacanthus chaetodon* (blackbanded sunfish), p. 318
3b. No black blotch on front of spinous dorsal fin (Plate 17 b, c); side marking variable (may be horizontal rows of spots or 5–8 vertical bars), but always on an olive background; shallow notch present between spinous and soft (rayed) dorsal fins, spines nearly the same length or only slightly shorter after the fourth spine (Plate 17 b, c) **4**

[3] **4a.** Black spot on ear tab two-thirds the size of eye pupil (Plate 18 a); generally shallow and thin caudal peduncle with (14)16–18(20) scales around the narrowest portion; shallow and more slender/laterally compressed body; side of body marked with rows of blue or greenish iridescent spots that are round with rather distinct outer margins and are more numerous and prominent in nuptial males; smaller juveniles and an occasional larger individual may have faint bars on side . . . . *Enneacanthus gloriosus* (bluespotted sunfish), p. 321
4b. Black spot on ear tab larger than eye pupil (Plate 18 b); generally deep and thick caudal peduncle with (17)19–22(24) scales around the narrowest portion; deep and chubby or chunky body; side of body generally marked with 5–8 dark to faint vertical bars, also frequently marked with scattered greenish or copperish spots that are crescent shaped with diffuse outer margins . . . . *Enneacanthus obesus* (banded sunfish), p. 324

[1] **5a.** Anal spines 5 or more (rarely 5) **6**
5b. Anal spines (2)3(4) **8**

[5] **6a.** Dorsal spines 11–13; anal spines 7–8(9) (Plate 17 n, o) . . . . *Centrarchus macropterus* (flier), p. 315
6b. Dorsal spines 5–8; anal spines (5)6–7 . . . . **7**

[6] **7a.** Dorsal spines 5–6(7); dorsal fin base length short, shorter than distance from dorsal fin origin to top of head above posterior margin of eye (Plate 18 c); 4–5 rows of scales on cheek; mottling on sides forming discernible vertical bars . . . . *Pomoxis annularis* (white crappie), p. 361
7b. Dorsal spines (6)7–8; dorsal fin base long, as long as distance from dorsal fin origin to top of head above posterior margin of eye (Plate 18 d); 6 rows of scales on cheek; no distinctive pattern in often heavily mottled sides . . . . *Pomoxis nigromaculatus* (black crappie), p. 364

[5] **8a.** Body shallow (elongate) and moderately compressed (see Plate 2 f, g for examples), with standard length usually 3 times greater than body depth; 55 or more lateral line scales; deep or moderately deep notch between spinous and soft dorsal fins (Plate 17 d, e) . . . . *Micropterus* **9**
8b. Body deep (see Plate 2 d for example) and usually strongly compressed (Plate 2 h), with standard length less than 3 times body depth; fewer than 55 lateral line scales; shallow or no notch between spinous and soft dorsal fins (Plate 17 f, g, j, k) . . . . *Lepomis* **10**

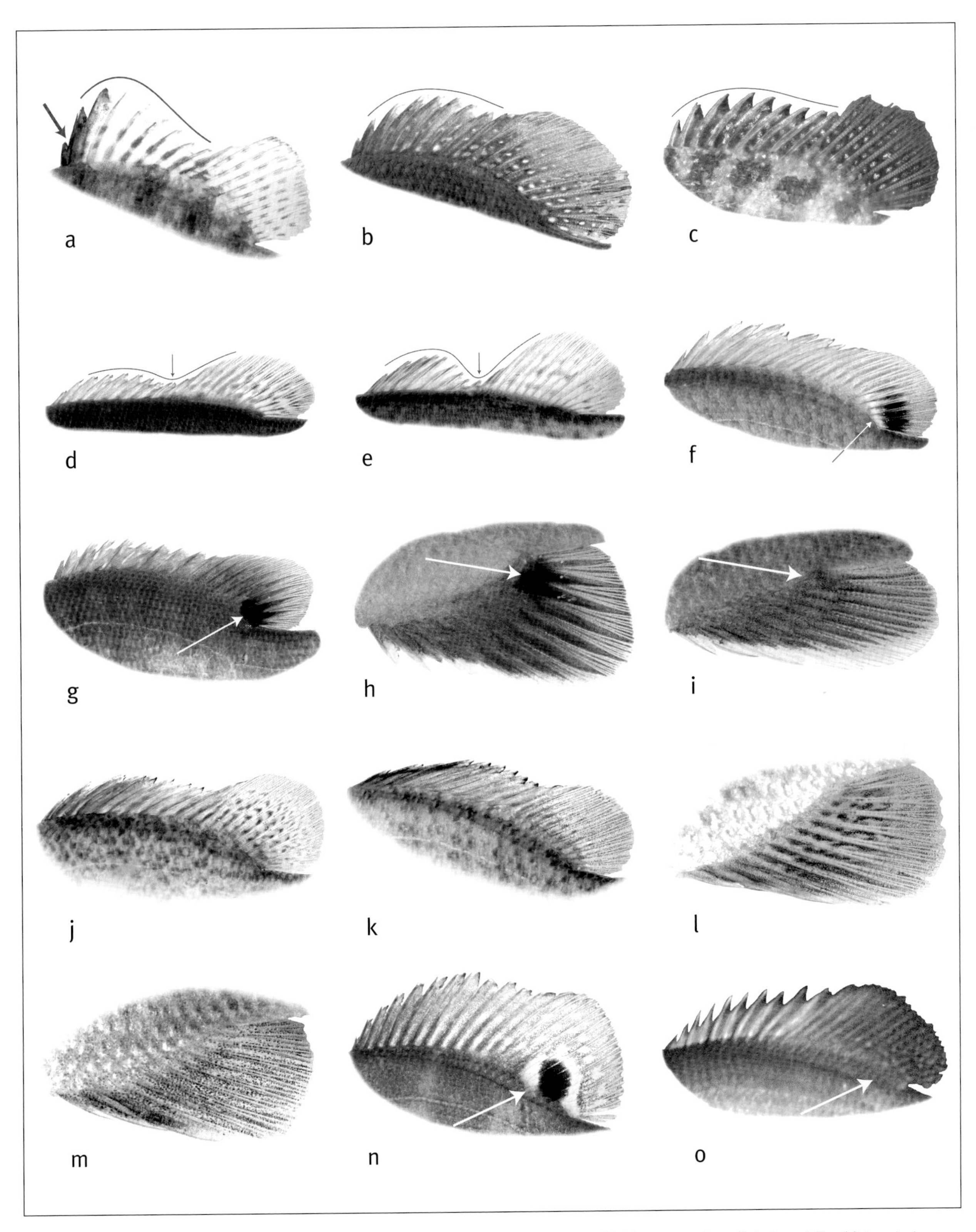

PLATE 17. Fin characteristics of Centrarchidae: (a) blackbanded sunfish dorsal fin, (b) bluespotted sunfish dorsal fin, (c) banded sunfish dorsal fin, (d) redeye bass dorsal fin, (e) largemouth bass dorsal fin, (f) bluegill dorsal fin, (g) adult green sunfish dorsal fin, (h) adult green sunfish anal fin, (i) juvenile green sunfish anal fin, (j) pumpkinseed dorsal fin, (k) redear sunfish dorsal fin, (l) pumpkinseed anal fin, (m) redear sunfish anal fin, (n) juvenile flier dorsal fin, (o) adult flier dorsal fin.

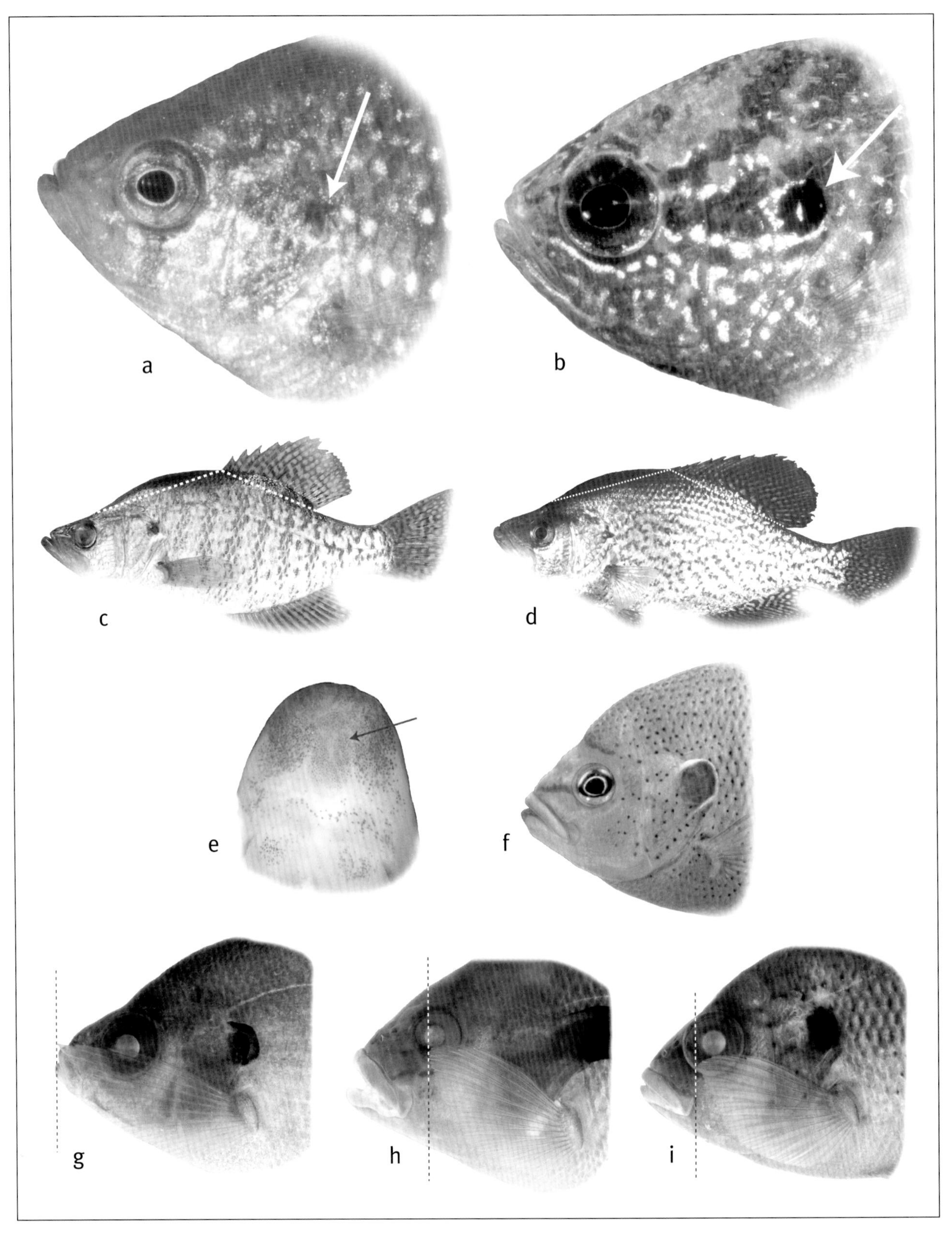

PLATE 18. Morphological characters of Centrarchidae: (a) bluespotted sunfish ear tab relative size, (b) banded sunfish ear tab relative size, (c) white crappie dorsal fin base length, (d) black crappie dorsal fin base length, (e) tooth patch on tongue of warmouth, (f) color pattern on spotted sunfish head and operculum, (g) long pointed pectoral fin, (h, i) shorter, rounded pectoral fin.

[8] **9a.** Ventrolateral side marked with horizontal rows of small dark spots; midlateral side may be marked with faint black blotches, but these do not form a distinct broad horizontal stripe; sides of young may be heavily mottled; median tooth patch present on tongue; upper and lower tips of caudal fin marked with white; notch between spinous and soft dorsal fins only moderately deep so that fins are more distinctly joined (Plate 17 d) . . . . *Micropterus coosae* (redeye bass), p. 354

9b. Ventrolateral side marked with irregularly arranged small, dark spots; midlateral side generally marked by distinct broad horizontal stripe made up of joined or nearly joined black blotches; no median tooth patch present on tongue; upper and lower tips of caudal fin not marked with white; notch between spinous and soft (rayed) dorsal fins deep, with the fins nearly separated (Plate 17 e) . . . . *Micropterus salmoides* (largemouth bass), p. 357

[8] **10a.** Large mouth with upper jaw (posterior edge of maxilla) extending to near center of pupil of eye when mouth is closed **11**

10b. Small mouth with upper jaw (posterior edge of maxilla) not extending to pupil of eye when mouth is closed **12**

[10] **11a.** Tongue with median patch of teeth (Plate 18 e); 3 dark, relatively deep stripes radiating from posterior margin of eye; no thin, wavy iridescent blue lines on side of head; no large black blotch on posterior base of soft dorsal fin . . . . *Lepomis gulosus* (warmouth), p. 337

11b. Tongue only occasionally with teeth; no dark stripes radiating from posterior margin of eye; wavy iridescent blue stripes on sides of head; large black blotch at posterior base of soft dorsal fin (Plate 17 g) (faint in young); also dark blotch at posterior base of anal fin in adults, may be faint or absent in juveniles (Plate 17 h, i) . . . . *Lepomis cyanellus* (green sunfish), p. 330

[10] **12a.** Pectoral fins long with pointed tips, extending beyond anterior margin of eye when folded forward toward eye (Plate 18 g), and extending beyond origin of anal fin when pressed against body **13**

12b. Pectoral fins short with rounded tips, not extending beyond anterior margin of eye when folded forward (Plate 18 h, i), and not extending beyond origin of anal fin when pressed against body. **15**

[12] **13a.** Large black blotch at posterior base of soft dorsal fin (faint in young) (Plate 17 f); no red/orange spot on posterior margin of ear tab; gill rakers long and slender, about 5–10 times longer than wide . . . . *Lepomis macrochirus* (bluegill), p. 340

13b. No large black blotch on posterior base of soft dorsal fin; red/orange spot on mid-posterior margin of ear tab in adults; gill rakers short and thick, about 2–3 times longer than wide **14**

[13] **14a.** Soft portions of dorsal and anal fins with spots or wavy mottling (Plate 17 j, l); wavy iridescent blue lines on side of head; posterior edge of ear tab (excluding membranous flap) stiff . . . . *Lepomis gibbosus* (pumpkinseed), p. 333

14b. Soft portions of dorsal and anal fins with no spots or wavy mottling (Plate 17 k, m); no wavy blue lines on cheek; posterior edge of ear tab (excluding membranous flap) flexible . . . . *Lepomis microlophus* (redear sunfish), p. 348

[12] **15a.** Distinct spots on side of head, especially opercle, but no iridescent blue wavy lines or markings (Plate 18 f); long and moderately slender gill rakers about 5–6 times longer than wide; posterior edge of ear tab (excluding membranous flap) usually stiff . . . . *Lepomis punctatus* (spotted sunfish), p. 351

15b. Blue iridescent wavy lines or markings on side of snout and/or cheek and opercle; gill rakers 2–3 times longer than mid-width; posterior edge of ear tab (excluding membranous flap) flexible **16**

[15] **16a.** Cheek scale rows 6–8; posterior edge of ear tab with white or light margin; no palatine teeth . . . . *Lepomis marginatus* (dollar sunfish), p. 345

16b. Cheek scale rows 3–4; no light margin on posterior edge of ear tab; palatine teeth present. . . . . *Lepomis auritus* (redbreast sunfish), p. 327

Note: Teeth on the tongue can be easily felt by rubbing a blunt probe, finger, or pen across the tongue.

Acantharchus pomotis (Baird, 1855)

Mud sunfish

DESCRIPTION Body compressed but chunky and relatively elongate (compared with most *Lepomis*). Eyes large. Mouth large and oblique with posterior edge of maxilla extending to posterior margin of pupil. Tongue may have a median tooth patch. Gill rakers long, but fewer than 10 present on lower arch. Pectoral fins broadly rounded and short; when folded forward their tips extend only to operculum, well behind eye. Caudal fin rounded. Lateral line complete and scales cycloid.

Three lines cross the forehead and extend posteriorly across the side of the light tan face and operculum. Dorsum brown, sides marked with chocolate brown mottling on yellowish tan background. Ventral head and breast yellowish tan, becoming increasingly mottled posteriorly on belly to flanks. Mottling on sides mostly irregular, but sometimes forming three to four moderately deep parallel horizontal lines. Dorsal and anal rays dusky brown and may be mottled, especially in young individuals. Pectoral fins largely translucent. Pelvic fins with light anterior margin, then dark brown, fading in posterior rays. Ear tab short and deep, dark brownish black with light dorsal and ventral margins.

Sexual dimorphism is not pronounced during the breeding season, but the chocolate brown mottling and ear tab are darker in males than females. Mottling is not complete on small juveniles, which have large, diffuse blotches on lateral scales producing many rows of parallel lines.

Acantharchus pomotis, mud sunfish

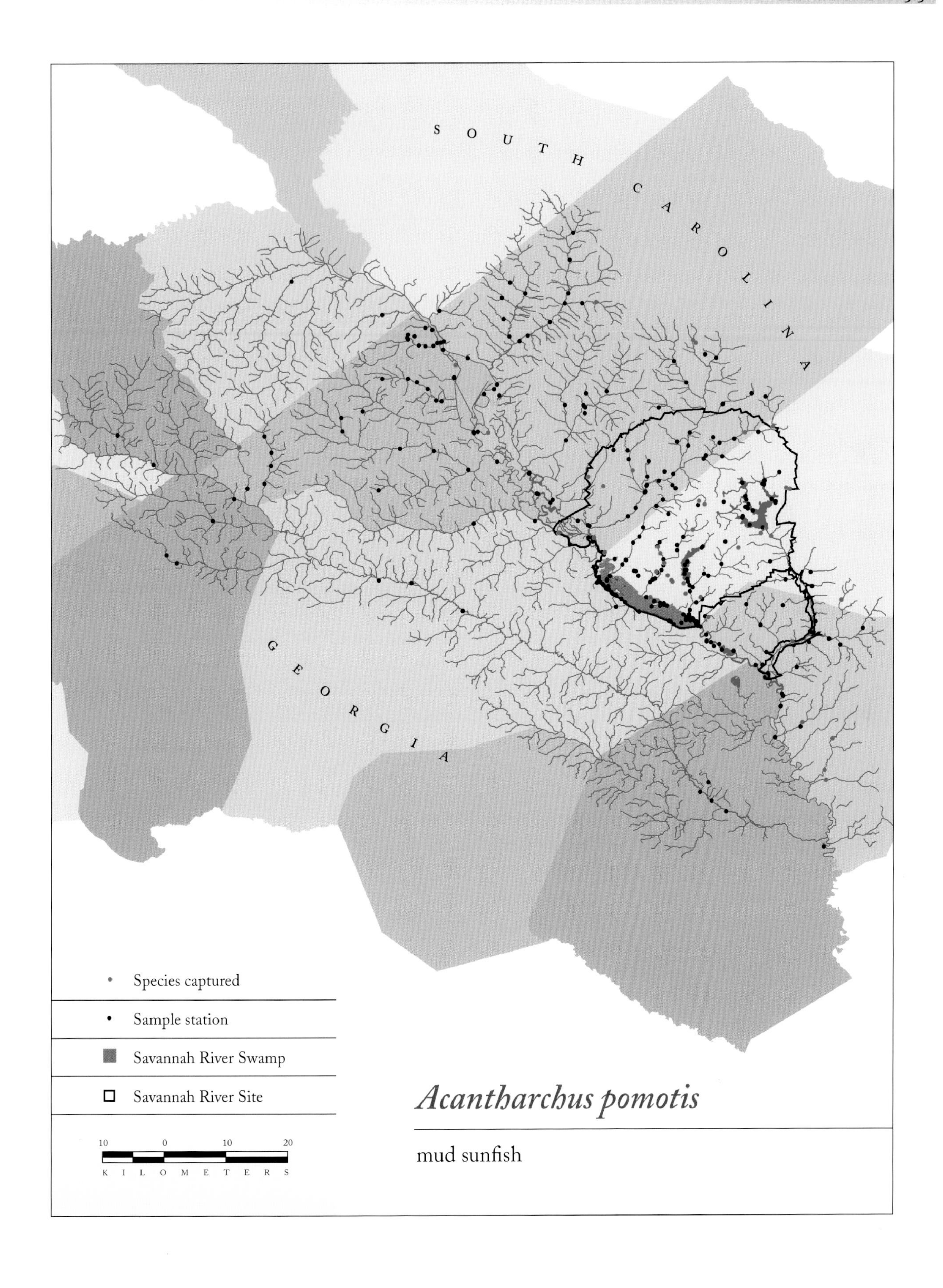

Acantharchus pomotis

mud sunfish

SIZE May reach 170 mm, but most adults are 100–145 mm SL (Cashner et al. 1989; Jenkins and Burkhead 1993).

MERISTICS Dorsal spines (10)11(12), dorsal rays (9)10–12(13), anal spines (4)5(6), anal rays (9)10(11), pectoral rays 14–15; lateral line scales (34)37–43(45) (Jenkins and Burkhead 1993).

DISTRIBUTION The Atlantic Coastal Plain from New York to northern Florida and westward in the Florida Panhandle to the St. Marks River, but common nowhere (Cashner *in* Lee et al. 1980). In Georgia, found on the Coastal Plain from the Savannah River to the Ochlockonee River (Dahlberg and Scott 1971a). Widely distributed but seldom common in tributary streams of the MSRB, most common in headwater streams. Abundant at times in some Carolina bays.

HABITAT Sluggish streams, ponds, and swamps (Cashner et al. 1989) over a pH range from about 4 to nearly 9 (Graham 1993). In small tributary streams, frequently found in undercut banks or woody debris in pools. More frequently found in pools than runs and more frequently over detritus than mud bottoms (Pardue 1993).

BIOLOGY The biology of this nocturnal and secretive fish is poorly known (Mansueti and Elser 1953; Cashner et al. 1989). The spawning season appears to vary with latitude: gravid females were collected in May and June in Delaware, indicating late spring/summer spawning (Wang and Kernehan 1979); spawning may occur from early fall to late winter in the Okefenokee Swamp of Georgia and Florida (Laerm and Freeman 1986), and in early June in New Jersey (Breder 1936). Females with mature eggs have been recorded from January to May, but few gravid females were examined; males had flowing milt from December to May in North Carolina; water temperatures ranged from 7 to 20 °C (Pardue 1993). Adults appear in reproductive condition in early to mid-spring in the MSRB (DEF pers. obs.). Fowler (1923) reported a male mud sunfish guarding a saucer depression nest, about 30 cm in diameter, in shallow water near the shore among aquatic vegetation. In the MSRB, males have been captured on small nests, about 15–20 cm in diameter; generally these were in protected areas near the bank in pools in very small headwater streams (DEF pers. obs.). Details of spawning and nest guarding are not well known. Wang and Kernehan 1979 includes a drawing of a juvenile. Annuli formed on scales in late March in North Carolina swamp streams, where mud sunfish lived to 4 years of age, with most individuals not exceeding 3 years (Pardue 1993). Based on examination of scales, mud sunfish may live to 7–8 years (Breder and Redmond 1929; Mansueti and Elser 1953). Mud sunfish studied by Pardue (1993) matured in 1 year, and total egg counts within ovaries ranged from 5,508 to 11,838, with 1,515–3,812 eggs measuring at least 0.7 mm diameter.

Adults frequently rest head-down in weeds and are most active at night (Laerm and Freeman 1986). Mud sunfish are reported to be rather sedentary with small home ranges; more than 70% of all recaptures in a North Carolina swamp stream were within 200 m of the original site of capture (Whitehurst 1981). Movements most frequently occur in the spring around spawning season while temperatures are low and water levels are high (Pardue 1993). In the MSRB, the mud sunfish is a good invader of isolated wetlands that dry infrequently, particularly those located in the lower parts of drainage basins (Snodgrass et al. 1996); they also frequently invade ephemeral reaches of tributaries.

Mud sunfish are reported to eat predominantly crustaceans and insects, but include fish in the diet occasionally (Pardue 1993). Mud sunfish may produce grunting sounds (Abbott 1884 *in* Breder and Rosen 1966).

SCIENTIFIC NAME *Acantharchus* = "spined or thorned anus" (Greek), in reference to the large number of well-developed anal spines; *pomotis*, from *poma* = "lid or gill cover" and *otis* = "ear" (Latin).

Centrarchus macropterus (Lacépède, 1801)
Flier

DESCRIPTION Body deep, strongly compressed, and slab sided. Eyes large. Mouth small, supraterminal, and oblique, with posterior edge of maxilla extending to anterior margin of pupil. Two elongate median tooth patches present on tongue. Gill rakers long and slender. Pectoral fin relatively rounded and short, when folded forward extending to between anterior margin of pupil and anterior margin of eye. Caudal fin emarginate. Anal fin base very long. Dorsal and anal fins tall and with broad lobes posteriorly. Lateral line complete; scales ctenoid. As in crappies (*Pomoxis* sp.), the posterior edge of the preopercle is distinctly serrate (Pflieger 1975).

Dorsum dark olive fading to brassy green on lateral sides and to brassy white ventrally. Sides marked with parallel stripes formed by many scales having a black-olive central spot. On ventral side stripes/mottling fade anteriorly onto breast. Cheek brassy with black suborbital bar. Dorsal and anal rays dusky with black, tan, and orange mottling (light in preservative). Mottling most distinctive on soft portion of fins. Caudal fin frequently with dark mottling. Ear tab short, deep, and dark olive-black.

Juveniles have a prominent black ocellus with reddish orange margin on the posterior portion of the soft dorsal fin that fades with size. Small juveniles are also marked with four to six dark olive-black bars that fade into dark saddles as the brassy sides take on a mottled appearance and dark spots gradually fill in the longitudinal rows typical of adults.

Centrarchus macropterus, flier; adult

SIMILAR SPECIES The flier closely resembles the crappies, but differs by having 10–13 dorsal spines.

SIZE Adults are generally 127–178 mm TL (Mettee et al. 1996).

MERISTICS Dorsal spines (11)12(13), dorsal rays (12)13–14, anal spines 7–8, anal rays (14)15–16(17), pectoral rays (11)13–14; lateral line scales (36)37–41(42) (Jenkins and Burkhead 1993).

DISTRIBUTION From eastern Virginia and extreme southern Maryland south to north-central Florida, westward along the Gulf Coast to eastern Texas, and north in the Mississippi River basin to southern Illinois and Indiana; the population in southern Maryland may be the result of introduction (Lee and Gilbert *in* Lee et al. 1980). In Georgia, restricted to the Coastal Plain (Dahlberg and Scott 1971a). Common but not abundant in the MSRB, inhabiting primarily the lowlands of the modern Savannah River floodplain and occasionally tributary streams, especially on the fluvial terraces of the Savannah River Valley.

HABITAT Swampy backwaters of large rivers, oxbow lakes, outflow pools, and slow-moving creeks (Robison and Buchanan 1984; Mettee et al. 1996), generally in quiet, clear, and heavily vegetated areas with mud and detritus bottoms (Robison and Buchanan 1984). Gunning and Lewis (1955) found this species in a green alga–yellow pond lily–duckweed community in a spring-fed portion of an Illinois swamp, and stressed the role of the water temperature around the springs in maintaining populations near the northern extent of the range. In Arkansas, larvae were exclusively found in flooded bottomland oak-hickory forest as opposed to the main channel and tupelo swamp habitats (Killgore and Baker 1996).

BIOLOGY The spawning season seems to vary with longitude. Spawning occurs in April in southeast Missouri and in March–May at 17 °C in Alabama (Carlander 1977). Males build and guard nests, frequently in colonies (Carlander 1977; Laerm and Freeman 1986). Fliers in Mis-

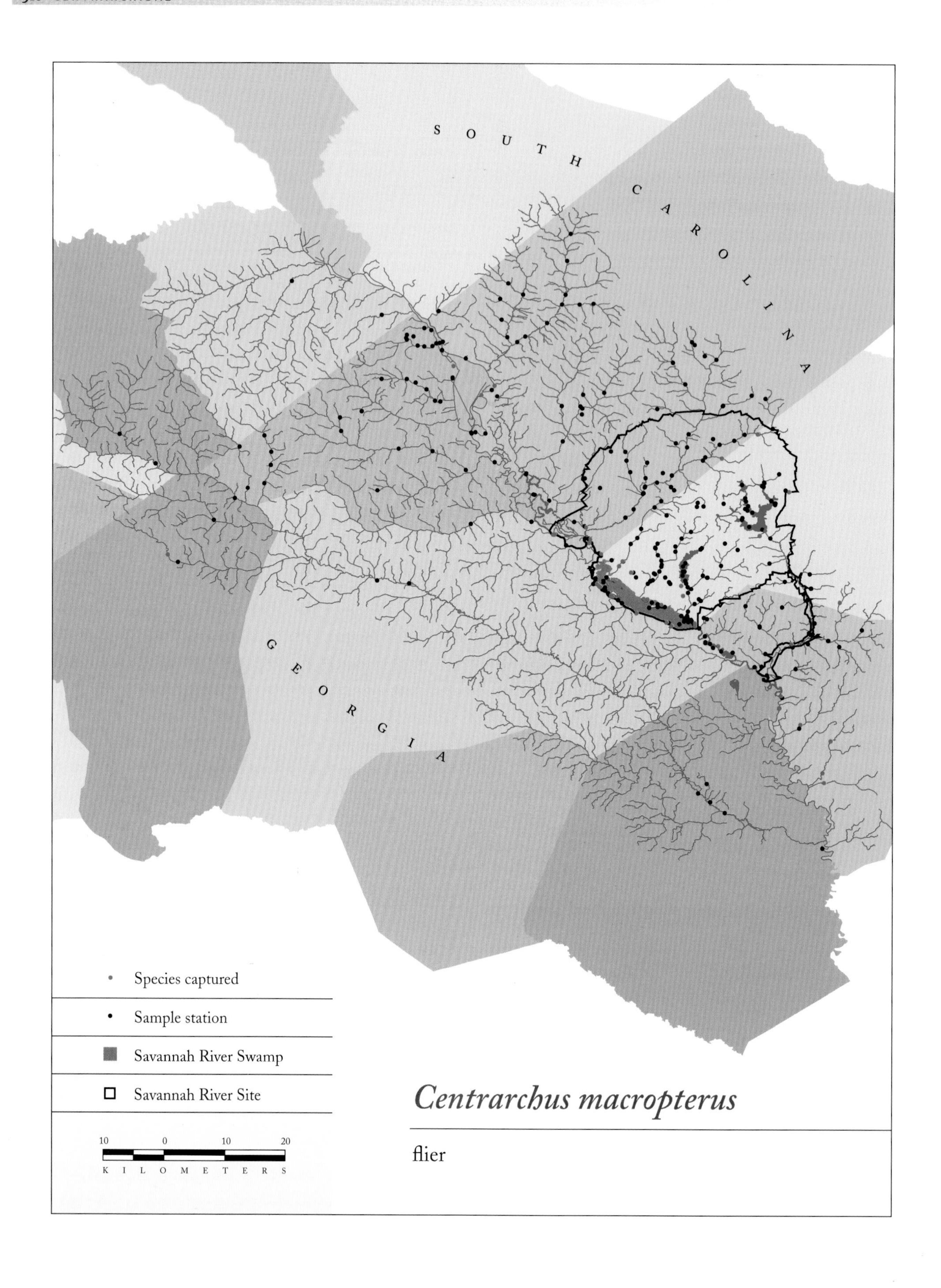

Centrarchus macropterus

flier

Centrarchus macropterus, flier; juvenile

souri may mature in 1 year at 71 mm, and reach lengths of 56, 99, 133, 157, 180, 193, and 198 mm at ages 1–7, respectively (Conley 1966 *in* Pflieger 1975).

Fliers feed primarily on invertebrates. The diet varies geographically. Food consists generally of aquatic and terrestrial insects, cladocerans, amphipods, copepods, and arachnids, but may also include crayfish, filamentous algae, and fish; juveniles eat large numbers of copepods (Gunning and Lewis 1955; Flemer and Woolcott 1966; Gatz 1979; Etnier and Starnes 1993).

In the Okefenokee Swamp, fliers do not aggregate in the summer but may do so in the winter (Laerm and Freeman 1986). Although reported to be a mostly sedentary species with a small home range (generally less than 200 m), a portion of the population in a North Carolina swamp stream moved an average of 4.5 km in 68 days, with individuals moving up to 12.7 km (Whitehurst 1981). These movements were greatest from March through May.

SCIENTIFIC NAME *Centrarchus* = "spined anus" (Greek), in reference to the large number of anal spines; *macropterus* = "large fin" (Greek).

Enneacanthus chaetodon (Baird, 1855)

Blackbanded sunfish

DESCRIPTION Body short, deep, and strongly compressed. Mouth terminal and small with posterior edge of the maxilla barely anterior to or barely reaching eye when mouth is closed. Teeth present on vomer and palatines. Deep notch between the spinous and soft portions of the dorsal fin, best observed when fins are expanded. Median spines of dorsal fin markedly longer than anterior and posterior spines (Carr and Goin 1959). Caudal fin rounded.

Dorsum brown to black. Sides lighter, but variable in color and fading to whitish ventrally; sides often have a pearly sheen and may have a rose or pink blush. Sides of juveniles and adults marked with six distinctive black bars, the anteriormost passing through the eye and the third bar extending into the anterior portion of the dorsal and pelvic fins. Leading edges of pelvic fins reddish orange or pink, as is the third membrane of the dorsal fin. Iris coppery to orange.

Spawning females are more brilliantly colored than males (Breder and Rosen 1966). Spawning males have a prominent black spot on the operculum.

SIMILAR SPECIES The distinctive color pattern distinguishes the blackbanded sunfish from other fish species in the MSRB.

SIZE To at least 100 mm TL.

MERISTICS Dorsal spines (8)10(11), dorsal rays (10)11–12(13), anal spines 3(4), anal rays (10)11–12(14), pectoral rays (8)9–11(12) (Jenkins and Burkhead 1993), pelvic spine 1, pelvic rays 4–5 (Hardy 1978); midlateral scales (23)25–29(32) (Jenkins and Burkhead 1993); gill rakers on upper limb 2, on lower limb 10–11 (Hardy 1978).

CONSERVATION STATUS Listed by Georgia as imperiled or critically imperiled (S1/S2) in the state because of extreme rarity, but the global rank is "apparently secure globally" (G4). Listed as rare in Florida by Gilbert (1992).

Enneacanthus chaetodon, blackbanded sunfish

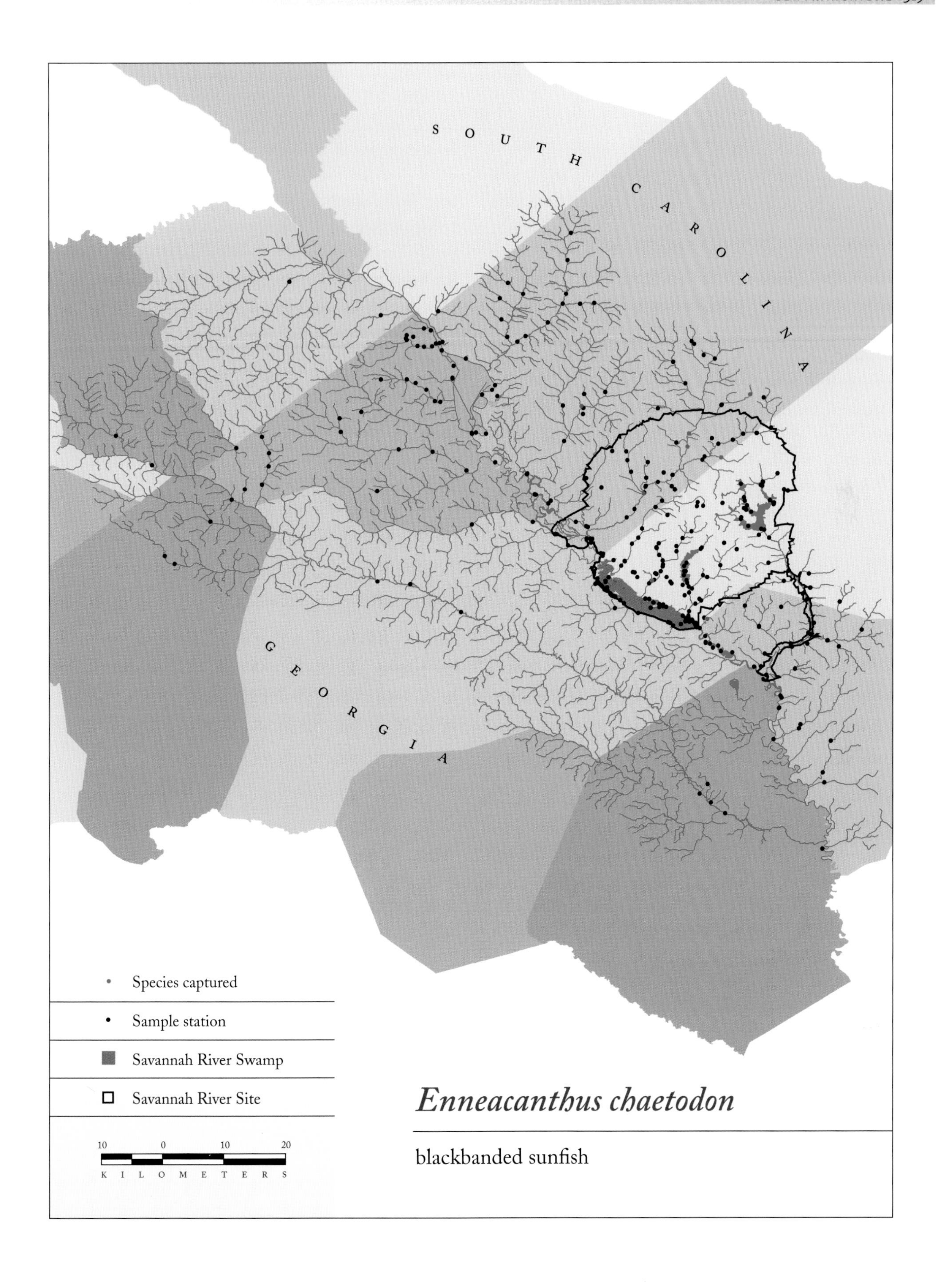

Enneacanthus chaetodon

blackbanded sunfish

DISTRIBUTION The Atlantic Coastal Plain from the Raritan River, New Jersey, to the St. Johns River, Florida, and in the Gulf of Mexico drainages of Georgia (Hardy 1978); also in some Gulf of Mexico drainage rivers in northern Florida (Lee *in* Lee et al. 1980). The range is fragmented, and this species is often missing from areas where appropriate habitat is present (Rohde et al. 1994). It has been introduced into Germany (Hardy 1978). In the MSRB, generally found in upland tributary streams, above the Savannah River fluvial terraces.

HABITAT Quiet, shallow, frequently densely vegetated margins of lakes, ponds, swamps, roadside ditches, and streams; largely restricted to stained, acidic water with pH 4.0–5.0 (Smith 1907; Breder 1936; Jenkins et al. 1975), but some populations inhabit water with pH up to 6.6 (Graham and Hastings 1984) or 7.0 (Graham 1993). Frequently found over sand substrate overlain by variable amounts of detritus (Gilbert 1992). Females in reproductive condition have been collected in ephemeral water bodies (Abbott 1870). In lotic systems of the MSRB, adults seasonally migrate into beaver ponds to spawn. Beaver ponds are also an important nursery habitat as many more young-of-the-year remain in these ponds than in the adjacent streams (Snodgrass and Meffe 1999).

BIOLOGY Spawning was observed in April in New Jersey (Abbott 1870) and in March in North Carolina (Smith 1907). In the MSRB, spawning probably takes place in mid-spring, at the same time the bluespotted sunfish (*E. gloriosus*) spawn. Spawning behavior appears to be variable. Breder and Rosen (1966) summarized nest-building behavior. Some males construct nests by forming suspended hollows in vegetation or among filamentous algae or macrophytes. Alternatively, saucer depressions may be formed in the bottom substrate; these depressions average about 10 cm in diameter and are usually placed in water about 30 cm deep (see also Abbott 1870). Nests may be built in a clear area or among aquatic macrophytes (Jenkins and Burkhead 1993). Males guard territories around the nest. The female lays adhesive eggs within the nest. Males may cover the eggs with sand (Bade 1931). Blackbanded sunfish may live up to 3 years in nature and 8 years in aquaria (Sternburg 1986).

The diet consists mostly of insects and small crustaceans taken from the surface of vegetation (Jenkins and Burkhead 1993).

SCIENTIFIC NAME *Enneacanthus* = "nine-spined" (Greek); *chaetodon* = "hairlike, fine narrow teeth" (Latin), referring to a characteristic of the marine butterfly fish genus *Chaetodon*, which blackbanded sunfish superficially resemble.

Enneacanthus gloriosus (Holbrook, 1855)
Bluespotted sunfish

DESCRIPTION Body moderately deep and compressed. Eyes moderate in size. Mouth small, terminal or superior, and oblique, with posterior edge of maxilla barely reaching anterior edge of eye when mouth is closed. Pectoral fins relatively short and rounded. Caudal fin rounded. No deep notch between spinous and soft portions of dorsal fin. Lateral line generally lacking on several posterior scales; western populations are more likely than eastern populations to have an incomplete lateral line (Peterson and Ross 1987; Ross 2001).

Body brownish to olive or very dark midnight blue with iridescent blue, green, silver, or gold spots, more distinctly round than those of banded sunfish (*E. obesus*). Mississippi populations may lack spots (Peterson and Ross 1987). Pupil dark blue or black, iris dull red or gold. Dark, somewhat oblique bar below eye. Opercle with small black or pearly blue spot that is two-thirds to three-fourths of the eye diameter (Plate 20 a). Soft dorsal, anal, and caudal fins may be pink or reddish. Pelvic fins colorless or pink.

Spots on sides are brightest and most developed on nuptial males, which have a nearly black background coloration; median fins mostly black (may be some orange hue) with bright iridescent spots; and soft portions of soft dorsal and anal fins enlarged. Breeding males are also a little deeper bodied and have longer fins than breeding females (Breder and Redmond 1929). Breeding females are drabber than breeding males, generally with an olive background color (Breder and Rosen 1966). Young and nonreproductive adults may have indistinct dark bars on sides.

SIMILAR SPECIES The bluespotted sunfish is very similar to the banded sunfish in appearance. Individuals collected in the MSRB with intermediate characteristics are suspected to be hybrids. Hybridization has been documented in New Jersey, which appears to be an old, stable hybrid zone (Graham and Felley 1985). Peterson and Ross 1987 includes a detailed comparison of the meristic characters of bluespotted sunfish and banded sunfish.

SIZE To at least 85 mm TL.

MERISTICS Dorsal spines (7)8–9(11), dorsal rays (9)10–12(13), anal spines (2)3(4), anal rays (8)9–10(13), pectoral rays (9)12–13 (Jenkins and Burkhead 1993), pelvic spine 1, pelvic rays 5 (Hardy 1978); pored lateral line scales (20)28–32(35) (Jenkins and Burkhead 1993), lateral line scales 25–35, scales around caudal peduncle 14–20; gill rakers on upper limb 0–3, on lower limb 8–13; branchiostegals 6 (Hardy 1978).

DISTRIBUTION Mainly found on the Coastal Plain from southeastern New York through peninsular Florida with a disjunct, probably introduced, population in the Great Lakes drainage of New York (Werner 1972) that may have been extirpated (Cudmore-Vokey and Crossman 2000). Also found in Gulf of Mexico drainage rivers in Florida,

Enneacanthus gloriosus, bluespotted sunfish; (left) male, (right) female

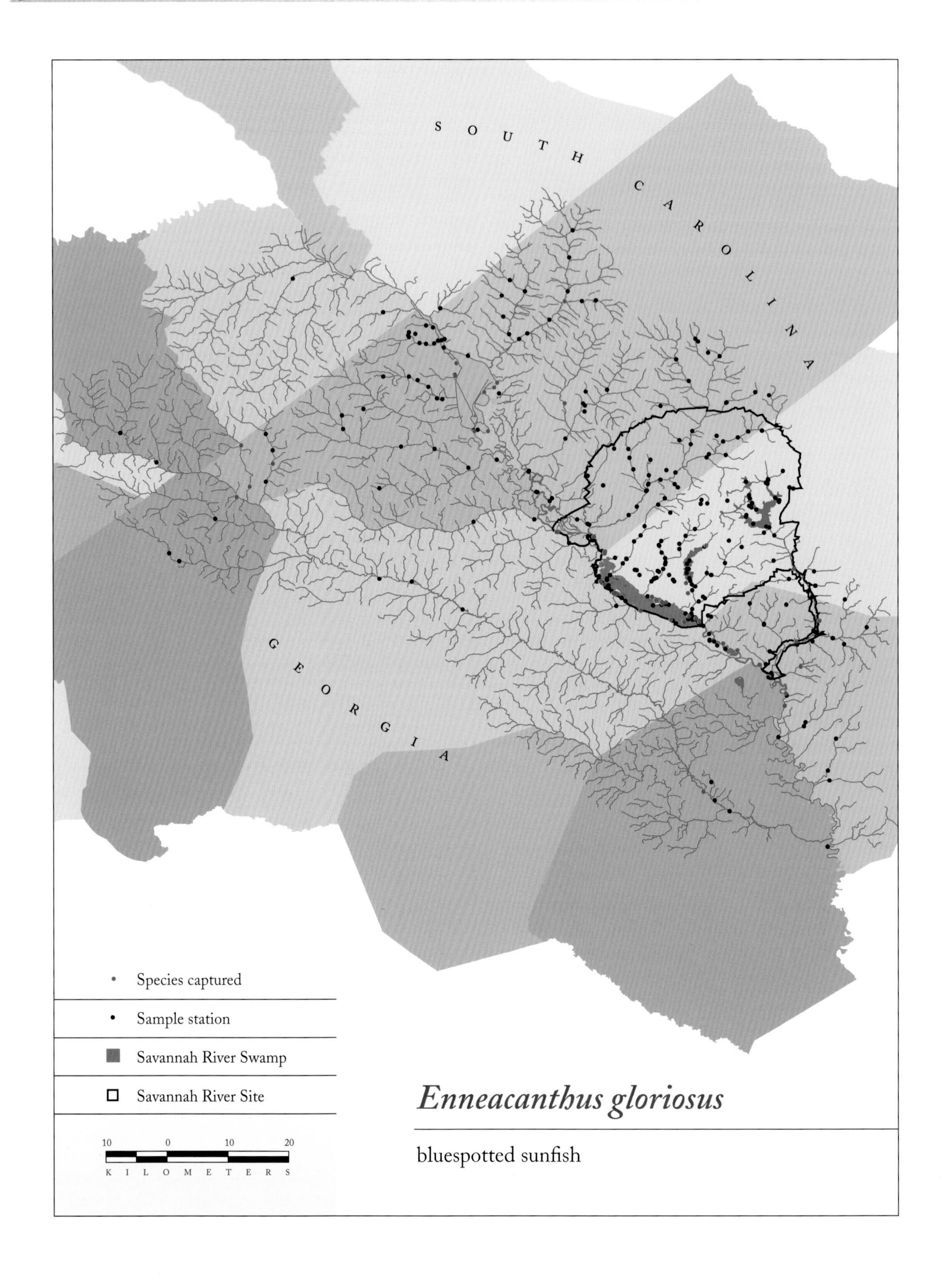

Enneacanthus gloriosus

bluespotted sunfish

Georgia, and Alabama (Lee and Gilbert *in* Lee et al. 1980). In Mississippi, populations, probably native, extend from the Pascagoula River westward to the Jourdan River (Clemmer et al. 1975 as reported in Peterson and Ross 1987; Miller and Clemmer 1980; Peterson and VanderKooy 1997); also introduced into ponds off the Big Black River (Peterson and Ross 1987). In Georgia, this species is restricted to the Middle and Lower Coastal Plain (Dahlberg and Scott 1971a). Its attractive appearance makes it a popular aquarium fish, and consequently it has been introduced outside its native range (Werner 1972; Peterson and Ross 1987). In the MSRB, most frequently found in the lower portions of tributary streams, especially those portions on the fluvial terraces or modern floodplain of the Savannah River, and in backwaters and swamps on the Savannah River floodplain.

HABITAT Quiet, shallow, weedy areas of ponds, oxbows, swamps, sluggish backwaters of streams and canals, and roadside ditches. Habitats include densely vegetated areas over generally a soft substrate, but substrates may include mud, rock, organic debris, and sand (Anderson 1964; Guillory 1979c; Peterson and VanderKooy 1997; Rakocinski et al. 1997) in water less than 1.2 m deep (Breder and Redmond 1929; Guillory 1979c). In Mississippi, bluespotted sunfish were reported from a temperature range of 12–32 °C and in water with dissolved oxygen as low as 1.0 ppm (Peterson and Ross 1987). They were more abundant in oxbow side ponds than stream channels, and their presence was correlated with fine substrate, abundant submergent and emergent vegetation, slightly acidic waters, high conductivity, low turbidity, and low velocity (Peterson and VanderKooy 1997; Rakocinski et al. 1997). Acidity down to pH 4.0 and alkalinity up to pH 9.0 are tolerated (Graham and Hastings 1984; Graham 1993), and unlike in the banded sunfish, pH does not seem to influence distribution patterns (Graham and Hastings 1984). The banded sunfish is tolerant of lower pH levels than bluespotted sunfish (Gonzalez and Dunson 1989). Bluespotted sunfish may inhabit tidal fresh water but are not usually found in brackish water (Wang and Kernehan 1979).

BIOLOGY Adults spawn multiple clutches over a protracted spawning season that varies geographically and is shorter in more northern latitudes (Snyder and Peterson 1999). Spawning occurs from July through September in Mississippi, and in May–June in Maryland at about 20 °C (Lippson and Moran 1974). Populations in the MSRB spawn in mid-to-late spring (DEF pers. obs.). Males construct and guard a small circular nest (only 102–127 mm in diameter), mostly in sandy substrate in water about 30 cm deep (Breder and Rosen 1966). The adhesive eggs are attached to rootlets in the nest (Breder and Rosen 1966) or laid among the vegetation (Breder and Redmond 1929; Breder 1936). The eggs hatch in about 57 hours (2.4 days) at 23 °C (Breder and Redmond 1929). Breder and Redmond (1929) and Lippson and Moran (1974) described the larvae.

Geographic variation in age at maturation has been noted, and males often mature at a smaller size than females (Peterson and VanderKooy 1997). Individuals mature at 2 years of age and around 40 mm SL in New Jersey (Breder and Redmond 1929), and at lengths less than 20 mm SL in coastal Mississippi (Peterson and VanderKooy 1997). Bluespotted sunfish may live up to 6 years (Breder and Redmond 1929); however, longevity was only 4 years in a Mississippi population located at the westernmost periphery of the range, and even 4-year-old individuals were rare (Snyder and Peterson 1999). In the same population, clutch size averaged 117 ova (Snyder and Peterson 1999).

Bluespotted sunfish feed diurnally, often among submergent aquatic vegetation (Snyder and Peterson 1999). They appear to be opportunistic feeders; the dominant prey items vary from one site to another, but terrestrial and aquatic insects (especially chironomids), amphipods, cladocerans, ostracods, copepods, and gastropods are eaten (Breder and Redmond 1929; Flemer and Woolcott 1966; Gatz 1979; Snyder and Peterson 1999). Size also influences diet; individuals smaller than 12 mm SL feed mostly on planktonic copepods (Ross 2001), while fish larger than 20 mm SL add gastropods and amphipods to the diet (Snyder and Peterson 1999). Bluespotted sunfish exhibit a crepuscular activity pattern (Casterlin and Reynolds 1980).

SCIENTIFIC NAME *Enneacanthus* = "nine-spined" (Greek); *gloriosus* = "glorious" or "handsome" (Latin).

Enneacanthus obesus (Girard, 1854)
Banded sunfish

DESCRIPTION Body short, deep, and compressed, but robust. Eyes moderate in size. Mouth relatively small, with posterior edge of maxilla barely reaching anterior margin of eye. Pectoral fins relatively short and rounded; caudal fin also rounded. No deep notch between spinous and soft dorsal fins. Lateral line usually incomplete. Posterior edge of dorsal and anal fins of breeding males elongated.

Body color variable, ranging from olive green to pale brown with belly lighter than upper sides. Sides of body marked with five to eight usually dark vertical bars that may vary in distinctness. Operculum with black spot (sometimes with gold flecks) as large as or slightly larger than diameter of eye, bordered in gold. Dark bar extends below eye. Body and fins with scattered small spots ranging from gold or copper to green, purple, or blue. Spots are less well developed than on bluespotted sunfish (*E. gloriosus*) and are generally crescents or flecks, as opposed to the more rounded spots of the bluespotted sunfish.

Breeding adults have white spines in the dorsal and anal fins. Radiating lines of the same color as the spots may be present on the head.

SIMILAR SPECIES The banded sunfish is most similar in appearance to the bluespotted sunfish, with which it may hybridize (see the *E. gloriosus* account for details).

SIZE May reach 70–80 mm SL. The largest reported specimen from Virginia is 74.5 mm SL (Jenkins and Burkhead 1993).

MERISTICS Dorsal spines (7)8–9(11), dorsal rays (9)11–12(13), anal spines 3 (3–4 in Hardy 1978), anal rays (9)10–11(14), pectoral rays (10)11–13 (Jenkins and Burkhead 1993), pelvic spine 1, pelvic rays 5; lateral line scales 27–35; gill rakers upper limb 1–4, lower limb 9–13 (Hardy 1978); in Virginia, the number of pored lateral line scales is bimodal, (15)18–23(25) and (26)28–32(34) (Jenkins and Burkhead 1993).

DISTRIBUTION The Atlantic Coastal Plain from Florida north to extreme southern New York, with a disjunct population in eastern Massachusetts, Rhode Island, and

Enneacanthus obesus, banded sunfish

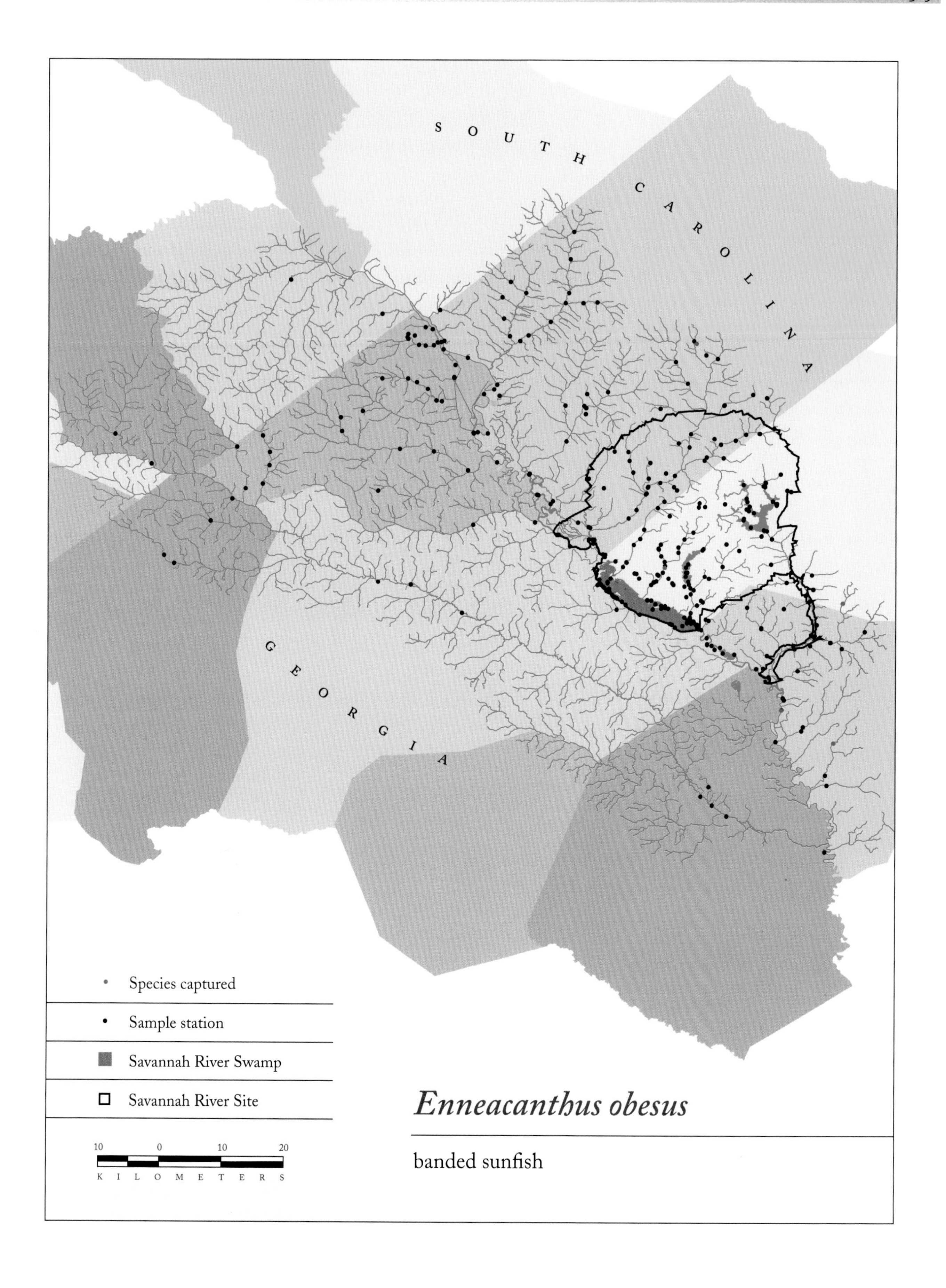

Enneacanthus obesus

banded sunfish

Connecticut; extending into western Florida and barely entering Alabama in the Perdido River system (Mettee et al. 1996). Introduced into ponds off the Big Black River in Mississippi (Peterson and Ross 1987). New England populations may occur above the Fall Line. In Georgia, banded sunfish are restricted to the Lower and Middle Coastal Plain (Dahlberg and Scott 1971a). Generally found in upland tributaries in the MSRB, where they are apparently uncommon.

HABITAT Lakes, ponds, swamps, roadside ditches, and streams, but more common in lakes and ponds around weedy areas; they are largely restricted to blackwater habitats (Jenkins and Burkhead 1993) and are found over mud, muck, organic debris, or sand bottoms. Distribution is highly correlated with and nearly restricted to low pH (usually pH 4.1–6.6); populations have been found at pHs as low as 3.7 (Graham and Hastings 1984) and as high as 7.0 (Graham 1993). Banded sunfish tolerate lower pH levels than the other two *Enneacanthus* species (Gonzalez and Dunson 1989).

BIOLOGY Spawning occurs in the spring when water temperatures reach 18–20 °C. Photoperiod cues seem to be important in initiating the spawning period (Harrington 1956). In the Okefenokee Swamp, juveniles 7–9 mm SL appear only in early June (Freeman and Freeman 1985); in Virginia, adults in reproductive condition have been collected from April through July (Jenkins and Burkhead 1993). Males defend territories and construct and guard a circular nest in the substrate in which adhesive eggs are laid. Juveniles grow relatively rapidly, reaching 34–45 mm SL by October of their first year (Freeman and Freeman 1985). Males are sexually mature by 45 mm (Harrington 1956). Females may be mature at 1 year, and maximum longevity for the species is 6 years (Jenkins and Burkhead 1993). Cohen (1977 *in* Jenkins and Burkhead 1993) estimated fecundity to be 802–1,400 eggs in females 1–6 years old.

Banded sunfish feed mostly on insects, insect larvae, crustaceans, and some plant material. In the Okefenokee Swamp, they eat mostly cladocerans and dipteran larvae (Freeman and Freeman 1985).

SCIENTIFIC NAME *Enneacanthus* = "nine-spined" (Greek); *obesus* = "fat" (Latin).

Lepomis auritus (Linnaeus, 1758)

Redbreast sunfish

DESCRIPTION Body laterally compressed and deep, but rather elongate for *Lepomis*, particularly the females and juveniles. Mouth moderate in size, terminal, and oblique; posterior edge of maxilla generally reaching anterior edge of pupil. Palatine teeth present (Etnier and Starnes 1993), but tooth patches on tongue absent. Gill rakers moderately long—longer and more slender than those of the dollar sunfish (*L. marginatus*). The nape of breeding males can be notably convex; that of females and juveniles is more gently sloping. Pectoral fins relatively short and rounded, with tips extending to pupil but not beyond front of eye when bent forward. Caudal fin emarginate. Lateral line complete; scales ctenoid.

Caudal fin of both sexes marked with orange on upper and lower margins. Cheeks and sides of snout marked with bluish green stripes radiating from mouth. Belly and breast of juveniles olive. Small juveniles under 30 mm SL elongate with narrow olive bars on sides.

Breeding males have a bright red or orange breast fading onto the belly; the sides are light powder blue with orange spots. Dorsum and head olive, although cheeks of very large males can be orange. The elongate, upward-angled ear tab is slender and entirely bluish black. Breeding females have a yellow olive breast and belly, olive brown dorsum, and sides lighter olive yellow with coppery spots; ear tab short, black, and unmargined.

SIZE To at least 190 mm SL in the MSRB, but more commonly less than 160 mm SL.

MERISTICS Dorsal spines (9)10(11), dorsal rays 11–12, anal spines 3, anal rays (8)9–10, pectoral rays (13)14–15; lateral line scales (39)42–46(54) (Jenkins and Burkhead 1993).

DISTRIBUTION The native range is east of the Appalachian Mountains from southern New Brunswick, Canada, to central Florida, and west to the Choctawhatchee River of Florida and Alabama; Mobile Bay drainage populations are probably introduced (Lee *in* Lee et al. 1980). Introduced in several places outside the native range, including northern Italy (Lee *in* Lee et al. 1980), Puerto Rico (Erdman 1984), and eastern Tennessee, where it is established (Etnier and Starnes 1993). Mettee et al. (1996) considered the populations in the Coosa and Tallapoosa drainages of Alabama, Georgia, and Tennessee to be native; those of the Black Warrior and Choctawhatchee drainages of Alabama may be introduced. Widespread and abundant in the MSRB in many permanent lentic waters (ponds, large reservoirs, swamps, oxbows). Although they occur in the Savannah River and nearly all streams but the smallest headwaters, they are most abundant in intermediate-size streams.

HABITAT Generally found in deeper pools, often associated with woody debris, stumps, or undercut banks. Substrates can range from coarse sand in runs to mud in slow or still waters. In small to intermediate-size streams,

Lepomis auritus, redbreast sunfish; (left) male, (right) female

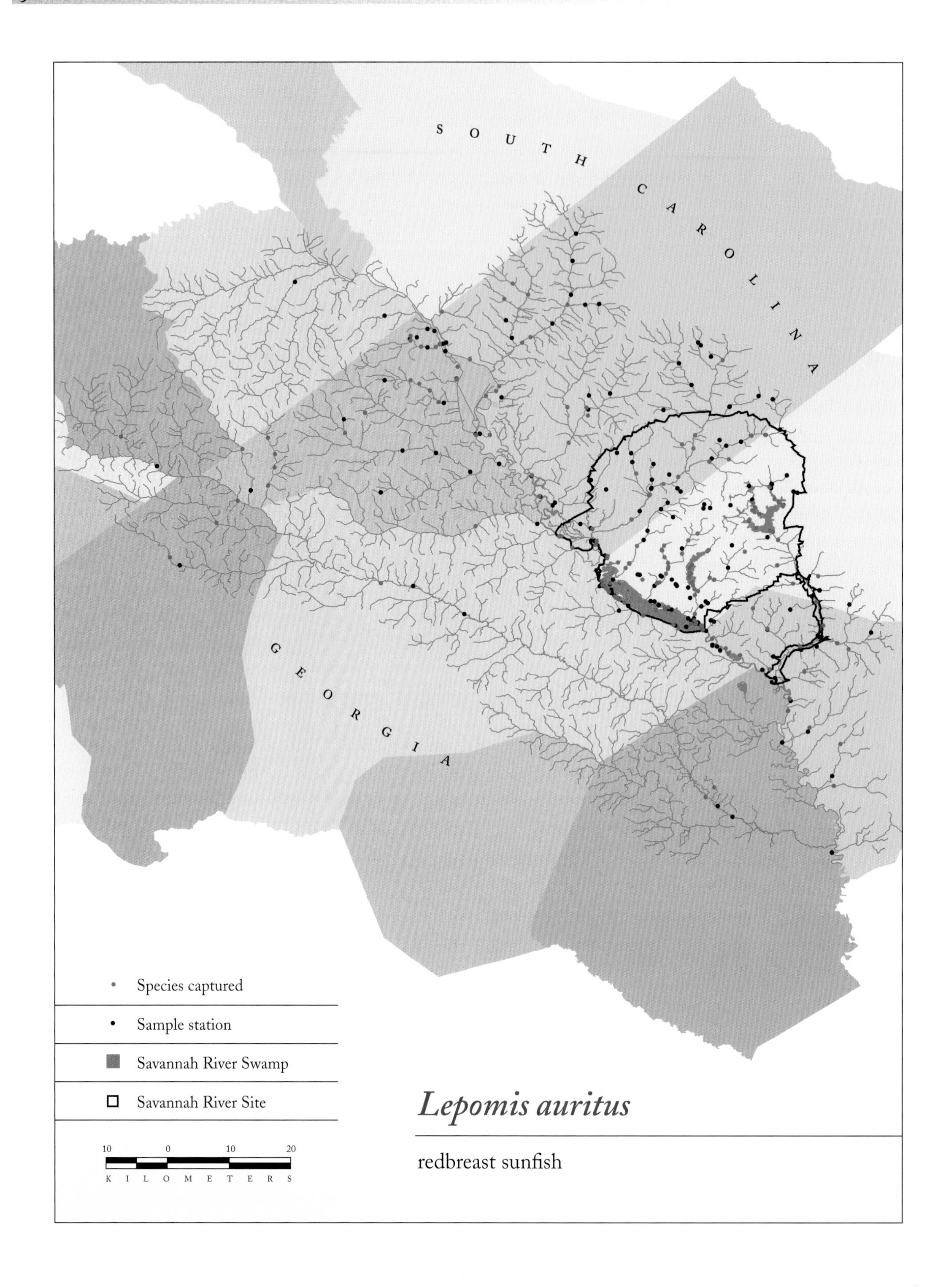
S O U T H C A R O L I N A
G E O R G I A
Species captured
Sample station
Savannah River Swamp
Savannah River Site
10 0 10 20
K I L O M E T E R S
Lepomis auritus
redbreast sunfish

abundance generally increases with decreasing water velocity and increased depth and cover (Aho et al. 1986b; Meffe and Sheldon 1988). In some lotic systems, especially headwaters, beaver ponds may be important spawning and nursery habitat (Snodgrass and Meffe 1999). The pH range tolerated is about 6–9 (Graham 1993).

BIOLOGY In tributary streams of the MSRB, spawning generally occurs from late May through the end of July, although it begins earlier in some warmer lentic habitats. Water released from reservoirs can significantly expedite or delay spawning downstream depending on its temperature; for example, spawning was delayed until July in Steel Creek during cold-water releases from L Lake. Spawning occurs at water temperatures ranging from 20 to 31 °C (DEF pers. obs.). Males build large saucer nests by sweeping a depression in the substrate with the tail (Breder and Rosen 1966; Lukas and Orth 1993). Redbreast sunfish build larger nests than most other *Lepomis* species of about the same size. Nests average 80 cm in diameter, more than five times the standard length of the males that construct and defend them. The nest substrate is almost always fine gravel. Females are courted to the nest, and the male guards the developing offspring until the fry swim up and leave the nest.

Nesting habitat is described in Breder and Rosen 1966; Davis 1971; Aho 1986b; Helfrich et al. 1991; Lukas and Orth 1993. Redbreast sunfish living in streams spawn in beaver ponds, backwaters, and coves or main flowing channels. When spawning occurs in swift channels, nests are generally built near the bank in eddies behind obstructions such as woody debris, stumps, or man-made structures. Nests in tributaries are typically built in water less than 1 m deep, but nests in large reservoirs can be in water as deep as 2 m. Nests may be solitary, in loose aggregations, or in dense colonies of more than 80 nests (Fletcher 1993). Generally, the distance from the bank and the aggregation size increase as habitats become more lentic. Nests in large reservoirs (e.g., L Lake) may be more than 100 m from the bank. Genetic analyses indicated that multiple females spawn in each nest, and low rates of cuckoldry occur in Fourmile Branch (DeWoody et al. 1998). Larvae are described in Buynak and Mohr 1978; and Lippson and Moran 1974.

Dusky shiners (*Notropis cummingsae*), which are among the most abundant fish in some streams in the MSRB, spawn on redbreast sunfish nests (Fletcher 1993). Consequently, the redbreast sunfish may be a keystone species in some habitats. Redbreast sunfish are relatively sedentary with small home ranges, although a few individuals within a population may move long distances (Gatz and Adams 1994). Movements are greatest in the spring preceding the spawning season (Hudson and Hester 1975; Gatz and Adams 1994). Redbreast sunfish live up to 5 years in Virginia (Saecker and Woolcott 1988), and to 6 years in North Carolina (Davis 1971). Nest-guarding males in Fourmile Branch ranged from 2 to 7 years old, but most were between 3 and 5 (DEF unpub. data). Males averaged 66, 97, 133, 152, and 157 mm SL at ages 1–5, respectively, in North Carolina, while females were 65, 98, 126, 138, and 156 mm SL (Davis 1971). The number of ova ranges from 940 to almost 10,000 and increases with female size (Davis 1971; Bass and Hitt 1974).

Feeding occurs mostly during daylight hours (Johnson and Dropkin 1995). Redbreast sunfish are invertivores with a geographically variable diet that includes aquatic insects, terrestrial insects, microcrustaceans, crustaceans, ostracods, and molluscs (Davis 1971; Bass and Hitt 1974; Coomer et al. 1977; Cooner and Bayne 1982; Johnson and Johnson 1984; Thorp et al. 1989; Sheldon and Meffe 1993; Wiltz 1993). Juveniles feed primarily on benthic insect larvae (e.g., dipterans) and microcrustaceans (Coomer et al. 1977; Johnson and Johnson 1984; Etnier and Starnes 1993; Sheldon and Meffe 1993). Nest-guarding males may reduce the time spent foraging (Thorp et al. 1989) and may cannibalize offspring from their own nest (DeWoody et al. 2001).

SCIENTIFIC NAME *Lepomis* = "scaled gill cover" (Greek); *auritus* = "eared" (Latin).

Lepomis cyanellus (Rafinesque, 1819)
Green sunfish

DESCRIPTION Body laterally compressed, somewhat deep and oblong. Eye large. Mouth large, terminal or nearly terminal, and oblique, with the posterior edge of the maxilla extending to the center of the pupil when the mouth is closed. Palatine, vomer, and both jaws with fine teeth (Scott and Crossman 1973); small patch of teeth sometimes present on the tongue (Jenkins and Burkhead 1993). Gill rakers long. Pectoral fins short and rounded. Caudal fin emarginate. Lateral line complete; scales ctenoid.

Dorsum olive green fading to brassy, yellowish, or whitish ventrally. Wavy iridescent blue lines radiate from mouth across cheek and operculum. Similar iridescent patches on the body scales may form rows of horizontal lines across a darker background. Lines may be broken into irregular mottling, especially posteriorly. Soft dorsal, caudal, and anal fins may be mottled, and all three have light margins. Black blotch present at posterior base of soft dorsal fin, and sometimes at posterior base of anal fin. Ear tab dark with a wide, light margin.

Nuptial males are more brilliantly colored than females and have a prominent light margin on the median fins. Breeding females have dark vertical bars (Hunter 1963) and sides; young-of-the-year have uniformly grayish sides (Etnier and Starnes 1993). A whitish gold color morph exists in some populations (Dunham and Childers 1980).

SIZE Adults can exceed 270 mm TL, but few exceed 200 mm TL.

MERISTICS Dorsal spines (9)10(12) (maximum 11 in Etnier and Starnes 1993), dorsal rays 10–11(12), anal spines 3, anal rays (8)9–10 (possibly 11 in Etnier and Starnes 1993), pectoral rays 13–14(15); lateral line scales (43)46–50(53) (Jenkins and Burkhead 1993); gill rakers 11–14; lateral line scales 43–52 (Etnier and Starnes 1993); lateral line scales 40–59, scale rows above lateral line 6–11, scale rows on cheek 7–10 (Hardy 1978).

CONSERVATION STATUS Introduced into the MSRB.

DISTRIBUTION Native from the Great Lakes south through the Mississippi River drainage, including most of the area west of the Appalachian Mountains and east of the Rocky Mountains, south into northeastern Mexico (Lee *in* Lee et al. 1980). Through introduction this species is now found east of the Appalachians from Ontario and New York south into Georgia, and west of the Rocky Mountains in Arizona, New Mexico, California, Utah, Nevada, and Idaho, and in Sonora and Baja California Norte in Mexico (Lee *in* Lee et al. 1980). In the Great Lakes, not native to either Lake Superior or Lake Michigan (Cudmore-Vokey and Crossman 2000). While this species is abundant in the upper Savannah River basin (Dahlberg and Scott 1971a), it is rare in the MSRB. Collection records exist for at least six locations in the Savannah River and one location at the mouth of a tributary. Early collections on the SRS also yielded green

Lepomis cyanellus, green sunfish; (left) adult, (right) juvenile

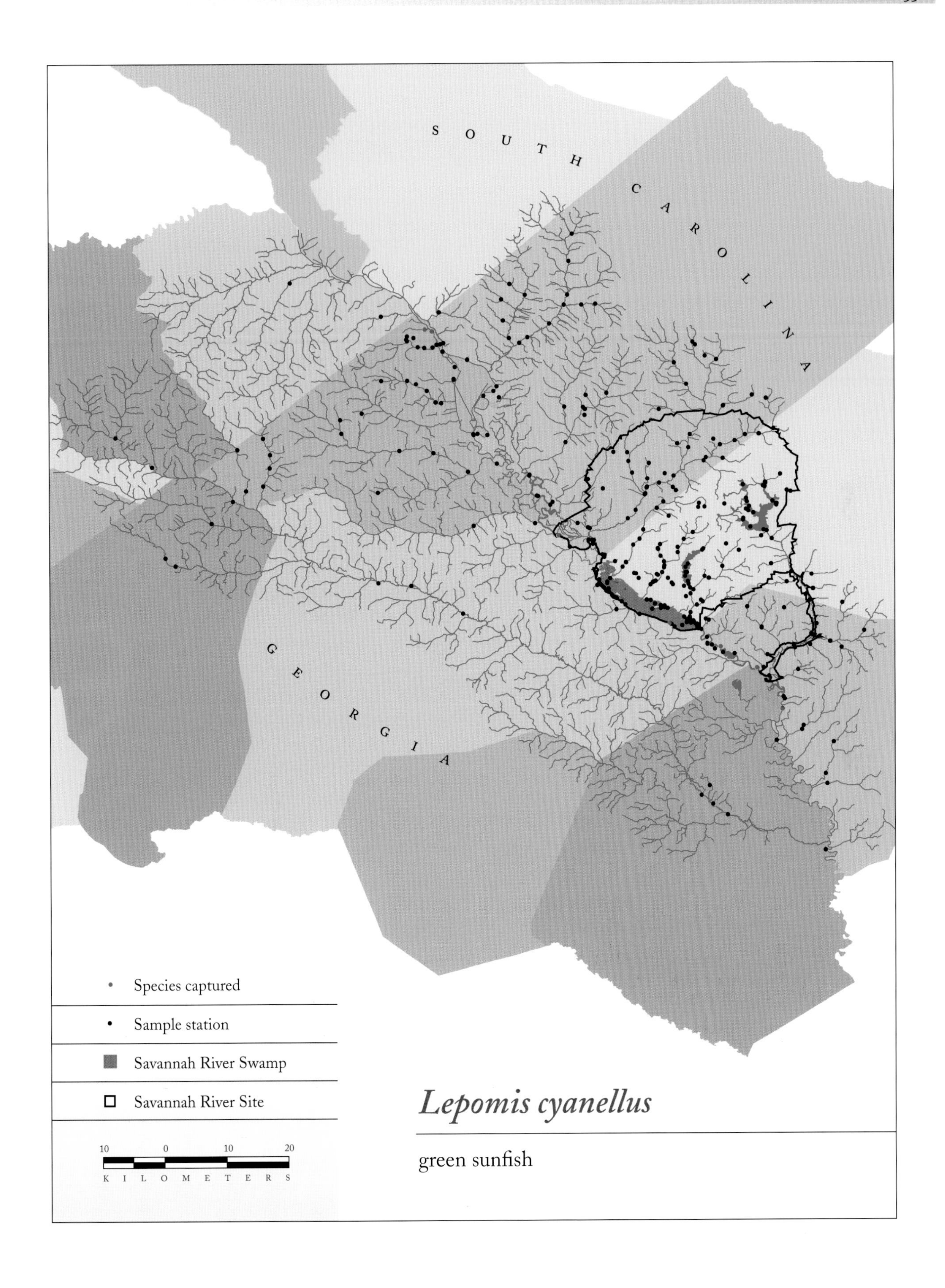

Lepomis cyanellus

green sunfish

sunfish in Upper Three Runs (Whitworth 1969) and Lower Three Runs (Freeman 1958).

HABITAT In the Carolinas, small rivers, sluggish creeks, and slow-moving areas of low or moderate gradient, usually in small, rocky Piedmont and mountain foothill creeks (Rohde et al. 1994); also in floodplains of streams and rivers (Ross and Baker 1983; Kwak 1988). Green sunfish are more common in lotic systems but may be common in lakes, particularly those with turbid water (Cahn 1927). They are extremely tolerant of adverse conditions such as drought, and are rapid invaders and colonizers of new habitats (Etnier and Starnes 1993). The extreme range of tolerance may allow this species to thrive where few other sunfish species occur (Pflieger 1975). The recorded temperature range is 4–28 °C, but experimental subjects tolerated temperatures up to 34.4 °C (Hardy 1978). Introduced green sunfish populations suppressed native fish populations in headwater streams of Piedmont North Carolina (Lemly 1985).

BIOLOGY Prespawning gonadal development is primarily controlled by both photoperiod and water temperature; gonadal regression is controlled mainly by water temperature (Kaya and Hasler 1972; Kaya 1973). In Missouri, spawning begins when the water temperature reaches about 21 °C and continues over a protracted season from late May sometimes into August with a peak in June (Pfleiger 1975). Spawning occurs from April through August in Arkansas (Robison and Buchanan 1984), and from late spring to early summer in Wisconsin at water temperatures of 18–24 °C (Kaya and Hasler 1972; Kaya 1973). Spawning occurs episodically at intervals that may average about 8–9 days (Hunter 1963).

Hunter (1963) described spawning and nesting in detail. Males construct a depression nest by sweeping substrate away with the tail. Nests are constructed in shallow water—generally less than 35 cm and sometimes as shallow as 4 cm—and areas near cover and over gravel are preferred over soft muck substrates (Hunter 1963), although the use of clay or detritus bottoms has been reported (Robison and Buchanan 1984). Nests may be placed in compact colonies (Hunter 1963) or in large groups when nest sites are limited (Pflieger 1975). Males defend small territories. Females are courted to the nests, and spawning occurs during daylight hours, but males continue guarding the nests after dark (Hunter 1963). Males fan the eggs and vigorously protect the offspring from predators until the fry leave the nest. Females and nonnesting males may congregate around the spawning areas (Hunter 1963). Each female lays between 2,000 and 10,000 eggs over a period of 1–2 days (Beckman 1952), and multiple females may spawn in a given nest (Robison and Buchanan 1984). The eggs are yellow and adhesive (Scott and Crossman 1973). Hatching occurs in 35–55 hours and larvae swim up in 6 days at 24–27 °C (Taubert 1977). Eggs and larvae were described by Taubert (1977), and have been identified using allozyme analysis (Rettig 1998). Larvae often migrate from the spawning grounds into the limnetic zone of lakes (Rettig 1998); subsequently the juveniles return to the weedy littoral zones.

Sexual maturity occurs in 1–3 years, depending on the latitude (Jenkins and Burkhead 1993). A study in a southern Illinois stream indicated that green sunfish can live up to 6 years, and grow to 68, 94, 125, 147, 157, and 190 mm SL in each of the 6 year classes, respectively (Lewis and Elder 1956). Green sunfish are often relatively sedentary with very small home ranges (Pflieger 1975; Etnier and Starnes 1993), but a portion of the population may be highly motile and move longer distances (Smithson and Johnston 1999). Hybridization may occur with many other sunfish species (Jenkins and Burkhead 1993).

The green sunfish is an ambush predator whose large mouth allows it to feed on a wider size range of food items than small-mouthed *Lepomis* species can ingest (Werner and Hall 1977). The diet consists largely of aquatic insects, with smaller amounts of microcrustaceans, molluscs, and small fish (Sadzikowski and Wallace 1976; Jenkins and Burkhead 1993); crayfish are added after the fish reach a length of about 100 mm (Applegate et al. 1966). Green sunfish become more piscivorous at 50–99 mm TL (2–3 years of age), when 10–65% of their diet consists of fish (Mittelbach and Persson 1998).

SCIENTIFIC NAME *Lepomis* = "scaled operculum" (Greek); *cyanellus* = "blue" (Greek).

Lepomis gibbosus (Linneaus, 1758)

Pumpkinseed

DESCRIPTION Body deep, laterally compressed, but stout. Eyes large. Mouth small, terminal or nearly terminal, with the posterior edge of the maxilla barely reaching the anterior margin of the eye when the mouth is closed. Pectoral fins long and pointed, tips reaching beyond anterior margin of eye when folded forward. Caudal fin emarginate. Palatine teeth and tooth patches on tongue absent (Scott and Crossman 1973; Etnier and Starnes 1993). Gill rakers thick and relatively short.

Dorsum olive, body color lighter ventrally with yellow to orange breast. Brassy yellow/olive sides densely marked with dusky spots; orange to yellow spots may also be present. Caudal, anal, and pelvic fins tend to be yellowish. Dorsal fin may be dark. Soft dorsal and anal fins with dark to vague mottling or wavy lines. Sides of snout, cheeks, and opercles marked by light iridescent blue vermiculations. Ear tab mostly black with light posterior margin and pale orange to reddish orange spot on posteriormost tip.

Breeding males are more brightly colored with intensified cheek vermiculations, black pelvic fins, and bluish dorsal and caudal fins (Adams and Hankinson 1932); a distinct white border and bright red spot on the ear tab; bright red iris; bright yellow belly; and bright blue-green and yellow speckling on sides of body (Stacey and Chiszar 1978). Breeding females are duller, lack the white border on the ear tab, and have dark vertical bars on sides (Stacey and Chiszar 1978).

SIMILAR SPECIES The long, pointed pectoral fin, short gill rakers, reddish orange on the ear tab, and mottling or wavy lines on the soft dorsal and anal fins are particularly diagnostic for this species. The body scales, especially dorsolaterally, appear spotted. Redear sunfish (*L. microlophus*) lack the mottling in the median fins and have more uniformly colored body scales. At about 30 mm TL, juvenile pumpkinseeds develop eight or nine vertical bars similar to the bluegill's (*L. macrochirus*), but the pumpkinseed has dark spots between the bars, especially between the second and third anterior bars (Brown and Colgan 1981).

Lepomis gibbosus, pumpkinseed; male

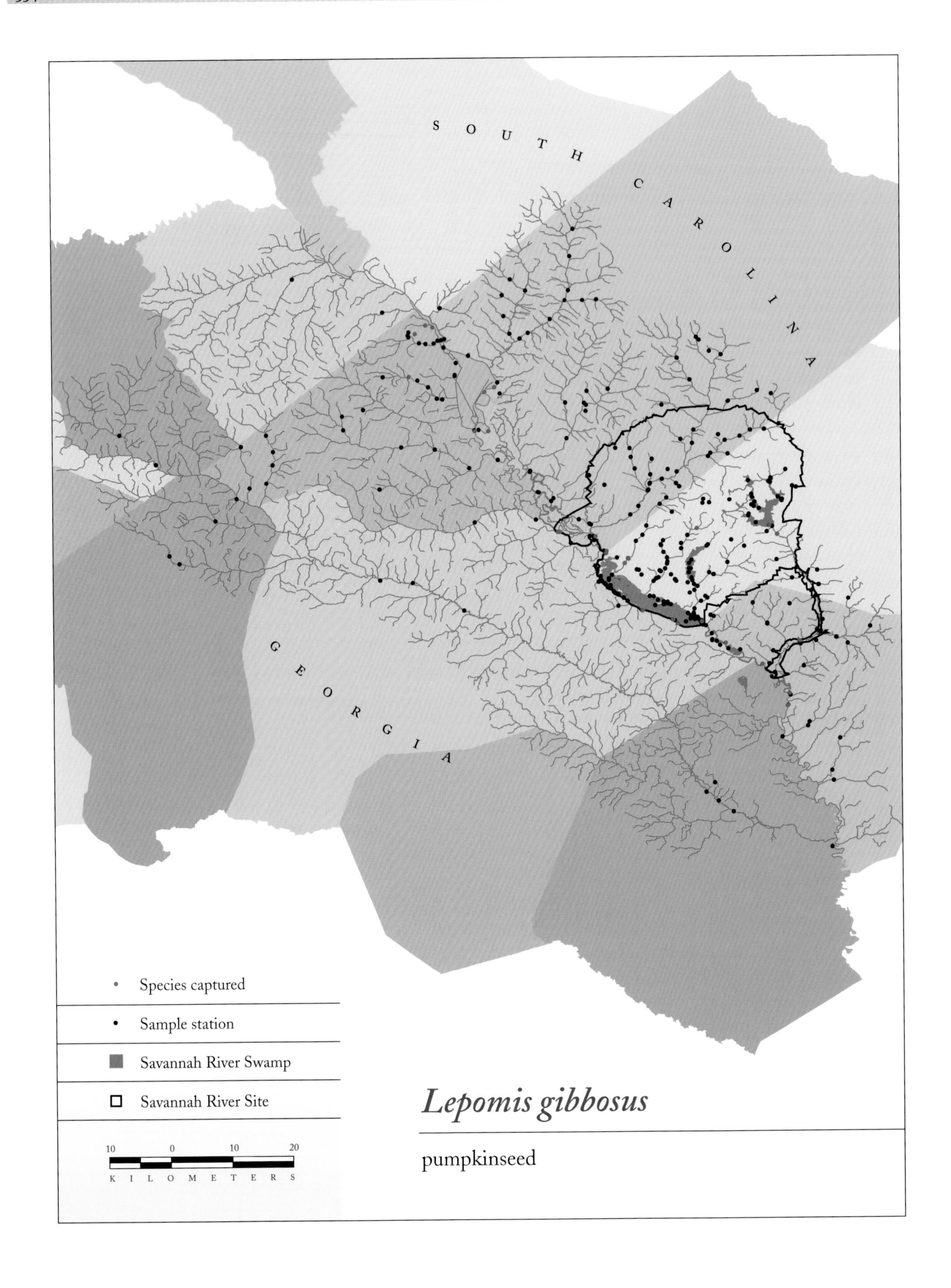

Lepomis gibbosus

pumpkinseed

SIZE The maximum size recorded is 381 mm (Rohde et al. 1994).

MERISTICS Dorsal spines (9)10(12), dorsal rays (10)11–12(13), anal rays (8)9–10(12), pectoral rays (11)12–13(14); lateral line scales (36)37–44(47) (Jenkins and Burkhead 1993); gill rakers 9–12 (Etnier and Starnes 1993).

DISTRIBUTION Native from New Brunswick east of the Appalachian Mountains south to northern Georgia, the Great Lakes drainages, the upper Mississippi Valley south to Missouri and South Dakota, and north into southern Ontario and Manitoba (Lee *in* Lee et al. 1980). Widely introduced and established in the northern tier of U.S. states and southern Canada westward to British Columbia, Washington, and Oregon, and south and west in California, with populations in Colorado and New Mexico; also established in Europe (Lee *in* Lee et al. 1980). Generally uncommon in the MSRB; most frequently found in backwaters of the Savannah River, occasionally found in large reservoirs, and sometimes found in ponds or sluggish sections of mostly larger tributaries.

HABITAT Quiet, sluggish waters of lotic systems; they thrive in farm ponds and quiet areas of lakes and reservoirs in heavy vegetation (Rohde et al. 1994). Known to winter in schools in deep water (Clark and Keenleyside 1967). Often found in cooler waters than other sunfishes, normally at 5.0–32.5 °C, becoming inactive below 10.0 °C (Hardy 1978). Pumpkinseeds occur over a broad pH range, from less than 5 to above 9 (Graham 1993), but larval mortality is high at pH 4.25 and lower (Graham and Hastings 1984).

BIOLOGY Gonad development prior to the spawning season depends primarily on water temperature and may not begin until the water temperature rises above 12.5 °C (Burns 1976). Spawning occurred from May through August at water temperatures between 16 and 27 °C in Maryland (Lippson and Moran 1974), and from May through July at water temperatures generally between 17 and 23 °C in various Ontario lakes (Danylchuk and Fox 1994a). In both studies, peak spawning occurred at water temperatures between 21 and 24 °C (Lippson and Moran 1974; Danylchuk and Fox 1994a). In Lake Opinicon, Ontario, pumpkinseeds spawned in June, earlier than the syn-

Lepomis gibbosus, pumpkinseed; female

topic bluegill (Keast 1978). Over the protracted spawning season, females spawn multiple times, averaging three episodes in northern latitudes (Fox and Crivelli 1998). Offspring from early-season spawnings may survive their first winter better than those hatched later (Bernard and Fox 1997).

Males construct nearly circular depression nests by sweeping substrate away with the tail. The nests are generally placed in shallow water, usually less than 60 cm deep, and the substrate type influences the nest dimensions (Breder 1940; Lippson and Moran 1974). Spawning is not colonial, but nests are generally solitary or scattered in loose aggregations (Keast 1978; Gross and MacMillan 1981). Each male defends a territory around his nest from rival males (Stacey and Chiszar 1978). Females are courted to the nest for spawning. The adhesive eggs attach to sediment particles, small stones, roots, or sticks (Adams and Hankinson 1932). Larvae have been identified using allozyme analysis (Rettig 1998) and were described by Lippson and Moran (1974). The male guards offspring in the nest from predators until the fry leave the nest (Adams and Hankinson 1932). Larvae often migrate from the shallow spawning grounds into the limnetic zone of lakes (Rettig 1998), but juveniles subsequently migrate back to the weedy littoral areas.

Some populations contain reproductive parasitic males (Taborsky 1994), which become sexually mature at a smaller size and allocate a larger proportion of their total body mass to testes than do nest-building males. These parasitic morphs do not build nests or court females, but spawn by sneaking onto nests of other males while the nest builders are spawning (Gross 1979). Rasping sounds, presumably made with the pharyngeal pads, are produced during agonistic and reproductive behaviors (Ballantyne and Colgan 1978a, 1978b).

Females in the Ontario populations studied by Bertschy and Fox (1999) generally matured at 2–3 years of age and 65–95 mm TL. Age at maturation likely varies with latitude and local conditions, and maturation might occur at earlier ages in the MSRB. In some populations, smaller, younger females may mature later in the spawning season than larger females (Danylchuk and Fox 1994b). Maturation may occur earlier in the season when high population densities produce stunted fish (Fox 1994). The general life span of the pumpkinseed is 8–10 years (Scott and Crossman 1973). Clutch size (batch fecundity) increases with female size and ranges from 1,800 to 14,100 ova per spawning (Fox and Crivelli 1998).

The diet varies seasonally and with age (Keast 1978). Juveniles feed largely on soft-bodied invertebrates that live on vegetation; the adults' morphology is well adapted for feeding on vegetation-dwelling gastropods (Mittelbach 1984; Osenberg et al. 1992). Adults studied in Lake Opinicon, Ontario, fed primarily on molluscs, isopods, and chironomid larvae, with smaller amounts of trichopterans and ephemeropterans; juveniles also ate Cladocera (Keast 1978). Because snails are an important food, low snail abundance may limit pumpkinseed population levels (Mittelbach 1984; Osenberg et al. 1992). Redear sunfish and pumpkinseeds are allopatric through most of their native distributions, but a small area of overlap in the Carolinas includes the MSRB; the adults of these two sunfish when syntopic may compete for molluscs (Huckins 1997; Huckins et al. 2000). Although pumpkinseeds are not as specialized for feeding on molluscs as the closely related redear sunfish (Lauder 1983), they have pharyngeal pads with large molariform teeth that are efficient at crushing shells (Keast 1978). These teeth may become more robust in habitats with higher mollusc densities (Mittelbach et al. 1992). Pumpkinseeds can adapt to a more planktivorous diet as well, especially in the absence of more open-water planktivores like bluegills (Osenberg et al. 1992). Within a lake, habitat-specific morphs that differ in body shape and diet can exist in open water and shallow water, especially when bluegills are absent (Robinson et al. 2000). This polymorphism is attributed largely to phenotypic plasticity, but can partly be attributed to genetic differences (Robinson and Wilson 1996).

Pumpkinseeds feed mostly during daylight hours, but some nocturnal feeding occurs, especially when planktonic organisms are the food (Collins and Hinch 1993). Pumpkinseeds are mostly sedentary, as reported from a North Carolina swamp stream (Whitehurst 1981).

SCIENTIFIC NAME *Lepomis* = "scaled operculum" (Greek); *gibbosus* = "formed like the full moon" (Latin).

Lepomis gulosus (Cuvier, 1829)

Warmouth

DESCRIPTION Body compressed, but stocky and relatively elongate. Eyes large. Mouth of adults large and oblique; posterior edge of maxilla extends to beyond posterior margin of pupil but does not reach posterior margin of eye. Mouth of juveniles smaller, with maxilla reaching only to mid-pupil. Anterior profile generally concave at junction of head and nape. Palatine teeth are present (Etnier and Starnes 1993), as are two elongate median tooth patches on tongue. Gill rakers relatively long, but thick. Pectoral fins short and rounded, when folded forward their tips extend to posterior edge of eye, but not to pupil. Caudal fin emarginate. Lateral line complete; scales ctenoid.

Body with extensive brown mottling, darkness of mottling variable and appears to fade or intensify rapidly. Dorsum background tan with coppery iridescence, especially on breeding fish. Background color of side more whitish, again highly iridescent on breeding fish (whitish gold to pale blue-green depending on direction of lighting). Three dark bars radiate posteriorly from the eye, with an occasional fourth bar located dorsally. Bars dark brown on tan iridescent background. Forehead usually dark, but sometimes with two faint dark bars between the eyes. Dorsal and anal fins dusky and marked with mottling or spots. Markings may range from dark brown to light tan to dull or bright orange, and may include iridescent bluish green, especially on posterior soft dorsal fin and basal one-third of anal fin (light in preservative). Similar markings or iridescence present at base of caudal fin. Bright red spot at posterior base of soft dorsal fin of breeding males. Pectoral fins translucent; pelvic fins dusky with light anterior margin. Ear tab short, dark brownish black, with light posterior margin that includes a small reddish spot. Juveniles have translucent, light dusky to slightly mottled fins, and no red on ear tab.

SIZE At least to 244 mm TL (Larimore 1957); up to 240 mm TL on the SRS.

MERISTICS Dorsal spines (9)10(11), dorsal rays (9)10(11), anal spines 3, anal rays 9–10, pectoral rays (12)14; lateral line scales (38)41–45(48) (Jenkins and Burkhead 1993).

Lepomis gulosus, warmouth

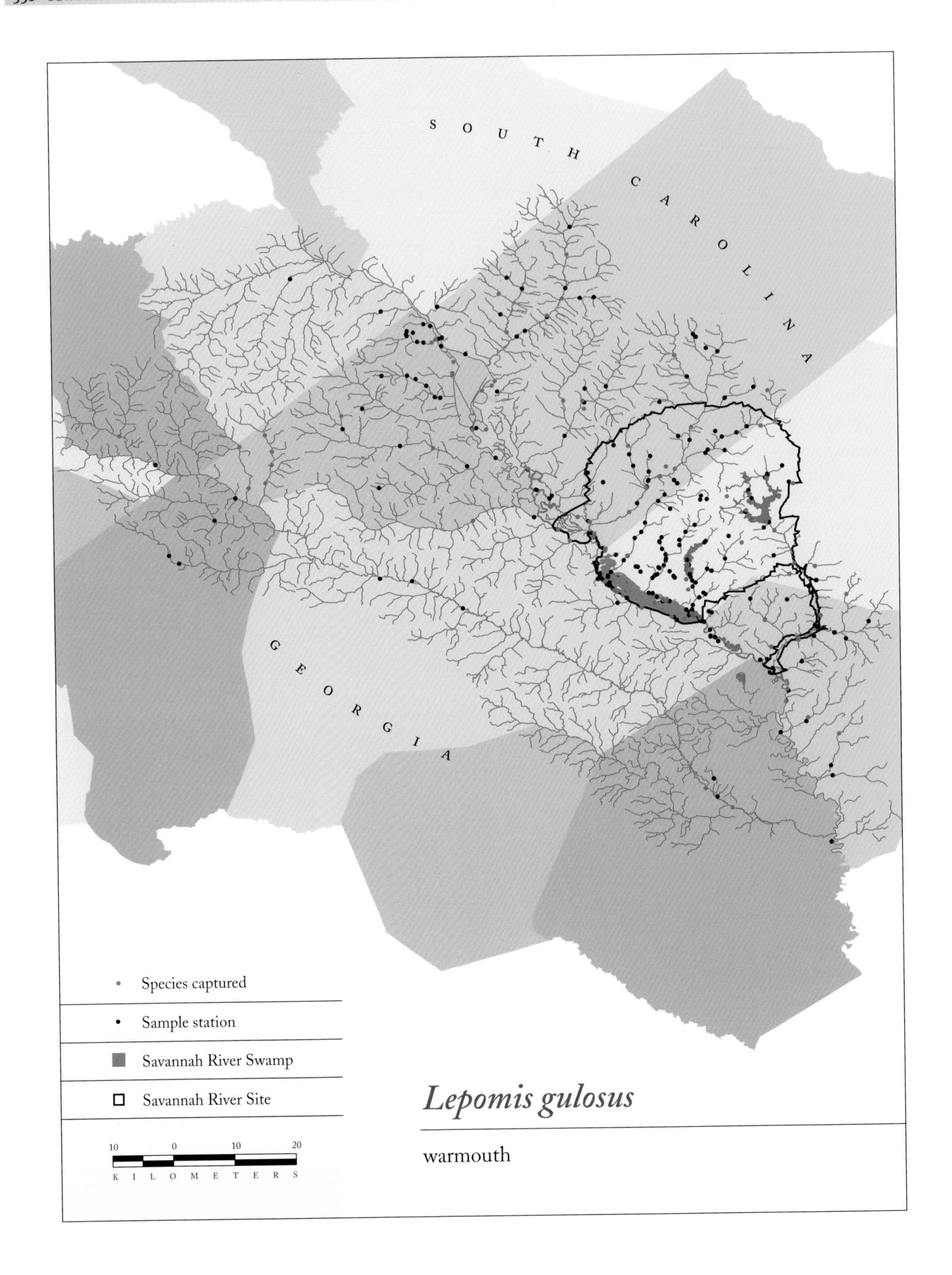

Lepomis gulosus

warmouth

DISTRIBUTION The Atlantic Coast from Maryland (where it may have been introduced) and Virginia south to southern Florida, west to the Rio Grande of Texas and New Mexico, north through Kansas and Iowa to southern Wisconsin and Michigan, and east to western Pennsylvania (Lee *in* Lee et al. 1980). Widely introduced west of the Rocky Mountains and on the East Coast (Lee *in* Lee et al. 1980), and in Puerto Rico, where it is possibly established (Erdman 1984). In the Great Lakes, native only to Lake Erie (Cudmore-Vokey and Crossman 2000). In the MSRB, widely distributed in lentic systems ranging from small ponds to large reservoirs, and with a similar wide distribution in lotic systems. Common in backwaters of the Savannah River. Also in larger tributaries, especially those with large reservoirs in their headwaters.

HABITAT Slow-flowing or still waters, including lentic habitats such as ponds, reservoirs, oxbows, and Carolina bays; and sluggish, low-gradient portions of lotic systems (e.g., Larimore 1957; Pflieger 1975). Often found along the shoreline around aquatic vegetation, submerged roots, and brush piles. On the SRS, warmouths are particularly abundant in rock rubble placed along the shorelines of dams, stilling basin channels, and banks of the Savannah River. In lentic systems, they frequently inhabit shallow littoral habitats 30–100 cm deep, around dense submergent vegetation and over soft organic substrates (Guillory 1979c); shoreline depth is an important indicator of abundance (Rutherford et al. 2001). While this species thrives in clear water, it will tolerate moderate levels of turbidity, and is also more tolerant of lower dissolved oxygen concentrations than the bluegill (Larimore 1957). It is occasionally abundant in tidal portions of rivers in Alabama at salinities up to 15 ppt (Mettee et al. 1996).

BIOLOGY The ovaries of females studied in central Illinois began to enlarge in March and April, and males became ripe before females (Larimore 1957). Spawning is reported to occur from late spring (April–May) through summer (July–August) in Arkansas, Tennessee, Missouri, and Illinois, with a peak in June (Larimore 1957; Pflieger 1975; Robison and Buchanan 1984; Etnier and Starnes 1993). Spawning begins at water temperatures around 21 °C in Illinois (Larimore 1957). In general, males construct solitary, circular depression nests in shallow water near structures such as stumps, wood debris, roots, rock rubble, or vegetation (Larimore 1957). Exposed substrates are avoided, and soft rather than sandy substrates are generally utilized (Larimore 1957). In Fire Pond on the SRS, warmouth nests are frequently hidden among submergent vegetation (DEF pers. obs.). Fletcher and Burr (1992) found warmouths building nests in cavities created by the buttressing roots of cypress trees. A female courted to a nest deposits a clutch of eggs during multiple spawning episodes, and may lay eggs in nests of more than one male (Larimore 1957). Males guard the developing offspring until the fry leave the nest. Embryos and larvae are described in Larimore 1957; and additional characters are provided in Taubert 1977.

Larimore (1957) described courtship and spawning, including rapid color changes, larval development, and growth, for a population in two Illinois lakes, as follows. Eggs hatched in 33–36 hours at 25–26.4 °C; yolk supplies were nearly exhausted in 4 days; swim-up occurred in 5 days; and the young often hid within or in small open pockets among submergent vegetation. Sexual maturity occurred at 1 or 2 years of age between 76 and 102 mm TL, and all fish were mature by 137 mm TL. Annulus formation on scales occurred in early May, and in one population, fish averaged 58, 89, 132, 168, and 196 mm TL at ages 2–6, respectively. The life span was 8 or more years (Larimore 1957). In a northern South Carolina lake, fecundity estimates increased with female size, ranging from 798 to 34,257 for fish between 96 and 222 mm TL (Panek and Cofield 1978).

The diet generally consists of aquatic insects, crustaceans including amphipods and shrimp, and fish (Hunt 1953; Larimore 1957; Flemer and Woolcott 1966; Wiltz 1993; Mittelbach and Persson 1998). The warmouths (65–215 mm TL) Wiltz (1993) studied in the MSRB ate aquatic insects, microcrustaceans, decapods, and gastropods, and a few individuals ate fish. Warmouths feed on fish more than most other *Lepomis* species (Robison and Buchanan 1984) and become increasingly piscivorous at 125 mm TL and 3–4 years of age, when 6–20% of the diet may consist of fish (Mittelbach and Persson 1998). Larimore (1957) made a detailed analysis of a warmouth population's diet in an Illinois lake, including an assessment of ontogenetic, seasonal, and habitat influences. While the diets of those warmouths and largemouth bass overlapped, the feeding areas differed, with warmouths feeding on soft bottoms in shallow waters or along the bank (Larimore 1957).

Warmouths are generally solitary outside of spawning season, but may aggregate in desirable cover (Larimore 1957). In Tennessee streams, their home range is frequently less than 100 m (Gatz and Adams 1994), but in a North Carolina swamp, most have sizable home ranges of more than 200 m (Whitehurst 1981).

SCIENTIFIC NAME *Lepomis* = "scaled operculum" (Greek); *gulosus* = "large mouthed" (Latin) (Mettee et al. 1996).

Lepomis macrochirus Rafinesque, 1819

Bluegill

DESCRIPTION Body strongly laterally compressed and deep. Eyes moderate in size. Mouth small, terminal, and oblique, with posterior margin of maxilla anterior to or barely reaching anterior edge of eye when mouth is closed. Pectoral fins long and pointed, reaching beyond anterior margin of eye when folded forward. Caudal fin emarginate. Palatine teeth (Etnier and Starnes 1993) and tooth patches on tongue generally absent. Gill rakers long and relatively slender.

The classic bluegill character is the black blotch near the posterior base of the soft dorsal fin. Fish of all ages lack mottling on the fins. Large juveniles and females are olive dorsally, fading to white ventrally on the flanks and belly, and goldish yellow or olive on the breast. The sides are generally marked with vertical bars. Ear tab short and black. Small juveniles have an olive dorsum and silvery sides, usually with a purplish sheen. At 20–25 mm TL, juveniles develop distinctive dark vertical bars on the sides and have no spots between bars (Brown and Colgan 1981). Juveniles are whitish ventrally with fins mostly translucent except for the black spot in the soft dorsal fin, which may be vague in small individuals.

Nuptial males are olive dorsally fading to lighter olive with a bluish or purplish sheen on the sides; ventral head from mouth to opercle light powder blue. Ear tab elongate and black.

SIMILAR SPECIES In the MSRB, only green sunfish (*L. cyanellus*) and juvenile fliers (*Centrarchus macropterus*) share the black blotch near the posterior base of the soft dorsal fin. Juvenile pumpkinseeds (*L. gibbosus*) develop spots between the dark bars by about 30 mm TL; these are absent in the bluegill (Brown and Colgan 1981).

SIZE Bluegills may reach close to 2.5 kg and 381 mm TL.

MERISTICS Dorsal spines (9)10(12), dorsal rays (9)10–11(13), anal spines 3, anal rays (9)11–12, pectoral rays (12)13–14(15); lateral line scales (38)41–46(50) (Jenkins and Burkhead 1993); gill rakers 13–16, the longest four–five times as long as the basal width (Etnier and Starnes 1993).

Lepomis macrochirus, bluegill; (above, left) aberrantly colored breeding male from Par Pond, (above, right) breeding male, (facing page, left) adult or large juvenile, (facing page, right) juvenile

DISTRIBUTION Native from western New York through the Great Lakes to the Dakotas, south through the Mississippi Valley to the Gulf of Mexico, west through Texas into northeastern Mexico, and eastward to Florida and northward east of the Appalachian Mountains to southern Virginia (Lee *in* Lee et al. 1980). Widely introduced and established in all 48 contiguous U.S. states and in northwestern Mexico in Sonora and Baja California Norte; also established in Europe, South Africa (Lee *in* Lee et al. 1980), Hawaii (Brock 1960; Maciolek 1984), Puerto Rico (Erdman 1984), Korea, and Japan. Widely distributed in the MSRB, but most common in reservoirs, ponds, and backwaters on the modern Savannah River floodplain. Generally rare in tributary streams that lack impounded habitat (Meffe 1991).

HABITAT Bluegills are a versatile species whose broad environmental tolerance allows them to use nearly all inshore lentic habitats (Guillory 1979c; Keast et al. 1978). They are commonly associated with clear, nonflowing water with dense vegetation. Most abundant in lentic habitats such as reservoirs, lakes, and ponds, but also found in lotic systems, especially in backwaters, coves, or oxbows of rivers. When bluegills were stocked into L Lake on the SRS, they invaded but did not successfully colonize Meyers Branch, a largely undisturbed tributary stream (Meffe 1991).

Offshore movement after sunrise and onshore movement at night have been noted in lentic habitats (Baumann and Kitchell 1974). They tend to overwinter in deeper water (Keast 1978), often in aggregations. Bluegills are highly gregarious and are commonly found in loose schools of 20–30 (Pflieger 1975). Schooling behavior may be enhanced by predator activity, and bluegills may reduce their use of deeper water in the presence of predators like largemouth bass (DeVries 1990). Ontogenetic habitat changes include migration of larvae from the spawning grounds into the limnetic zone of lakes (Rettig 1998). Juveniles move to shallow littoral zones, and the adults subsequently switch back to more open-water feeding areas (Mittelbach and Osenberg 1993).

Bluegills can be found at water temperatures ranging from 5.0 to 33.5 °C, at salinities up to about 14 ppt (Hardy 1978), and over a broad pH range of 5–9 (Graham 1993); larval mortality is extensive below pH 4.5 (Graham and Hastings 1984). Bluegills from Pond C, where populations may have declined by as much as 90% during high-temperature effluent releases (Taylor et al. 1991), had a critical thermal maximum of 37–41 °C (Holland et al. 1974). Adult distribution is significantly correlated with pH and water color (Graham and Hastings 1984).

BIOLOGY Spawning occurs from May through August with a peak in June at water temperatures of 18.5–25.0 °C in Delaware (Wang and Kernehan 1979), and from late May through July or into August in Missouri (Pflieger 1975). The spawning season is controlled largely by water temperature, and spawning may stop if the water tem-

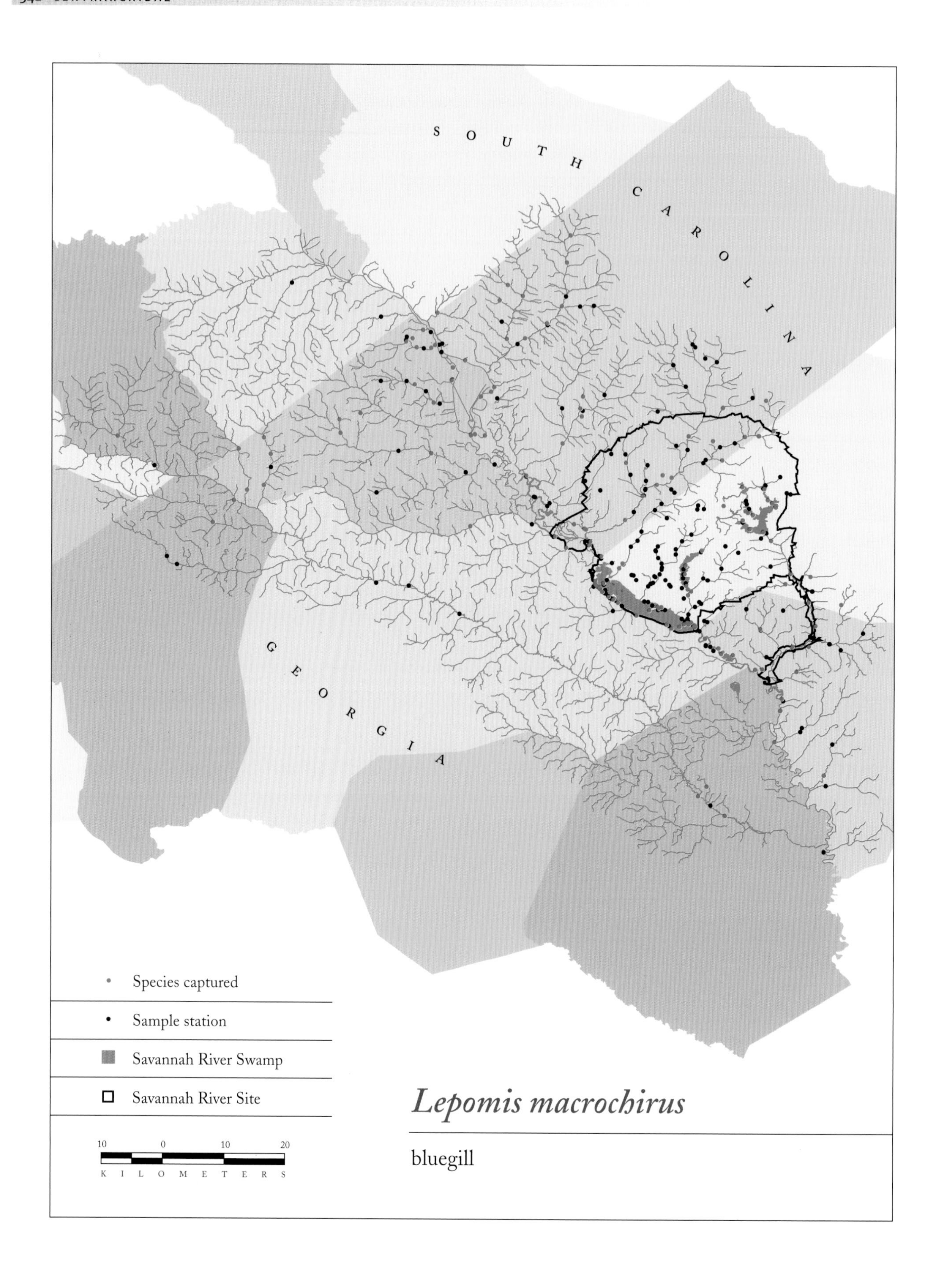

Lepomis macrochirus

bluegill

perature becomes too hot (Taylor et al. 1991) or too cold. Spawning was extended into fall or early winter by warm-water effluents into Pond C on the SRS; but cold-water releases delayed or prevented spawning in Steel Creek below the L Lake dam (DEF pers. obs.). Early-spawned young-of-the-year may be more likely to survive their first winter than later-spawned young (Cargnelli and Gross 1996). Larvae are described in Taber 1969; Wang and Kernehan 1979; and Tin 1982; and have been identified using allozyme analysis (Rettig 1998). Geographic and temporal variation in larval myomere counts was addressed by Bosley and Conner (1984).

Males construct saucer depression nests, usually about 30–40 cm in diameter, by sweeping away substrate with the tail. Bluegills are highly colonial nesters. While smaller numbers are more common, spawning colonies may exceed 250 nests, with nearly 3 nests/m^2 (Côté and Gross 1993). A male's ability to maintain a nest within a colony is at least partly related to his size (Dominey 1980, 1981a). Aggression among nest-building males decreases in denser colonies (Colgan et al. 1979). Male-male interactions within colonies include nest takeovers, which are common (Dominey 1981a). Colonial nesting may reduce predation of offspring in the centrally located nests (Dominey 1981b; Gross and MacMillan 1981). The guarding male courts a female to the nest, and after spawning enhances the survival of the offspring by guarding the nest from predators until the fry leave (Bain and Helfrich 1983), generally a period of 7–10 days (Dominey 1981a; Gross and MacMillan 1981). Spawning and agonistic behaviors among nest-guarding males, including rasping sounds presumably produced with the pharyngeal pads, are described in detail in Avila 1976; and Ballantyne and Colgan 1978a, 1978b. Kindler et al. (1991a, 1991b) studied hormonal control of reproductive behavior.

Under certain environmental conditions, some males parasitize nests built by other males rather than building their own nests (Dominey 1980; Gross 1984; Chan and Ribbink 1990; Taborsky 1994). "Sneaker" males dart into another male's nest and release sperm while the nest builder is spawning (Gross 1982, 1984). Other males mimic females in appearance and behavior, and stealthily join a spawning event between a nest-building male and a female to fertilize eggs (Dominey 1980, 1981a; Gross 1984). Such males are generally too small to successfully build a nest and attract females (Dominey 1980; Dominey 1981a), and parasitic spawning secures successful reproduction via sperm competition rather than through female monopolization (Taborsky 1994). These males mature at a smaller size and a younger age (2 years versus 7 years in some populations), and allocate more of their body weight to gonad tissue than do nest-building males (Dominey 1980; Gross 1982, 1984; Ehlinger 1997). Nest builders attempt to defend their nests from these cuckolding males, but must balance nest defense with courtship (Gross 1982; Chan and Ribbink 1990). Genetic analyses confirmed successful spawning of cuckolding males and multiple females in a single nest (Colbourne et al. 1996). Lakes with lower fishing pressure produce larger, later-maturing males—and consequently fewer cuckolder morphs (Drake et al. 1997); no cuckolder morphs were detected in unfished Par Pond on the SRS (Belk and Hales 1993). In stunted populations, nest-building males mature at a younger age and smaller size while cuckolders increase in abundance (Ehlinger 1997). Increasing the number of large nest-building males in a population causes young males to delay sexual maturation and consequently to grow faster (Jennings et al. 1997). Bluegills in postthermal Par Pond on the SRS matured at total lengths between 175 and 215 mm at 2–4 years of age, about 80 mm longer and 2 years older than populations in other southern U.S. reservoirs (Belk and Hales 1993).

Otolith annulus formation occurs from February through June in Par Pond on the SRS (Hales and Belk 1992) and in Florida (Schramm 1989). Hatching time within the protracted spawning season influences the opaqueness of the otolith center (Hales and Belk 1992). Most annual growth of juveniles occurs in midsummer, when lipid reserves later utilized for winter survival are accumulated (Booth and Keast 1986). Larger adults survive winter in better condition than smaller fish (Cargnelli and Gross 1997). While northern bluegills may live longer, latitude does not appear to affect maximum size, as bluegills in southern populations tend to be larger at a given age (Belk and Houston 2002). Bluegills may reach 11 years of age (Hardy 1978; Rohde et al. 1994). Bluegills grew faster and reached a larger size in Par Pond than in other reservoirs, reaching average total lengths, respectively, of 66, 123, 188, 248, 257, 259, 263, 264, and 260 mm at ages 1–9, the maximum age (Belk and Hales 1993). Mortality rates in Par Pond were higher for young bluegills and lower for older bluegills than in other reservoirs as well (Belk and Hales 1993). These life history differences and the above-mentioned maturation rates appear to be due to the abundance of large largemouth bass (*Micropterus salmoides*) in unfished Par Pond (Belk 1993; Belk and Hales 1993) and the lack of fishing pressure on larger bluegills, and appear to represent phenotypic variation rather than genetically fixed differences (Belk 1995). Populations exposed to low or no predation for a long time lose this plasticity (Belk 1998).

The diet and feeding behavior of bluegills have been extensively studied. Although mostly invertivorous, bluegills feed opportunistically, and diet composition is influenced by season, fish size (Taylor et al. 1991), the co-occurrence of other species of sunfish (Werner and Hall 1976, 1977), and habitat (Flemer and Woolcott 1966; Mittelbach and Osenberg 1993). The plant material frequently found in bluegill stomachs was probably secondarily ingested along with insects that live on aquatic vegetation (Sadzikowski and Wallace 1976). Small fish are only sporadically eaten (Keast 1985b). Larval bluegills deplete their yolk and begin active feeding about 8–9 days after fertilization at 23.5 °C (Toetz 1966). Larvae less than 8 mm in TL feed on rotifers, copepod nauplii, and cladocerans; larvae over 9 mm TL switch to ostracods, cladocerans, copepods, and dipterans (Partridge and DeVries 1999). Juveniles feed largely on vegetation-dwelling invertebrates, but the adults' morphology is well adapted for feeding on open-water zooplankton (Mittelbach 1984; VanderKooy et al. 2000); the long, fine gill rakers (Werner and Hall 1976) and the fine needlelike teeth on the pharyngeal pads in the throats (Keast 1978) retain small prey. As adults move from shallow littoral zones to more open water, the diet switches from vegetation-dwelling invertebrates to zooplankton (Mittelbach and Osenberg 1993).

Bluegills are "semi-motile"; excursions of more than 200 m were uncommon in a North Carolina swamp (Whitehurst 1981), but bluegills seemed more mobile than warmouths (*L. gulosus*) or redbreast sunfish (*L. auritus*) in Tennessee streams (Gatz and Adams 1994).

Juveniles may clean parasites off adults that assume a vertical head-down display (Sulak 1975). In Hawaii, where bluegills and largemouth bass have been introduced, the two species have produced an intergeneric hybrid that appears to be fertile and has persisted since 1963 (Maciolek 1984). Disruption of the substrate by bluegill nest building affects aquatic invertebrate distributions (Pierce et al. 1987).

SCIENTIFIC NAME *Lepomis* = "scaled operculum" (Greek); *macrochirus* = "large hand" (Greek), in reference to the body size and shape.

Lepomis marginatus (Holbrook, 1855)

Dollar sunfish

DESCRIPTION Body short, deep, and compressed, with a lateral profile that tends to be relatively more rounded than in other MSRB *Lepomis*. Mouth small and oblique, with the posterior edge of the maxilla extending to the anterior edge of the eye, but not reaching the pupil. Palatine teeth (Etnier and Starnes 1993) and tooth patches on the tongue absent. Gill rakers stubby, short, and thick. Lateral line complete; scales ctenoid. Pectoral fin relatively short and rounded; when folded forward the tip extends to the pupil, but not to the anterior edge of the eye. Caudal fin emarginate. Forehead and nape of breeding males with distinctly convex arch; nape gently sloping and not rounded in females and juveniles.

Wavy blue iridescent lines on sides of snout and cheeks, sometimes extending to operculum, more pronounced in breeding males. Dark reddish orange spots on sides, especially on the lateral line. Similar spots frequently on posterior base of soft dorsal fin.

Breeding males are quite colorful with dorsum, dorsolateral sides, and top of head dark olive, fading ventrally. Dorsolateral and lateral scales dark basally, distal edges with metallic blue edging (lighter in preservative), fading to goldish iridescence ventrally. Ventrolateral and ventral body burnt orange. Chin metallic blue. Dorsal and anal rays metallic blue with dark reddish orange membranes, color more intense in basal and posterior portions of anal fin (dusky in preservative). Pectoral fins largely translucent; pelvic fins metallic blue. The blue iridescence is more visible when observed underwater, especially with bright light. Ear tab moderately elongate, angled dorsally, and dark bluish black with a distinctive white border.

Breeding females have a goldish olive dorsum fading to yellow-orange ventrally. Scales on lateral sides generally dark basally and marked with iridescent gold/pale bluish patches. Fins largely translucent, but membranes of dorsal and anal fins dusky in life and preservative. Ear tab bluish black with light margin, sometimes with gold iridescent flecks.

SIMILAR SPECIES The dollar sunfish is one of four local *Lepomis* species with wavy blue iridescent lines on the sides of the snout and cheeks; the others are the green sunfish (*L. cyanellus*), pumpkinseed (*L. gibbosus*), and redbreast sunfish (*L. auritus*). See the key and species accounts for identification characters.

SIZE Reported up to 127 mm TL in Louisiana (Douglas 1974); in the MSRB, most are less than 90 mm SL.

MERISTICS Dorsal spines (9)10(11), dorsal rays (10)11(12), anal spines 3, anal rays 9–10, pectoral rays (11)12(13); lateral line scales 34–41, cheek scale rows 3–4; gill rakers 9–10, stubby, short, and thick (Etnier and Starnes 1993; Mettee et al. 1996).

DISTRIBUTION North Carolina south to southern Florida, west to east Texas, north through the Mississippi

Lepomis marginatus, dollar sunfish; (left) male, (right) female

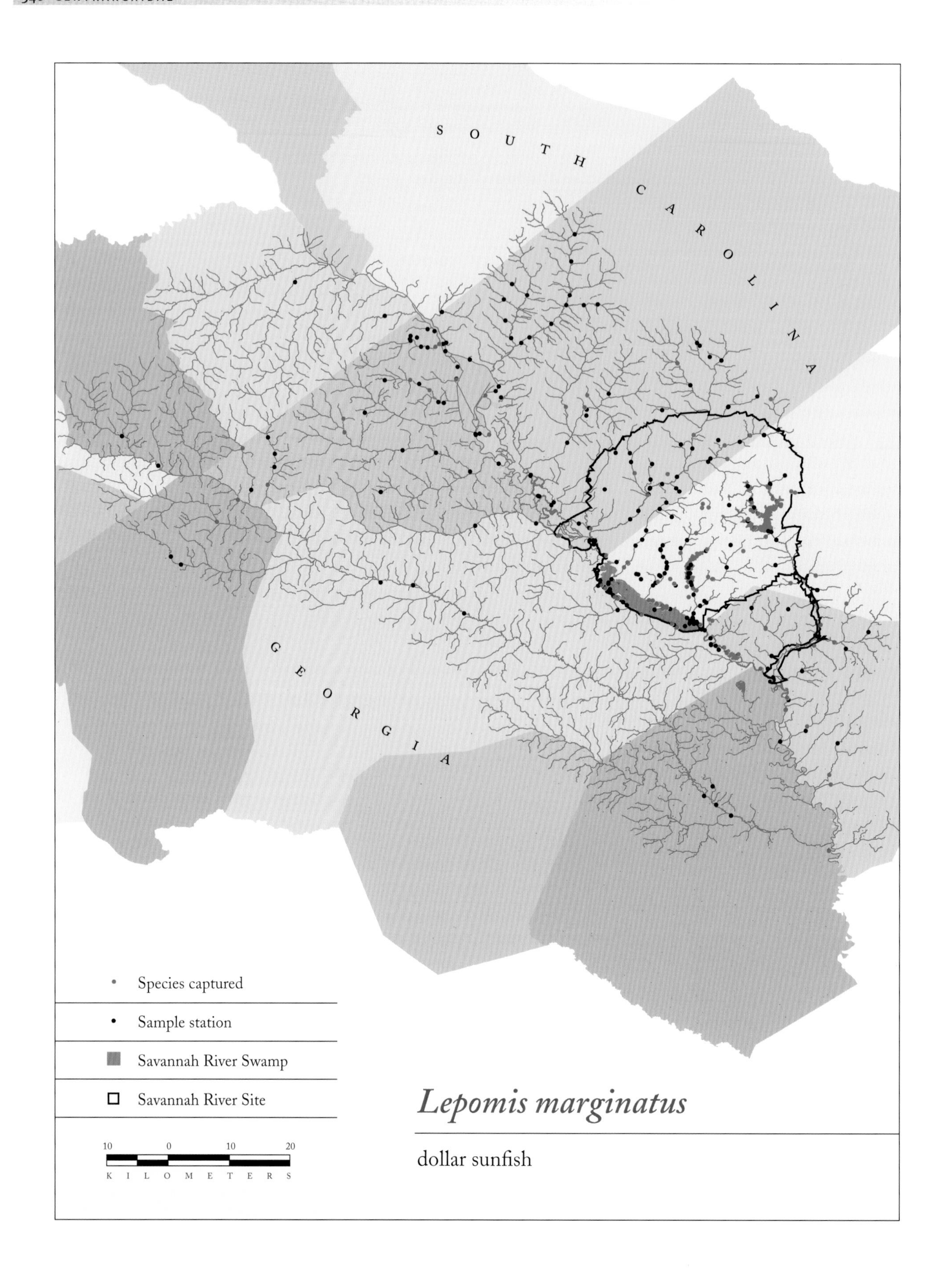
SOUTH CAROLINA
GEORGIA
Species captured
Sample station
Savannah River Swamp
Savannah River Site
10 0 10 20
KILOMETERS
Lepomis marginatus
dollar sunfish

River basin through southeastern Oklahoma and Arkansas to eastern Tennessee and southeastern Kentucky (Bauer *in* Lee et al. 1980). In Georgia, mostly on the Coastal Plain (Dahlberg and Scott 1971a). Reports of the longear sunfish (*L. megalotis*) from Augusta probably represent misidentified dollar sunfish (Dahlberg and Scott 1971a). In the MSRB, the dollar sunfish is common where favorable habitat is available.

HABITAT In the MSRB, small headwater streams or side channels and backwaters of braided streams such as those on the fluvial terraces or on the modern floodplain of the Savannah River. Isolated wetlands (wetlands connected to other surface water bodies only during flooding conditions) in the MSRB such as Carolina bays are also frequently inhabited, particularly those in the headwaters of drainage systems where source populations are readily available (Snodgrass et al. 1996). In headwaters, dollar sunfish are generally found in shallower and narrower channels than are other common centrarchids (Meffe and Sheldon 1988; Paller 1994); in lotic systems, they are more common in beaver ponds than in adjacent streams (Snodgrass and Meffe 1999). In a Mississippi stream, dollar sunfish left the main channels and exploited the floodplains during floods (Ross and Baker 1983). In western Tennessee they inhabit unaltered, good-quality sluggish streams, vegetated swamps, and natural lakes (Etnier and Starnes 1993). Dollar sunfish can tolerate a temperature range of at least 4–36 °C (Winkelman 1993).

BIOLOGY Dollar sunfish spawn over a protracted season. In lentic systems in the MSRB, they may spawn from May through August, with peak spawning from mid-May to the end of June (Winkelman 1993). In Fourmile Branch, spawning generally extends from late May through the end of July at water temperatures around 24–27 °C. Over this time, spawning episodes occur every 1–3 weeks, with females spawning multiple times per season. Males sweep away silt and fine organic matter from a variety of substrates to form a shallow depression about 30 cm in diameter; nests are generally located near the bank in shallow water 10–50 cm (Winkelman 1996). Adhesive eggs are attached to a variety of substrates including submergent vegetation, detritus, twigs, sand, gravel, and roots (Winkelman 1993; DEF unpub. data). The male guards the nest until the fry leave. During field trials, Winkelman (1996) showed that threats of avian predation reduced the time a male spent guarding a nest, but the presence of offspring in the nest increased the male's willingness to risk his own death to promote the survival of his offspring.

Although nests are primarily solitary or in loose aggregations in lotic habitats (Mackiewicz et al. 2002), dense colonies of more than 20 nests may form in lentic habitats such as Par Pond (DEF pers. obs.). In the largely solitary lotic nests studied in Fourmile Branch, nest-guarding males sired more than 90% of the offspring contained within the nests they guarded (Mackiewicz et al. 2002). Nest takeovers by rival males and cuckoldry by neighboring males were also documented. No evidence of a parasitic cuckolding morph male was found in an examination of 97 males. The genetic analyses also indicated that multiple females spawned in a single nest.

Winkelman (1993) studied the life history of dollar sunfish in three ponds (Fire, Dick's, and Par) on the SRS. Maximum size and age varied substantially in the ponds; on average, males were larger than females: at the maximum ages of 4, 3, and 2 years, males were about 120, 80, and 75 mm TL, respectively, and females were about 100, 70, and 65 mm TL. Dollar sunfish may live as long as 6 years (Etnier and Starnes 1993). An analysis of annual lipid storage cycles indicated that females utilize lipid stores to initiate spawning (ovarian development) in the spring while males expend their stores guarding nests (Winkelman 1993).

In tributary streams of the MSRB, dollar sunfish feed mostly on aquatic insects, especially chironomids, terrestrial insects, snails, and oligochaetes (Sheldon and Meffe 1993); a similar diet has been reported from Tennessee (Etnier and Starnes 1993).

SCIENTIFIC NAME *Lepomis* = "scaled operculum" (Greek); *marginatus* = "edged" (Latin), probably in reference to white lining of the ear tab (Mettee et al. 1996). The common name is derived from the small, round body, which is often about the size of a silver dollar.

Lepomis microlophus (Günther, 1859)
Redear sunfish

DESCRIPTION Body distinctly compressed. Mouth relatively small, terminal or nearly terminal, with the posterior edge of the maxilla barely reaching the anterior edge of the eye when the mouth is closed. Pectoral fins long and pointed, reaching beyond anterior margin of eye when folded forward. Caudal fin emarginate. Lacking palatine teeth (Etnier and Starnes 1993) and tooth patches on tongue. Gill rakers relatively short and thick. Pharyngeal arches with well-developed molar surfaces (Lauder 1983).

Redear sunfish are less colorful than most local *Lepomis* species. Dorsum olive fading to silvery gray or brassy straw yellow on sides, marked with irregular dusky spots. Whitish or yellowish ventrally. Nuptial coloration may include orange tinges on breast and fins, and red iris. Ear tab flexible, black, with a red posterior margin on males, orange on females (Etnier and Starnes 1993); may be only pale orange on juveniles.

SIMILAR SPECIES Juvenile redear sunfish can be distinguished from similar-appearing pumpkinseeds by the length of the depressed pectoral fin, which reaches beyond the base of the dorsal fin in redear sunfish (Casey Huckins pers. comm.).

SIZE To more than 300 mm TL (Carlander 1977); may weigh in excess of 2 kg, but generally less than 1 kg (Jenkins and Burkhead 1993; Rohde et al. 1994).

MERISTICS Dorsal spines 10(11) (sometimes 9 in Etnier and Starnes 1993), dorsal rays (10)11–12, anal spines 3, anal rays (9)10(11), pectoral rays 13–15(16) ([12]13–14 in Etnier and Starnes 1993); lateral line scales (40)41–44(47) (Jenkins and Burkhead 1993); gill rakers 9–11, length of longest rakers about twice their basal width in adults (Etnier and Starnes 1993).

DISTRIBUTION Native in the southeastern United States from Texas north to southern Illinois and east to the Atlantic Ocean; introductions have expanded the range into New Mexico and north to Michigan, Ohio, and Pennsylvania (Lee *in* Lee et al. 1980). Established in

Lepomis microlophus, redear sunfish; male

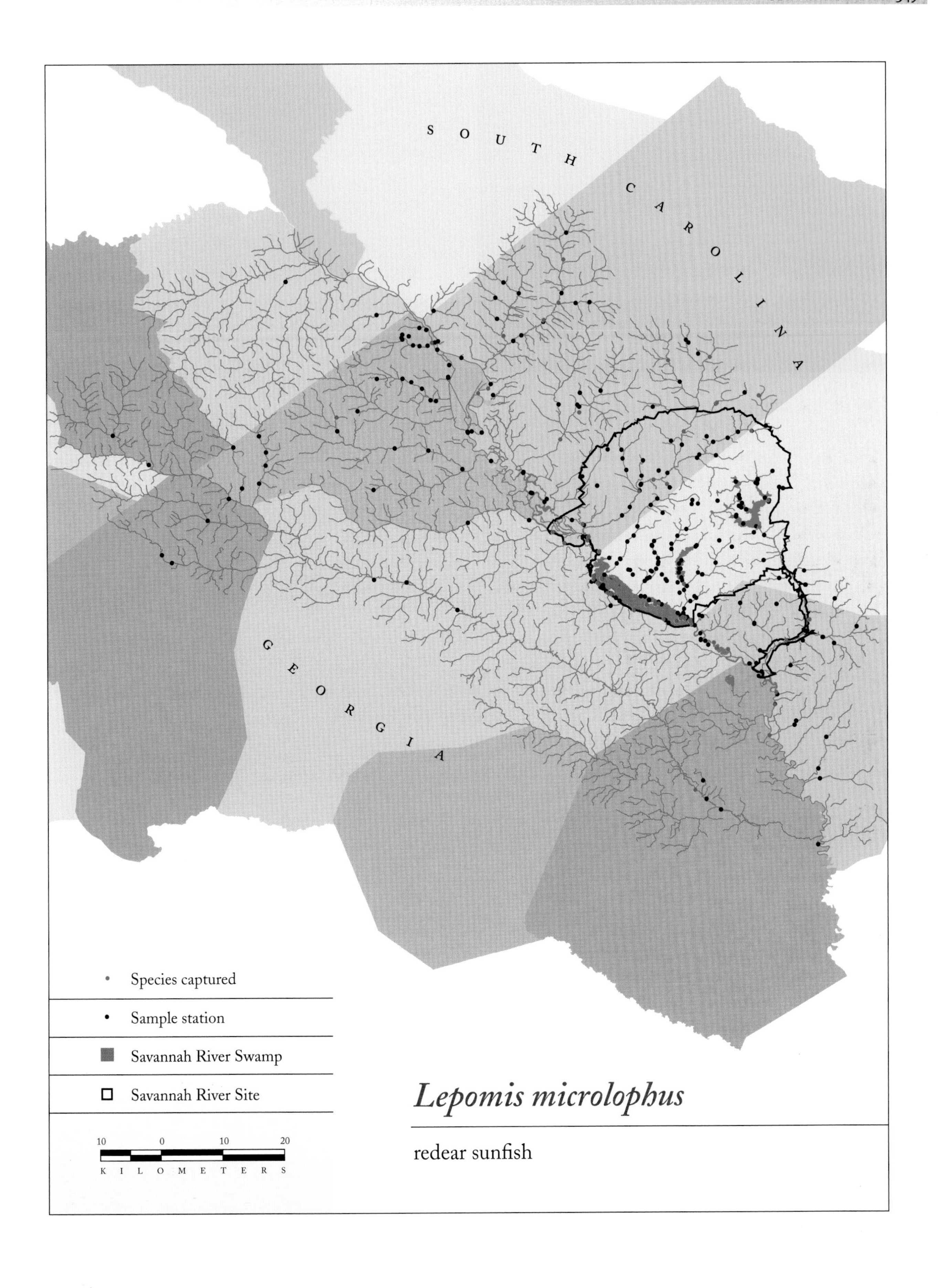

Lepomis microlophus

redear sunfish

Lepomis microlophus, redear sunfish; adult or large juvenile

Puerto Rico, where it was introduced both as a sport fish and to control the snail host of the parasite *Schistosoma mansoni* (Erdman 1984). Not widespread in the mid-Atlantic area, but can be locally abundant (Rohde et al. 1994). The MSRB is at the northern extent of its native geographic range along the Atlantic seaboard. In this area, redear sunfish are most commonly found in the Savannah River and in backwaters on the modern Savannah River floodplain. Present but generally rare in larger tributaries. Also found in some ponds and large reservoirs.

HABITAT Lowland swamps, ponds, reservoirs, and sluggish sections of lotic systems (Burr and Warren 1986). More frequently found in backwaters than main channels of streams (Pflieger 1975). In a Florida study, redear sunfish were nearly 10 times more abundant in an oxbow lake than in the nearby river (Beecher et al. 1977). Found in salinities up to 12.3 ppt (Hardy 1978).

BIOLOGY Spawning occurs in May and June in Missouri (Pflieger 1975), in April to early June in Alabama (Mettee et al. 1996), and may begin in spring when water temperatures reach 20–21 °C (Jenkins and Burkhead 1993). Nests may be placed in shallow water but are frequently placed in water 1 to more than 2 m deep (Wilbur 1969). Redear nests are generally in deeper water than those of the sympatric pumpkinseed (*L. gibbosus*) (Casey Huckins pers. comm.) and are sometimes constructed in dense colonies (Clugston 1966; DEF pers. obs.). A male sweeps out a saucer depression nest to which he attracts females. During courtship, males make popping sounds by snapping their jaws together (Gerald 1971). After spawning, the male continues to guard the nest until the fry depart. Sexual maturity occurs in 1–2 years, and the life span may reach 8 years (Carlander 1977). Ovaries of mature females from Florida ranging in size from 190 to 242 mm TL contained 12,196–25,965 eggs (Carlander 1977).

Redear sunfish are commonly called "shell crackers" because they feed on molluscs. While juveniles eat soft-bodied invertebrates and zooplankton, larger individuals (>55 mm SL) feed primarily on sediment-dwelling prey, and their functional morphology is well suited for molluscivory (Huckins 1997; Huckins et al. 2000; VanderKooy et al. 2000). In the MSRB, *L. microlophus* feed primarily on pelecypod molluscs, with gastropods and chironomids next in frequency (Wiltz 1993), but seasonal diet shifts have been noted (VanderKooy et al. 2000). Redear sunfish and pumpkinseed are generally allopatric throughout most of their native distributions, but a small area of overlap in the Carolinas includes the MSRB. When syntopic, the adults of the two species may compete for molluscs (Huckins et al. 2000). Redear sunfish are more specialized molluscivores than pumpkinseeds; they switch to molluscivory at a smaller size, can crush harder-shelled snails, and their diet contains a larger proportion of molluscs (Huckins 1997).

SCIENTIFIC NAME *Lepomis* = "scaled operculum" (Greek); *microlophus* = "small nape" (Greek).

Lepomis punctatus (Valenciennes, 1831)
Spotted sunfish

DESCRIPTION Body deep and laterally compressed, moderately stocky. Eye moderate in size. Mouth small, oblique, and terminal, with posterior edge of maxilla extending to anterior edge of eye but not to pupil. Palatine teeth (Etnier and Starnes 1993) and tooth patches on tongue absent. Gill rakers relatively thick and of moderate length, length-to-width ratio on lower limb of first arch about 4:1 (Warren 1992). Outline of forehead and nape of breeding males with a distinctly convex arch; those of females and juveniles more gently sloping. Pectoral fins relatively short and rounded, when folded forward the tip extends to, but not past, pupil of eye. Caudal fin emarginate. Lateral line complete; scales ctenoid.

Dorsum dark olive fading to lighter shades of olive on sides and ventral surface. Scales on sides variably marked with black spots. Head olive with dark spots on posterior portion of cheek and opercle. Dorsal and anal rays dusky; spots present on basal one-third of soft dorsal, anal, and caudal fins (may appear as mottling if faded in preservative). Pectoral fins translucent; pelvic fins dusky.

Pelvic fins in breeding males become intense bluish black with an iridescent blue anterior margin and light posterior margin. Dorsal, anal, and caudal fin membranes also darken to dusky bluish black (spots sometimes obscured), with reddish orange on branched portions (outer portions) and a light distal margin that is most distinct on soft rays of unpaired fins. Breast and belly with coppery butterscotch yellow and orange cast. Sides marked with about five to seven narrow, dark vertical bars. Spots on sides noticeably faded compared with dorsolateral and ventrolateral spots. Sides sometimes with rosy iridescent sheen, especially posteriorly. Ear tab of males moderate in size, angled dorsally, and dark bluish black with a light margin. May have a poorly defined butterscotch yellow blotch (pale in preservative) above opercular tab (Warren 1992). Orbit of eye develops brilliant iridescent turquoise blue on ventral curve.

Breast of female olive with distinctive yellow cast. Ear tab small and black. Caudal fin of juvenile reddish orange, flanks with purple or blue iridescent cast.

SIZE May reach 200 mm TL (Etnier and Starnes 1993). In SRS tributary streams, most individuals are less than 120 mm SL, with a few between 120 and 160 mm SL (Reichert 1995).

MERISTICS Dorsal spines 10(11), dorsal rays 10–11(12), anal spines 3, anal rays 10(11), pectoral rays 13(15); gill rakers 8–11 (Etnier and Starnes 1993); lateral line scales (38) 39–42(45), caudal peduncle scale rows 19–21(23), cheek scale rows (4)5–6(7) (Warren 1992).

DISTRIBUTION From southeastern North Carolina south to southern Florida and west through most of Texas, north in the Mississippi River basin to Illinois and Mis-

Lepomis punctatus, spotted sunfish; (left) male, (right) female

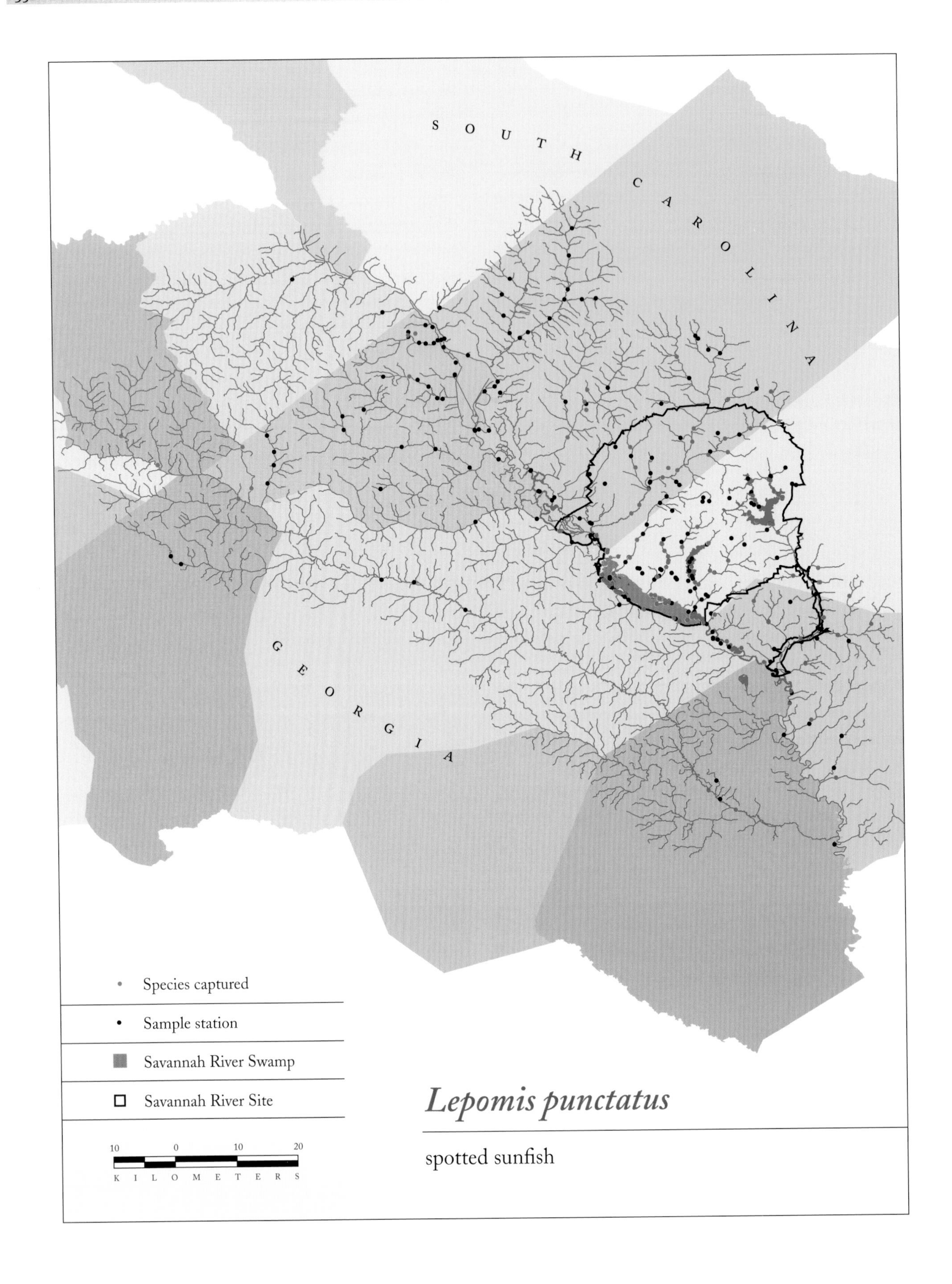

Lepomis punctatus

spotted sunfish

souri (Lee *in* Lee et al. 1980). Spotted sunfish are the most common and widely distributed centrarchids in tributary streams of the MSRB. They are nearly equally common in first–fourth-order streams (Paller 1994). While most common in tributary streams, they may also be found in the Savannah River and in many lentic habitats.

HABITAT Most common in clear, slow-to-moderate-velocity streams, rivers, and oxbow lakes. In tributary streams, often associated with woody debris, stumps, or undercut banks of pools. Distribution in tributary streams in the MSRB is associated with slower-flowing water and softer substrates (Meffe and Sheldon 1988). Spotted sunfish, especially young-of-the-year, may also be common in beaver ponds in the MSRB (Snodgrass and Meffe 1999). In Tennessee, the species is abundant only in natural lakes of the Mississippi floodplain (Etnier and Starnes 1993). Spotted sunfish have a strong tolerance for salinity (Bailey et al. 1954).

BIOLOGY In tributary streams of the MSRB, spawning typically occurs from late May through the end of July at water temperatures around 24–27 °C. Spawning occurs from April to August in Texas and Arkansas (Robison and Buchanan 1984), and from May to August in Tennessee (Etnier and Starnes 1993). Males excavate small nests, generally 15–30 cm in diameter, near and frequently against the bank in shallow water (Carr 1946; DEF pers. obs.). Nests in tributary streams on the SRS are generally solitary (DeWoody et al. 2000a). Carr (1946) reported that most nests in a Florida stream were solitary, but groups of two to three aggregated nests were found as well. The spawning act was described by Carr (1946). Genetic analyses indicated that at least three, and an estimated average of seven females spawn in each nest (DeWoody et al. 2000a). After spawning, males remain on the nest, defending it from potential predators and fanning the offspring to keep them free of silt. The eggs are small, adhesive (frequently attached to fine roots along the shoreline side of the nest), demersal, and dark brownish olive in color. Filial cannibalism was genetically confirmed by DeWoody et al. (2001).

Two morphological types of reproductively mature males have been reported in Fourmile Branch on the SRS. Larger, sexually dimorphic males build nests and court females for spawning, while a smaller parasitic morph displays no secondary sexual characters, has greatly enlarged testes, and sneaks onto nests of other males to fertilize eggs (DeWoody et al. 2000a). These parasitic cuckolding morphs were very rare in Fourmile Branch, and genetic analyses indicated that nest-guarding males sired most of the offspring they guarded (DeWoody et al. 2000a).

Spotted sunfish are invertivores. In the Savannah River, they feed largely on aquatic insects, terrestrial insects, microcrustaceans, and decapods, occasionally ingesting gastropods, pelecypods, algae, vascular plants, and an occasional fish (Wiltz 1993). In tributaries of the MSRB, spotted sunfish over 75 mm SL feed primarily on terrestrial insects, followed by smaller amounts of aquatic insects, snails, and decapods. Smaller spotted sunfish eat more chironomids, along with other aquatic and terrestrial insects and a few water mites and crustaceans (Sheldon and Meffe 1993). We have watched large spotted sunfish hiding beneath undercut banks watching for prey drifting in the current. They dart out and capture the prey, then return to their refuge (DEF pers. obs.).

SCIENTIFIC NAME *Lepomis* = "scaled gill cover" (Greek); *punctatus* = "spotted" (Latin).

Micropterus coosae Hubbs and Bailey, 1940
Redeye bass

DESCRIPTION Body elongate and slender, more so than other black basses (Gwinner et al. 1975). Mouth large, with posterior edge of maxilla reaching to or barely behind posterior margin of eye in adults. Pectoral fin short and rounded. Tongue usually (in more than 70% of specimens) with a circular, median patch of teeth (King and Parsons 1951). Caudal fin strongly emarginate. Only a slight notch is present between the spinous and soft dorsal fins, in contrast to the deep cleft in largemouth bass (*M. salmoides*). Lateral line complete; scales ctenoid.

Dorsum dark olive fading on the mottled sides to a generally whitish ventrum. Mid-side with black blotches or vertical bars, faint on all sizes and usually obscured on large specimens, but never forming a distinct horizontal stripe as in largemouth bass. Ventrolateral side marked with distinct horizontal rows of small dark spots in adults, but may be more irregular in juveniles. Entire side of juveniles may appear mottled with dark olive. Three (2–4) dark bars radiate from eye onto cheek at all sizes. An indistinct black basicaudal spot may be present. White present on upper and lower margins of caudal fin. Fins, especially median fins, may have orange or reddish tinge in life. Dark mottling sometimes in soft dorsal and soft anal fins. As the common name implies, eyes typically reddish, with a white fleck in the posterior portion. Like most *Micropterus* species, the redeye bass is monomorphic, with almost no external differences between the sexes.

SIZE Reaches at least 432 mm TL. Adults average around 225 g (Parsons 1954; Etnier and Starnes 1993); specimens above 1 kg are rare.

MERISTICS Dorsal spines (9)10(11), dorsal rays (11)12(13), anal spines 3, anal rays (9)10(11) (Etnier and Starnes 1993; Mettee et al. 1996), pectoral rays (14)15–16(17); lateral line scales 57–77 (Etnier and Starnes 1993) or 63–74 (Mettee et al. 1996).

DISTRIBUTION Native to Alabama and Georgia drainages feeding into Mobile Bay, the upper Chattahoochee River, and the Altamaha River, and in the Savannah River. Also native to two small drainages in Tennessee proximate to the Georgia border (Parsons 1954). Introduced in Arkansas (Robison and Buchanan 1984), Tennessee, Kentucky (Burr and Warren 1986; Mettee et al. 1996), California (Moyle 2002), and Puerto Rico, where it is thriving (Erdman 1972, 1984). In the MSRB, redeye

Micropterus coosae, redeye bass, juvenile

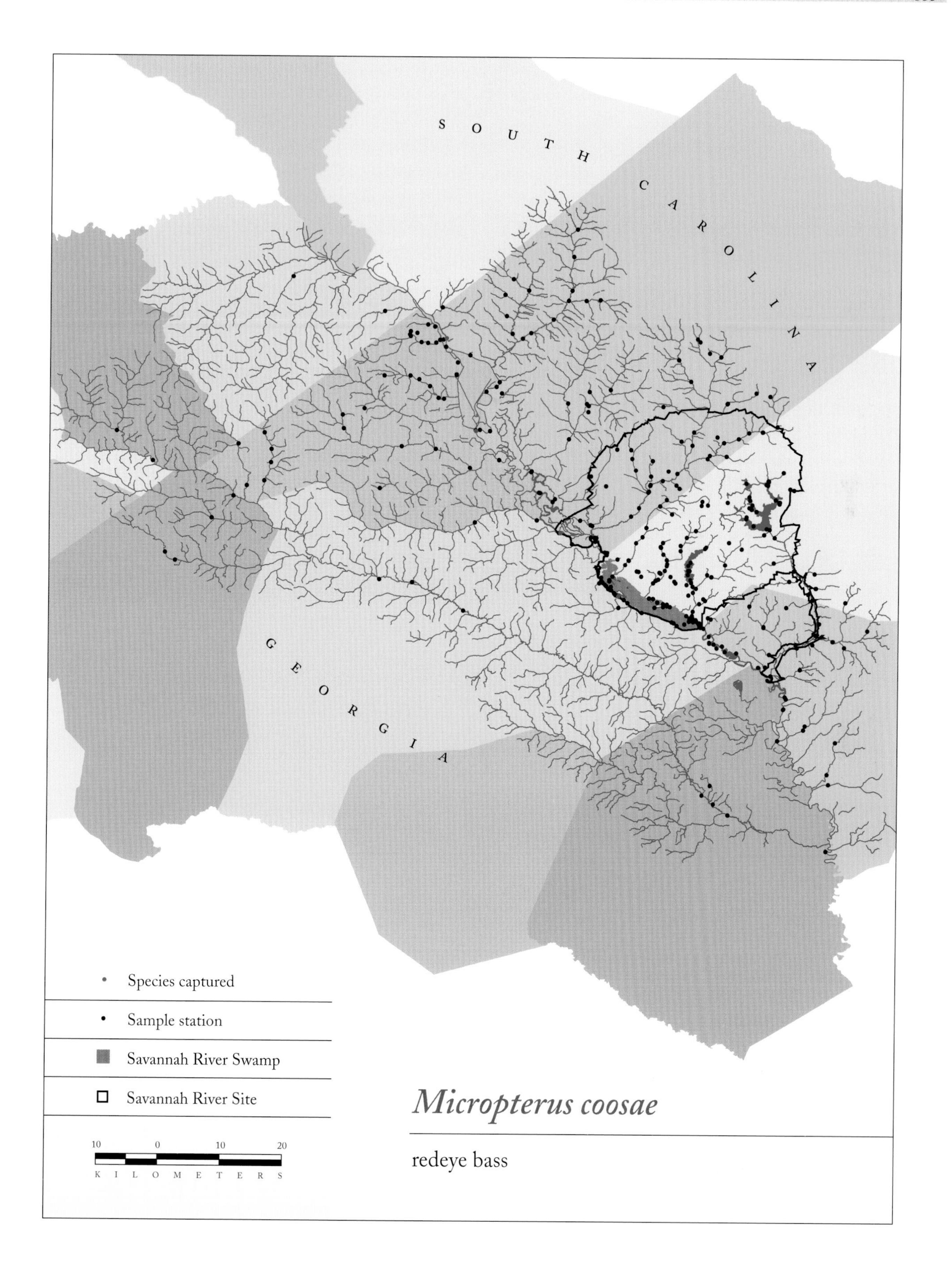
SOUTH CAROLINA
GEORGIA
Species captured
Sample station
Savannah River Swamp
Savannah River Site
10 0 10 20
KILOMETERS
Micropterus coosae
redeye bass

bass are primarily found in the Savannah River, most often in the rapids area, but juveniles have been collected as far downstream as the Steel Creek confluence.

HABITAT Primarily a lotic species that inhabits upland streams and usually remains near the cover of boulders, logs, vegetation, or undercut banks. More abundant in pools with noticeable current in pool and riffle stretches of streams. Occurs mostly in streams too warm for trout and cooler than preferred by largemouth bass. In Tennessee, most commonly established in high-gradient, third-order or smaller streams with extensive canopy coverage and substrate characterized by bedrock, boulders, cobbles, or sand (Pipas and Bulow 1998). Not usually tolerant of impoundments, but has been successfully introduced in some South Carolina reservoirs. In reservoirs, generally found along the shoreline near cover (Burr and Warren 1986).

BIOLOGY Spawning occurs in May, June, and early July in Tennessee (Parsons 1954) and from April to June in Alabama (Mettee et al. 1996). The male constructs a nearly circular nest by fanning a depression in a coarse gravel bottom in shallow water. Redeye bass move into larger rivers in the winter and back into small streams in spring (Parsons 1954). Females may be sexually mature by 122 mm SL (Parsons 1954). Longevity is about 10 years (Etnier and Starnes 1993; Parsons 1954), and 9- and 10-year-old fish from cool streams average 170–211 mm SL (Parsons 1954). Redeye bass appear to grow faster in reservoirs than in streams (Moyle 2002).

This predaceous carnivore feeds on terrestrial insects, crayfish, small fishes, salamanders, and aquatic insects. Insects taken from the surface make up the bulk of the diet (Parsons 1954; Gwinner et al. 1975). When introduced into waters where smallmouth bass occur, the two species sometimes hybridize (Pipas and Bulow 1998). Redeye bass also hybridize with spotted bass in California (Moyle 2002). In impoundments, this species seems unable to compete with other black bass species (Parsons 1954).

SCIENTIFIC NAME *Micropterus* = "small fin" (Greek), referring to torn fin of the type specimen; *coosae* = from the Coosa River.

Micropterus salmoides (Lacépède, 1802)
Largemouth bass

DESCRIPTION Body laterally compressed, shallow, and elongate for a sunfish; standard length usually three times greater than the body depth. Head long and eye large. Mouth very large, oblique, and supraterminal, with posterior edge of maxilla generally extending to or beyond posterior margin of eye. Mouth smaller in juveniles, with maxilla reaching only posterior margin of pupil. Relative mouth size increases with growth. Pectoral fins well rounded and short, when folded forward extending to operculum. Spinous and soft dorsal fins only narrowly joined, with a deep cleft between them. Caudal fin emarginate. Lateral line complete; scales ctenoid.

Dark above with a light or white ventral surface—a common color pattern for open-water fish. Dorsum olive green, fading ventrally. Sides marked with a dark greenish black stripe that is generally complete and is formed by contiguous dark blotches and mottling. Ventrolateral sides below stripe may be mottled with olive, but dark mottling does not form distinct horizontal rows. Breast, belly, and flanks bright white, often pearly. Two or three dark lines radiating from posterior margin of eye; dorsal or middle eye stripe generally darkest and at same elevation as lateral body stripe. Soft dorsal fin mottled in basal half, light dusky in distal half. Pectoral and pelvic fins mostly translucent. Faint mottling sometimes present in mostly translucent anal fin. Ear tab short and dark, but not always distinctive as it blends into the lateral stripe.

Unlike *Lepomis* species, which often develop distinctive sexually dimorphic characters, *Micropterus* species are largely monomorphic. Adult largemouth bass can be sexed by the shape of the scales around the vent, which are generally circular in males and pear shaped in females (Parker 1971), but practice is required to become skilled at using this character.

SIZE Maximum length well over 600 mm TL, and weight over 3 kg in the Savannah River.

MERISTICS Dorsal spines (9)10(11), dorsal rays (11)12–13(14), anal spines (2)3, anal rays (10)11(12), pectoral rays (13)14–15(17); lateral line scales (58)61–65(69) (Jenkins and Burkhead 1993).

DISTRIBUTION The native distribution is from central South Carolina south to southern Florida, westward into northeastern Mexico, and northward through most of the Mississippi Basin west of the Appalachian Mountains into southern Quebec and Ontario (Lee *in* Lee et

Micropterus salmoides, largemouth bass, juvenile

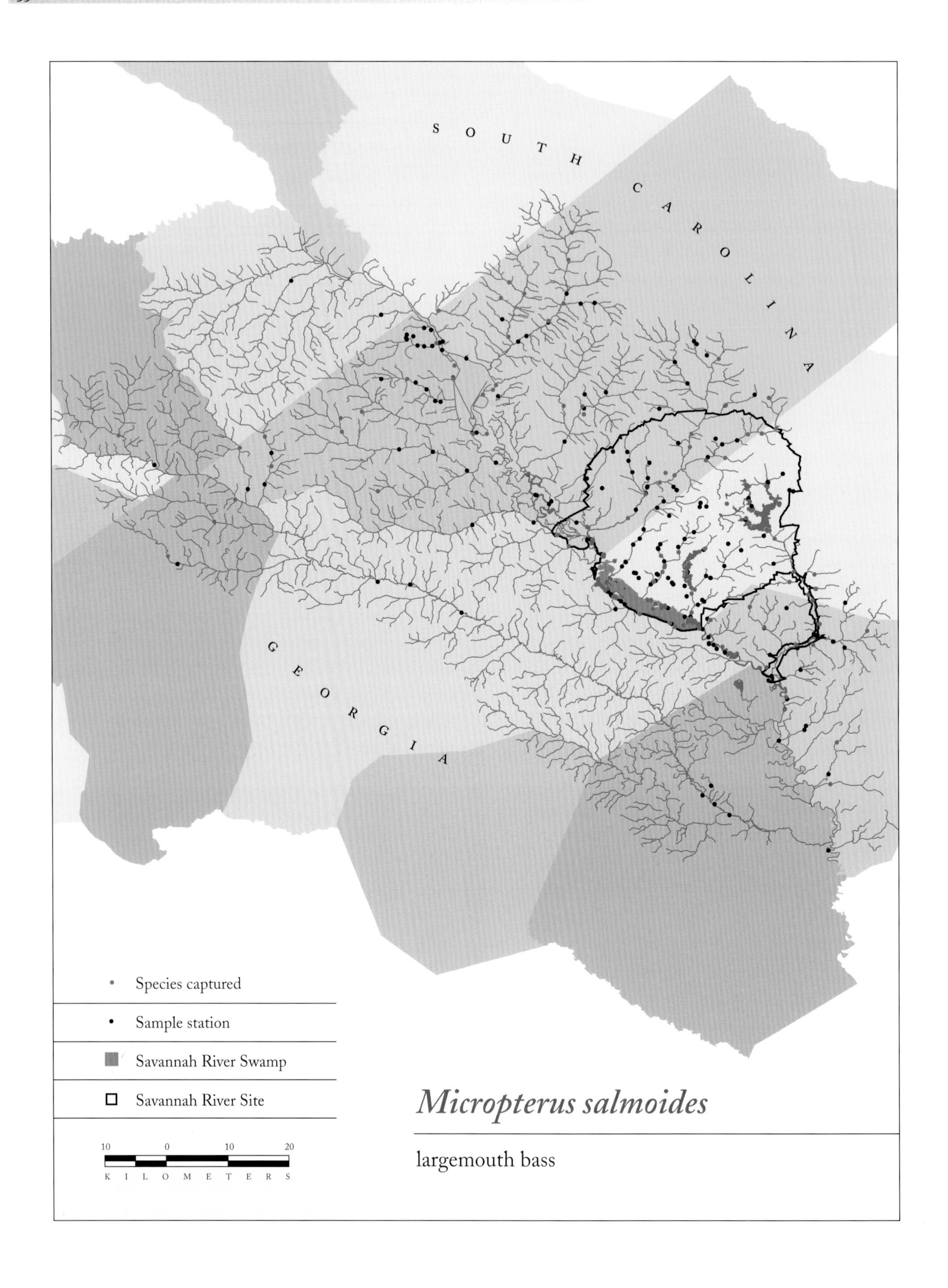

Micropterus salmoides

largemouth bass

al. 1980). Widely introduced into all 48 contiguous U.S. states and southern Canada, Mexico and Central America, South America, and most of Cuba (Lee *in* Lee et al. 1980). Also introduced and established in Hawaii (Brock 1960) and Puerto Rico (Erdman 1984). Worldwide the largemouth bass has been introduced into more than 35 countries (Heidinger 1976). In the MSRB, largemouth bass are most abundant in larger lentic waters including reservoirs and ponds as well as oxbows and backwaters of the Savannah River. Also common in the Savannah River and present in tributary streams, especially below reservoirs discharging warm water.

HABITAT Natural and man-made ponds and lakes, permanent deep pools of small streams, and quiet backwaters of large rivers; replaced by other *Micropterus* species where strong, continuous currents occur (Adams and Hankinson 1932; Pflieger 1975). More tolerant of turbidity and slack current than other *Micropterus* species (Mettee et al. 1996). May frequent deeper water just off bars or ledges bordered by weed beds (Cahn 1927). Occurs over a broad pH range from less than 5 to above 9 (Graham 1993). May occur in salinities up to 15.6 ppt (Swift et al. 1977). During thermal releases into Par Pond, bass tolerated temperatures as high as 36 °C for short periods (Zimmerman et al. 1989).

BIOLOGY In the MSRB, spawning occurs in March and April at 19–23 °C in Steel Creek, L Lake (DEF unpub. data), and Par Pond, but occurred earlier in heated areas of the reservoir (Bennett and Gibbons 1975). Spawning occurs in April and May at temperatures of 17–20 °C in Alabama (Mettee et al. 1996), from mid-April to June in Missouri (Pflieger 1975), and in February and March at 15.6–18.3 °C in Florida (Chew 1974).

The reproductive behavior of largemouth bass in Steel Creek and L Lake was described by DeWoody et al. (2000c). Nests are generally not as well formed as those of *Lepomis* species. Frequently, nests are swept in sand or gravel at the base of logs, stumps, boulders, or emergent macrophytes such as cattails. In the latter case, the fine roots are swept clean and the adhesive eggs are attached to the roots. Tops of boulders, logs, or stumps may be swept clean and used for spawning too. Nests, which are about 70 cm in diameter, are generally placed in water less than 1 m deep, within 2–3 m of the bank. The spawning act is detailed in Adams and Hankinson 1932; and Chew 1974; embryonic and larval development are described in Chew 1974.

Genetic analyses confirmed behavioral observations in Steel Creek and L Lake that largemouth bass were largely monogamous (DeWoody et al. 2000c); only a low level of cuckoldry by males or females was reported. Unlike *Lepomis* species, both parents care for their offspring (Smith 1907; Hankinson 1908; DeWoody et al. 2000c) and vigorously defend them from predators. Generally, the male remains directly over the nest and the female is about 1 m away facing the nest, but at times the female will position herself directly on the nest. The female may take over the primary guarding when a male is removed. Variability in parental care may exist as males have been described defending nests alone (Richardson 1913; Chew 1974; Isaac et al. 1998). Parental protection continues after the schooling juveniles leave the nest. Bass nests may be used as spawning sites by local cyprinids such as the golden shiner (*Notemigonus crysoleucas*) and the taillight shiner (*Notropis maculatus*) (Chew 1974).

Largemouth bass in streams grow more slowly than those in lakes (Robison and Buchanan 1984). Correspondingly, those in small tributaries of the MSRB, such as Fourmile Branch and Pen Branch, are generally smaller (rarely larger than 300 mm SL) than those in reservoirs and in the Savannah River. In Steel Creek below L Lake during the drought conditions of 1999–2003, bass rarely exceeded 350 mm SL. Due to lack of fishing pressure, largemouth bass are three to four times as abundant and 10–30% larger in postthermal Par Pond than in other reservoirs in the southeastern United States, and this large predator population is influencing life histories of smaller fish such as bluegills (Belk and Hales 1993). Largemouth bass in northern latitudes live longer than those in southern populations, but fish in southern populations tend to be larger at a given age (Belk and Houston 2002). Males may mature at a smaller size and younger age than females (Pardue and Hester 1967). Age at sexual maturity varies locally, but females generally mature at age 2 or 3 (Carlander 1977). In a Wisconsin population, largemouth bass averaged 9.9, 28.4, 35.8, 39.4, 46.0, 49.8, 51.6, and 53.3 cm TL at ages 1, 3, 5, 7, 9, 11, 13, and 15 years, respectively (Bennett 1937). Largemouth bass may live 15 years or more (Bennett 1937; Carlander 1977). The oldest largemouth bass on record had 23 annuli in its otolith

(Green and Heidinger 1994). In North Carolina, otoliths were found to be better indicators of age than scales (Besler 1999).

In Hawaii, where largemouth bass and bluegills (*L. macrochirus*) have been introduced in reservoirs, the two species produce an intergeneric hybrid, which appears to be fertile and has persisted since 1963 (Maciolek 1984).

The response of largemouth bass to heated effluents in Par Pond on the SRS has been the focus of extensive research, including studies of movements and activity patterns (e.g., Gibbons and Bennett 1971; Gibbons et al. 1972; Quinn et al. 1978; Block et al. 1984; Zimmerman et al. 1989), feeding and energetics (e.g., Bennett and Gibbons 1972; Rice et al. 1983; Kennedy et al. 1985), and parasitology (e.g., Eure 1976; Esch and Hazen 1978, 1980; Hazen et al. 1978; Huizinga et al. 1979). Largemouth bass in Tennessee streams were relatively sedentary but moved more than syntopic *Lepomis* species (Gatz and Adams 1994). Migrations of largemouth bass in response to water temperature have been recorded in postthermal Steel Creek (Jones 2001).

Largemouth bass are carnivorous predators. In addition to fish they eat crayfish, insects, other crustaceans, frogs, mice, and almost any other animal of appropriate size that has fallen in or is swimming in the water (Pflieger 1975; Wiltz 1993). In the Savannah River, cyprinids make up the bulk of the diet, but at least six additional families of fishes are eaten as well (Wiltz 1993). Stable isotope studies in a Florida lake indicated that largemouth bass there fed mainly on planktivorous fish (Gu et al. 1996). Habitat alteration can influence hunting behavior, and in some habitats largemouth bass hunt in groups around structural cover (Annett 1998). Groups of bass may herd prey such as mosquitofish into coves against the bank before attacking en masse (DEF pers. obs.). Young bass pass through an invertivorous phase, generally in their first summer (Keast 1985b). During this time, small juveniles (13–30 mm TL) feed largely on zooplankton and larger juveniles (31–75 mm TL) feed more on insects, amphipods, and decapods (Chew 1974). Juveniles as small as 23 mm may begin to eat larvae of other fish species (Keast 1985b), and by the time they reach 50–100 mm TL and an age of 0 or 1, 75–95% of their diet may consist of fish (Mittelbach and Persson 1998). Conditions causing slow growth in the invertivore phase, such as competition from high densities of bluegill juveniles, may delay the switch to piscivory until age 2 (Olson 1996; Olson et al. 1995), but growth may increase when piscivory begins and food is plentiful (Olson et al. 1995).

SCIENTIFIC NAME *Micropterus* = "small fin" (Greek), referring to the torn dorsal fin of the type specimen; *salmoides*, from *salmo* = "trout" (Greek), the vernacular name originally applied to this species in southern states, and *oides* = "like" (Latin), hence "troutlike."

Pomoxis annularis Rafinesque, 1818

White crappie

DESCRIPTION Body deep and extremely compressed. Mouth large and oblique, with the maxilla reaching nearly to the posterior margin of the pupil when the mouth is closed. Eye large. Length of dorsal fin base less than the distance from the dorsal fin origin to top of head above posterior margin of eye, and about equal to the anal fin base length. Spinous and soft dorsal fins broadly connected (Morgan 1954); pectoral fins relatively short and rounded; caudal fin emarginate. Teeth present on vomer, palatine, and ectopterygoids. Gill rakers long and slender. Lateral line complete; scales ctenoid.

Dorsum black to dark olive, but may be mottled, especially posteriorly. Sides flashy silver with iridescent sheen that may be bluish, greenish, or silvery (Trautman 1957). Black to dark olive mottling on sides, generally arranged in 8–12 narrow vertical bars; bars described as chainlike, dusky, or black double bands (Beckman 1952). Vague black spot present near edge of opercle (Scott and Crossman 1973). Adults from turbid waters may lack lateral markings. Lighter ventrally, especially on flanks. Dorsal, anal, and caudal fins generally with mottling, especially on soft (rayed) portions. Mottling may fade in preservative, especially in small specimens.

Breeding males are much darker than breeding females (Hansen 1943; Whiteside 1964). Like *Micropterus*, *Pomoxis* species are not as sexually dimorphic as the male nest-guarding *Lepomis*.

Pomoxis annularis, white crappie

SIMILAR SPECIES The black crappie (*P. nigromaculatus*) has a similar general body form but lacks distinct bands on the sides. The dorsal fin of the white crappie generally has fewer spines (5–6[7] versus [6]7–8) than that of the black crappie, and the length of the dorsal fin base is less than the distance from the dorsal fin origin to the eye in the white crappie, but equal to it in the black crappie. Hybridization between black and white crappies has been well documented, and mixing of these species through introduction should be avoided (Travnichek et al. 1996, 1997; Epifanio et al. 1999). Morphological characters are insufficient to accurately identify hybrids (Smith et al. 1995).

SIZE Up to about 530 mm TL (Scott and Crossman 1973). The common size range in South Carolina is between 203 and 355 mm (SCDNR).

MERISTICS Dorsal spines 5–6 rarely 7 (Jenkins and Burkhead 1993) or (4)6(8) (Trautman 1957), dorsal rays (13)14–15 (Jenkins and Burkhead 1993) or rarely 16 (Whitworth et al. 1968; Etnier and Starnes 1993), anal spines 6–7 (Jenkins and Burkhead 1993) or (5)6(7) (Etnier and Starnes 1993; Jenkins and Burkhead 1993), anal rays 16–18 (Etnier and Starnes 1993), pectoral rays usually 13 (Jenkins and Burkhead 1993) or 14–16 (Etnier and Starnes 1993); lateral line scales usually 34–44 (Jenkins and Burkhead 1993) or 34–47; gill rakers 25–32; branchiostegal rays 7 (Etnier and Starnes 1993).

DISTRIBUTION Native from southeastern Ontario and the southern drainages of all Great Lakes except Lake Superior west to Minnesota and south through the Mississippi Valley to the Gulf of Mexico, east to the Appalachian Mountains and west into South Dakota, Nebraska, Kansas, and Oklahoma; along the Gulf Coast west through Texas into northeastern Mexico, possibly native farther west in Kansas and Nebraska and in Georgia (Lee *in* Lee et al. 1980). Introduced and established in parts of all 48 contiguous U.S. states except Idaho and most of New England (Lee *in* Lee et al. 1980). Found in many waters in South Carolina and Georgia, especially in large reservoirs, natural lakes, and backwaters, but not in mountain streams. Most MSRB records are from the Savannah River, but the species appears to be very rare and not well established.

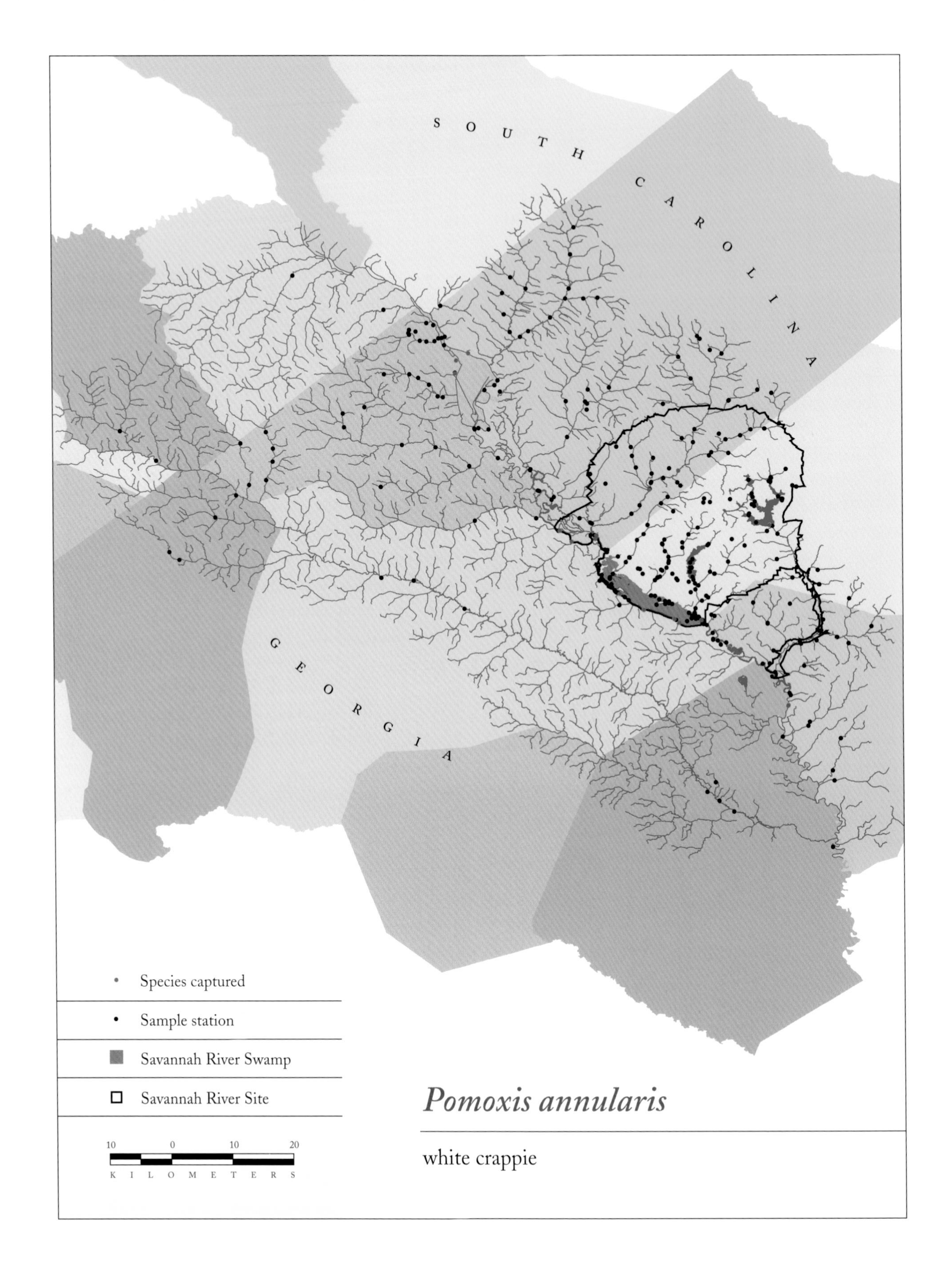

Pomoxis annularis

white crappie

HABITAT Small white crappies may enter small streams, but adults are generally restricted to rivers and reservoirs as reported for Alabama (Mettee et al. 1996). They tolerate turbid water and soft bottoms (Cahn 1927) better than most other sunfishes, including black crappies. Adults are often associated with abundant cover such as submerged brush and heavy vegetation (Webster 1942; Whitworth et al. 1968), and with deep water. Larvae under 10 mm TL are pelagic; larger juveniles move into shallow littoral weed beds (Cahn 1927; Sammons and Bettoli 1998). In the summer, white crappies may be attracted to areas with a steep bottom gradient and an abundance of structure such as stumps, logs, or boulders (Markham et al. 1991). White crappies have been recorded at salinities as high as 6.0 ppt and at a maximum temperature of 29 °C.

CONSERVATION STATUS Probably introduced into the MSRB (Dahlberg and Scott 1971b; Lee *in* Lee et al. 1980; Etnier and Starnes 1993; Rohde et al. 1994).

BIOLOGY Spawning generally begins in early April in South Carolina (Stevens 1959a), and occurs from April through June in Tennessee (Etnier and Starnes 1993) and Arkansas, where nest construction begins at water temperatures above 13.3 °C (Robison and Buchanan 1984). Males make poorly defined, shallow depression nests at water depths between 0.6 and 2.4 m (Siefert 1968), sometimes in aggregations; the adhesive eggs may attach to vegetation (Hansen 1943). Descriptions of eggs, larvae, and juveniles can be found in Ward and Leonard 1954; Faber 1963; and Siefert 1969. Siefert (1969) compared larvae of white and black crappies. Males guard the nests until the fry leave (Etnier and Starnes 1993). Fecundity increases with female size, and egg production was estimated to range from 9,000 to 145,000 in a Pennsylvania pond (Mathur et al. 1979). High water levels prior to spawning enhance reproductive success in tributary impoundments but have less effect on success in main-stem river impoundments (Sammons et al. 2002). Females appear to begin maturing at age 2 (Mathur et al. 1979; Hale 1999). The life span in South Carolina is about 9 or 10 years (SCDNR), although a maximum age of 7 years at 378 mm TL was reported in Lake Moultrie, South Carolina (Stevens 1959a). White crappies lived to 10 years in an Oklahoma lake, but most individuals were less than 5 years old; fish reached 87, 126, 159, 201, 244, 273, 261, 329, 309, 355 mm TL at ages 1–10, respectively (Muoneke et al. 1992). Otoliths are a better indicator of age than scales in Oklahoma and Mississippi (Boxrucker 1986; Hammers and Miranda 1991).

While not a schooling species per se, white crappies may occur in loose aggregations near brush piles and other cover (Robison and Buchanan 1984). They frequently have a crepuscular or nocturnal movement pattern, moving to shallow water at dusk and to deeper water at dawn; the greatest seasonal movements occur in spring before spawning (Markham et al. 1991; Guy et al. 1994).

White crappies in Lake Moultrie eat insects and fish, predominantly threadfin shad (*Dorosoma petenense*) and gizzard shad (*D. cepedianum*) (Stevens 1959a). Individuals less than 150 mm TL feed largely on invertebrates, but adults switch to piscivory, feeding on shad and other forage fish (Mathur and Robbins 1971; Mathur 1972; Muoneke et al. 1992).

SCIENTIFIC NAME *Pomoxis* = "sharp opercle" (Greek), referring to the two flat points on the operculum rather than an ear tab; *annularis* = "having rings" (Latin), referring to the bars on the body.

Pomoxis nigromaculatus (Lesueur, 1829)

Black crappie

DESCRIPTION Body deep and extremely compressed. Mouth oblique and large, posterior edge of maxilla reaching nearly to posterior edge of pupil. Eye large. Length of dorsal fin base long, equal to distance from dorsal fin origin to top of head above posterior margin of eye. Pectoral fins relatively short and narrowly rounded, tip extending to, but not past, pupil of eye when folded forward. Caudal fin emarginate. Fine teeth present on both jaws and palatine (Scott and Crossman 1973). Generally with two small tooth patches on posterior portion of tongue; these may be difficult to see in small individuals but can be felt with a probe. Gill rakers very long, slender, and pointed. Lateral line complete; scales ctenoid.

Dorsum black to dark olive, but may be mottled, especially posteriorly. Dorsolateral and midlateral sides flashy silver, with iridescent shine and generally heavy but variable black to dark olive mottling but no vertical bars. Flanks lighter in color with less mottling than dorsal areas. Dorsal, anal, and caudal fins generally with heavy mottling, especially the soft (rayed) portions. Mottling may fade in preservative, especially in small specimens.

In breeding males, the black, particularly on the head, becomes darker and velvety (Scott and Crossman 1973). Like the *Micropterus* species, *Pomoxis* are not as sexually dimorphic as the male nest-guarding *Lepomis*.

SIMILAR SPECIES The white crappie (*P. annularis*) has a similar general body form but generally has a distinct banded pattern on the sides. The dorsal fin of the white crappie generally has fewer spines (5–6[7] versus [6]7–8) than that of the black crappie, and the length of the dorsal fin base is shorter in the white crappie than in the black crappie (see ratio in key). Hybridization between black and white crappies has been well documented, and mixing of these species through introduction should be avoided (Travnichek et al. 1997; Epifanio et al. 1999). Morphological characters are insufficient to accurately identify hybrids (Smith et al. 1995).

SIZE To more than 350 mm TL (Carlander 1977). Large adults may reach 1.4 kg, but the average is more often less than 0.5 kg (Smith 1907; Stevens 1959a). A black crappie measuring 488 mm TL and weighing 2.3 kg was caught in Lake Moultrie, South Carolina, in the 1950s (Stevens 1959a).

MERISTICS Dorsal spines (rarely 6)7–8, dorsal rays 14–16, anal spines 6–7 (Jenkins and Burkhead 1993), rarely

Pomoxis nigromaculatus, black crappie

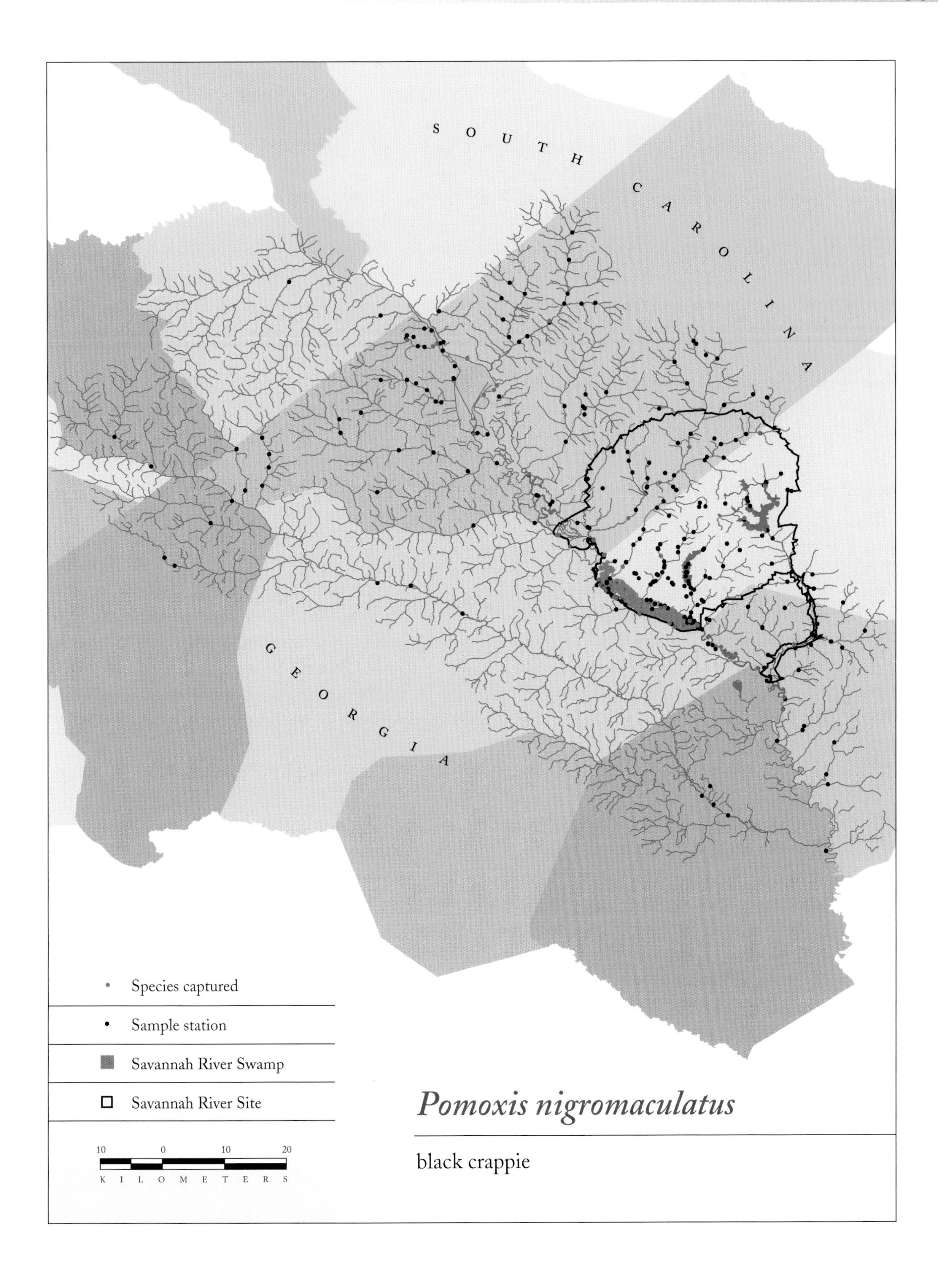
SOUTH CAROLINA
GEORGIA
Species captured
Sample station
Savannah River Swamp
Savannah River Site
10 0 10 20
KILOMETERS
Pomoxis nigromaculatus
black crappie

5 (Etnier and Starnes 1993), anal rays 16–18 (Jenkins and Burkhead 1993) or 16–19 (Etnier and Starnes 1993), pectoral rays 13–15 (Etnier and Starnes 1993; Jenkins and Burkhead 1993); lateral line scales 36–44 (Jenkins and Burkhead 1993) or 35–41; gill rakers 27–32; branchiostegal rays 7 (Etnier and Starnes 1993).

DISTRIBUTION Native east of the Appalachian Mountains from southern Virginia south through peninsular Florida, west along the Gulf Coast in Texas, and north through the Mississippi River drainages into southeastern Manitoba and southwestern Ontario and then east through the Great Lakes into the St. Lawrence Valley; possibly native in western Kansas and Nebraska; introduced and established in parts of all 48 contiguous U.S. states (Lee *in* Lee et al. 1980). In the MSRB, most common in large reservoirs or backwaters of the Savannah River; occasionally found in larger tributaries.

HABITAT Lentic environments more commonly than flowing-water habitats, usually in quiet backwaters of streams or rivers, reservoirs, and larger ponds; less commonly found in muddy sloughs (Adams and Hankinson 1932; Rohde et al. 1994). Found in clear, deeper water over a hard bottom, often near aquatic vegetation or structure (Cahn 1927; Adams and Hankinson 1932). In Florida, found more frequently in oxbows than river habitats (Beecher et al. 1977). In Arkansas, larvae were taken in flooded bottomland oak-hickory forest and in tupelo swamp, but not in the river channel (Killgore and Baker 1996). Larvae below 10 mm TL are pelagic; larger juveniles move into littoral zones (Sammons and Bettoli 1998). Black crappies require clearer and cooler water than do white crappies (Carlander 1977; Robison and Buchanan 1984; Mettee et al. 1996). They may migrate into deeper water in the autumn, returning to the shallows in the spring (Adams and Hankinson 1932). Black crappies occur over a pH range of less than 5 to above 8.5 (Graham 1993). The recorded temperature range is 4.0–32.5 °C (Hardy 1978), and the maximum salinity recorded is 4.9 ppt (Smith 1971). SCDNR constructs and marks bush piles to attract black crappies for recreational fisheries.

BIOLOGY Throughout its range, the black crappie generally spawns at water temperatures of about 15–20 °C, peaking at 18–20 °C (Schneberger 1972 *in* Wang and Kernehan 1979; Jenkins and Burkhead 1993). Spawning begins in March in North Carolina (Smith 1907) and occurs mostly in April in Virginia. Males move into the spawning area and establish and guard territories, but nest construction does not begin until females arrive (Ginnely 1971 *in* Carlander 1977). Nests may be constructed and guarded by the male, as in *Lepomis* (Richardson 1913; Scott and Crossman 1973), or by both parents, as in largemouth bass (*M. salmoides*) (Smith 1907; DeWoody et al. 2000c). The possibility of biparental care should be more closely examined in this species, which, like *Micropterus*, lacks strong sexual dimorphism.

Nests are similar to *Lepomis* nests but not as well defined. They are usually built in shallow water near vegetation, and sometimes are placed in dense groups. Nests are constructed on a variety of substrates, including sand, gravel, and solid objects (Richardson 1913; Adams and Hankinson 1932). Nests in lentic habitats are generally constructed in shallow water (0.4–0.8 m) in areas sheltered from wind and waves (Pope and Willis 1997). The adhesive eggs lie on the substrate or are attached to roots or leaves of aquatic vegetation in the nest (Richardson 1913; Adams and Hankinson 1932). Descriptions of eggs, larvae, and juveniles can be found in Merriner 1971; and Ward and Leonard 1954. Siefert (1969) compared larvae of black and white crappies.

Both sexes generally mature in 2 years in Florida lakes (Huish 1954), and in 2–4 years in Iowa (Erickson 1953). Each female spawns several times and can lay 11,000–188,000 eggs (Jenkins and Burkhead 1993). In the Keowee Reservoir of the upper Savannah River drainage, fecundity increased with female size, and females 3–5 years old measuring 159–281 mm TL contained 6,100–109,900 ova (Barwick 1981). Scott and Crossman (1973) reported an average fecundity of 37,700 for black crappies in Canada.

Few individuals live more than 8 years (Stevens 1959a; Carlander 1977). Fish in Lake Marion and Lake Moultrie, South Carolina, reached 340 and 381 mm TL, respectively, at the maximum age of 8 (Stevens 1959a). In a study of three Georgia reservoirs, black crappies lived up to 5 years, but few were older than 3 (Larson et al. 1991). See Kruse et al. 1993 for a comparison of otoliths versus scales as techniques for aging black crappies, along with a discussion of geographic variation. In Lake Wylie, a lake on the North Carolina–South Carolina border that receives heated effluents, growth rates vary spatially and are positively correlated to water temperature in young-of-the-year, and to prey densities (especially shad) in 1-year-old fish (McInerny and Degan 1991). Reproductive success may be enhanced in some habitats by higher water levels (Miller et al. 1990). Although high water level prior to spawning enhances reproductive success in tributary impoundments, it has less of an effect on reproduction in main-stem river impoundments (Sammons et al. 2002).

Black crappies are nocturnal or crepuscular feeders (Adams and Hankinson 1932; Keast et al. 1978). Larvae feed primarily on copepods, with smaller larvae preying more on nauplii (Schael et al. 1991). Young-of-the-year continue to feed primarily on zooplankton such as copepods and small cladocerans, but add dipteran larvae and pupae on reaching 50 mm TL (Hansen and Qadri 1984). After 1 year the diet switches largely to aquatic invertebrates such as chaoborids or chironomid larvae, but may include an occasional fish as well (Keast 1985b). Black crappies become more piscivorous at 140–180 mm TL and 3–4 years, when 25–60% of the diet may consist of fish (Mittelbach and Persson 1998). A stable isotope study in a Florida lake determined that black crappies fed mostly on planktivorous fish (Gu et al. 1996) such as shads (*Dorosoma cepedianum* and *D. petenense*), as has also been reported for Lake Moultrie, South Carolina (Stevens 1959a).

SCIENTIFIC NAME *Pomoxis* = "sharp opercle" (Greek), referring to the two flat points on the operculum rather than an ear tab; *nigromaculatus* = "black spotted" (Latin).

Percidae (darters and perches)

Most species of the perch family (176) are found in fresh waters of North America, although a few (14) occur in Eurasia. A number of species have yet to be described, especially among the darters. There are 9 or 10 recognized genera, 4 of which occur only in Eurasia (see details in Craig 2000). Three genera are known from the MSRB: *Perca* and *Percina*, each with 1 species, and *Etheostoma*, with 6 species. *Perca* species are generally medium to large fish; the members of the other 2 genera are relatively small darters.

The general percid body form is roughly cylindrical or compressed and moderately long to elongate. The two dorsal fins are separated or only slightly joined, and the first is spinous. The anal fin has one or two spines, and the pelvic fin has one spine. Many species, in particular the darters, exhibit striking sexual dimorphism, especially during the breeding season. Darters can be spectacularly colored, some year-round, others only during the brief breeding season.

All percids are visual predators, but in terms of habitat and life history they can be divided into two groups: the larger perches and pikeperches, and the smaller darters. The first group is found predominantly in larger rivers and lakes. Their morphology is adapted to life in the water column: they have a swimbladder, forked tail, and moderate-sized paired fins. They eat fish or large invertebrates and spawn by broadcasting a large number of eggs individually (*Stizostedion*) or in strands (e.g., *Perca*). The darters are more adapted to life on or near the substrate. They are small (usually <15 cm, often much less) and have only a rudimentary swimbladder and large paired fins. Their diet consists of small aquatic invertebrates. Darters utilize a variety of spawning strategies. Some species bury their eggs in loose sand or gravel; others attach eggs to aquatic vegetation, roots, rocks, or other structures. Species with this reproductive strategy abandon the eggs after spawning. Some species employ an egg-clustering strategy, attaching the eggs in a single layer on the flat underside of an object such as rock or wood. This is the most complex strategy, and it has arisen independently in several subgenera (Page 1983). Males clean the spawning substrate and aggressively defend the nest, to which females are courted. At least in some species, males continue to guard the nest, and clean and aerate the developing offspring within it (Winn 1958; Page 1983; Constantz 1985). Tessellated darters (*Etheostoma olmstedi*) maintain a clean saucer under the nest into which the larvae fall after hatching.

Recreational and commercial fishermen target larger perches. Many species are also sought after by aquarists and photographers because of their bright colors. Their narrow tolerance for environmental disturbance has made darters widely recognized as environmental indicators. See Craig 2000 for further details on many aspects of the biology and exploitation of the Percidae. Kuehne and Barbour (1983) and Page (1983) have published comprehensive information on the American darters in particular.

KEY TO THE SPECIES OF PERCIDAE

1a. Origin of first dorsal fin directly above origin of pectoral fins; mouth large, upper jaw extending to under middle of eye; posterodorsal edge of maxillary not concealed by suborbital (Plate 19 a); caudal fin moderately forked . . . . *Perca flavescens* (yellow perch), p. 390

1b. Origin of first dorsal fin well behind origin of pectoral fins; mouth small, upper jaw not extending beneath middle of eye; posterodorsal edge of maxillary obscured under suborbital (Plate 19 b); caudal fin straight, rounded, or emarginate; TL no larger than 15 cm . . . . darters **2**

[1] **2a.** Ten to 15 elliptical or diamond-shaped (square on caudal peduncle and generally decreasing in width anteriorly) solid black bars on sides, bars often contiguous forming a black lateral stripe; males have enlarged, scutelike, often spiny scales on the belly, 1–3 between the pelvic fins (Plate 19 c) and a single row along the belly (Plate 19 d) . . . . *Percina nigrofasciata* (blackbanded darter), p. 393

2b. No elliptical blotches; bars may be present or absent, but are not elliptical and solid black; no enlarged scales on belly . . . . *Etheostoma* **3**

[2] **3a.** Middle portion of lateral line arching strongly upward (lateral line incomplete) (Plate 19 e, f) **4**

3b. Lateral line straight or only slightly arched at anterior end (may be complete or incomplete) (Plate 19 g, h) **5**

[3] **4a.** Entire posterior margin of preoperculum serrate (Plate 19 i); 2 small, intensely black, vertically aligned basicaudal spots present (Plate 19 l); 13–14 dorsal rays . . . . *Etheostoma serrifer* (sawcheek darter), p. 387

4b. Edge of preoperculum sometimes partially but never entirely serrate (Plate 19 j, k); caudal spot(s) variable, but never 2 intensely black, vertically aligned spots (Plate 19 m); 10–11 dorsal rays . . . . *Etheostoma fusiforme* (swamp darter), p. 376

[3] **5a.** Frenum absent, upper jaw (premaxilla) protractile (Plate 20 a); straw-colored (black in nuptial males) dorsolateral sides with sparsely spaced dark brown W and X markings . . . . *Etheostoma olmstedi* (tessellated darter), p. 384

5b. Frenum present, upper jaw (premaxillary) not protractile (Plate 20 b, c); dorsolateral side with dark brown stripe, bars, or dense mottling **6**

[5] **6a.** Branchiostegal membranes broadly joined at isthmus (Plate 20 d); laterally marked with thin, parallel horizontal lines formed by red or brown (dark in preservative) spots . . . . *Etheostoma inscriptum* (turquoise darter), p. 382

6b. Branchiostegal membranes narrowly or moderately joined at isthmus (Plate 20 e); no narrow horizontal lines formed by red spots **7**

[6] **7a.** Spinous dorsal fin with a red (dark in preservative) band near the margin, band thick in male (Plate 20 h) and thin in female (Plate 20 j); usually a broad dark brown/black dorsolateral stripe extends the entire length of each side of body; right and left stripes are separated dorsally by a lighter tan or gold stripe that extends the length of the dorsum (Plate 20 f); if lateral stripes are faded, dorsal stripe still evident . . . . *Etheostoma fricksium* (Savannah darter), p. 373

7b. Spinous dorsal fin with a red (dark in preservative) band distinctly near the middle of the fin, band thick in male (Plate 20 i) and thin in female (Plate 20 k); no dark dorsolateral stripes and no light dorsal stripe; instead dorsum marked with dark saddles including (in the MSRB) 2 dark rectangular predorsal blotches (Plate 20 g) . . . . *Etheostoma hopkinsi* (Christmas darter), p. 379

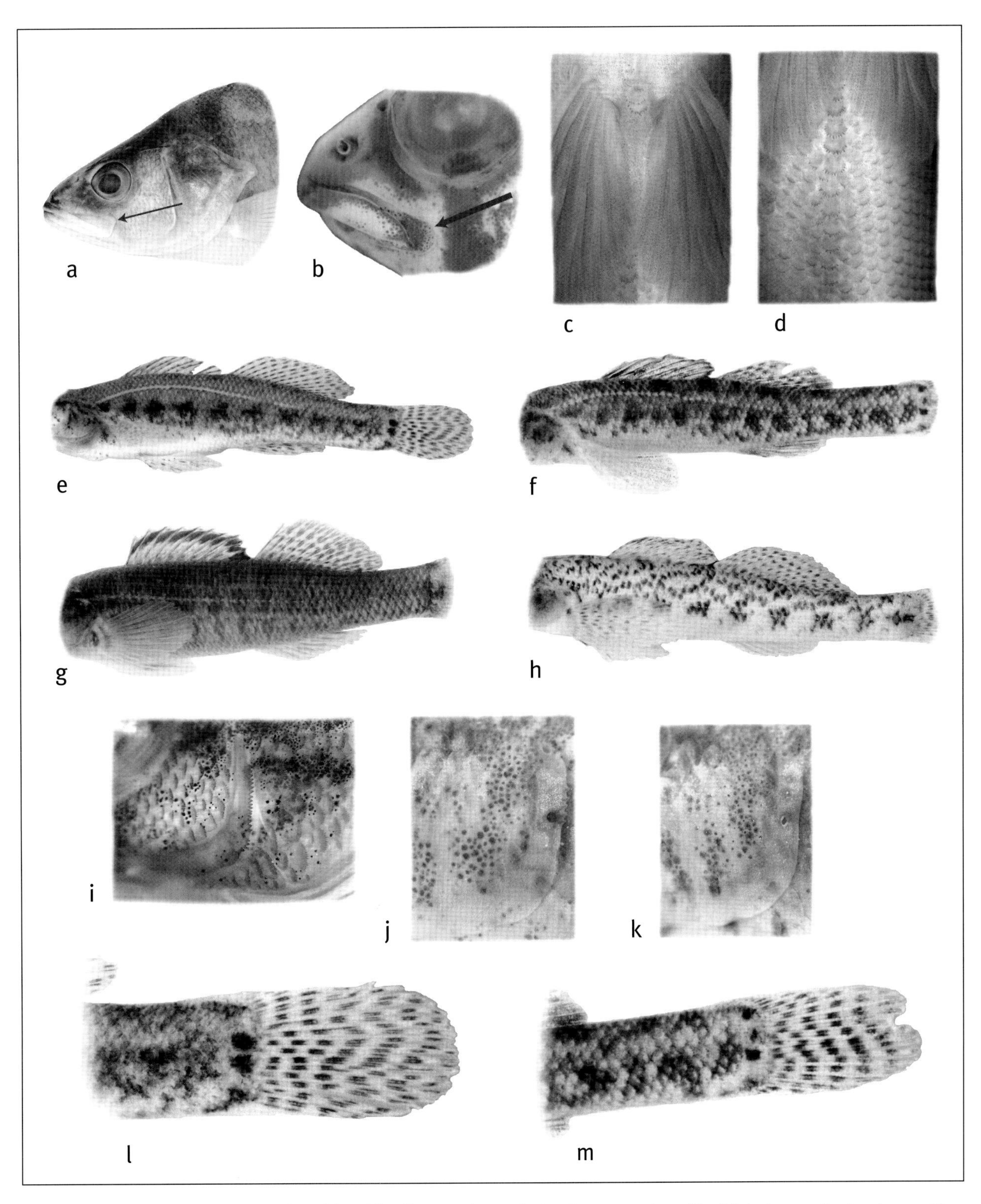

PLATE 19. Morphological characters of Percidae: (a) yellow perch mouth showing exposed maxilla, (b) darter mouth showing maxilla mostly hidden by suborbital, (c, d) ventral scutes and enlarged scales of blackbanded darter, (e, f) darter incomplete lateral line strongly arched in the middle, (g, h) darter lateral lines that are straight or only slightly curved, (i) sawcheek darter cheek showing strongly serrate preopercle, (j) swamp darter with partially serrate preopercular margin, (k) swamp darter with smooth preopercular margin, (l) sawcheek darter tail showing two vertically aligned median spots, (m) swamp darter tail lacking two vertically aligned median spots.

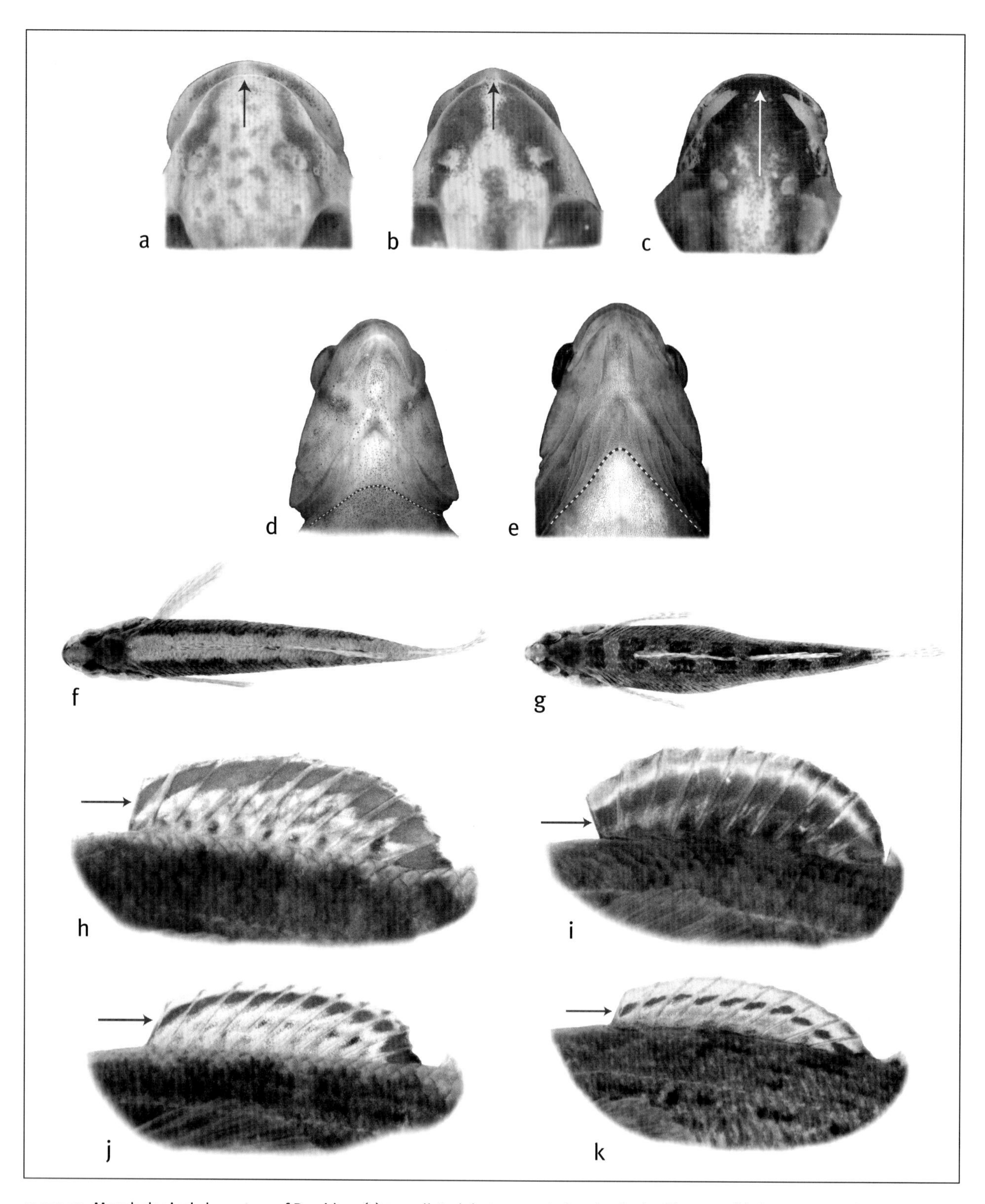

PLATE 20. Morphological characters of Percidae: (a) tessellated darter snout showing lack of frenum, (b) darter snout showing presence of a narrow frenum, (c) darter snout showing presence of a wide frenum, (d) turquoise darter throat showing branchiostegal membrane broadly joined across isthmus, (e) darter throat showing branchiostegal membrane moderately joined across isthmus, (f) light stripe of Savannah darter dorsum, (g) Christmas darter dorsum with dark saddles, (h) male Savannah darter first dorsal fin, (i) male Christmas darter first dorsal fin, (j) female Savannah darter first dorsal fin, (k) female Christmas darter first dorsal fin. For h–k the arrow indicates the location of the red band.

Etheostoma fricksium Hildebrand, 1923

Savannah darter

DESCRIPTION Head small and conical; upper lip with a frenum; two anal spines (Kuehne and Barbour 1983). Gill membranes moderately connected (Page 1983). Lateral line straight and nearly complete or complete, and light tan in life. Cheeks, opercle, and prepectoral area fully or only partially scaled. Belly often scaled but may be unscaled, breast usually unscaled (Kuehne and Barbour 1983; Page 1983).

Sides marked with 8–12 dark vertical bands, bars, or blotches; bar width variable; generally more defined posteriorly; bars may be obscured to varying degrees by the following described dark horizontal stripe. Both sexes with dark horizontal stripe on side; stripe highly variable and sometimes faded; often most distinct dorsolaterally; stripe may be broken by the vertical bars. Distinctive broad tan, light brown, or gold stripe running the length of the dorsum; stripe is sandwiched between the right and left dark lateral stripes; even when the lateral stripes are faded, the dorsal stripe is evident, and is thus an excellent diagnostic character for this species; edges of dorsal stripe scalloped by the dorsal edges of the dark vertical bars (Plate 20 f). Three to four black, irregularly shaped, vertically aligned basicaudal spots. Brown or black suborbital bar usually distinctive but may be faded. Cheek color variable; ranges from intensely mottled with dark markings to markings being mostly absent. Spinous dorsal fin with thin, clear, dusky or light green edge underlain by a thick red band near the fin's margin; interior of fin bluish; the red line is thinner in females and located slightly lower in the fin but is distinctly closer to the edge of the fin than the middle; note that the position of the red band is diagnostic for this species (Plate 20 h, j). Other fins marked with brown wavy lines or mottling. Caudal fin with four to five vertical rows of dark brown spots, forming bands.

Dorsal fin colors intensify in breeding males; caudal and pectoral fins develop a yellow wash whereas pelvic and anal fins acquire green coloration (Layman 1993). Breeding males are also characterized by bright alternating dark or green and orange-red bars on side, dark/green bars extending more dorsally than red bars; width of red bars variable; often posteriorly located red bars encircle belly, but not always; the dark lateral stripe may be superimposed over the colored bars or it may be faded and indistinct. Underside of head and breast light green; general amount of green variable, with some individuals having a green wash over much of the body, especially individuals that have narrow red bars. Coloration of males in general is less intense outside the breeding season. Male genital papilla shaped as triangular flap (Kuehne and Barbour 1983). Coloration of breeding females much more subdued; green largely absent from fins, head, and breast, and red vertical bars on sides are faint or absent (Layman 1993). The female develops a long tubular genital papilla, almost the length of the first anal spine (Kuehne and Barbour 1983; Layman 1993).

SIMILAR SPECIES The variable coloration of Savannah darters has led to their frequent misidentification on the SRS; most frequently misidentified as the Christmas

Etheostoma fricksium, Savannah darter; (left) breeding male, (right) breeding female

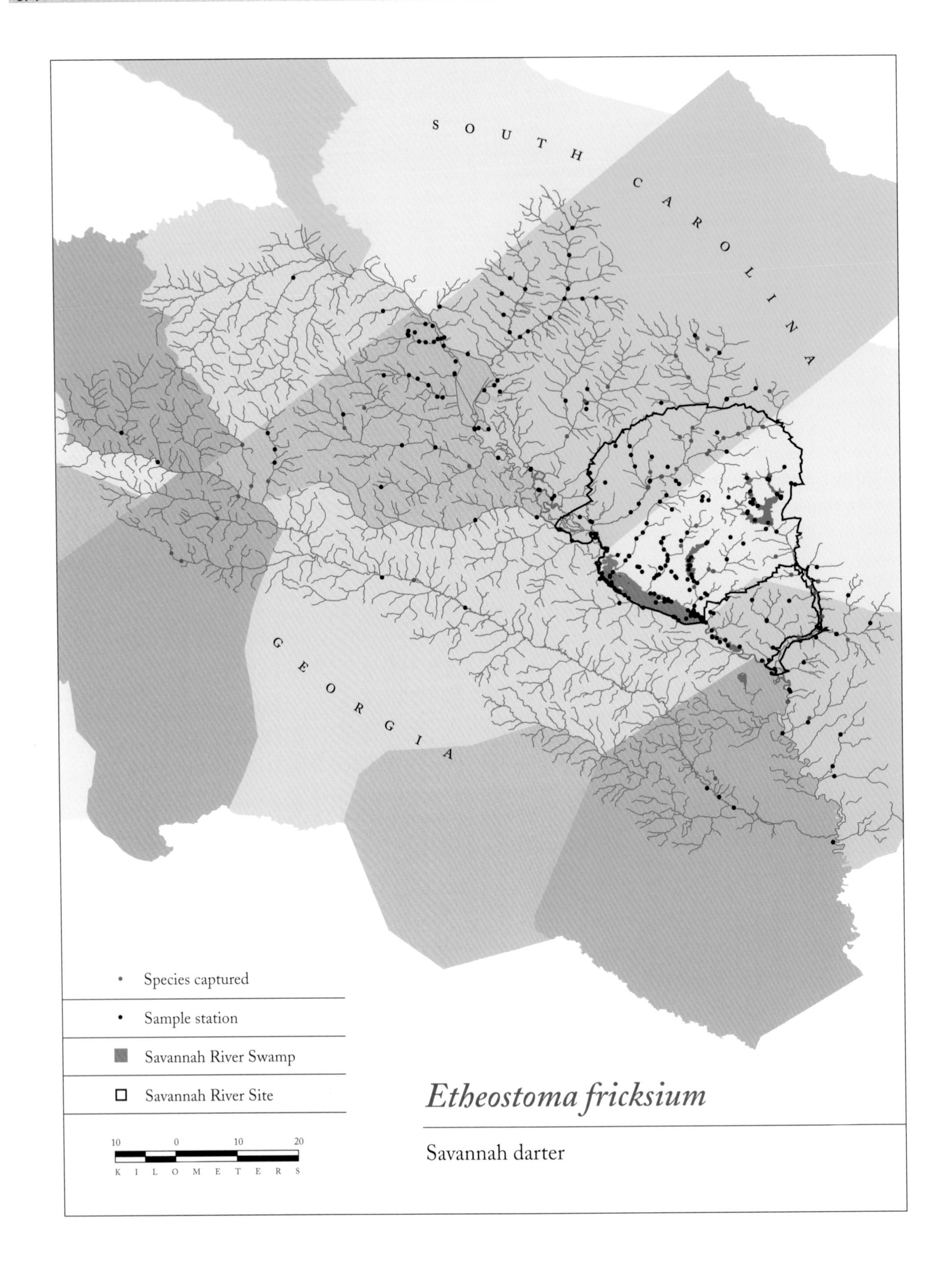

Etheostoma fricksium

Savannah darter

darter (*E. hopkinsi binotatum*). We, like Layman (1993), have been unable to confirm any records of Christmas darters from the SRS. Layman (1993) was also unable to confirm records anywhere in the Savannah River drainage below the Fall Line, but our information is inconclusive. Christmas darters differ in having no dark dorsolateral stripes and no light stripe running the length of the dorsum (Plate 20 f). The dorsum of Christmas darters is instead marked with dark saddles, in the MSRB, including two dark rectangular predorsal blotches (Plate 20 g). Also diagnostic, the red stripe of the spinous dorsal fin is located in the middle of the fin instead of near the edge of the fin as in the Savannah darter (Plate 20 h–k).

SIZE May reach 64 mm SL or more in the MSRB, but most adults are 35–60 mm SL. The males measured by Layman (1993) in Tinker Creek on the SRS were larger than the females.

MERISTICS Dorsal spines (8)10–11(12), dorsal rays (10)12 (13), anal spines 2, anal rays 8–9(10), pectoral rays (12)13–14(15); lateral line scales (35)38–42(45), scales around caudal peduncle (12)15–16(18) (Page 1983).

DISTRIBUTION Restricted to below the Fall Line in the Edisto, Combahee, Broad, and Savannah river drainages of South Carolina and Georgia (Rohde *in* Lee et al. 1980; Kuehne and Barbour 1983). Widespread in smaller tributaries of the MSRB.

HABITAT Clear, small to medium-size creeks, usually with a pronounced current. The preferred sediment type is sand and gravel where logs, sticks, and leafy detritus are present (Bennett and McFarlane 1983; Layman 1993; Rohde et al. 1994).

BIOLOGY The spawning season is from February until the end of May, when water temperatures are between 11 and 23 °C (Layman 1993). The adhesive eggs are buried in fine gravel or sand, and are deposited in clutches of 10–46 eggs; it is unclear how many clutches an individual female produces during a single spawning season (Layman 1993). The females outnumber the males by 1.5 to 1 (Layman 1993). Sexual maturity is reached at the end of the first year, and few individuals live more than 3 years (Layman 1993).

The Savannah darter forages among woody debris and leafy vegetation, preying predominantly on aquatic insects and their larvae (especially chironomids); but terrestrial insects, zooplankton, small snails, and worms are also included in the diet (Bennett and McFarlane 1983; Layman 1993; Sheldon and Meffe 1993).

SCIENTIFIC NAME *Etheostoma*, etymology unclear; literally "strain" and "mouth" (Greek), but stated by Rafinesque to mean "various mouth" because the species known to him at the time had mouths of various shapes (Jordan and Evermann 1896); *fricksium* = after L. D. Fricks, former surgeon of the U.S. Public Health Service, who was in charge of a malaria control program using mosquitofish when the Savannah darter was discovered (Kuehne and Barbour 1983).

Etheostoma fricksium, Savannah darter; (left) nonbreeding individual, (right) nonbreeding individual

Etheostoma fusiforme (Girard, 1845)
Swamp darter

DESCRIPTION Body slender, elongate, and slightly compressed. Head and mouth small, and frenum fairly broad (Kuehne and Barbour 1983). Lateral line incomplete, curving upward anteriorly (see Plate 19 f). Belly, breast, cheek, opercle, and nape fully scaled (Page 1983). Edge of preoperculum sometimes partially but never entirely serrate (see Plate 19 j, k).

Color pattern quite variable, but ground color generally pale whitish or yellowish. Dorsum dark brown to greenish brown. Ventrolateral side often mottled with brown, mottling becoming more intense dorsally. Ventral side light with black and brown specks. Seven to 12 often indistinct dorsal saddles and 6–13 blotches on side, midlaterally. Blotches on adults can be obscured by mottling, and breeding males may be dark throughout (Collette 1962). Clearly defined dark suborbital bar (teardrop) present. Variable number of weakly colored basicaudal spots present, usually three, sometimes four. Fins marked with dark wavy lines or mottling.

Breeding females have conical genital papillae. Breeding males generally darken in color and sometimes have large, dark melanophores in the first dorsal fin that can form a band near the middle of the fin.

Geographic variation within this species is discussed in Collette 1962. Two subspecies, *E. f. fusiforme* and *E. f. barratti*, are recognized. *E. f. barratti*, the local form, ranges north to the Pee Dee River (Collette 1962; Kuehne and Barbour 1983).

SIMILAR SPECIES The swamp darter belongs to the closely related group of "swamp darters" that includes the sawcheek darter (*E. serrifer*) and the Carolina darter (*E. collis*) (Collette 1962). The sawcheek darter has an entirely serrated preopercle and two distinct dark spots on the caudal fin base.

SIZE Can reach up to 40 mm TL in Virginia (Etnier and Starnes 1993), and 46 mm TL in the northern part of its distribution (Schmidt and Whitworth 1979).

MERISTICS In the Savannah River and MSRB, dorsal spines 9–11, elsewhere 8–12(13); dorsal rays (8)9–12, elsewhere (8)9–12(13); anal spines 2, anal rays 6–9, elsewhere (5)6–9(10); pectoral rays 13, elsewhere (12)13–14(15); lateral series scales (45)48–57(63), elsewhere (40)44–57(63); pored lateral scales (11)16–27(37), elsewhere 0–27(37)

Etheostoma fusiforme, swamp darter

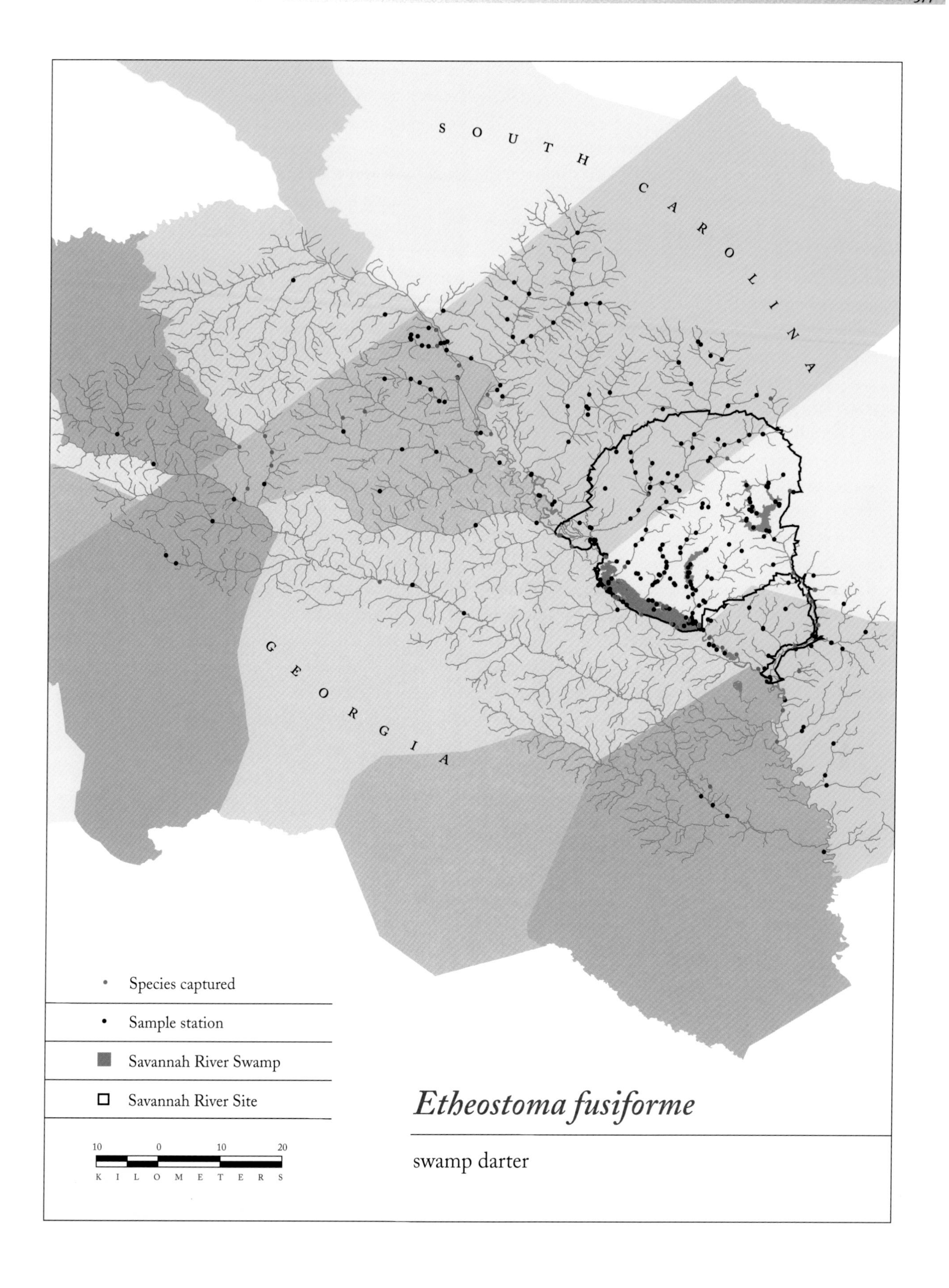

Etheostoma fusiforme

swamp darter

(Collette 1962); scales around caudal peduncle 18–22 (Page 1983). See Collette 1962 for detailed meristic information.

DISTRIBUTION From southeastern Maine southward along the Coastal Plain through peninsular Florida and westward on the Coastal Plain into Louisiana; extending northward and farther westward in the former Mississippi Embayment to Texas, Oklahoma, Missouri, and Kentucky. Introduced populations are present in eastern Tennessee (Norden *in* Lee et al. 1980). In the MSRB, locally abundant in tributary streams and reservoirs, where appropriate habitat abounds.

HABITAT Warm ponds, swamps, sluggish streams, and rivers over mud and sometimes sand, usually in or near heavy vegetation; rarely found in faster-flowing waters (Collette 1962; Rohde et al. 1994). The swamp darter tolerates very low pH conditions but is also found in alkaline waters (Kuehne and Barbour 1983). When found with the sawcheek darter, the swamp darter is more common further back in backwaters, and the sawcheek darter is usually found in the middle of the stream in clumps of underwater vegetation (Collette 1962).

BIOLOGY In the mid-Atlantic region, reproduction probably occurs in February, March, and April, possibly through September (Clemmons and Lindquist 1983). Neither sex establishes a territory (Collette 1962; Rohde et al. 1994). Spawning probably takes place at the surface near floating vegetation or algae, and eggs are deposited singly on leaves of aquatic vegetation (Collette 1962).

The swamp darter has been reported to be an annual species that rarely survives a second fall season (Collette 1962; Schmidt and Whitworth 1979). However, the presence of only a single adult size class may be unusual; most populations probably have some adults that survive a second breeding season (Kuehne and Barbour 1983).

The diet consists of small aquatic invertebrates like copepods and aquatic insects, but may also include larvae and juveniles of other fish species (Collette 1962; Gatz 1979; Schmidt and Whitworth 1979).

SCIENTIFIC NAME *Etheostoma*, etymology unclear; literally "strain" and "mouth" (Greek), but stated by Rafinesque to mean "various mouth" because the species known to him at the time had mouths of various shapes (Jordan and Evermann 1896); *fusiforme* = "spool or coil shaped" (Latin), referring to the long, cylindrical body; *barratti* = after John Barratt, a South Carolina naturalist.

Etheostoma hopkinsi (Fowler, 1945)

Christmas darter

DESCRIPTION Lateral line straight; frenum present; two anal spines. Gill membranes moderately connected. See head and body squamation below.

Body yellow to green dorsally and ventrally. Dorsum marked with eight dark saddles, in the MSRB, including two dark, rectangular predorsal blotches. Side with 10–12 dark bars that are green or bluish green in breeding males. Dark bars separated by orange-red bars in mature males (more distinct in the breeding season), and by yellow bars that may include some reddish orange in females. Spinous dorsal fin with a thick (males) or thin (females) red band located distinctly in the middle of the fin; fin dusky green basally; note that the position of the red band is diagnostic for this species (Plate 20 i, k). Three to four vertically aligned, often irregularly shaped basicaudal spots (Page 1983). Suborbital bar black. Humeral spot well developed (Page 1983). Breeding males develop tubercles on ventral body scales and fins, and the red or yellow coloration intensifies (Kuehne and Barbour 1983; Page 1983). The genital papilla of the male is a short tube; that of the female is slender and moderately long and slightly crenulate (Kuehne and Barbour 1983).

Two subspecies—*E. h. hopkinsi* and *E. h. binotatum*—have been recognized, with *E. h. binotatum* being the form in the Savannah River (Kuehne and Barbour 1983; Page 1983). Although variable, the subspecies can be distinguished by the number of lateral scales (see below) and by *E. h. hopkinsi* usually having fully scaled opercles and prepectoral areas, while the opercles of *E. h. binotatum* are unscaled and the prepectoral areas are unscaled or have only a few scales (Page 1983). Also, *E. h. binotatum* has two well-developed dark, rectangular predorsal blotches that are absent or poorly developed in *E. h. hopkinsi* (Page 1983).

SIMILAR SPECIES The Savannah darter (*E. fricksium*) is most likely to be confused with the Christmas darter; see the Savannah darter account for discussion and distinguishing characters.

Etheostoma hopkinsi, Christmas darter; males

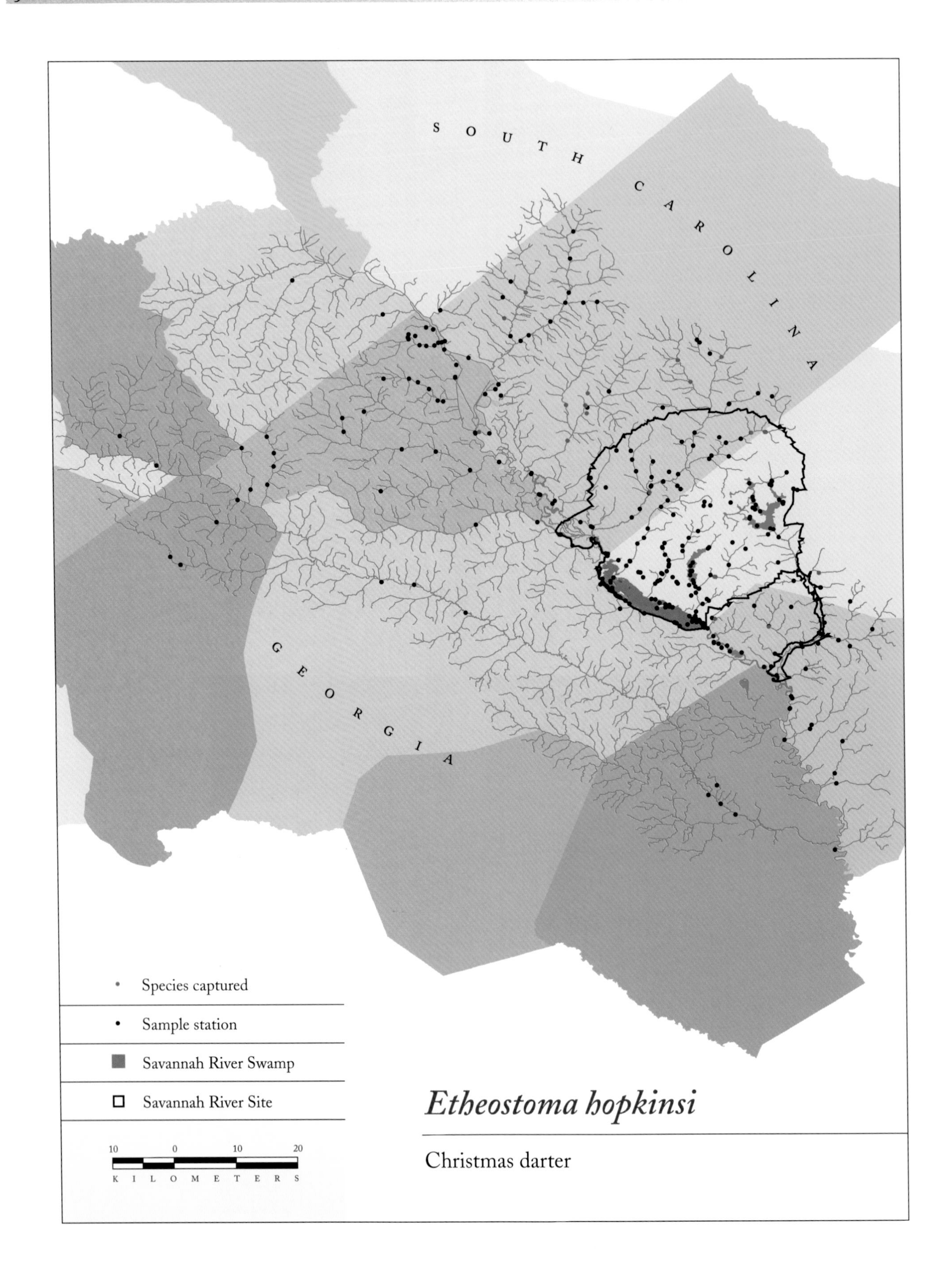

Etheostoma hopkinsi

Christmas darter

Etheostoma hopkinsi, Christmas darter; female

SIZE Adult size 30–60 mm (Rohde *in* Lee et al. 1980; Kuehne and Barbour 1983).

MERISTICS Dorsal spines (8)10–11(12), dorsal rays 11–13(14), pectoral rays (11)12–13(14), anal spines 2, anal rays 7–9(10) (Page 1983). Lateral series scales for *E. h. binotatum* (39)41–49(52) with the last 8–13 scales unpored, and for *E. h. hopkinsi* (39)40–47(50) and (0)2–11(14), respectively (Kuehne and Barbour 1983). Scales around caudal peduncle (14)16–19(20).

CONSERVATION STATUS Listed by the state of South Carolina as a species of concern.

DISTRIBUTION *E. h. hopkinsi* is found in the Altamaha and Ogeechee river drainages of Georgia, while *E. h. binotatum* is found in the Savannah River drainage of Georgia and South Carolina (Rohde *in* Lee et al. 1980). Found above and below the Fall Line (Page 1983). In the MSRB, which lies mostly below the Fall Line, it appears to be more commonly collected in upland creeks near the Fall Line. However, the distribution of this species is problematic because of frequent misidentification of Savannah darters as Christmas darters (Layman 1993). We, like Layman (1993), could confirm no records of Christmas darters from the SRS. Layman also could confirm no records anywhere in the Savannah River drainage below the Fall Line, but our information is inconclusive.

HABITAT Generally found in shallow riffles of clear, cool creeks and medium-size streams with a medium to fast current, sandy substrate, and aquatic vegetation and stones that provide hiding places (Kuehne and Barbour 1983; Page 1983; Rohde et al. 1994). Occasionally found in springs (Rohde *in* Lee et al. 1980).

BIOLOGY The biology is largely unknown. Ripe females were found in the Ocmulgee River in mid-March (Kuehne and Barbour 1983). The Christmas darter feeds predominantly on larvae of small aquatic insects (Rohde et al. 1994).

SCIENTIFIC NAME *Etheostoma*, etymology unclear; literally "strain" and "mouth" (Greek), but stated by Rafinesque to mean "various mouth" because the species known to him at the time had mouths of various shapes (Jordan and Evermann 1896); *hopkinsi* after N. M. Hopkins Jr., the original collector; *binotatum* = "two spotted" (Latin), referring to the two saddles on the back (Kuehne and Barbour 1983).

Etheostoma inscriptum (Jordan and Brayton, 1878)
Turquoise darter

DESCRIPTION Snout blunt (Bennett and McFarlane 1983). Lateral line straight. Gill membranes broadly connected. Six (rarely five) dark black-brown saddles along the back and six often less distinct dark blotches on the side; these may be darker in males.

Ground color light brown dorsally fading to pale ventrally. Side marked with thin horizontal brown lines, more distinctive in males. First dorsal fin with red marginal band and dusky base. Tips of dorsal spines and immediately posterior membranes white. Suborbital bar (teardrop) black and generally prominent.

Breeding males have prominent red spots on the side, red on the opercle and cheek (Page 1983); green bars on the sides, between the dark blotches, that extend to but do not encircle the belly (Richards 1966); and blue on the lower part of the head and pelvic and anal fins (Rohde et al. 1994). The male's genital papilla is small, the female's is a moderately long tube that is grooved at the tip (Kuehne and Barbour 1983).

SIZE May reach up to 66 mm SL (Page 1983) or 78 mm TL (Rohde et al. 1994).

MERISTICS Dorsal spines (8)9–11(12), dorsal rays (9)10–12(13), pectoral rays (13)14–15(16), anal spines 2, anal rays (6)7–9(10) (Kuehne and Barbour 1983; Page 1983); lateral line scales (39)44–53(61), usually all pored; scales around caudal peduncle (13)16–17(20) (Page 1983).

CONSERVATION STATUS Listed in North Carolina as a species of special concern.

DISTRIBUTION Common in medium-size Piedmont and lower Blue Ridge streams of the Altamaha and Savannah river drainages of North Carolina, South Carolina, and Georgia. Uncommon in the upper Ogeechee and Edisto drainages, and rare on the Coastal Plain (Starnes *in* Lee et al. 1980). The spotty distribution reflects habitat availability. In the MSRB, rare below the Fall Line (Kuehne and Barbour 1983) and reported from only a few streams, such as the middle reaches of Upper Three Runs and Tinker Creek on the SRS.

HABITAT Riffles of clear, larger creeks and small rivers with gravel and stones (Kuehne and Barbour 1983; Baker 2002). In streams on the SRS, regularly collected from around woody debris on the bottom of sandy runs of intermediate-size streams.

BIOLOGY The biology is poorly known. Spawning appears to occur from March to early June. Males have breeding tubercles from late March to early June (Richards 1966). Females collected in May near Athens, Georgia (Kuehne and Barbour 1983), and in a tributary of the Alcovy River of Georgia (Baker 2002) had maturing or yolky eggs.

The diet consists predominantly of aquatic insect lar-

Etheostoma inscriptum, turquoise darter; (left) male, (right) female

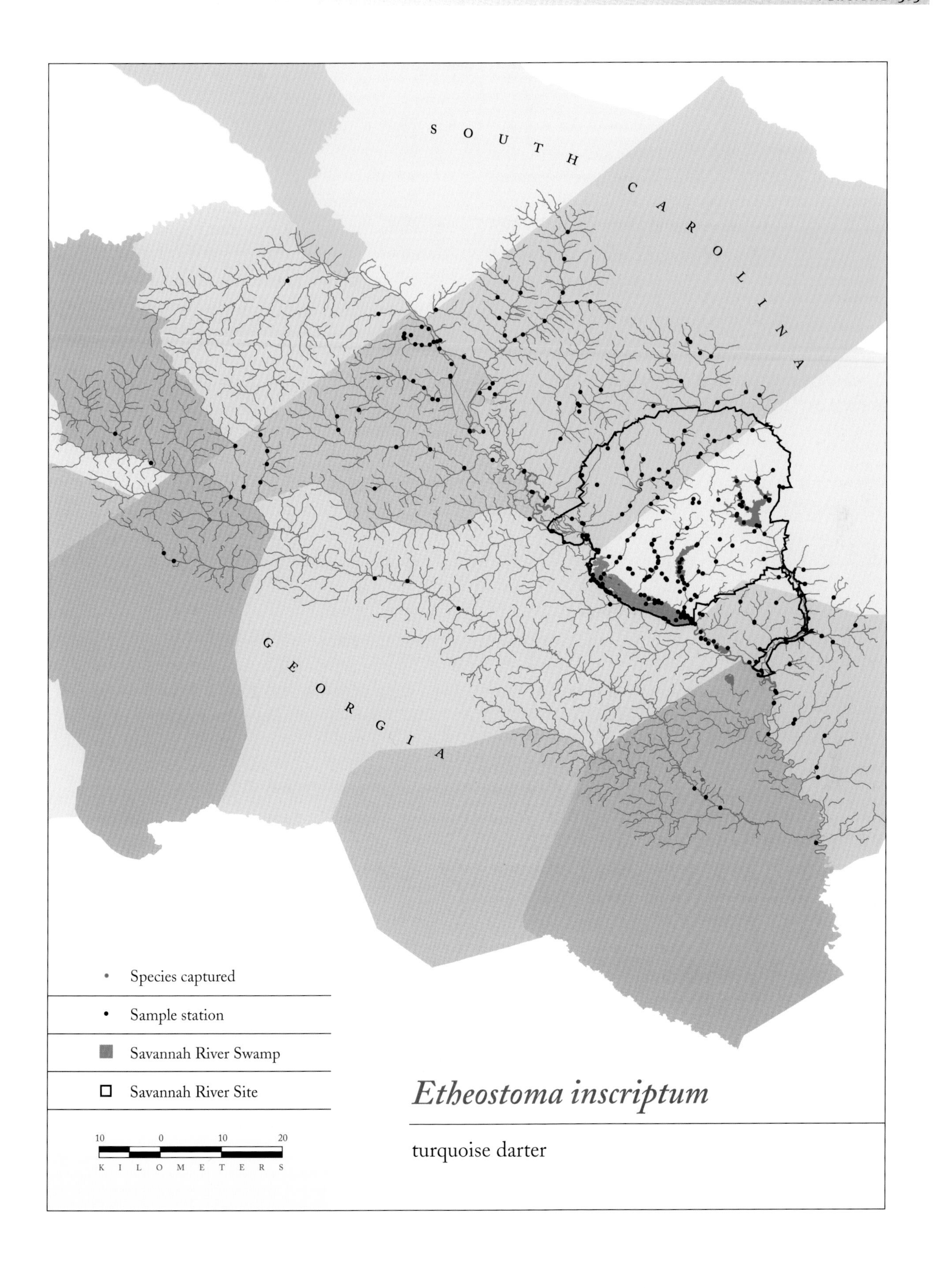

Etheostoma inscriptum

turquoise darter

vae and nymphs, with chironomid larvae being the most important food item (Baker 2002).

SCIENTIFIC NAME *Etheostoma*, etymology unclear; literally "strain" and "mouth" (Greek), but stated by Rafinesque to mean "various mouth" because the species known to him at the time had mouths of various shapes (Jordan and Evermann 1896); *inscriptum* = "written on" (Latin), referring to the lines on the body of mature males (Kuehne and Barbour 1983).

Etheostoma olmstedi Storer, 1842

Tessellated darter

DESCRIPTION Body elongate and not very compressed except toward the caudal fin. Snout relatively blunt. Mouth subterminal, lacking a frenum. Lateral line straight and complete. Operculum with a prominent spine. Gill membranes narrowly connected (Page 1983).

Straw-colored above fading to white ventrally. Dorsum marked with six dark brown saddles. Sides with dark brown vermiculations and X- or W-shaped markings. Midlateral markings may be arranged into 9–11 blotches. Preorbital bar moderately developed, extending downward and forward but failing to meet its counterpart on the lip (Kuehne and Barbour 1983). Suborbital bar prominent; first dorsal fin with a dark spot at the front.

Breeding males become very dark, particularly the head, anal fin, and pelvic fins, obscuring the markings on the body. The dark males may be difficult to see on the silty substrate that generally surrounds their nests. Males lack breeding tubercles (Kuehne and Barbour 1983).

SIMILAR SPECIES The tessellated darter has uninterrupted infraorbital canals, while the Johnny darter (*E. nigrum*) has interrupted infraorbital canals. Other characters that distinguish the two species are given in Cole 1967 (table 1). We are not aware of any Johnny darters being collected in the MSRB, but this species has been introduced outside its native range (Page and Burr 1991). Tessellated darters in the geographic range that includes the MSRB differ from those in other geographic areas by having two anal spines and virtually no scales on the belly (Cole 1967).

SIZE Up to 110 mm TL (Rohde et al. 1994), 88 mm SL (Raney and Lachner 1943). Although adults in the MSRB are generally smaller than 70 mm SL, we have collected individuals as large as 89 mm SL.

MERISTICS For *E. o. maculaticeps* from the Savannah River: dorsal spines (5)8–10(11), dorsal rays (10)12–14(16), pectoral rays (11)12–13(14), anal spines (1)2 (1 in other parts of the range), anal rays (6)7–8(9) (Cole 1967); lateral line scales (40)46–57(58), scales around caudal peduncle (13)15–21(24).

DISTRIBUTION Atlantic slope drainages from southern Quebec south to the Altamaha River of Georgia, with a disjunct population in the lower Oklawaha River, Florida. South of the Susquehanna River, confined to the

Etheostoma olmstedi, tessellated darter; (left) male, (right) adult or large juvenile

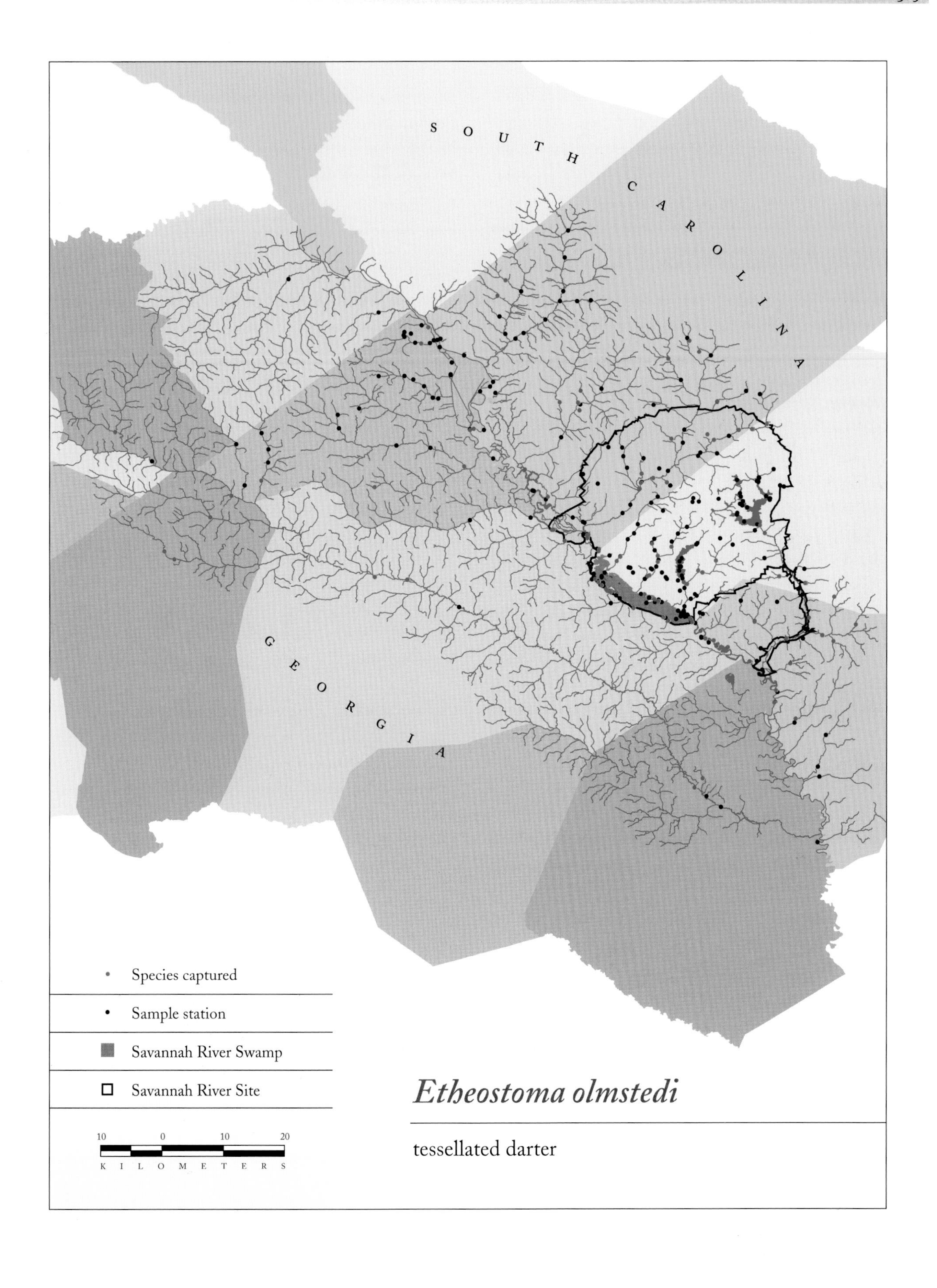

Etheostoma olmstedi

tessellated darter

Coastal Plain except in the Potomac, Rappahannock, York, Roanoke, upper Cooper, and Pee Dee river drainages (Lee and McAllister *in* Lee et al. 1980). Occurs in Lake Ontario but not in the other Great Lakes (Cudmore-Vokey and Crossman 2000). *Etheostoma o. maculaticeps* occurs from Cape Fear, North Carolina, to the St. Johns River, Florida (Cole 1967). It is one of the most common darters in lotic waters of the MSRB.

HABITAT Flowing small streams over a variety of sediments; may be particularly abundant in shallow, sandy runs with a moderate current.

BIOLOGY Reproduction occurs during a 7-week period from late March through June (Gale and Deutsch 1985). In the MSRB, peak spawning appears to be in March and April. Males guard a territory with a nest that is usually located under woody debris, rocks, or other objects in a pool or relatively still eddy. Nests are most easily located by searching for a clean saucer depression below an object with a flattened underside. The edges of the saucer often stand out against the surrounding silty substrate. The saucer is swept clean by the male, as is the undersurface (spawning surface) on which the eggs are to be laid. Gale and Deutsch (1985) summarized the spawning behavior based on their observations and those of others: spawning occurs during the day; eggs are placed in a single layer on the undersurfaces of stones and other submerged objects where they are guarded, aerated, and cleaned by the male. Nest-guarding males eat apparently healthy embryos from their nests, and genetic analyses confirmed filial cannibalism (DeWoody et al. 2001). The hatchlings drop into the cleaned saucer. Nesting and non-nesting males may make excursions to other nests where they attempt, frequently successfully, to fertilize eggs (DeWoody et al. 2000b). Multiple females spawn in each nest (DeWoody et al. 2000b). Some breeding male tessellated darters have white nobs on the tips of the pectoral and pelvic fin rays. While the breeding male is inverted during prespawning displays and spawning, these fins are in close proximity to the nest surface. The nobs are thought to have developed to protect the eggs from the sharp points of the rays, but may have later developed into egg mimics that may stimulate females to spawn (Page and Bart 1989). Females can lay up to eight clutches of eggs, each clutch consisting of 19–324 eggs (Gale and Deutch 1985). See Hardy 1978 for a description of eggs and larvae. Tessellated darters reach maturity after 1 year and typically live for 2–3 years; rarely they may live up to 4 years (Tsai 1972; Layzer and Reed 1978; Schultz 1999a).

Tessellated darters feed during the day, predominantly in the morning hours, preying on zooplankton and small aquatic insects and their larvae (Layzer and Reed 1978). In the MSRB, they feed primarily on aquatic insects, especially chironomid larvae (Sheldon and Meffe 1993).

SCIENTIFIC NAME *Etheostoma*, etymology unclear; literally "strain" and "mouth" (Greek), but stated by Rafinesque to mean "various mouth" because the species known to him at the time had mouths of various shapes (Jordan and Evermann 1896); *olmstedi* = after the naturalist Charles Olmstead, who discovered the species.

Etheostoma serrifer (Hubbs and Cannon, 1935)

Sawcheek darter

DESCRIPTION Body somewhat compressed. Snout blunt with a frenum. Preopercle with completely serrated posterior edge (hence the common name). Lateral line pale yellow and incomplete, arching slightly upward anteriorly. Gill membranes narrowly joined. Cheek, opercle, and breast fully scaled.

Brownish above; may have 7–11 generally vague dorsal blotches, more distinct in very small individuals (20 mm SL). Sides brownish and marked with a midlateral row of 10–12 irregular dark blotches that may be partially fused. Belly lighter, greenish with black flecks. Four vertically aligned dark spots at the base of the tail, the middle pair—in some individuals partially fused—vertically centered on the tail, intensely black, and noticeably more distinct than the outer two spots. Suborbital bar often present in specimens from the MSRB, but may be faint or absent (Page 1983). Soft dorsal, caudal, and anal fins marked with brown tessellations.

Nonbreeding males are more intensely colored than nonbreeding females. Breeding females have flattened bilobed genital papillae. The basal portion of the first dorsal fin of breeding males is almost solid black, and the fin membranes may have large dark or reddish melanophores (Collette 1962). Breeding males have tubercles on their anal and pelvic rays (Jenkins and Burkhead 1993).

SIMILAR SPECIES The sawcheek darter belongs to the closely related group known as "swamp darters" that also includes the swamp darter (*E. fusiforme*) and the Carolina darter (*E. collis*) (Collette 1962). The preopercle of the swamp darter may be partially serrate or lack serrations but is never completely serrated; the swamp darter also usually lacks the two distinct, vertically aligned median caudal spots.

SIZE May reach 57 mm in length (collected in South Carolina); females grow to larger sizes than males (Collette 1962).

MERISTICS Dorsal spines (9)10–12(13), dorsal rays (11)13–14(17), anal spines 2, anal rays 5–7(9), pectoral rays (11)12 (13) (Collette 1962); lateral series scales (44)50–58(66), pored scales (20)28–38(45); scales around caudal peduncle (23–27) (Page 1983).

DISTRIBUTION The Atlantic Coastal Plain from the Dismal Swamp of Virginia south to the Altamaha River of Georgia. Occurs above the Fall Line only in the Cape

Etheostoma serrifer, sawcheek darter

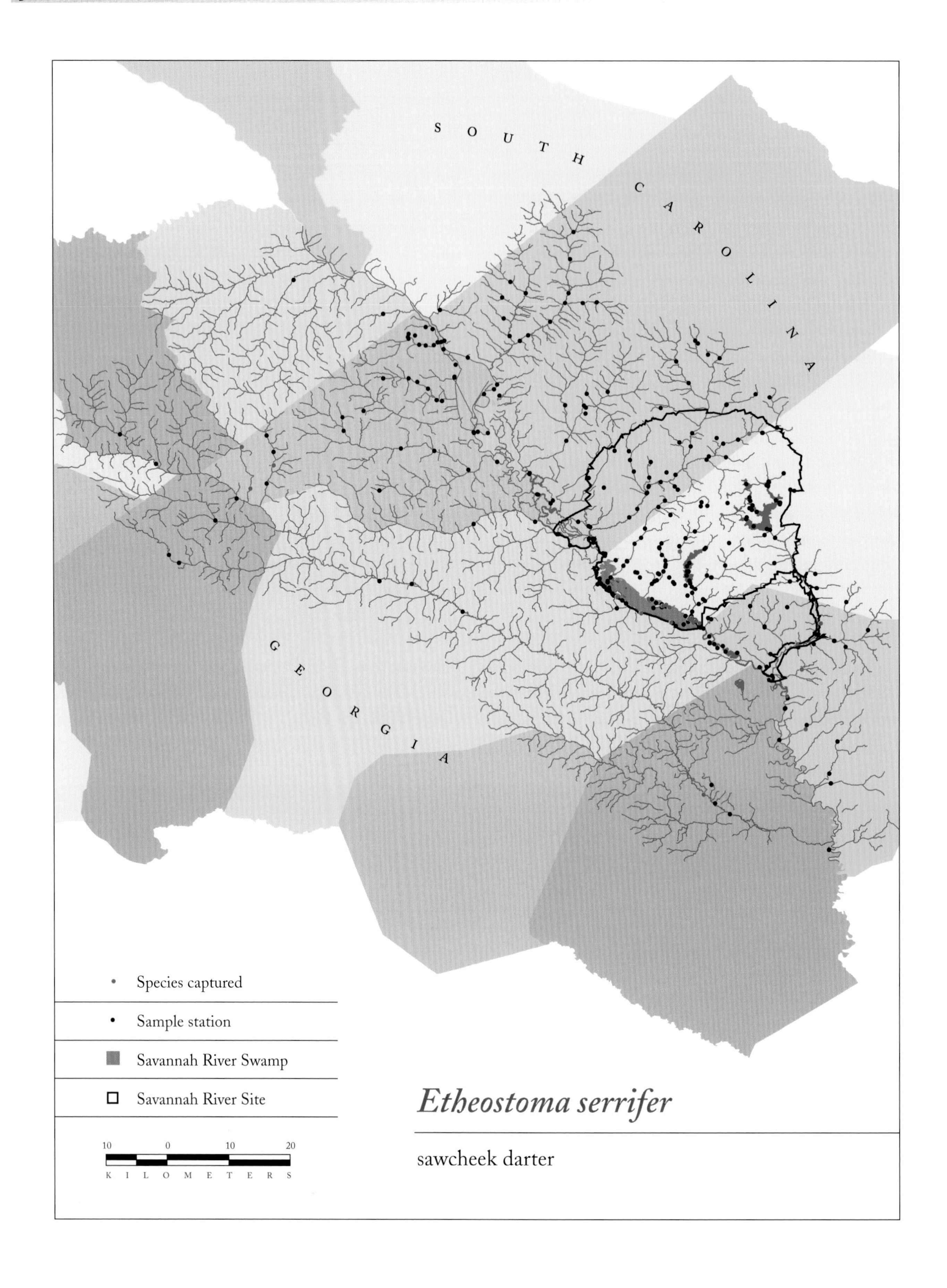
SOUTH CAROLINA
GEORGIA
Species captured
Sample station
Savannah River Swamp
Savannah River Site
10 0 10 20
KILOMETERS
Etheostoma serrifer
sawcheek darter

Fear drainage of North Carolina (Hocutt *in* Lee et al. 1980). Although not widespread in the MSRB, it is regularly found where appropriate habitat occurs. Because of its habitat specificity, it may have a patchy distribution in tributary streams.

HABITAT Warm, slow-flowing streams, swamps, and ponds; typically associated with darkly stained Coastal Plain waters with sediments consisting of sand, mud, and detritus, often littered with sticks and stumps (Jenkins and Burkhead 1993). In the MSRB, most often associated with backwaters, beaver ponds, swamps, and sluggish braids of tributaries. Throughout its range it may inhabit small to medium-size creeks with a more moderate current over sand to gravel with submerged vegetation (Collette 1962; Rohde et al. 1994). When it occurs with the swamp darter, the sawcheek darter is usually found in the middle of the stream in clumps of aquatic vegetation, while the swamp darter prefers the backwaters (Collette 1962).

BIOLOGY Reproduction probably takes place in late March and April, based on the presence of breeding tubercules (Collette 1962) and gravid females in mid-March (Kuehne and Barbour 1983). Gravid females have been observed in Fourmile Branch in the SRS in early to mid-spring. No territorial behavior has been observed (Collette 1962).

The sawcheek darter preys voraciously on a variety of small aquatic invertebrates (Collette 1962; Jenkins and Burkhead 1993).

SCIENTIFIC NAME *Etheostoma*, etymology unclear; literally "strain" and "mouth" (Greek), but stated by Rafinesque to mean "various mouth" because the species known to him at the time had mouths of various shapes (Jordan and Evermann 1896); *serrifer* = "serrated" or "saw bearing" (Latin), referring to the edge of the operculum. Robins et al. (1991) rejected the recently used species name *serriferum* and restored the original spelling: *serrifer*.

Perca flavescens (Mitchill, 1814)

Yellow perch

DESCRIPTION Body deep and slab sided with noticeably separated dorsal fins. Mouth relatively large; preopercle strongly serrated.

Ground color brownish green dorsally fading to bronze on sides and to whitish or yellowish bronze ventrally. Dorsum marked with seven dark saddles that extend down onto sides forming vertical bars that gradually narrow and fade toward the belly. Additional saddles restricted to the dorsum are generally present at the posterior base of the spinous dorsal fin and at the posterior margin of the head. The first (spinous) dorsal fin has a dark blotch on the posterior end. Pectoral, pelvic, and anal fins red to yellow-orange; dorsal and caudal fins generally dusky.

SIZE Can reach 400 mm TL but is much smaller (290 mm) on the SRS (Bennett and McFarlane 1983).

MERISTICS First dorsal spines (11)13–15, second dorsal spines 1–2(3), dorsal rays 12–15(16), anal spines 2, anal rays 6–8(9), pectoral rays (13)14–15(16), pelvic spine 1, pelvic rays 4–5; lateral line scales 51–70 (Thorpe 1977; see also Craig 2000).

CONSERVATION STATUS Introduced in the Savannah River drainage.

DISTRIBUTION The native range is from the Santee River of South Carolina north along the Atlantic Coast to Labrador; the Great Lakes; most of Canada to Saskatchewan, Northwest Territories, and Yukon; and south through the Dakotas to northern Missouri and southern Illinois. Widely introduced in most of the 48 contiguous U.S. states, though no self-sustaining populations are known in Louisiana, Mississippi, Arkansas, or Oklahoma (Lee *in* Lee et al. 1980). A small, apparently relict population occurs in the Mobile Delta of Alabama (Mettee et al. 1996). Introduced into the MSRB, where it is most common in the Savannah River and larger reservoirs such as Par Pond; it may occasionally be found in the Savannah River Swamp and larger tributaries, especially those portions on the Savannah River floodplain.

HABITAT Lakes and streams with clear, slowly flowing water; often found near aquatic vegetation.

BIOLOGY Spawning occurs March–May along sand and gravel shorelines in areas with vegetation and submerged brush. Females are pursued by as many as 25 males and typically produce 15,000–25,000 eggs, although large fe-

Perca flavescens, yellow perch

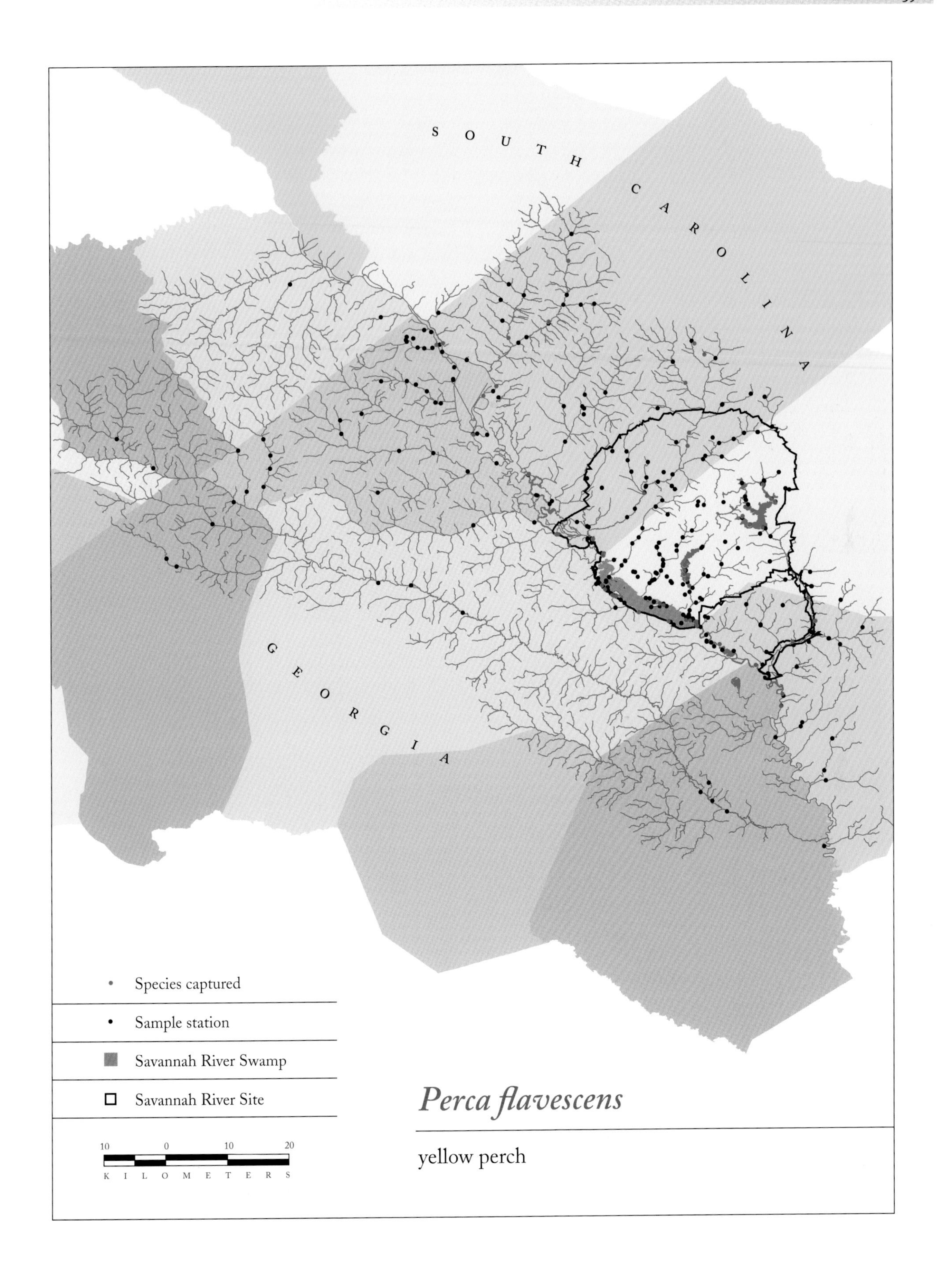

Perca flavescens

yellow perch

males can spawn up to 100,000 eggs (Dahlberg 1971; Clugston et al. 1978; Etnier and Starnes 1993; Craig 2000). Eggs are laid in long strands garlanded over underwater objects (Harland 1947). A description of the eggs and larvae and an overview of their development are given in Hardy 1978 and Craig 2000.

The life span is between 6 and 21 years (Craig 2000). Since mortality may be temperature related, with northern populations living longer than southern ones, the life span in the MSRB is probably close to the lower end of this range (Craig 2000). Females seem to live longer and grow faster than males (Scott and Crossman 1973).

Yellow perch are visual daytime hunters with a diverse diet consisting of worms, crustaceans, insects, molluscs, and fish (including other smaller perch) (Tharratt 1959; Muncy 1962; Thorpe 1977; Craig 2000). At 100–170 mm TL and 4–5 years of age, the diet may consist of 25–60% fish (Mittelbach and Persson 1998). Yellow perch are frequently taken by anglers and commercial fishers in northern states (Etnier and Starnes 1993), and the species has been widely cultured because of its popularity (see details and overview in Craig 2000).

SCIENTIFIC NAME *Perca* = "perch" (old Latin), possibly from the Greek word for "dusky" (Jordan and Evermann 1896); *flavescens* = "yellowish" (Latin).

Percina nigrofasciata (Agassiz, 1854)
Blackbanded darter

DESCRIPTION Lateral line complete; opercle and cheek fully scaled. Frenum present.

Body coloration, particularly overall darkness, may be influenced by the color of the bottom substrate (Mathur 1973a). Ground color generally yellowish brown to olive dorsally and whitish to light olive brown ventrally. Body sides marked with 10–15 solid black vertical bars mostly visible as elliptical or diamond-shaped blotches (square on caudal peduncle and generally decreasing in width anteriorly); bars often contiguous, forming a black lateral stripe. Dorsum marked with six to eight darker, but sometimes vague, saddles or blotches. Three dark brownish to black caudal spots are often fused, especially the ventral two. Postorbital bar (teardrop) variable and ranges from prominent to faint or absent.

Sexual dimorphism is especially apparent in fish larger than 35 mm TL. Females have dark spots and blotches in the spaces between the portions of the dark vertical bars below the lateral line; these are absent on males (Crawford 1956; Mathur 1973a). Breeding females have numerous large chromatophores in all fins (Mathur 1973a). Breeding males become much darkened with black and develop a greenish iridescence, especially on the opercle and ventral half of the body. The genital papilla is long, conical, and viliform in females, and short, rounded, and indiscriminately viliform in males. Modified enlarged scales (usually two or three) are present between the posterior bases of the pelvic fins (Plate 19 c) of males and often females.

Geographic variation, particularly in meristic characters, has been noted between populations in the upper Savannah River and those in the MSRB (Crawford 1956).

SIMILAR SPECIES The blackbanded darter is the largest of the darters found in the MSRB.

SIZE Reaches at least 95 mm SL in the MSRB, but only a few exceed 80 mm, and most are less than 70 mm SL.

MERISTICS Dorsal spines (9)11–13(15), dorsal rays (10)11–12(13), anal spines 2, anal rays (7)9(10), pectoral rays 13–15; lateral line scales (46)50–62(71) (Page 1983). In the MSRB, lateral line scales 55–69 (Crawford 1956).

DISTRIBUTION From the Edisto River of South Carolina south to the St. Johns and Kissimmee rivers of Florida, west to the lowermost tributaries of the Mississippi River in eastern Louisiana and southern Mississippi. Found from headwaters down onto the Coastal Plain (Burgess *in* Lee et al. 1980). In the MSRB, widespread and common in the Savannah River and tributary streams, more common in intermediate-size streams (third and fourth order) than in headwaters (Paller 1994).

Percina nigrofasciata, blackbanded darter; male

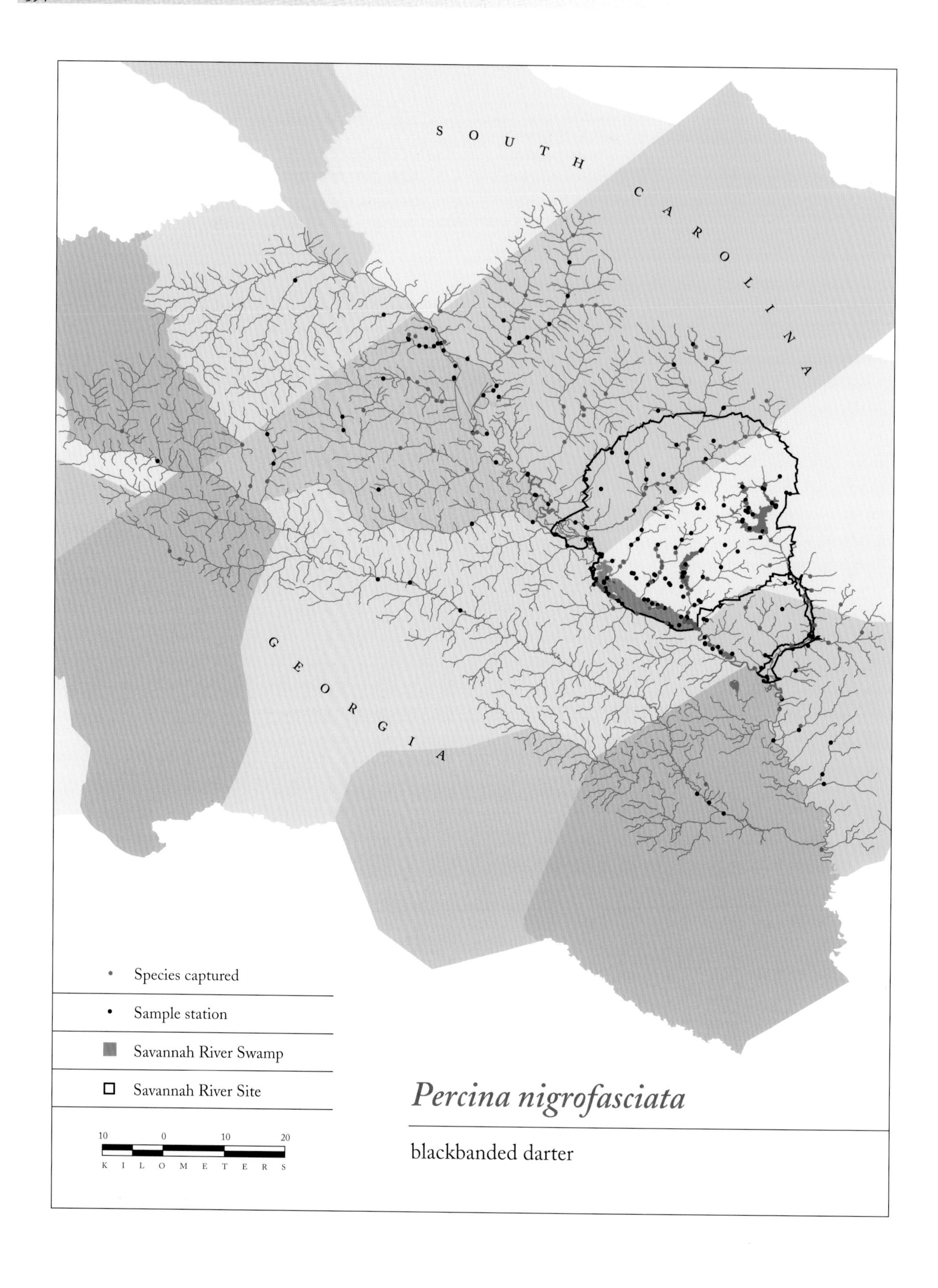

Percina nigrofasciata

blackbanded darter

Percina nigrofasciata, blackbanded darter; adult or large juvenile

HABITAT Occupies a wide range of habitats with at least a moderate water flow (Etnier and Starnes 1993). Found over a variety of sediments, but seems to prefer gravel and is more common over sandy, gravel, rock, or rubble substrates than over silt or mud (Crawford 1956; Mathur 1973a).

BIOLOGY Spawning takes place from early May through June in Alabama (Mathur 1973a), possibly starting as early as February (Suttkus and Ramsey 1967). The eggs are probably buried in loose sand and gravel, as is the case with other *Percina* (see Page 1985). Clutch size is small and increases with the female's size; a female 43 mm TL had 38 eggs, and a female 69 mm TL had 250 eggs (Mathur 1973a). The ripe ova are between 1.0 and 1.9 mm in diameter and contain two or three oil globules (Mathur 1973a). Sexual maturity is reached at age 1, and most individuals do not live more than 2 years; only rarely do blackbanded darters live longer than 3 years (Rohde et al. 1994).

The blackbanded darter is an insectivore that feeds during the daytime, predominantly on aquatic insect larvae (Mathur 1973b). Individuals generally stay within a small stream area, but some occasionally make excursions of more than 100 m, especially from November through June (Freeman 1995). Blackbanded darters may be hosts for glochidia of at least two species of freshwater mussels: *Ptychobranchus greeni* and *Medionidus acutissimus* (Haag and Warren 1997).

SCIENTIFIC NAME *Percina* = diminutive, "little perch" (Latin); *nigrofasciata* = "black banded" (Latin).

Mugilidae (mullets)

Mullets are primarily marine fishes, although a few species spend their lives in fresh water. They are characterized by a torpedo-shaped body that is round in cross section anteriorly and more laterally compressed posteriorly. Scales are cycloid or, rarely, weakly ctenoid, and the lateral line is absent. An adipose eyelid is present, and the first of the two widely separated dorsal fins has four spines. The pectoral fins are placed high on the sides, and the caudal fin is forked. The very long intestinal tract is related to a diet consisting primarily of detritus and algae. Most mullets have an olive, bluish, or greenish back and silvery lower sides. Many species form very large schools. The family—with approximately 17 genera and 80 species—is distributed worldwide in tropical and subtropical waters, with most genera and species in the Indo-Pacific. Two species are reported from the MSRB: the mountain mullet (*Agonostomus monticola*) and the striped mullet (*Mugil cephalus*). Mullets in the genus *Mugil*, and perhaps some in other genera as well, have a prejuvenile stage called the "queremana" that is characterized by a strongly laterally compressed and bright silvery body, large eyes, and only two anal spines. The marine larvae metamorphose into queremanas, which migrate inshore to bays and estuaries to metamorphose into juveniles with more rounded bodies, adipose eyelids, and three anal spines. While most species lay their eggs in full-strength seawater or have larvae that migrate into marine areas, at least some species can go through the full life cycle in brackish water. Mullets are important commercially, used both for human food and for bait to catch larger predatory species.

KEY TO THE SPECIES OF MUGILIDAE

1a. Lower lip with thin front edge; lower jaw angular; adipose eyelid, clear in living individuals but thick and well formed; no distinct spot on caudal peduncle . . . . *Mugil cephalus* (striped mullet), p. 401

1b. Lower lip thick, without thin edge; lower jaw rounded; adipose eyelid thin and not obvious or even missing; distinct spot on caudal peduncle . . . . *Agonostomus monticola* (mountain mullet), p. 398

Agonostomus monticola (Bancroft, 1834)

Mountain mullet

DESCRIPTION Body moderately robust. Head length 25–27% of standard length; lips thick, upper lip very high; no symphysial knob on lower jaw. Several rows of teeth in both jaws; outer row unicuspid, inner rows either bicuspid or tricuspid. Posterior end of maxilla reaching to beneath the eye. No lateral line. Side of head scaly. Second dorsal fin lightly scaled anteriorly and along base. No axillary scale associated with pectoral fin.

Back dark with scales outlined in black. Side gray, grading ventrally to whitish or silvery with scales having darker edges. Unpaired fins may be bright yellow, especially at their bases, and the second dorsal fin has a darkened area along the bases of the first few rays. Dark stripe running from snout to posterior edge of operculum. A black spot present at bases of pectoral fins and another usually on the caudal peduncle.

SIMILAR SPECIES The only similar species in our area is the striped mullet. *Mugil* species have a gizzard for processing food (Loftus and Gilbert *in* Gilbert 1992); *Agonostomus* does not. Also, the axillary process is present in *Mugil* and absent in *Agonostomus*.

SIZE Reaches 360 mm (Thomson 1978).

MERISTICS First dorsal fin 4 spines, second dorsal fin 1 spine and (8)9 (Dizon et al. 1973) rays; anal fin 2 spines and 10 rays; pectoral fin rays 13–16 (Suttkus 1956); total caudal rays 32–34 (Dizon et al. 1973); gill rakers on lower gill arch 17–20; branchiostegals 5; lateral scale rows 38–47.

DISTRIBUTION Tropical and subtropical waters from North Carolina to northern South America, including the West Indies and the Caribbean Coast of Mexico and Central America (Rohde *in* Lee et al. 1980); also found in the Pacific Coast drainages of Mexico (Lyons and Navarro-Perez 1990). The patchy distribution within the United States reflects habitat availability near sources of recruitment (Loftus and Gilbert *in* Gilbert 1992). Only a few individuals in two collections are reported from the MSRB, all from the main channel of the Savannah River.

HABITAT Adults inhabit fresh water, often high-gradient streams (Carr and Goin 1955; Suttkus 1956; Erdman 1972; Cruz 1987), but may also occur in lentic waters (Loftus et al. 1984). Larvae live in marine water both nearshore and offshore; juveniles enter and inhabit the lower stretches of rivers. Fall collections in Louisiana and Florida indicate that this species enters North American rivers more than was previously thought (Swift et al. 1977).

Agonostomus monticola, mountain mullet

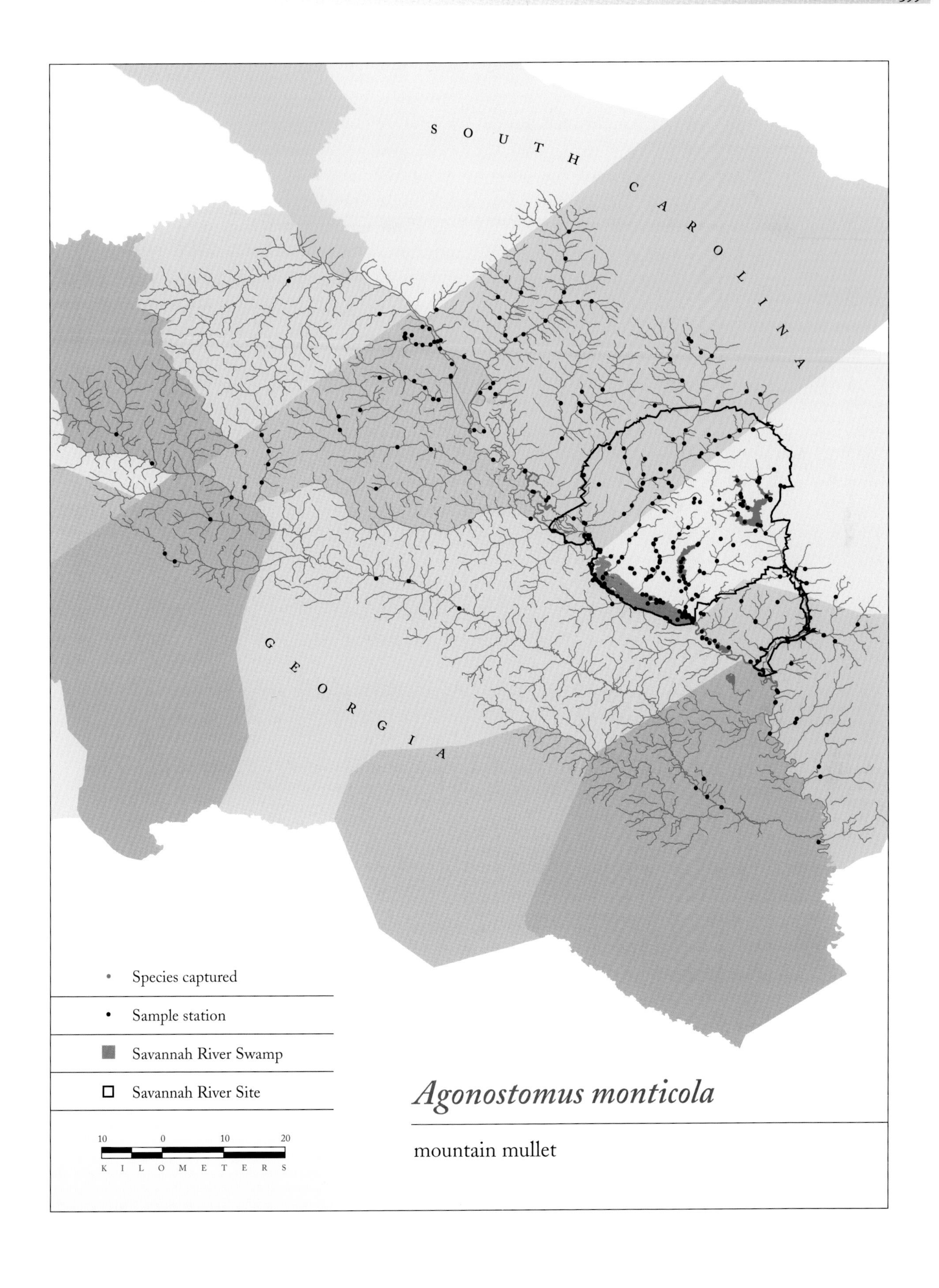

Agonostomus monticola

mountain mullet

BIOLOGY The adults are found mainly in the upper reaches of rivers but may migrate toward the sea for spawning. The spawning season appears to coincide with the local rainy season throughout the range (Loftus and Gilbert *in* Gilbert 1992). Larvae have been found primarily in seawater or in brackish water. This species is often thought to be catadromous but may be amphidromous (Gilbert and Kelso 1971; Erdman 1972). Gilbert (1992) discussed circumstantial evidence supporting amphidromy, and Phillip (1993) discussed evidence supporting catadromy. Cruz (1987) suggested that adults may migrate to estuaries to spawn. Perhaps several strategies are used (W. Loftus pers. comm.). Larvae have been found as far north as Louisiana and Texas in August (Ditty and Shaw 1996). The mountain mullet is probably a frequent stray in South Carolina and Georgia waters but not a resident. The population in southern Florida is supplemented by recruits from Cuba and the West Indies (Loftus et al. 1984). Parrot Jungle in Miami Beach, Florida, had an artificial stream that was home for a possibly reproducing population (Loftus et al. 1984; Loftus and Kushlan 1987). This species may be more common than records indicate as it is difficult to capture with most collecting gear (Pezold and Edwards 1983; Lyons and Navarro-Perez 1990) and is susceptible to capture only with dynamite (Carr and Giovannoli 1950) and rotenone-based fish toxins (FDM pers. obs.). They are readily caught by electrofishing when conditions are suitable (W. Loftus pers. comm.)

In a Trinidad population, the sex ratio of adult females to males was 20.4:1 (Phillip 1993); in Jamaica the ratio was only 1.2:1.0 (Aiken 1998). Both sexes are sexually mature by 135 mm FL (Phillip 1993).

Mountain mullet feed on fruits, seeds, algae, aquatic insects, land plants, detritus, molluscs, small fishes, and crustaceans (Erdman 1972; Loftus et al. 1984; Cruz 1987; Phillip 1993; Aiken 1998).

SCIENTIFIC NAME *A* (prefix) = "without," *gony* = "knee" and, by extension, "angle," and *stoma* = mouth (Greek): "mouth without angle"; *monticola* = "inhabiting mountains" (Latin).

Mugil cephalus (Linnaeus, 1758)

Striped mullet

DESCRIPTION Body somewhat compressed and stout. Caudal peduncle stout. Head short and broad; side of head scaly; mouth small and terminal; lips thin and terminal, the lower lip with a high symphysial knob. If a vertical line were drawn through the head in front of the eye, the upper jaw would barely reach this line. Teeth labial, one to six rows in upper lip and one to four in lower lip with outer rows unicuspid and inner rows usually bicuspid. Well-developed adipose eyelids present. Dorsal fins well separated. No lateral line. Gill arches have numerous close-set gill rakers. Scales extend well onto the caudal fin and somewhat onto the dorsal and anal fins; scales cycloid and at least some small scales are present on the soft dorsal and anal fins. Axillary process present at the base of the pectoral fins. Gizzard present.

Back olive green to bluish gray, side silvery, belly silvery white. A large dark spot is present at the base of the pectoral fin, bluish in life. Scales on most of the side have dusky center spots giving the appearance of rows of spots. Juveniles less than 150 mm are bright silvery, and very small juveniles are strongly compressed with small eyes.

SIZE Up to 910 mm (Robison and Buchanan 1984).

MERISTICS First dorsal fin 4 spines, second dorsal fin 1 spine, 6–9 rays, usually 8; anal fin 3 spines and (7)8 rays (in juveniles less than 35–45 mm TL, the third anal element is a ray so that there are only 2 anal spines [Ditty and Shaw 1996]); pectoral fin rays (14)16–17(18); pelvic fin 1 spine, 5 rays; total caudal rays 28–30 (Dizon et al. 1973); principal caudal rays 14; lateral series scale rows 37–42; total gill rakers about 80; branchiostegals 5–6.

DISTRIBUTION Marine and estuarine waters worldwide in the tropics and subtropics. In North America, most common south of the Chesapeake Bay on the Atlantic Coast and south of Los Angeles on the Pacific Coast (Burgess *in* Lee et al. 1980). Absent from the Bahamas and most of the Caribbean and West Indies (Robins and Ray 1986). Common in the MSRB, mostly in the main channel of the Savannah River, although there are some records from mouths of creeks that enter the Savannah River.

Mugil cephalus, striped mullet; adult

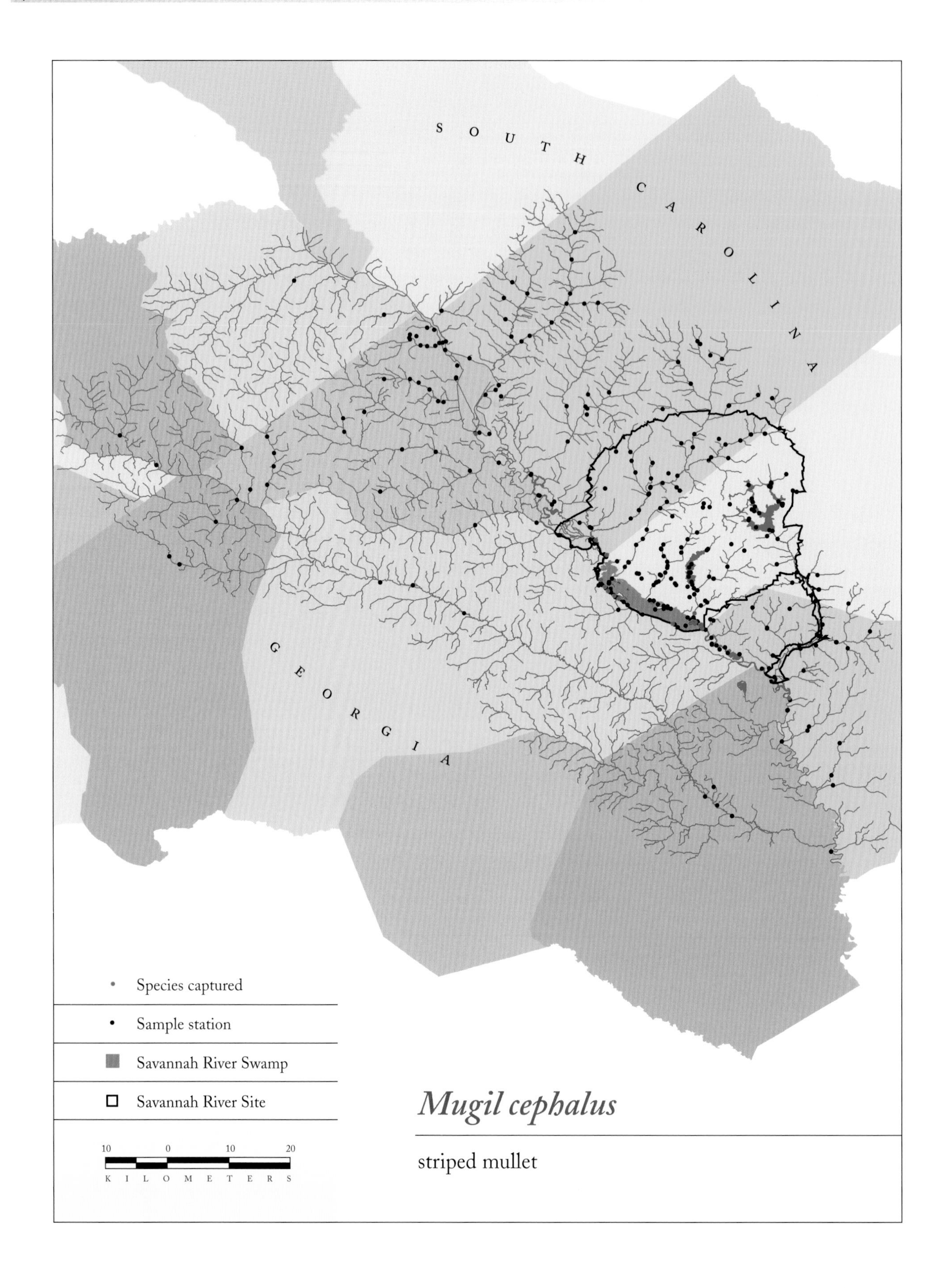

Mugil cephalus

striped mullet

Mugil cephalus, striped mullet; closeup of head showing adipose eyelid

HABITAT Adults are most often found in open water of bays and large rivers, where they often form large schools. They occur at salinities ranging from fresh water to up to 75 ppt (Hoese and Moore 1977). They are absent from rivers during the late fall and early winter spawning season (Beecher and Hixson 1982). Common in fresh and brackish canals throughout South Florida (Loftus and Kushlan 1987), moving far inland. In Alabama, found as far as 603 km upstream and 69.3 m elevation above sea level (Boschung and Hemphill 1960). Spawning occurs offshore, near the edge of the continental shelf. Juveniles use estuarine areas almost exclusively.

BIOLOGY This schooling species forms much larger aggregations during spawning, which occurs well offshore in late fall and winter, as late as March in the northern Gulf of Mexico (Ditty and Shaw 1996). Small juveniles enter estuaries in the early spring. Adults move inland in late spring and summer. Sexual maturity is achieved at 2 years of age or older, and individuals may live up to 13 years.

Juveniles eat microcrustaceans and insect larvae; adults eat plant material, plankton, and detritus, which they filter through their close-set gill rakers and process in the gizzardlike stomach. They may also feed by ingesting sediment.

SCIENTIFIC NAME *Mugil*, from *mulgeo* = "to suck" or "to milk" (Latin), the old vernacular name for mullet; *cephalus*, from *céfalo*, the old name for mullet, derived from *cephalus* = "head" (Latin).

Achiridae (American soles)

The Achiridae are flatfishes that occur in both salt and fresh water. The nine genera and 28 species live mainly in shallow waters off the Atlantic and Pacific shores of North and South America. In the past, the Achiridae was considered one of two subfamilies (Achirinae and Soleinae) of the Soleidae. Jordan (1923), and more recently Chapleau and Keast (1988) and Chapleau (1993), suggested elevating the Achiridae to family status, a view generally accepted to date. The hogchoker (*Trinectes maculatus*) is the only American sole found in the MSRB.

The American soles are laterally flattened, and the general body shape is rounded or oval. The larvae have eyes on each side of the head like other fishes, but during metamorphosis the left eye moves over the head; after metamorphosis both eyes are located on the right side of the head. At this time, the juvenile starts its demersal life and the eyed side takes on the adult color pattern. The blind side is usually white or mottled. All species are carnivores preying largely on benthic invertebrates.

The southern flounder (*Paralichthys lethostigma*, family Bothidae, left-eyed flounders) has been reported in the MSRB, but little information exists on its distribution within this area.

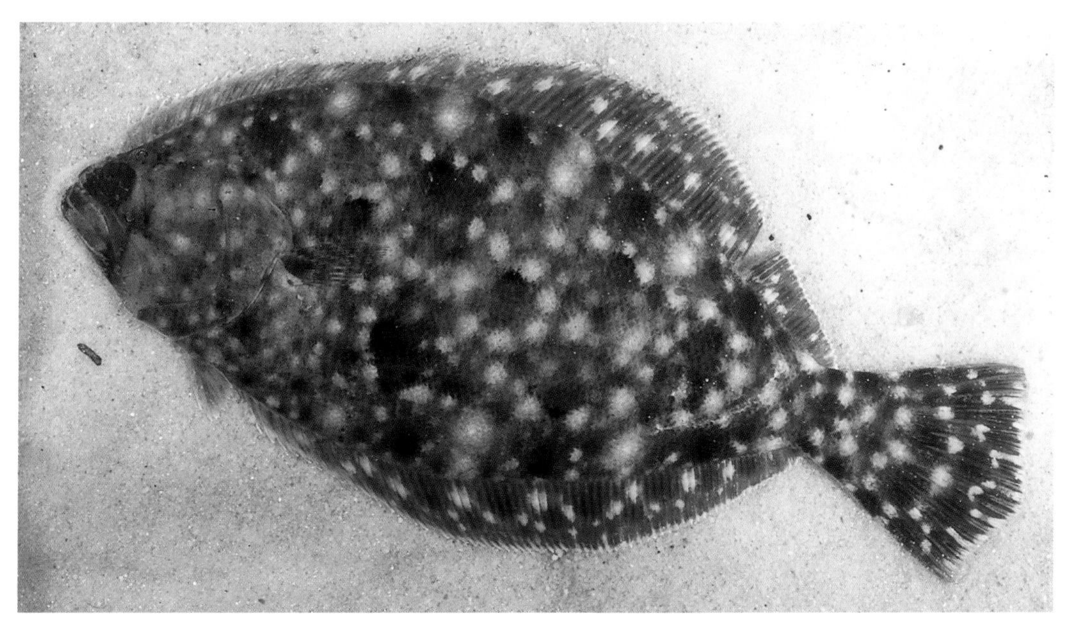

Paralichthys lethostigma, southern flounder. This species is thought to occur in the MSRB, but we have seen neither specimens nor verifiable records.

Trinectes maculatus (Bloch and Schneider, 1801)
Hogchoker

DESCRIPTION Body laterally flattened. Eyes small, both on right side of body after metamorphosis is completed. Mouth small and curved.

Color on eyed side light to dark brown with seven or eight dark, narrow vertical lines; row of spots along lateral line. Two large, diffuse blotches at about midbody on base of dorsal and anal fins. Blind side whitish, sometimes with darker spots. Pectoral fins absent. Right pelvic fin continuous with anal fin. Dorsal, anal/pelvic, and caudal fins profusely mottled with dark blotches. Tiny ctenoid scales present on both sides.

SIMILAR SPECIES Since the southern flounder (*Paralichthys lethostigma*) is also a flatfish, it may be confused with the hogchoker. However, the southern flounder looks entirely different: both eyes on the left side; a big, straight mouth with clearly visible teeth; a prominent pectoral fin; very small scales; and no lines on the eyed side (for more information on the southern flounder, see Reagan and Wingo 1985; Blandon et al. 2001). Reports of the southern flounder in the MSRB usually involve adults larger than 20 cm, the maximum reported length for the hogchoker.

SIZE Can reach 200 mm TL, but usually no larger than 170 mm. Females grow larger than males.

MERISTICS Dorsal rays 50–56, anal rays 36–42, pelvic rays on eyed side 3, on blind side 1 (Hildebrand and Schroeder 1928), total caudal rays 16 (Dizon et al. 1973), pectoral rays absent.

DISTRIBUTION From Cape Ann, Massachusetts, along the Atlantic and Gulf of Mexico coasts to Panama, but not in the West Indies (Burgess *in* Lee et al. 1980). Relatively common and locally abundant in the MSRB,

Trinectes maculatus, hogchoker

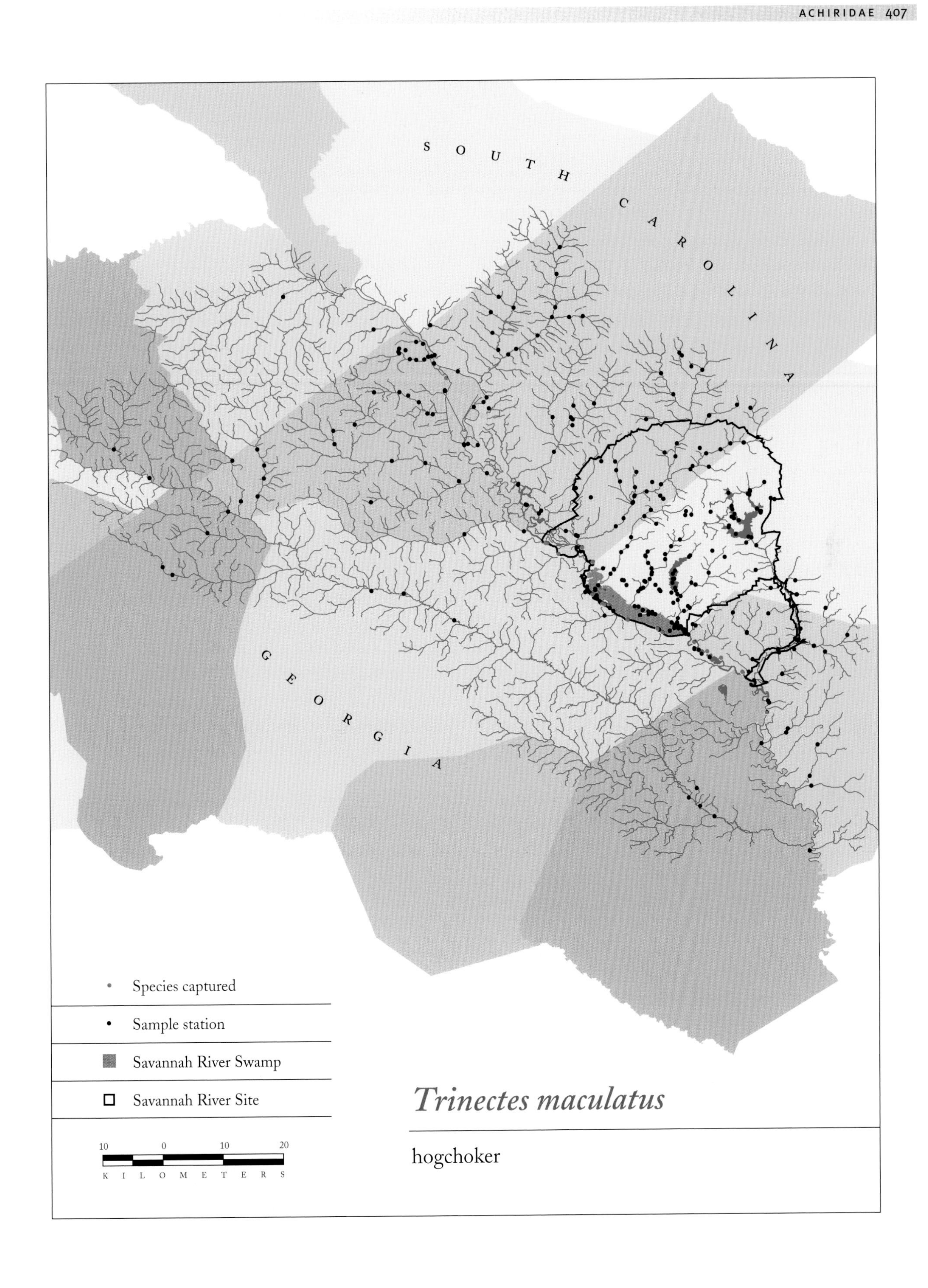

Trinectes maculatus

hogchoker

where it is found almost exclusively in the Savannah River but may enter lower portions of streams on the modern Savannah River floodplain.

HABITAT Coastal marine and fresh waters. Juveniles prefer tidal fresh water and seagrass beds around tidal creek mouths (Mettee et al. 1996). In the MSRB, locally abundant on sandbars along the inside bank of curves or bars at the mouths of tributary streams. Smaller individuals (ca. 40–65 mm SL) are generally found in water less than 1 m deep; larger individuals can be found in deeper water on sandy bottoms. In the Chesapeake Bay, typically found on mud bottoms (Murdy et al. 1997).

BIOLOGY Spawning takes place in the summer in marine coastal waters and estuaries (Dovel et al. 1969; Koski 1978). The eggs are pelagic, but less buoyant in lower salinities (Lippson and Moran 1974). Eggs and larvae have been collected from May, when water temperatures reached 20 °C, through September in Maryland (Dovel et al. 1969), through October in North Carolina (Hettler and Chester 1990), and at least through August in Georgia (Reichert and Van der Veer 1991). Temperature and day length may be important environmental factors initiating peak spawning activity in July (Dovel et al. 1969). Analyses of migration patterns in Maryland (Dovel et al. 1969) and in the Gulf of Mexico (Peterson 1996) suggested that after spawning in the summer, most adults migrate upriver into low-salinity waters, where they remain from November to March. In spring, they migrate back to the estuarine spawning grounds. In summer and fall, the developing larvae and early juveniles migrate upstream into fresh water. Eggs and larvae are described in Lippson and Moran 1974; and Wang and Kernehan 1979.

Small larvae have two pectoral fins and one eye on each side of the body. When the larvae are between 1.6 and 6.0 mm in length, the pectoral fins degenerate to small flaps and ultimately disappear. At lengths between 6.0 and 10.0 mm the left eye migrates over the head to the adult position on the right side (Able and Fahay 1998), and juveniles begin their demersal life. With increasing size, juveniles gradually start following the adult migration patterns, but do not seem to migrate into the high-salinity spawning grounds until full sexual maturity is reached between the ages of 2 and 4 (see details in Dovel et al. 1969; Koski 1978). The maximum life span may be 5 years for males and 7 years for females, but most do not live beyond 3 and 4 years, respectively (Koski 1978). Growth is not very fast, and average total lengths for hogchokers in the Patuxent River of Maryland (Mansueti and Pauly 1956) and the Hudson River of New York (Koski 1978) were 50–59, 81–83, 99–103, 115–122, and 128–139 mm at ages 1–5, respectively. Peterson-Curtis (1996) reported much slower growth rates for the Gulf of Mexico, where southern hogchokers reached average total lengths of 25, 50, 75, 95, and 115 mm at the same respective ages.

The diet consists of benthic organisms, the prey composition depending largely on salinity. In fresh water, hogchokers predominantly eat small aquatic crustaceans and insects, but molluscs and annelid worms are also included. In more saline waters the diet consists largely of polychaete and oligochaete worms, and crustaceans and molluscs become less important as prey (Derrick and Kennedy 1997).

SCIENTIFIC NAME *Trinectes* = "three swimming" (Greek), referring to using dorsal, anal, and caudal fins to swim; *maculatus* = "spotted" (Latin). The common name originates from the times when hogs were fed on discarded fishes. They had difficulty swallowing hogchokers because of the hard, rough scales (Hildebrand and Schroeder 1928).

Appendix: Sources Utilized for Distribution Data for Maps

Academy of Natural Sciences of Philadelphia [ANSP]. 1993. Spawning and nursery use of the Savannah River Swamp by fishes and effects of the 1992 K-Reactor power ascension test. Report 93-7F for Westinghouse Savannah River Co.

Academy of Natural Sciences of Philadelphia [ANSP]. 1995. 1994 Savannah River biological survey in the vicinity of Georgia Power and Light's Vogtle Nuclear Power Plant site. Report 95-11F for Westinghouse Savannah River Co.

Academy of Natural Sciences of Philadelphia [ANSP]. 2002. Online catalog of species.

Aho, J. M., C. S. Anderson, K. B. Floyd, and M. T. Negus. 1986. Patterns in fish species assemblage structure and dynamics in waters of the Savannah River Plant. SREL-27 UC-66e.

American Museum of Natural History. 2002. Online catalog of species.

Auburn University. 2002. Online catalog of species.

Bart, Hank. 2002. Tulane University, New Orleans, Louisiana, personal communication.

Clugston, J. P. 1973. The effects of heated effluents from a nuclear reactor on species diversity, abundance, reproduction, and movement of fish. Doctoral dissertation. University of Georgia, Athens.

Cornell University. 2002. Information obtained via e-mail query concerning collection holdings.

Dinkins, G. R. 2002. Letter to Barton C. Marcy Jr. Subject: Savannah River drainage records.

Dudley, R. G. 1980. Additional studies of striped bass movements in the Savannah River. University of Georgia, Department of Fisheries and Wildlife, Final Report, Anadromous Fish Project, Georgia AFS-12.

Dudley, R. G., Q. W. Mullis, and J. W. Terrell. 1975. Movements of striped bass (*Morone saxatilis*) in the Savannah River, Georgia. University of Georgia, School of Forest Resources, Final Report, Anadromous Fish Project, Georgia AFS-10.

Fletcher, Dean. 2002. University of Georgia, Savannah River Ecology Laboratory, personal communication.

Florida Museum of Natural History [FMNH]. 2002. Online catalog of species.

Friday, Gary P. 2002. Westinghouse Savannah River Co., Savannah River National Laboratory, personal communication.

Harvard University, Museum of Comparative Zoology [MCZ]. 2002. Online catalog of species.

Hogan, D. C. 1977. Distribution and relative abundance of prey fish in a reservoir receiving heated effluent. Master's thesis. University of Georgia, Athens.

Hoover, J. J. 2002. U.S. Army Corps of Engineers, Waterways Experiment Station, Vicksburg, Mississippi, personal communication.

Hoover, J. J., S. G. George, and N. H. Douglas. 1998. The bluebarred pygmy sunfish (*Elassoma okatie*) in Georgia. Southeast. Fishes Counc. Proc. 36:7–9.

Illinois Natural History Survey [INHS]. 2002. Information obtained via e-mail query concerning collection holdings.

Jenkins, Bob. 2002. Roanoke College, Salem, Virginia, personal communication.

Law Environmental. 1991. Data report for Indian Grave Branch and Pen Branch electrofishing, Barnwell County, Savannah River Site, South Carolina, 17 December 1990–8 January 1991, revision 1.

Marcy, Barton C. Jr. 2002. Westinghouse Savannah River Co., personal communication.

Martin, F. D., M. H. Paller, and L. D. Wike. 2002. Westinghouse Savannah River Co., personal communication regarding fishes taken incidental to seepline invertebrate studies.

Matthews, R. A., and C. F. Musca. 1983. Shortnose and Atlantic sturgeon in the Savannah River. DPST-83-753. E. I. du Pont de Nemours and Co.

McFarlane, R. W., F. F. Frietsche, and R. D. Miracle. 1978. Impingement and entrainment of fishes at the Savannah River Plant: an NPDES 316b demonstration. Report DP-1494. E. I. du Pont de Nemours and Co.

Mitchell, C. 2002. Savannah River Ecology Laboratory, University of Georgia, personal communication.

North Carolina State Museum [NCSM], cited in F. C.

Rohde and R. G. Arndt. 1987. Two new species of pygmy sunfishes (Elassomatidae, *Elassoma*) from the Carolinas. Proc. Acad. Nat. Sci. Phila. 139:65–85.

O'Hara, J., and V. Osteen. 1983. Interim report on adult fisheries studies in the Savannah River, September 1982–February 1983. ECS-SR-4, DPST-83-737.

Paller, M. 1984. Summary of the ichthyoplankton sampling data from the creeks and swamps of the Savannah River Plant. Interim report. ECS-SR-10. Environmental and Chemical Science, Inc., to E. I. du Pont de Nemours and Co.

Paller, M. H., B. H. Saul, and D. W. Hughes. 1986. The distribution of ichthyoplankton in thermal and non-thermal creeks and swamps on the Savannah River Plant, February–July 1985. ECS-SR-25. Environmental and Chemical Science, Inc., to E. I. du Pont de Nemours and Co.

Reichert, M. J. M. 1995. Index of biotic integrity: results of sampling the fish assemblage of 13 creeks at the Savannah River Site area in the fall and winter of 1994/95. Report to Westinghouse Savannah River Co.

Samsel, J. 2002. Augusta's backyard catfish. Ga. Sportsman (July).

Saul, B. H. 2002. Augusta State University, Augusta, Georgia, personal communication.

Sayers, R. E. Jr., H. G. Mealing, and K. K. Patterson. 1990a. L-Lake fish: L-Lake/Steel Creek biological monitoring program January 1986–December 1989. Report NAI-SR-113. Normandeau Associates, Inc.

Sayers, R. E. Jr., H. G. Mealing, and K. K. Patterson. 1990b. Steel Creek fish: L-Lake/Steel Creek biological monitoring program January 1986–December 1989. Report NAI-SR-122. Normandeau Associates, Inc.

Shealy Environmental. 1992. Stream fisheries characterization study at the Savannah River Site. PR no. D18518, PRCN no. 18433, May–July 1992.

Snodgrass, J. W., A. L. Bryan Jr., R. F. Lide, and G. M. Smith. 1996. Factors affecting the occurrence and structure of fish assemblages in isolated wetlands of the upper Coastal Plain. Can. J. Fish. Aquat. Sci. 53:443–454.

South Carolina Fisheries Information Network System [SC-FINS]. 2002. Information obtained from Jim Bulak of South Carolina Department of Natural Resources.

Specht, W. L. 2002. Westinghouse Savannah River Co., Savannah River National Laboratory, personal communication.

Thomason, C. 2002. South Carolina Department of Natural Resources, personal communication.

United States National Museum [USNM, Smithsonian]. 2002. Information obtained from online catalog of species.

University of Georgia [UGA]. 2002. Information obtained from Lee Hartle.

University of Michigan, Museum of Zoology [UMMZ]. 2002. Information obtained from online catalog of species.

Wike, Lynn D. 2002. Westinghouse Savannah River Co., personal communication.

Wilde, E. W., editor and compiler. 1987. Comprehensive cooling water study final report. Volume 7. Ecology of Par Pond. Report DP-1739-7. E. I. du Pont de Nemours and Co.

Wiltz, J. W. 1981. Savannah River fish population study and impingement prediction for Plant Vogtle, Burke County, Georgia. Report to Georgia Power Co.

Glossary

abdomen The lower surface of the body, especially the part between the pectoral fins and the anus.
abdominal Pertaining to the belly; pelvic fins are abdominal when inserted far behind the base of the pectoral fins.
adipose eyelid An immovable transparent outer covering or partial covering of the eye of some groups of bony fishes, such as mullets and jacks, which serves protective and streamlining functions. It may cover the eye in front and in back, leaving a vertically elongated central opening, or may cover the entire eye except for a small opening in the middle.
adipose fin A small, fleshy, rayless fin located on the midline of the back between the dorsal and caudal fins of catfishes, salmonids, and some other fishes.
aestivation See "estivation."
air bladder (also "gas bladder" or "swimbladder") A membranous, gas-filled sac in the upper part of the body cavity beneath the vertebral column.
allopatric Not occurring in the same area.
amphidromy Living and spawning in fresh water, but eggs and/or larvae wash out to sea and develop there.
anadromous A fish species, such as the American shad in the Savannah River, that spawns and spends its early life in fresh water, but spends most of its life and grows to sexual maturity in the ocean.
anal fin An unpaired median fin on the underside of a fish, usually just behind the anus.
anterior Toward the head; regarding the head.
anus The exterior opening of the digestive tract; the vent.
axillary process A long, thin, membranous flap or modified scale at the anterior base of the pelvic fins in the Hiodontidae and Salmonidae, and at the anterior bases of both the pectoral and pelvic fins in the Clupeidae and Mugilidae.
axillary scales Elongate scales at the insertion of the pelvic fin (and the pectoral fins in the Clupeidae).
barb A hooklike serration such as the pectoral and dorsal fin spines of carp and catfishes.
barbel A slender, elongate, flexible process located near the mouth, snout, and chin that is tactile and gustatory in function.
basicaudal Referring to the area at the base of the caudal fin.
basioccipital The posteriormost bone on the underside of the skull; it articulates with the centrum of the first vertebra.
belly The ventral surface posterior to the base of the pelvic fins and anterior to the anal fin.
bicuspid Having two points (applicable especially to teeth).
branched ray A soft ray that forks or branches away from its base.
branchiostegal ray One of the elongate, flattened bones supporting the gill membranes ventral to the operculum.
breast The ventral surface in front of the pelvic fins and anterior to the belly.
breeding tubercles Small, raised, often pointed epidermal structures on the head, body, or fin rays of adult males and females. Tubercles may consist of aggregations of nonkeratinized, superficially keratinized, or fully keratinized epidermal cells that are organized to form a discrete, usually conical, cap. They develop, often in large numbers and often in males only, just before the breeding season and remain present during the breeding season. They are usually located at body points where two individuals come into contact and are used in territorial aggression, to stimulate females during courtship, to help maintain body contact during spawning, and possibly in species and sex recognition. They are found in 15 families within Salmoniformes, Gonorhynchiformes, Cypriniformes, and Perciformes. Also called "nuptial tubercles" and "pearl organs."
caducous Readily shed (as belly scales in genus *Percina*).
caecum (pl. caeca) A blind pouch or other saclike evagination of the digestive tract, especially at the pylorus (junction of the stomach and small intestine).
catadromous A fish, such as the American eel, that grows to sexual maturity in fresh water but migrates to the ocean to spawn.
caudal Regarding the tail.

caudal fin Tail fin.
caudal peduncle The narrow region of the body between the posterior end of the base of the anal fin and the base of the caudal fin.
ceratohyal The anterior of the two bones to which the bases of the branchiostegial rays are attached.
chromatophores Pigment cells that form the color patterns of fish.
cleithrum The major bone of the pectoral girdle, extending upward from the fin base and forming the posterior margin of the gill chamber.
cloaca The common chamber into which the digestive tract and urogenital ducts discharge in some fishes and other vertebrates.
compressed Narrow from side to side (flattened laterally); deeper than broad.
concave Curving inward.
convex Curving outward.
crepuscular Active at dawn and/or dusk.
ctenoid scales Thin scales that bear a patch of tiny spine-like prickles (ctenii) on the exposed (posterior) surface.
cusp A point or projection on a tooth.
cycloid scales More or less rounded scales that are flat and bear no ctenii.
decurved Curving downward.
demersal eggs Eggs that are denser than water and sink.
dentate With toothlike notches.
depressed Flattened dorsoventrally (from top to bottom); wider than deep.
dimorphism Featuring two body forms in the same species, often referring to differences between male and female.
disjunct Widely separated; refers mostly to separated populations.
distal Farthest from the point of attachment (e.g., the free edge of fins farthest from their bases).
diurnal Active by day.
dorsal Referring to the back; an abbreviation for dorsal fin.
dorsal fin The unpaired median fin on the back of a fish, often consisting of two parts.
dorsum The upper part of the body; the back.
ear tab A flattened, flexible extension of the posterior part of the operculum; well developed in some sunfishes. Also called "opercular flap" and "ear flap."
ectopterygoid One of four paired bones in the roof of the mouth; it articulates anteriorly with the palatine, posteriorly with the quadrate, and mesially with the entopterygoid (if the latter is present). It may be called the pterygoid when there is no entopterygoid (see Helfman et al. 1997 for skeletal details).
elevated scales Scales on the sides that are much higher than they are long (i.e., the distance from top to bottom is much greater than the distance from front to rear).
elongate Long and narrow.
emarginate Having a distal margin notched, indented, or slightly forked.
endangered In danger of extinction.
endemic Found only in, or limited to, a particular geographic region or locality.
entopterygoids Thin paired bones that roof the mouth. One of four paired bones in the roof of the mouth. Also called mesopterygoids (see Helfman et al. 1997 for skeletal details).
epihyal The posterior of the two bones to which the bases of the branchiostegal rays are attached.
epilimnion The upper layer of water in a lake that is characterized by a temperature gradient of less than 1 °C per meter of depth.
estivation Dormancy during summer (also "aestivation").
estuarine Occupying estuaries.
euryhaline Able to tolerate a wide range of osmotic pressures in the environment; a fish that can move freely between saltwater and freshwater environments.
extirpated Exterminated on a local basis (i.e., from a political or geographic part of the range).
eye diameter The horizontal diameter of the eyeball—in contrast to iris diameter, the distance across the black aperture in the center of the eye; or orbit diameter, the horizontal distance between the anterior and posterior margins of the socket.
falcate Deeply concave or sickle shaped, as a falcate fin.
fecundity Used here to refer to the number of ripening eggs produced by a female fish in preparation for the next spawning period.
filament A threadlike process usually associated with the fins.
fimbriate Fringed at the margin with slender, elongate processes.
fingerling A young fish, usually late in the first year.
fin ray A bony or cartilaginous rod supporting the fin membrane. Soft rays usually are segmented (cross-striated), branched, and flexible near their tips; spines are not segmented, are never branched, and usually are stiff to their sharp distal tips.
fork length Distance from the anteriormost margin of the head to the tip of the middle ray of the caudal fin.
frenum A fold of skin that limits the movements of an organ (e.g., the membrane across the snout of some darters).
frontal spines Spines on the head between the posterior margins of the eyes.
fry A young fish at the age when the yolk has been con-

sumed and the fish is actively feeding (alternate term, "larva," preferred).

fulcrum (pl. fulcra) Spinelike structures bordering the anterior rays of the fins of some fishes.

fusiform Spindle shaped; referring to the form of fishes whose body tapers both anteriorly and posteriorly, and is slightly or not at all compressed.

gape In fishes, width of gape is the transverse distance between the two ends of the mouth cleft when the mouth is closed; length of gape is the diagonal distance from the anterior (median) end of the lower lip to one end of the mouth cleft.

gas bladder (also "air bladder" or "swimbladder") A membranous gas-filled sac in the upper part of the body cavity of some fishes that regulates buoyancy and aids in respiration, sound production, and sound reception.

genital papilla A fleshy projection adjacent to the anus, as in darters.

gill cover The bony covering of the gill cavity composed of opercular bones.

gill filaments Respiratory structures projecting posteriorly from the gill arches.

gill membranes Membranes that close the gill cavity ventrolaterally, supported by the branchiostegals.

gill rakers Projections (knobby or comblike) from the concave anterior surface of the gill arches.

gizzard Extra stomach for the digestion of difficult-to-digest materials such as plants and algae; present in *Mugil* species and others.

gonopodium The modified, rodlike anal fin of male livebearers (only *Gambusia* and *Heterandria* in the MSRB) used to transfer sperm to the female's genital pore.

gravid Used in this book to refer to a female fish containing ripe or nearly ripe eggs.

gular fold A transverse fold of soft tissue across the throat.

gular plate A large, median dermal bone on the throat, as in the bowfin or tarpon.

heterocercal A type of tail in which the vertebral column turns upward into the dorsal lobe.

homocercal A type of tail in which the vertebrae terminate at the caudal fin base in a hypural plate; neither lobe of the caudal fin is invaded by the vertebral column.

humeral "spot" or "scale" A scalelike bone, often dark, behind the gill opening and above the base of the pectoral fin (in darters).

hybrid Used here to refer to the offspring resulting from a cross between individuals of two different species.

hyoid teeth Teeth on the tongue.

hypersaline Water with a salinity higher than 31–37 parts per thousand, the average salinity of seawater.

hypolimnion The cool bottom layer of water in a stratified lake, below the thermocline characterized by a temperature gradient of less than 1 °C per meter of depth.

hypural plate The expanded terminal vertebral process at the caudal base in fishes having a homocercal tail, formed by fusion of the last few caudal vertebrae. The end of the hypural plate usually appears as a crease or line across the end of the caudal peduncle.

ichthyofauna The assemblage of fishes inhabiting a specific body of water or geographic area.

ichthyology The science of the study of fishes.

immaculate Without spots or pigment pattern; usually white or colorless.

included Used in reference to the upper or lower jaw when one is shorter than the other.

inferior Beneath, lower, or on the ventral side; for example, inferior mouth.

infraorbital canal A segment of the lateral line canal in the suborbital bones that curves beneath the eye and extends forward onto the snout; the canal may be complete or interrupted.

insertion The anteriormost end of the bases of the paired fins.

intermuscular bones Fragile, branched bones that are isolated in the connective tissue between body muscles.

interopercle A small bone of the operculum situated between the preopercle and the subopercle.

interorbital The region on top of the head between the eyes.

interradial membranes Membranes between fin rays.

introduced Refers to nonnative species that have been transplanted outside their natural area of distribution.

isthmus The narrow part of the breast that projects forward between (and separates) the gill chambers.

jugular Refers to the throat region; pelvic fins are jugular when they are inserted in front of the pectoral fin bases.

keel Scales or tissue forming a sharp edge or ridge.

labial teeth Teeth present on the lips of some species such as mullets.

lachrymal A dermal bone of the skull located in front of the orbit; the anteriormost bone in the suborbital series, and often the largest.

larva (pl. larvae) The early immature form of an animal that changes structurally as it becomes an adult.

lateral Regarding the side.

lateral line A system of sensory tubules communicating to the body surface by pores; referring most often to a longitudinal row of scales that bear tubules and visible pores. The lateral line is considered incomplete if only the anterior scales have pores and complete if all scales in that row (to the base of the caudal fin) have pores.

lateral scales The scales on the side of the fish along the lateral line; or the series of scales along the midline of the side if no lateral line is present. When lateral line is incomplete, may be only those scales with a lateral line pore or tube. In the text, the latter series of scales is called "scales in lateral line."

lentic Standing-water habitat, such as a lake or pond.

leptocephalus Larval form found in eels, tarpons, ladyfishes, and bonefishes characterized by a clear, flattened, and ribbonlike body that shrinks in length during metamorphosis to the next developmental stage.

limnetic zone The region of open water in a lake from the shore extending downward to the maximum depth at which there is sufficient sunlight for photosynthesis.

lingual Pertaining to the tongue.

littoral Referring to the area along the shore of bodies of water.

lotic Running-water habitat, such as a river.

macrohabitat A habitat scale larger than the mesohabitat. In the context of this book, macrohabitat generally corresponds to the scale of individual creeks, rivers, or lakes.

mandible The main bone of the lower jaw.

mandibular pores Small openings along a tube that traverses the underside of each lower jaw (part of the lateral line system).

maxilla (maxillary) A bone of each upper jaw that lies immediately above (or behind) and parallel to the premaxilla.

melanophore A black pigment cell.

meristic Pertaining to the number of serial parts; for example, fin rays or lateral line scales.

mesionally In or toward the middle.

mesohabitat A habitat scale intermediate between microhabitat and macrohabitat. In streams, mesohabitat generally corresponds to the scale of individual pools, runs, and riffles.

metamorphosis Period of transformation from larval to adult form or to another larval or juvenile stage in species that have distinct stages of development.

microhabitat A habitat scale smaller than mesohabitat. In streams, microhabitat generally corresponds to the scale of a magnitude larger than the size of individual fish.

middorsal Pertaining to the midline of the back.

molar teeth Teeth with broad, flat surfaces adapted for grinding.

monomorphic With no external differences between males and females.

morphology The form and structure of organisms or parts of organisms.

morphometric Morphological characters that can be measured with a millimeter scale such as length of body parts; often expressed as proportions or percentages of the fish's standard length (SL).

mottled Marked with spots or blotches of different colors but not in a regular pattern.

mouth Inferior mouth: below snout, snout obviously overhanging mouth; oblique mouth: line of the mouth (when closed) at an angle of 45 degrees or greater; subterminal mouth: mouth slightly overhung by snout, not quite terminal; terminal mouth: tips of upper and lower jaw forming foremost part of the head; ventral mouth: mouth on ventral surface of head, as in sturgeons.

myomere A muscle segment.

myotome A muscle plate; a section of the repeated muscle units corresponding to the flakes of a cooked fish.

naked Without scales.

nape Dorsal part of the body from the occiput to the dorsal fin origin.

nares Nostrils; in fishes, each nostril usually has an anterior and a posterior narial opening, usually located above and in front of the eyes, respectively.

nonprotrusible Not capable of being protruded, extended, or thrust out; for example, the upper jaw or both jaws cannot project forward to form a tube or scoop.

nostril External opening of the olfactory organ.

nuptial tubercles See "breeding tubercles."

occiput The back part of the head or skull.

ocellus An eyelike spot, usually round with a light or dark border.

oligotrophic lake A nutrient-poor lake characterized by low production of plankton and fish and having considerable dissolved oxygen in the bottom waters (due to low organic content).

opercle The large posterior bone of the gill cover; it may be spiny, serrate, or entire (smooth).

opercular flap A posterior flaplike extension of the operculum, especially in sunfishes (the "ear tab" in *Lepomis* species).

opercular gill A rudimentary gill on the inner face of the operculum as in gars and sturgeons (= hemibranch).

operculum Bony flap covering the gills of fishes, also called the "gill cover."

orbit Eye socket.

origin (of fins) Anterior end of the base of a dorsal fin or anal fin.

otoliths Calcareous bodies with incremental structure (annuli), sometimes called "ear stones," found in the inner ear of bony fishes; the three otoliths in each inner ear respond to gravitational forces and changes in orientation of the fish as well as to vibrations caused by sound waves.

palatines A pair of cartilaginous bones in the roof of the mouth, frequently with teeth. One of four paired bones in the roof of the mouth (see Helfman et al. 1997 for skeletal details).

palatine teeth Teeth on the paired palatine bones that lie on the roof of the mouth.

papilla A small, soft, and rounded projection on the skin.

papillose Covered with papillae (as contrasted with plicate when referring to lips of suckers).

parietals Paired dermal bones in the roof of the skull, located between the frontal and occipital bones.

parr marks Nearly square or oblong blotches found along the sides of young trout before they have developed the adult color pattern.

pearl organ See "breeding tubercles."

pectinate Having teeth like a comb.

pectoral Referring to the anterior ventral portion of a fish; the breast.

pectoral fins The anteriormost paired fins, on the sides or on the breast, behind the head.

pelagic Of open water; usually referring to nonflowing water.

pelvic fins The ventral paired fins lying below the pectoral fins or between those and the anal fin.

peritoneum Membranous lining of the body cavity.

pharyngeal teeth Bony projections from the fifth pharyngeal (gill) arch.

physoclistous Having the swimbladder isolated from the esophagus.

physostomous Having the swimbladder connected to the esophagus by an open duct.

piscivorous Feeding on fish.

plicate Having parallel folds of soft ridges; for example, the grooved lips that are especially prominent in certain catostomids.

polymorphism The presence in a population of two or more distinct forms.

posteriad In a posterior direction.

posterior Regarding the tail end.

predorsal scales Scales on the midline of the dorsum between the back edge of the head and the dorsal fin origin.

premaxilla (premaxillary) A paired bone at the front of the upper jaw. The right and left premaxillae join anteriorly and form all or part of the border of the jaw.

preopercle The sickle-shaped bone lying behind and below the eye; may be serrated or smooth.

preoperculomandibular canal A cephalic portion of the lateral line system that extends along the preopercle and mandible.

preorbital The bone forming the anterior rim of the eye socket and extending forward on the side of the snout.

primary feeder (primary consumer) A herbivorous fish occupying the second trophic level.

principal rays Fin rays that extend to the distal margin of median fins; enumerated by counting only one unbranched ray anteriorly plus subsequent branched rays.

procurrent (rudimentary) rays Small, contiguous rays at the anterior bases of the dorsal, caudal, and anal fins of many fishes; not included in the count of principal fin rays.

protractile Capable of being thrust out; used herein when the upper jaw is completely separated from the snout by a continuous groove (frenum absent).

proximal Nearest the body; center or base of attachment.

pseudobranch An accessory gill-like structure on the inner surface of the operculum.

pustulose Having small blisterlike projections or elevations.

pyloric caeca Fingerlike blind tubes at the junction of the stomach and intestine in most fishes.

ray An articulated or jointed rod that supports the membrane of a fin.

recurved Curved upward and inward.

redd The gravel nest of salmonid fishes.

reservoir Used herein to refer to a man-made impoundment of approximately 200 ha or more at normal pool level.

retrorse Turned backward (of maxillary).

rhombic (or rhomboid) scale A heavy, bony, diamond-shaped, nonoverlapping scale found in gars.

robust Strongly or stoutly built; husky.

rough fish Nongame fishes that are generally considered to be undesirable in certain aquatic habitats from the standpoint of angling or fisheries management.

rudimentary rays See "procurrent rays."

sacculus A saclike chamber forming a small portion of the inner ear of a fish and containing the otolith known as the sagitta.

saddle Rectangular or linear bars or bands that cross the back and extend partially or entirely downward across the sides.

salinity The amount of dissolved salts in water. Note that we continue the use of ppt (parts per thousand or grams per liter) as the unit for salinity because it is still widely used and recognized. Currently, salinity is measured indirectly as the ratio of electrical conductivity of (sea)-water and that of the standard potassium chloride sample. This yields a nondimensional number, but many use "practical salinity units" (psu). If a unit is given, psu should be used when salinity is measured using conductivity, and ppt can be used if salinity is measured directly.

sawtooth scales Sharply pointed scales present on the belly of herring/shads.

scale Flattened, stiff, bony or horny structure on the skin, embedded in the scale pocket.

scale radius A line radiating from the focus, or center, of a scale to the margin.

scute A modified scale in the form of a horny or bony plate, often spiny or keeled.

semelparous Referring to a fish that ordinarily spawns once and subsequently dies (e.g., the American and European eels).

serrae Teeth of a sawlike organ or structure.

serrate Toothed or notched on the edge, like a saw.

snout Part of the head anterior to the eye but not including the lower jaw.

snout length The distance from the most anterior point of the head or upper jaw to the front margin of the eye socket.

soft ray Soft fin ray.

spatulate Paddle shaped or spoon shaped.

species The fundamental taxonomic category; subdivision of a genus; a group of organisms that naturally or potentially interbreed, are reproductively isolated from other such groups, and are usually morphologically separable from them.

spine A stiff, bony supporting structure in the fin membrane. Unlike rays, spines are never branched or segmented.

spiracle An opening on the back part of the head (above and behind the eye) of paddlefish and some sturgeons representing a primitive gill cleft.

spiral valve A spiral fold of mucous membrane projecting into the intestines.

standard length (SL) The distance from the tip of the snout to the end of the hypural plate.

stream order The relative position of a stream segment in a drainage network. The smallest, unbranched, intermittent tributaries are first order; the joining of two first-order streams produces a second-order stream; the joining of two second-order streams produces a third-order stream; and so on.

striate Streaked or striped by narrow parallel lines or grooves.

subopercle The bone immediately below the opercle in the operculum.

suborbital The thin bone forming the lower part of the orbital rim.

subspecies A taxonomic category, subdivision of a species; a group of local populations inhabiting a geographic subdivision of the species' range and differing taxonomically from other populations of the species.

subterminal mouth A mouth that is positioned slightly ventrally rather than straight forward and from the front of the head; the upper jaw slightly exceeds the lower jaw.

subtriangular Almost triangular.

supramaxilla The small, wedge-shaped, movable bone adherent to the upper edge of the maxilla near its posterior tip.

supraorbital canal A paired branch of the lateral line system that extends along the top of the head between the eyes and forward onto the snout.

supratemporal canal A branch of the lateral line system that extends across the top of the head at the occiput, connecting the lateral canals.

swimbladder (also "air bladder," "gas bladder") The gas-filled sac in the dorsal portion of the body cavity of most fishes that aids in buoyancy, sound reception and production, and respiration in some species.

sympatric Two or more populations occupying identical or broadly overlapping geographical areas.

symphysis The articulation of two bones in the median plane of the body, especially that of the two halves of the lower jaw (mandibles) at the chin.

syntopic Occurring at the same locations.

teardrop A vertical dark bar under the eye; synonym for subocular bar.

teleost A name applied to fishes usually having the skeleton fully ossified; in other words a "bony fish," in contrast to a shark, which is a cartilaginous fish.

terminal mouth Mouth located at the extreme anterior tip of the head.

thermocline The stratum of water between the epilimnion and hypolimnion of a stratified lake. It is generally characterized by a water temperature gradient of more than 1 °C per meter of depth.

threatened A form or forms likely to become endangered within the foreseeable future if certain conditions continue to deteriorate.

thoracic Pertaining to the chest region in fishes; pelvic fins are thoracic when inserted below the pectoral fins.

tooth formula The numerical expression of the number of teeth, usually pharyngeal, in each of one or more rows.

total length Length from the tip of the snout to the tip of the caudal fin (TL).

tricuspid Having three points (applicable especially to teeth).

trophic level A feeding level; for example, a herbivore or a first-level carnivore.

truncate fin shape A fin with the posterior edge squared off.

tubercle A small projection or lump; refers herein to keratinized or osseus structures developed during the breeding period; that is, breeding tubercles.

tuberculate Having tubercles.

unicuspid Having one point (applicable especially to teeth)

vent The anus; the external opening of the alimentary canal.

venter The belly or lower sides of a fish.

ventral Pertaining to the lower surface.

vertical fins The fins (dorsal, anal, and caudal) on the median (center) line of the body, in contrast to the paired fins (pectoral and pelvic fins).

vermiculate Wormlike; marked with irregular or wavy lines.

vermiform Wormlike in shape.

villiform In the form of villi (fingerlike projections); said of teeth that are slender and crowded closely together in bands.

vomer A median bone, usually bearing teeth, at the anterior extremity of the roof of the mouth.

Weberian apparatus A series of ossicles (small bones) that conduct vibrations or pressure changes from the swim-bladder to the ear, involving the first four or five fused vertebrae behind the head; in North American fresh water, restricted to species of the families Cyprinidae, Catostomidae, Characidae, and Ictaluridae.

Literature Cited

Abbott, C. C. 1861. Notes on the habits of *Aphredoderus sayanus*. Proc. Acad. Nat. Sci. Phila. 1861:95–96. Cited in D. A. Etnier and W. C. Starnes. 1993. The fishes of Tennessee. University of Tennessee Press, Knoxville.

Abbott, C. C. 1868. List of vertebrate animals of New Jersey. Pages 807–808 *in* G. C. Cook, editor. The geology of New Jersey. Board of Managers, Newark, N.J.

Abbott, C. C. 1870. Notes on fresh-water fishes of New Jersey. Am. Nat. 4:107.

Abbott, C. C. 1884. A naturalist rambles about home. D. Appleton, New York.

Able, K. W., and M. P. Fahay. 1998. The first year in the life of estuarine fishes in the Middle Atlantic Bight. Rutgers University Press, New Brunswick, N.J.

Adams, C. C., and T. L. Hankinson. 1932. Ecology and economics of Oneida Lake fish. Roosevelt Wildl. Ann., Syracuse University, Syracuse, N.Y.

Aho, J. M., C. S. Anderson, K. B. Floyd, and M. T. Negus. 1986a. Patterns in fish species assemblage structure and dynamics in waters of the Savannah River Plant. Savannah River Ecology Laboratory Report SREL-27 UC-66e, Savannah River Ecology Laboratory, Aiken, S.C.

Aho, J. M., C. S. Anderson, and J. W. Terrell. 1986b. Habitat suitability index models and instream flow suitability curves: redbreast sunfish. U.S. Fish Wildl. Serv. Biol. Rep. 82(10.119).

Aiken, K. 1998. Reproduction, diet and population structure of the mountain mullet, *Agonostomus monticola*, in Jamaica, West Indies. Environ. Biol. Fishes 53:347–352.

Albanese, B. 2000a. Reproduction behavior and spawning microhabitat of the flagfin shiner *Pteronotropis signipinnis*. Am. Midl. Nat. 143:84–93.

Albanese, B. 2000b. Life-history of the flagfin shiner, *Pteronotropis signipinnis*, from a blackwater stream in southeastern Mississippi. Ecol. Freshw. Fish 9:219–228.

Alberts, J. J., J. W. Bowling, J. E. Schindler, and D. E. Kyle. 1988. Seasonal dynamics of physical and chemical properties of a warm monomictic reservoir. Int. Verein. theoret. ange. Limnol. Verhand. 23:176–180.

Allen, D. E., and C. S. Thomason. 1993. Survey and inventory of fisheries resources in the Combahee River. Study completion report F-32. South Carolina Wildlife and Marine Resources Department, Columbia.

Anderson, A. A., C. Hubbs, K. O. Winemiller, and R. J. Edwards. 1995. Texas freshwater fish assemblages following three decades of environmental change. Southwest. Nat. 40:314–321.

Anderson, W. D. Jr. 1964. Fishes of some South Carolina Coastal Plain streams. Q. J. Fla. Acad. Sci. 27:31–54.

Andrews, A. H., E. J. Burton, K. H. Coale, G. M. Cailliet, and R. E. Crabtree. 2001. Radiometric age validation of Atlantic tarpon, *Megalops atlanticus*. Fish. Bull. 99:389–398.

Angermeier, P. L. 1985. Spatio-temporal patterns of foraging success for fishes in an Illinois stream. Am. Midl. Nat. 114:342–359.

Angus, R. A., and W. M. Howell. 1996. Geographic distributions of eastern and western mosquitofishes (Poeciliidae: *Gambusia*): delineations of ranges using fin ray counts. Southeast. Fishes Counc. Proc. 33:1–6.

Annett, C. A. 1998. Hunting behavior of Florida largemouth bass, *Micropterus salmoides floridanus*, in a channelized river. Environ. Biol. Fishes 53:75–87.

Applegate, R. L., J. W. Mullan, and D. I. Morris. 1966. Food and growth of six centrarchids from shoreline areas of Bull Shoals Reservoir. Proc. Annu. Conf. Southeast. Assoc. Game Fish Comm. 19:469–482.

Askerov, T. A. 1975. Survival rate and oxygen consumption of juvenile wild carp maintained under different conditions. J. Hydrobiol. (Gidrobio. Zh.) 11:67–68.

Astanin, L. P., and L. M. Trofimova. 1969. Comparative study of the food, growth, and fecundity of common carp and domesticated carp (*Cyprinus carpio* L.) in Yegorlyk Reservoir. J. Ichthyol. 9:354–363.

Atkinson, C. E. 1951. Feeding habits of adult shad (*Alosa sapidissima*) in fresh water. Ecology 32:556–557.

Atlantic States Marine Fisheries Commission. 1999. Amendment 1 to the Interstate Fishery Management Plan for Shad and River Herring. Fish Manage. Rep. 35:77.

Auer, N. A., editor. 1982. Identification of larval fishes of the Great Lakes basin with emphasis on the Lake Michigan drainage. Great Lakes Fish. Comm. Spec. Pub. 82-3, Ann Arbor, Mich.

Avila, V. L. 1976. A field study of nesting behavior of male bluegill sunfish (*Lepomis macrochirus* Rafinesque). Am. Midl. Nat. 6(1):195–206.

Avise, J. C., and M. J. Van Den Avyle. 1984. Genetic analysis of reproduction of hybrid white bass × striped bass in the Savannah River. Trans. Amer. Fish. Soc. 113:563–570.

Bachmann, R. W., B. L. Jones, D. D. Fox, M. Hoyer, L. A. Bull, and D. E. Canfield Jr. 1996. Relations between trophic state indicators and fish in Florida (U.S.A.) lakes. Can. J. Fish. Aquat. Sci. 53:842–855.

Bade, E. 1931. [No title.] Pages 261–820 *in* Das Süsswasser-Aquarium. Die Flora und Fauna des Süsswassers und ihre Pflege im Zimmer-Aquarium [in German]. Teil 2: Die Süsswasser-Fauna, 5th edition. Fritz Pfenningstorff, Berlin.

Bailey, R. M., H. E. Winn, and C. L. Smith. 1954. Fishes from the Escambia River, Alabama and Florida, with ecologic and taxonomic notes. Proc. Acad. Nat. Sci. Phila. 106:109–164.

Bain, M. B. 1997. Atlantic and shortnose sturgeons of the Hudson River; common and divergent life history attributes. Environ. Biol. Fishes 48:347–358.

Bain, M. B., and J. L. Bain. 1982. Habitat suitability index models: coastal stocks of striped bass. FWS/OBS-82/10.1:29. U.S. Fish and Wildlife Service, Washington, D.C.

Bain, M. B., and L. A. Helfrich. 1983. Role of male parental care in survival of larval bluegills. Trans. Am. Fish. Soc. 112:47–52.

Baker, S. 2002. Food habits and food electivity of the turquoise darter, *Etheostoma inscriptum*, in a Georgia Piedmont stream. J. Freshw. Ecol. 17:385–390.

Ballantyne, P. K., and P. W. Colgan. 1978a. Sound production during agonistic and reproductive behaviour in the pumpkinseed (*Lepomis gibbosus*), the bluegill (*L. macrochirus*), and their hybrid sunfish: I. Context. Biol. Behav. 3:113–135.

Ballantyne, P. K., and P. W. Colgan. 1978b. Sound production during agonistic and reproductive behavior in the pumpkinseed (*Lepomis gibbosus*), the bluegill (*L. macrochirus*), and their hybrid sunfish: II. Recipients. Biol. Behav. 3:207–220.

Ballek, M. S. 1994. Reproduction and early life history of the redfin pickerel (*Esox americanus americanus*). Master's thesis, University of South Carolina, Columbia.

Barney, R. L., and B. J. Anson. 1920. Life history and ecology of pygmy sunfish, *Elassoma zonatum*. Ecology 1:241–256.

Barr, B. R., and J. J. Ney. 1993. Predation by flathead catfish (*Pylodictus olivaris*) on black bass and sunfish populations in a New River impoundment. Va. J. Sci. 44:145.

Barret, T. 2000. History of the robust redhorse in the Savannah River. Presented at the 2000 Southern Division of the American Fisheries Society Midyear Meeting, Savannah, Ga.

Barwick, D. H. 1981. Fecundity of the black crappie in a reservoir receiving heated effluent. Progr. Fish-Cult. 43(3):153–154.

Bass, D. G., and V. G. Hitt. 1974. Ecological aspects of the redbreast sunfish, *Lepomis auritus*, in Florida. Proc. Annu. Conf. Southeast. Assoc. Game Fish Comm. 28:296–396.

Battle, H. I. 1940. The embryology and larval development of the goldfish (*Carassius auratus* L.) from Lake Erie. Ohio J. Sci. 40:82–93.

Baumann, P. C., and J. F. Kitchell. 1974. Diel patterns of distribution and feeding of bluegill (*Lepomis macrochirus*) in Lake Wingra, Wisconsin. Trans. Am. Fish. Soc. 2:255–260.

Beach, M. L. 1974. Food habits and reproduction of the taillight shiner, *Notropis maculatus* (Hay), in central Florida. Fla. Sci. 37:5–16.

Becker, G. C. 1983. Fishes of Wisconsin. University of Wisconsin Press, Madison.

Becker, H. R. 1923. The habitat of *Aphredoderus sayanus* in Kalamazoo County, Michigan. Occas. Pap. Mus. Zool. Univ. Mich. 138.

Beckman, W. C. 1952. Guide to the fishes of Colorado. Univ. Colo. Mus. Leafl. 11:110.

Beecher, H. A. 1979. Anomalous occurrence of lip projection on *Carpiodes cyprinus*. Fla. Sci. 42:62–63.

Beecher, H. A. 1980. Habitat segregation of Florida carpsuckers (Osteichthyes: Catostomidae: *Carpiodes*). Fla. Sci. 43:92–97.

Beecher, H. A., and W. C. Hixson. 1982. Seasonal abundance of fishes in three northwest Florida rivers. Fla. Sci. 45:145–171.

Beecher, H. A., W. C. Hixson, and T. S. Hopkins. 1977. Fishes of a Florida oxbow lake and its parent river. Fla. Sci. 40:140–148.

Behnke, R. J. 1992. Native trout of western North America. Am. Fish. Soc. Monogr. 6:275.

Belk, M. C. 1993. Growth and mortality of juvenile sunfishes (*Lepomis* sp.) under heavy predation. Ecol. Freshw. Fish 2:91–98.

Belk, M. C. 1995. Variation in growth and age at maturity in bluegill sunfish: genetic or environmental effects? J. Fish Biol. 47:237–247.

Belk, M. C. 1998. Predator-induced delayed maturity in

bluegill sunfish (*Lepomis macrochirus*): variation among populations. Oecologia 113:203–209.

Belk, M. C., and L. S. Hales Jr. 1993. Predation-induced differences in growth and reproduction of bluegills (*Lepomis macrochirus*). Copeia 1993(4):1034–1044.

Belk, M. C., and D. D. Houston. 2002. Bergmann's rule in ectotherms: a test using freshwater fishes. Am. Nat. 160:803–808.

Benke, A. C., K. A. Parsons, and S. M. Dhar. 1991. Population and community patterns of invertebrate drift in an unregulated Coastal Plain river. Can. J. Fish. Aquat. Sci. 48:811–823.

Benke, A. C., and J. B. Wallace. 1990. Wood dynamics in Coastal Plain blackwater streams. Can. J. Fish. Aquat. Sci. 47:92–99.

Bennett, D. H., and J. W. Gibbons. 1972. Food of largemouth bass (*Micropterus salmoides*) in a South Carolina reservoir receiving heated effluent. Trans. Am. Fish. Soc. 101(4):650–654.

Bennett, D. H., and J. W. Gibbons. 1975. Reproductive cycles of largemouth bass (*Micropterus salmoides*) in a cooling reservoir. Trans. Am. Fish. Soc. 104(1):77–82.

Bennett, D. H., and R. W. McFarlane. 1983. The fishes of the Savannah River Plant: National Environmental Research Park. University of Georgia, Savannah River Ecology Laboratory, Aiken, S.C.

Bennett, G. W. 1937. The growth of the largemouthed black bass, *Huro salmoides*, in the waters of Wisconsin. Copeia 1937:104–118.

Bentzen, P., G. C. Brown, and W. C. Leggett. 1989. Mitochondrial DNA polymorphism, population structure, and life history variation in American shad (*Alosa sapidissima*). Can. J. Fish. Aquat. Sci. 46:1446–1454.

Berg, L. S. 1964. Freshwater fishes of the USSR and adjacent countries, fourth edition, volume 2. Israeli Program for Scientific Translations, Jerusalem.

Berkman, H. E., and C. F. Rabeni. 1987. Effects of siltation on stream fish communities. Environ. Biol. Fishes 18:285–294.

Bernard, G., and M. G. Fox. 1997. Effects of body size and density on overwinter survival of age-0 pumpkinseeds. N. Am. J. Fish. Manage. 17:581–590.

Berry, F. H., and L. R. Rivas. 1962. Data on six species of needlefishes (Belonidae) from the western Atlantic. Copeia 1962:152–160.

Bertschy, K. A., and M. G. Fox. 1999. The influence of age-specific survivorship on pumpkinseed sunfish life histories. Ecology 80(7):2299–2313.

Besler, D. A. 1999. Utility of scales and whole otoliths for aging largemouth bass in North Carolina. Proc. Annu. Conf. Southeast. Assoc. Fish Wildl. Agencies 53:119–129.

Bettross, E. 2000. Savannah River trout stocking evaluation. Presented at the 2000 Southern Division of the American Fisheries Society Midyear Meeting, Savannah, Ga. [Abstract only.]

Bigelow, H. B., and W. C. Schroeder. 1953. Fishes of the Gulf of Maine. U.S. Fish Wildl. Serv. Fish. Bull. 53.

Bilkovic, D. M. 2000. Assessment of spawning and nursery habitat suitability for American shad (*Alosa sapidissima*) in the Mattaponi and Pamunkey rivers. Doctoral dissertation. College of William and Mary, Williamsburg, Va.

Black, D. A., and W. M. Howell. 1979. The North American mosquitofish, *Gambusia affinis*: a unique case in sex chromosome evolution. Copeia 1979:509–513.

Blandon, M. D., R. Ward, T. L. King, W. J. Karel, and W. J. Monaghan Jr. 2001. Preliminary genetic population structure of southern flounder, *Paralichthys lethostigma*, along the Atlantic Coast and Gulf of Mexico. Fish. Bull. 99(4):671–678.

Block, C. J., J. R. Spotila, E. A. Standora, and J. W. Gibbons. 1984. Behavioral thermoregulation of largemouth bass, *Micropterus salmoides*, and bluegill, *Lepomis macrochirus*, in a nuclear reactor cooling reservoir. Environ. Biol. Fishes 11(1):41–52.

Blumer, L. S. 1985. Reproductive natural history of the brown bullhead *Ictalurus nebulosus* in Michigan. Am. Midl. Nat. 114(2):318–330.

Blumer, L. S. 1986. The function of parental care in the brown bullhead *Ictalurus nebulosus*. Am. Midl. Nat. 115:234–238.

Böhlke, J. 1956. A new pygmy sunfish from southern Georgia. Notulae Naturae. Proc. Acad. Nat. Sci. Phila. 294:11.

Boltz, J. M., and J. R. Stauffer Jr. 1986. Branchial brooding in the pirate perch, *Aphredoderus sayanus* (Gilliams). Copeia 1986:1030–1031.

Booth, D. J., and J. A. Keast. 1986. Growth energy partitioning by juvenile bluegill sunfish, *Lepomis macrochirus* Rafinesque. J. Fish Biol. 28:37–45.

Borodin, N. 1925. Biological observations on the Atlantic sturgeon (*Acipenser sturio*). Trans. Am. Fish. Soc. 55:184–190.

Boschung, H. T., and A. F. Hemphill. 1960. Marine fishes collected from inland streams of Alabama. Copeia 1960:73.

Bosley, T. R., and J. V. Conner. 1984. Geographic and temporal variation in numbers of myomeres in fish larvae from the lower Mississippi River. Trans. Am. Fish. Soc. 113:238–242.

Boughton, D. A., B. B. Collette, and A. R. McCune. 1991. Heterochrony in jaw morphology of needlefishes (Teleostei: Belonidae). Syst. Zool. 40:329–354.

Bowers, J. A. 1992. Synoptic surveys of regional reservoirs in

South Carolina, 1988–1989. Savannah River Laboratory WSRC-92-RP-368, Westinghouse Savannah River Co., Aiken, S.C.

Boxrucker, J. 1986. A comparison of the otolith and scale methods for aging white crappies in Oklahoma. N. Am. J. Fish. Manage. 6:122–125.

Bozeman, E. L., and M. J. Van Den Avyle. 1989. Species profiles: life histories and environmental requirements of coastal fishes and invertebrates (South Atlantic)—alewife and blueback herring. Biol. Rep. 82(11.111), TR EL-82-4. U.S. Fish and Wildlife Service, National Wetlands Research Center, Washington D.C., and U.S. Army Corps of Engineers, Coastal Ecology Group—Waterways Experimental Station, Vicksburg, Miss.

Braccia, A., and D. P. Batzer. 2001. Invertebrates associated with woody debris in a southeastern U.S. forested floodplain wetland. Wetlands 21:18–31.

Brayton, S. L. 1981. Reproductive biology, energy content of tissues, and annual production of rainbow trout (*Salmo gairdneri*) in the South Fork of the Holston River, Virginia. Master's thesis, Virginia Polytechnic Institute and State University, Blacksburg.

Breder, C. M. Jr. 1936. The reproductive habits of the North American sunfishes (family Centrarchidae). Zoologica 21(1):1–47.

Breder, C. M. 1940. The nesting behavior of *Eupomotis gibbosus* (Linnaeus) in a small pool. Zoologica 23:353–360.

Breder, C. M. Jr. 1962. Effects of a hurricane on the small fishes of a shallow bay. Copeia 1962:459–462.

Breder, C. M. Jr., and A. C. Redmond. 1929. The bluespotted sunfish: a contribution to the life history and habits of *Enneacanthus* with notes on other Lepominae. Zoologica 9(10):379–401.

Breder, C. M. Jr., and D. E. Rosen. 1966. Modes of reproduction in fishes. Natural History Press, Garden City, N.Y.

Brenneman, W. M. 1992. Ontogenetic aspects of upper and lower stream reach cyprinid assemblages in a south Mississippi watershed. Doctoral dissertation, University of Southern Mississippi, Hattiesburg.

Brill, J. S. 1977. Notes on the abortive spawning of the pirate perch, *Aphredoderus sayanus*, with comments on sexual distinctions. Am. Currents 5(4):10–16.

Brim, J. 1991. Coastal plain fishes: floodplain utilization and the effects of impoundments. Master's thesis, University of South Carolina, Columbia.

Brock, V. E. 1960. The introduction of aquatic animals into Hawaiian waters. Int. Rev. ges. Hydrobiol. 45:463–480.

Brogensky, R. Y. 1960. Early development of the carp. Pages 129–149 *in* C. G. Krevanoski, editor. Works on the early development of bony fishes. Stud. A. N. Stevertsova Inst. Anim. Morphol., Sov. Acad. Sci. 28.

Brown, J. A., and P. W. Colgan. 1981. The use of lateral-body bar markings in identification of young-of-year sunfish (*Lepomis*) in Lake Opinicon, Ontario. Can. J. Zool. 59:1852–1855.

Brown, J. L. 1958. Geographic variation in southeastern populations of the cyprinodont fish *Fundulus notti* (Agassiz). Am. Midl. Nat. 59:477–488.

Bryant, B. 1981. Chemosensory social behavior in catfish: responses in a social context. Biol. Bull. Mar. Biol. Lab. Woods Hole 161:340–341.

Burbidge, R. G. 1974. Distribution, growth, selective feeding, and energy transformations of young-of-the-year blueback herring, *Alosa aestivalis* (Mitchill), in the James River, Virginia. Trans. Am. Fish. Soc. 1974:297–311.

Burgess, G. H., C. R. Gilbert, V. Guillory, and D. C. Taphorn. 1977. Distributional notes on some north Florida freshwater fishes. Fla. Sci. 40:33–41.

Burns, J. R. 1976. The reproductive cycle and its environmental control in the pumpkinseed, *Lepomis gibbosus* (Pisces: Centrarchidae). Copeia 1976(3):449–455.

Burr, B. M., and L. M. Page. 1975. Distribution and life history notes on the taillight shiner *Notropis maculatus* in Kentucky. Ky. Acad. Sci. 36:71–74.

Burr, B. M., and M. L. Warren Jr. 1986. A distributional atlas of Kentucky fishes. Ky. Nat. Preserv. Comm. Sci. Tech. Ser. 4:398.

Burr, B. M., M. L. Warren Jr., and K. S. Cummings. 1988. New distributional records of Illinois fishes with additions to the known fauna. Trans. Ill. Acad. Sci. 81:163–170.

Burr, B. M., M. L. Warren Jr., and R. C. Heidinger. 1989. Reproductive potential and habitat of the ironcolor shiner, *Notropis chalybaeus*, in Illinois. Final report submitted to Illinois Endangered Species Board.

Buynak, G. L., and H. W. Mohr. 1978. Larval development of the redbreast sunfish (*Lepomis auritus*) from the Susquehanna River. Trans. Am. Fish. Soc. 107(4):600–604.

Cahn, A. R. 1927. An ecological study of southern Wisconsin fishes: the brook silverside (*Labidestes sicculus*) and the sisco (*Leucichthys artedi*) in their relations to the region. Ill. Biol. Monogr. 11:1–151.

Cargnelli, L., and M. R. Gross. 1997. The temporal dimension in fish recruitment: birth date, body size, and size-dependent survival in a sunfish (bluegill: *Lepomis macrochirus*). Can. J. Fish. Aquat. Sci. 53:360–367.

Carlander, K. D. 1953. Handbook of freshwater fishery biology with the supplement. Brown, Dubuque, Iowa.

Carlander, K. D. 1969. Handbook of freshwater fishery biology. Volume 1: Life history data on freshwater fishes of the United States and Canada, exclusive of the Perciformes. Iowa State University Press, Ames.

Carlander, K. D. 1977. Handbook of freshwater fishery biology. Volume 2: Life history data on centrarchid fishes of the United States and Canada. Iowa State University Press, Ames.

Carlson, D. M., and K. W. Simpson. 1987. Gut contents of juvenile shortnose sturgeon in the upper Hudson estuary. Copeia 1987:796–802.

Carlson, J. E., and M. J. Duever. 1977. Seasonal fish population fluctuations in a south Florida swamp. Proc. Annu. Conf. Southeast. Assoc. Fish Wildl. Agencies 31:603–611.

Carr, A. F., and L. Giovannoli. 1950. The fishes of the Choloteca drainage of southern Honduras. Occas. Pap. Mus. Zool. Univ. Mich. 523:1–38.

Carr, A. F., and C. J. Goin. 1955. Guide to the reptiles, amphibians, and freshwater fishes of Florida. University of Florida Press, Gainesville.

Carr, A. F. Jr., and C. J. Goin. 1959. Reptiles, amphibians and freshwater fishes of Florida. University of Florida Press, Gainesville.

Carr, M. G., E. J. Carr, and S. R. Johnson. 1987. Seasonal changes in territoriality and frequency of agonistic behavior in two densities of juvenile brown bullhead, *Ictalurus nebulosus*. Environ. Biol. Fishes 19:175–181.

Carr, M. H. 1946. Notes on the breeding habitats of the eastern stumpknocker, *Lepomis punctatus punctatus* (Cuvier). J. Fla. Acad. Sci. 9:101–106.

Carscadden, J. E., and W. C. Leggett. 1975. Meristic differences in spawning populations of American shad, *Alosa sapidissima*: evidence for homing to tributaries in the St. John River, New Brunswick. J. Fish. Res. Board Can. 32:653–660.

Cartier, D., and E. Magnin. 1967. La croisance en longueuer et en poids de *Amia calva* L. de la region de Montreal. Can. J. Zool. 45:797–804.

Cashner, R. C., B. M. Burr, and J. S. Rogers. 1989. Geographic variation of the mud sunfish, *Acantharchus pomotis* (family Centrarchidae). Copeia 1989(1):129–141.

Cashner, R. C., and W. J. Matthews. 1988. Changes in the known Oklahoma fish fauna from 1973 to 1988. Proc. Okla. Acad. Sci. 68-1-7.

Cassani, J. R., and D. Maloney. 1991. Grass carp movement in two morphologically diverse reservoirs. J. Aquat. Plant Manage. 29:83–88.

Casterlin, M. E., and W. W. Reynolds. 1980. Diel activity of the bluespotted sunfish *Enneacanthus gloriosus*. Copeia 1980(2):344–345.

Castonguay, M., P. V. Hodson, and C. M. Couillard. 1994. Why is recruitment of American eel, *Anguilla rostrata*, declining in the St. Lawrence River and Gulf? Can. J. Fish. Aquat. Sci. 51:479–488.

Chan, T. Y., and A. J. R. Ribbink. 1990. Alternative reproductive behavior in fishes with particular reference to *Lepomis macrochira* and *Pseudocrenilabrus philander*. Environ. Biol. Fishes 28:249–256.

Chapleau, F. 1993. Pleuronectiform relationships—a cladistic reassessment. Bull. Mar. Sci. 52(1):516–540.

Chapleau, F., and A. Keast. 1988. A phylogenetic reassessment of the monophyletic status of the family Soleidae, with comments on the suborder Soleoidae (Pisces, Pleuronectiformes). Can. J. Zool. 66(12):2797–2810.

Chen, S. C. 1926. The development of the goldfish, *Carassius auratus*, as affected by being out of water, in distilled water, and solutions of alcohol. China J. Sci. Arts 4:294–303.

Chew, R. L. 1974. Early life history of the Florida largemouth bass. Fish Bull. 7:76. Florida Game and Freshwater Fish Commission, Tallahassee.

Chilton, E. W., and M. I. Muoneke. 1992. Biology and management of grass carp (*Ctenopharyngodon idella*, Cyprinidae) for vegetation control: a North American perspective. Rev. Fish Biol. Fish. 2:283–320.

Chittenden, M. E. Jr. 1976. Weight loss, mortality, feeding, and duration of residence of adult American shad, *Alosa sapidissima*, in fresh water. Fish. Bull. 74:151–157.

Christel-Rose, L. M. 1994. Historical wetlands mapping and GIS processing for the Savannah River Site database. EGG-11265-1018. EG&G Energy Measurements, Inc., Las Vegas, Nev.

Clark, F. W., and M. H. A. Keenleyside. 1967. Reproductive isolation between the sunfish *Lepomis gibbosus* and *L. macrochirus*. J. Fish. Res. Board Can. 24:495–514.

Clark, K. E. 1978. Ecology and life history of the speckled madtom, *Noturus leptacanthus* (Ictaluridae). Master's thesis, University of Richmond, Va.

Clemmer, G. H., R. D. Suttkus, and J. S. Ramsey. 1975. A preliminary checklist of endangered and rare fishes of Mississippi. Pages 6–11 *in* Preliminary list of rare and threatened vertebrates in Mississippi. Mississippi Game and Fish Commission, Jackson.

Clemmons, M. M., and D. G. Lindquist. 1983. Reproductive biology of the swamp darter in Lake Waccamaw, North Carolina. Page 12 *in* Abstracts of the 63rd Annual Meeting of the American Society of Ichthyologists and Herpetologists.

Cloutman, D. G., and R. D. Harrell. 1987. Life history notes on the whitefin shiner, *Notropis niveus* (Pisces: Cyprinidae), in the Broad River, South Carolina. Copeia 4:1037–1040.

Clugston, J. P. 1966. Centrarchid spawning in the Florida Everglades. Q. J. Fla. Acad. Sci. 29:137–143.

Clugston, J. P. 1973. The effects of heated effluents from a nuclear reactor on species diversity, abundance, reproduction, and movement of fish. Doctoral dissertation, University of Georgia, Athens.

Clugston, J. P., and E. L. Cooper. 1960. Growth of the common eastern madtom, *Noturus insignis*, in central Pennsylvania. Copeia 1960:9–16.

Clugston, J. P., J. L. Oliver, and R. Ruelle. 1978. Reproduction, growth, and standing crops of yellow perch in southern reservoirs. Am. Fish. Soc. Spec. Pub. 11:89–99.

Cochran, P. A. 1996. Cavity enhancement by madtoms (genus *Noturus*). J. Freshw. Ecol. 11:521–522.

Cohen, A. B. 1977. Life history of the banded sunfish (*Enneacanthus obesus*) in Green Falls Reservoir, Connecticut. Master's thesis, University of Connecticut, Storrs.

Colbourne, J. K., B. D. Neff, J. M. Wright, and M. R. Gross. 1996. DNA fingerprinting of bluegill sunfish (*Lepomis macrochirus*) using (GT)n microsatellites and its potential for assessment of mating success. Can. J. Fish. Aquat. Sci. 53:342–349.

Cole, C. F. 1967. A study of the eastern Johnny darter, *Etheostoma olmstedi storer* (Teleostei, Percidae). Chesapeake Sci. 8(1):28–51.

Colgan, P. W., W. A. Nowell, M. R. Gross, and J. W. Grant. 1979. Aggressive habituation and rim-circling in the social organization of bluegill sunfish (*Lepomis macrochirus*). Environ. Biol. Fishes 4:29–36.

Collette, B. B. 1962. The swamp darters of the subgenus *Hololepis* (Pisces, Percidae). Tulane Stud. Zool. 9:115–211.

Collette, B. B., and G. Klein-MacPhee, editors. 2002. Bigelow's and Schroeder's fishes of the Gulf of Maine. Smithsonian Institution Press, Washington, D.C.

Collins, M. R., S. G. Rogers, and T. I. J. Smith. 1996. Bycatch of sturgeon along the southern Atlantic coast of the USA. N. Am. J. Fish. Manage. 16:24–29.

Collins, M. R., S. G. Rogers, T. I. J. Smith, and M. L. Moser. 2000b. Primary factors affecting sturgeon populations in the southeastern United States: fishing mortality and degradation of essential habitats. Bull. Mar. Sci. 66:917–928.

Collins, M. R., and T. I. J. Smith. 1993. Characteristics of the adult segment of the Savannah River population of shortnose sturgeon. Proc. Annu. Conf. Southeast. Assoc. Fish Wildl. Agencies 47:485–491.

Collins, M. R., and T. I. J. Smith. 1997. Distributions of shortnose and Atlantic sturgeons in South Carolina. N. Am. J. Fish. Manage. 17:995–1000.

Collins, M. R., T. I. J. Smith, W. C. Post, and O. Pashuk. 2000a. Habitat utilization and biological characteristics of adult Atlantic sturgeon in two South Carolina rivers. Trans. Am. Fish. Soc. 129:982–988.

Collins, M. R., T. I. J. Smith, K. Ware, and J. Quatto. 1999. Culture and stock enhancement of shortnose and Atlantic sturgeons. Bull. Natl. Res. Inst. Agric., Suppl. 1:101–108.

Collins, N. C., and S. G. Hinch. 1993. Diel and seasonal variation in foraging activities of pumpkinseeds in an Ontario pond. Trans. Am. Fish. Soc. 122:357–365.

Congdon, B. C. 1995. Unidirectional flow and maintenance of genetic diversity in mosquitofish *Gambusia holbrooki* (Teleostei: Poeciliidae). Copeia 1995:162–172.

Conner, J. V., R. P. Gallagher, and M. F. Chatry. 1980. Larval evidence for natural reproduction of the grass carp (*Ctenopharyngodon idella*) in the lower Mississippi River. Pages 1–19 *in* L. A. Fuiman, editor. Proceedings of the Fourth Annual Larval Fish Conference. FWS/OBS-80/43. U.S. Fish and Wildlife Service, Biological Services Program, National Power Plant Team, Ann Arbor, Mich.

Conner, J. V., and R. D. Suttkus. 1986. Zoogeography of freshwater fishes of the western Gulf slope of North America. Pages 413–456 *in* C. H. Hocutt and E. O. Wiley, editors. The zoogeography of North American freshwater fishes. John Wiley and Sons, New York.

Conover, D. O., and B. E. Kynard. 1984. Field and laboratory observations of spawning periodicity and behavior of a northern population of the Atlantic silverside, *Menidia menidia* (Pisces: Atherinidae). Environ. Biol. Fishes 11:161–171.

Constantz, G. D. 1985. Allopaternal care in the tessellated darter, *Etheostoma olmstedi* (Pices: Percidae). Environ. Biol. Fishes 14:175–183.

Coomer, C. E. Jr., D. R. Holder, and C. D. Swanson. 1977. A comparison of the diets of redbreast sunfish and spotted sucker in a Coastal Plain stream. Proc. Annu. Conf. Southeast. Assoc. Fish Wildl. Agencies 31:587–596.

Cooner, R. W., and D. R. Bayne. 1982. Diet overlap in redbreast and longear sunfishes from small streams of east central Alabama. Proc. Annu. Conf. Southeast. Assoc. Fish Wildl. Agencies 36:106–114.

Cooper, G. P. 1936. Age and growth of the golden shiner (*Notemigonus crysoleucas auratus*) and its suitability for propagation. Mich. Acad. Sci. Arts 21:587–597.

Cooper, G. P. 1941. A biological survey of the Androscoggin and Kennebec river drainage systems in Maine. Fish Surv. Rep. 4:238. Maine Department of Inland Fish and Game.

Cooper, J. C., and T. T. Polgar. 1981. Recognition of year-class dominance in striped bass management. Trans. Am. Fish. Soc. 110:180–187.

Copp, G. H. 1989. The habitat diversity and fish reproductive function of floodplain ecosystems. Environ. Biol. Fishes 26:1–27.

Côté, I. M., and M. R. Gross. 1993. Reduced disease in offspring: a benefit of coloniality in sunfish. Behav. Ecol. Sociobiol. 33:269–274.

Courtenay, W. R., D. A. Hensley, J. N. Taylor, and J. A. McCann. 1984. Distribution of exotic fishes in the continental United States. Pages 41–77 *in* W. R. Courtenay and J. R. Stauffer, editors. Distribution biology and management of exotic fishes. Johns Hopkins University Press, Baltimore, Md.

Courtenay, W. R. Jr., D. P. Jennings, and J. D. Williams. 1991. Appendix 2: Exotic fishes. Pages 97–107 *in* C. R. Robins, R. M. Bailey, C. E. Bond, J. R. Brooker, E. A. Lachner, R. N. Lea, and W. B. Scott, editors. Common and scientific names of fishes from the United States and Canada, 5th edition. Am. Fish. Soc. Spec. Pub. 20. American Fisheries Society, Bethesda, Md.

Cowell, B. C., and B. S. Barnett. 1974. Life history of the taillight shiner, *Notropis maculatus*, in central Florida. Am. Midl. Nat. 91:282–293.

Cowell, B. C., and C. H. Resico Jr. 1975. Life history patterns in the coastal shiner, *Notropis petersoni*, Fowler. Fla. Sci. 38(2):113–121.

Crabtree, R. E., E. C. Cyr, D. C. Chaverri, W. O. McLarney, and J. M. Davis. 1997. Reproduction of tarpon (*Megalops atlanticus*) from Florida and Costa Rican waters and notes on their ages and growth. Bull. Mar. Sci. 61:271–285.

Crabtree, R. E., E. C. Cyr, and J. M. Dean. 1995. Age and growth of tarpon (*Megalops atlanticus*) from south Florida waters. Fish. Bull. 93:619–628.

Craig, J. F. 1996. Pike—biology and exploitation. Chapman and Hall Fish and Fisheries series 19. Chapman and Hall, New York.

Craig, J. F. 2000. Percid fishes—systematics, ecology and exploitation. Fish and Aquatic Resources series 3. Blackwell Science, Oxford, England.

Crance, J. H. 1984. Habitat suitability index models and instream flow suitability curves: inland stocks of striped bass. FWS/OBS-82/10.85. U.S. Fish and Wildlife Service, Washington, D.C.

Crawford, R. W. 1956. A study of the distribution and taxonomy of the percid fish, *Percina nigrofasciata* (Agassiz). Tulane Stud. Zool. 4:3–55.

Cross, D. G. 1969. Aquatic weed control using grass carp. J. Fish Biol. 1:27–30.

Crossman, E. J. 1960. Variations in number and asymmetry in branchiostegal rays in the family Esocidae. Can. J. Zool. 38:363–375.

Crossman, E. J. 1962. The redfin pickerel, *Esox americanus*, in North Carolina. Copeia 1962:114–123.

Crossman, E. J. 1978. Taxomony and distribution of North American esocids. Am. Fish. Soc. Spec. Pub. 11:13–26.

Crossman, E. J. 1996. Taxonomy and distribution. Pages 1–11 *in* J. F. Craig, editor. Pike—biology and exploitation. Chapman and Hall Fish and Fisheries series 19, Chapman and Hall, New York.

Crossman, E. J., and K. Buss. 1965. Hybridization in the family Esocidae. J. Fish. Res. Board Can. 22:1261–1292.

Crumpton, J. 1971. Food habits of longnose gar (*Lepisosteus osseus*) and Florida gar (*Lepisosteus platyrhincus*) collected from five central Florida lakes. Proc. Annu. Conf. Southeast. Assoc. Game Fish Comm. 24(1970):419–424.

Cruz, G. A. 1987. Reproductive biology and feeding habits of cuyamel, *Joturus pichardi*, and tepemechin, *Agonostomus monticola* (Pisces; Mugilidae) from Río Platano, Mosquitia, Honduras. Bull. Mar. Sci. 40:63–72.

Cudmore-Vokey, B., and E. J. Crossman. 2000. Checklists of the fish fauna of the Laurentian Great Lakes and their connecting channels. Can. Manuscr. Rep. Fish. Aquat. Sci. 2550.

Curran, H. W., and D. T. Ries. 1937. Fisheries investigations in the lower Hudson River. Pages 124–145 *in* A biological survey of the lower Hudson watershed. 26th Annu. Rep. New York State Conserv. Dep. (Suppl.), Albany.

Curry, K. D., and A. Spacie. 1984. Differential use of stream habitat by spawning catostomids. Am. Midl. Nat. 111:267–279.

Dadswell, M. J. 1976. Biology of the shortnose sturgeon, *Acipenser brevirostrum* LeSueur, in the St. John River estuary, New Brunswick. Trans. Can. Soc. Environ. Biol., Atlantic Chapter Annual Meeting 1975:20–72.

Dadswell, M. J., B. D. Taubert, T. S. Squires, D. Marchette, and J. Buckley. 1984. Synopsis of biological data on shortnose sturgeon, *Acipenser brevirostrum* Lesueur 1818. NOAA/NMFS Tech. Rep. 14.

Dahlberg, M. D. 1971. Fecundity of yellow perch, *Perca flavescens* Mitchill, in the Patuxent River, Maryland. Chesapeake Sci. 12:270–284.

Dahlberg, M. D. 1975. Guide to the coastal fishes of Georgia and nearby states. University of Georgia Press, Athens.

Dahlberg, M. D., and D. C. Scott. 1971a. The freshwater fishes of Georgia. Bull. Ga. Acad. Sci. 29:1–64.

Dahlberg, M. D., and D. C. Scott. 1971b. Introductions of freshwater fishes in Georgia. Bull. Ga. Acad. Sci. 29:245–252.

D'Amours, J., S. Thibodeau, and R. Fortin. 2001. Comparison of lake sturgeon (*Acipenser fulvescens*), *Stizostedion* spp., *Catostomus* spp., *Moxostoma* spp., quillback (*Carpiodes cyprinus*), and mooneye (*Hiodon tergisus*) larval drift in Des Prairies River, Quebec. Can. J. Zool./Rev. Can. Zool. 79:1472–1489.

Danylchuk, A. J., and M. G. Fox. 1994a. Seasonal reproductive patterns of pumpkinseed (*Lepomis gibbosus*) populations with varying body size characteristics. Can. J. Fish. Aquat. Sci. 51:490–500.

Danylchuk, A. J., and M. G. Fox. 1994b. Age and size-

dependant variation in the seasonal timing and probability of reproduction among mature female pumpkinseed, *Lepomis gibbosus*. Environ. Biol. Fishes 39:119–127.
Darr, D. P. 1987. A water quality assessment of Langley Pond, Aiken County, South Carolina. Tech. Rep. 010-86. South Carolina Department of Health and Environmental Control, Columbia.
Darr, D. P. 1988. Letter to South Carolina Wildlife and Marine Resources Department reporting on collection of fish during tissue analysis sampling on Langley Pond. South Carolina Department of Health and Environmental Control, Columbia.
Davis, B. J., and R. J. Miller. 1967. Brain patterns in minnows of the genus *Hybopsis* in relation to feeding habits and habitat. Copeia 1967:1–39.
Davis, B. M., and J. W. Foltz. 1991. Food of blueback herring and threadfin shad in Jocassee Reservoir, South Carolina. Trans. Am. Fish. Soc. 120:605–613.
Davis, J. R. 1971. The spawning behavior, fecundity rates, and food habits of the redbreast sunfish in southeastern North Carolina. Proc. Annu. Conf. Southeast. Assoc. Game Fish Comm. 25:556–560.
Davis, J. R., and D. E. Louder. 1971. Life history and ecology of the cyprinid fish *Notropis petersoni* in North Carolina waters. Trans. Am. Fish. Soc. 100:726–733.
Davis, S. M. 1980. American shad movement, weight loss and length frequencies before and after spawning in the St. Johns River, Florida. Copeia 1980(4):889–892.
Davis, W. L. 1980. A comparative food habit analysis of channel and blue catfishes in Kentucky and Barkley Lakes, Kentucky. Master's thesis, Murray State University, Murray, Ky. Published by Kentucky Department of Fish and Wildlife Resources.
Dean, B. 1895. Fishes and living fossils. Macmillan, New York.
Dean, B. 1899. On the dogfish (*Amia calva*), its habits and breeding. Annu. Rep. N.Y. Comm. Fish. Game For. 4(1898):246–256.
Dederen, L. H. T., R. S. E. Leuven, S. E. Wendelaar Bonga, and F. G. F. Oyen. 1986. Biology of the acid-tolerant fish species *Umbra pygmaea* (De Kay, 1842). J. Fish Biol. 28(3):307–326.
DeMeo, T. A. 2001. Report of the Robust Redhorse Conservation Committee annual meeting, 11–12 October 2000. Charlie Elliott Wildlife Center, Georgia.
DeMont, D. J. 1982. Use of *Lepomis macrochirus* Rafinesque nests by spawning *Notemigonus crysoleucas* (Mitchill) (Pisces: Centrarchidae and Cyprinidae). Brimleyana 8:61–63.
Derrick, P. A., and V. S. Kennedy. 1997. Prey selection by the hogchoker, *Trinectes maculatus* (Pisces: Soleidae), along summer salinity gradients in Chesapeake Bay, USA. Mar. Biol. 129:699–712.
Devaraj, K. V. 1974. Food of white catfish, *Ictalurus catus* (Linn.) (Ictaluridae) stocked in farm ponds. Int. Rev. ges. Hydrobiol. 59:147–151.
DeVries, D. R. 1990. Habitat use by bluegill in laboratory pools: where is the refuge when macrophytes are sparse and alternative prey are present? Environ. Biol. Fishes 29:27–34.
DeWoody, J. A., D. E. Fletcher, M. Mackiewicz, S. D. Wilkins, and J. C. Avise. 2000a. The genetic mating system of spotted sunfish (*Lepomis punctatus*): mate numbers and the influence of male reproductive parasites. Mol. Ecol. 9:2119–2128.
DeWoody, J. A., D. E. Fletcher, S. D. Wilkins, and J. C. Avise. 2000b. Parentage and nest guarding in the tesselated darter (*Etheostoma olmstedi*) assayed by microsatellite markers (Perciformes: Percidae). Copeia 2000:740–747.
DeWoody, J. A., D. E. Fletcher, S. D. Wilkins, and J. C. Avise. 2001. Genetic documentation of filial cannibalism in nature. Proc. Nat. Acad. Sci. USA 98:5090–5092.
DeWoody, J. A., D. E. Fletcher, S. D. Wilkins, W. S. Nelson, and J. C. Avise. 1998. Molecular genetic dissection of spawning, parentage, and reproductive tactics in a population of redbreast sunfish, *Lepomis auritus*. Evolution 52(6):1802–1810.
DeWoody, J. A., D. E. Fletcher, S. D. Wilkins, W. S. Nelson, and J. C. Avise. 2000c. Genetic monogamy and biparental care in an externally fertilizing fish, the largemouth bass (*Micropterus salmoides*). Proc. R. Soc. London B 267:2431–2437.
Dill, W. A., and A. J. Cordone. 1997. History and status of introduced fishes in California, 1871–1996. Manuscr. Fish Bull. Calif. Dep. Fish Game.
Dilts, E. W., and C. A. Jennings. 1999. Effects of fine sediment and gravel quality on survival to emergence of larval robust redhorse *Moxostoma robustum*. Presented at the 1999 Southern Division of the American Fisheries Society Midyear Meeting, Chattanooga, Tenn.
Dimmick, W. W. 1988. Ultrastructure of North American cyprinid maxillary barbels. Copeia 1988:72–80.
Ditty, J. G., and R. F. Shaw. 1996. Spatial and temporal distribution of larval striped mullet (*Mugil cephalus*) and white mullet (*M. curema*, family: Mugilidae) in the northern Gulf of Mexico, with notes on mountain mullet, *Agonostomus monticola*. Bull. Mar. Sci. 59:271–288.
Dixon, D. A., editor. 2003. Biology, management and protection of catadromous eels. Am. Fish. Soc. Symp. Ser. 33. American Fisheries Society, Bethesda, Md.
Dixon, D. A., J. D. Goins, J. E. Olney, and J. G. Loesch.

1998. Recapture of oxytetracycline (OTC) marked juvenile American shad (*Alosa sapidissima*) in the Pamunkey River, Virginia: 1995–1997. Proceedings of the 1998 Southern Division of the American Fisheries Society Midyear Meeting held in Lexington, Ky. [Abstract only.]

Dizon, A. E., R. M. Horrall, and A. D. Hasler. 1973. Meristic characters of some marine fishes of the western Atlantic Ocean. Fish. Bull. 71:301–317.

Dodson, J. J., and W. C. Leggett. 1974. Role of olfaction and vision in the behavior of American shad (*Alosa sapidissima*) homing to the Connecticut River from Long Island Sound. J. Fish. Res. Board Can. 31:1607–1619.

DOE [U.S. Department of Energy]. 1984. Final environmental impact statement: L-Reactor operation, Savannah River Plant, Aiken, S.C. DOE/EIS-0108. U.S. Department of Energy, Savannah River Operations Office, Aiken, S.C.

DOE [U.S. Department of Energy]. 1988. Impingement and entrainment at the river water intakes of the Savannah River Plant. CORR-860402. Savannah River Operations Office, Aiken, S.C.

DOE [U.S. Department of Energy]. 1990. Final environmental impact statement: continued operation of K-, L-, and P-Reactors at the Savannah River Site, Aiken, S.C. DOE/EIS-0147. U.S. Department of Energy, Savannah River Operations Office, Aiken, S.C.

DOE [U.S. Department of Energy]. 1993. Supplement 2 to Par Pond special environmental analysis, observed environmental impacts. Savannah River Operations Office, Savannah River Site, Aiken, S.C.

DOE [U.S. Department of Energy]. 1995. Environmental assessment for the natural fluctuation of water level in Par Pond and reduced water flow in Steel Creek below L-Lake at the Savannah River Site. DOE/EA/1070. Savannah River Operations Office, Aiken, S.C.

DOE [U.S. Department of Energy]. 1999. Environmental assessment for the Pond B Dam repair project at the Savannah River Site. DOE/EA-1285. Savannah River Operations Office, Aiken, S.C.

Dominey, W. J. 1980. Female mimicry in male bluegill sunfish—a genetic polymorphism? Nature 284:546–548.

Dominey, W. J. 1981a. Maintenance of female mimicry as a reproductive strategy in bluegill sunfish (*Lepomis macrochirus*). Environ. Biol. Fishes 6:59–64.

Dominey, W. J. 1981b. Anti-predator function of bluegill sunfish nesting colonies. Nature 290:586–588.

Douglas, N. H. 1974. Freshwater fishes of Louisiana. Louisiana Wildlife and Fish Commission, Baton Rouge.

Dovel, W. L., and T. J. Berggren. 1983. Atlantic sturgeon of the Hudson estuary, New York. N.Y. Fish Game J. 30:140–172.

Dovel, W. L., J. A. Mihursky, and A. J. McErlean. 1969. Life history aspects of the hogchoker, *Trinectes maculatus*, in the Patuxent River estuary, Maryland. Chesapeake Sci. 10:104–119.

Drake, M. T., J. E. Claussen, D. P. Philipp, and D. L. Pereira. 1997. A comparison of bluegill reproductive strategies and growth among lakes with different fishing intensities. N. Am. J. Fish. Manage. 17:496–507.

Drenner, R. W., F. deNoyelles Jr., and D. Kettle. 1982. Selective impact of filter-feeding gizzard shad on zooplankton community structure. Limnol. Oceanogr. 27:965–968.

Drenner, R. W., J. R. Mummert, F. deNoyelles Jr., and D. Kettle. 1984. Selective particle ingestion by a filter-feeding fish and its impact on phytoplankton community structure. Limnol. Oceanogr. 29:941–948.

Drenner, R. W., J. D. Smith, and S. T. Threlkeld. 1996. Lake trophic state and the limnological effects of omnivorous fish. Hydrobiologia 319:213–223.

Drenner, R. W., S. T. Threlkeld, and M. D. McCracken. 1986. Experimental analysis of the direct and indirect effects of an omnivorous filter-feeding clupeid on plankton community structure. Can. J. Fish. Aquat. Sci. 43:1935–1945.

Dudley, R. G., A. W. Mullis, and J. W. Terrell. 1977. Movements of adult striped bass (*Morone saxatilis*) in the Savannah River, Georgia. Trans. Am. Fish. Soc. 106:314–322.

Dunham, R. A., and W. F. Childers. 1980. Genetics and implications of the golden color morph in green sunfish. Progr. Fish-Cult. 42(3):160–163.

Dunham, R. A., C. Hyde, M. Masser, J. A. Plumb, R. O. Smitherman, R. Perez, and A. C. Ramboux. 1993. Comparison of culture traits of channel catfish, *Ictalurus punctatus*, and blue catfish, *I. furcatus*. J. Appl. Aquacult. 3:257–268.

Du Pont [E. I. du Pont de Nemours and Co.]. 1987. Comprehensive cooling water study. Volume 3: Radionuclide and heavy metal transport, M. W. Lower, editor. DP-1730-3. Savannah River Laboratory, Aiken, S.C.

Duryea, R., J. Donnelly, D. Gutherie, C. O'Malley, M. Romanowski, and R. Schmidt. 1996. *Gambusia affinis* effectiveness in New Jersey mosquito control. Proceedings of the Eighty-third Annual Meeting of the New Jersey Mosquito Control Association, Inc., 95–102.

Eberts, R. C. Jr., V. J. Santucci, and D. H. Wahl. 1998. Suitability of the lake chubsucker as prey for largemouth bass in small impoundments. N. Am. J. Fish. Manage. 18:295–307.

Echelle, A. A. 1968. Food habits of young-of-year longnose gar in Lake Texoma, Oklahoma. Southwest. Nat. 13:45–50.

Echelle, A. A., and C. D. Riggs. 1972. Aspects of the early life history of gars (*Lepisosteus*) in Lake Texoma. Trans. Am. Fish. Soc. 101:106–112.

Edds, D. R., W. J. Matthews, and F. P. Gelwick. 2002. Resource use by large catfish in a reservoir: is there evidence for interactive segregation and innate differences? J. Fish Biol. 60:739–750.

Edsall, T. A. 1964. Feeding by three species of fishes on the eggs of spawning alewives. Copeia 1964:226–227.

Ehlinger, T. J. 1997. Male reproductive competition and sex-specific growth patterns in bluegill. N. Am. J. Fish. Manage. 17:508–515.

Elder, H. S., and B. R. Murphy. 1997. Grass carp (*Ctenopharygnodon idella*) in the Trinity River, Texas. J. Freshw. Ecol. 12:281–289.

Eldridge, M. B., J. A. Whipple, D. Eng, M. J. Bowers, and B. M. Jarvis. 1981. Effects of food and feeding factors on laboratory-reared striped bass larvae. Trans. Am. Fish. Soc. 110:111–120.

Embody, G. C. 1914. The horned dace. Nat. Stud. Rev. 10:168–174.

EPA [U.S. Environmental Protection Agency]. 2000. Total maximum daily load (TMDL) development for total mercury in the middle/lower Savannah River, Ga. USEPA Region 4, Atlanta, Ga.

Epifanio, J. M., M. Hooe, D. H. Buck, and D. P. Philipp. 1999. Reproductive success and assortative mating among *Pomoxis* species and their hybrids. Trans. Am. Fish. Soc. 128:104–120.

EPRI [Electric Power Research Institute]. 1999. American eel (*Anguilla rostrata*) scoping study: a literature and data review of life history, stock status, population dynamics, and hydroelectric impacts. TR-111873. EPRI, Palo Alto, Calif.

Erdman, D. S. 1972. Inland game fishes of Puerto Rico. P.R. Fed. Aid Proj. F-1-20. Job 7, 4(2):1–96. Department of Agriculture, Commonwealth of Puerto Rico.

Erdman, D. S. 1984. Exotic fishes in Puerto Rico. Pages 162–176 *in* W. R. Courtenay Jr. and J. R. Stauffer Jr., editors. Distribution, biology, and management of exotic fishes. Johns Hopkins University Press, Baltimore, Md.

Erickson, J. G. 1953. Age and growth of the black and white crappies, *Pomoxis nigromaculatus* (Lesueur) and *P. annularis* Rafinesque, in Clear Lake, Iowa. Iowa State J. Sci. 26:491–505.

Esch, G. W., and T. C. Hazen. 1978. Thermal ecology and stress: a case history for red-sore disease in largemouth bass. Pages 331–363 *in* J. H. Thorp and J. W. Gibbons, editors. Energy and environmental stress in aquatic systems. DOE Symp. Ser. CONF-771114.

Esch, G. W., and T. C. Hazen. 1980. Stress and body condition in a population of largemouth bass: implications for red-sore disease. Trans. Am. Fish. Soc. 109:532–536.

Esch, G. W., T. C. Hazen, R. V. Dimock, and J. W. Gibbons. 1976. Thermal effluent and epizootiology of the ciliate *Epistylis* and the bacterium *Aeromonas hydrophila* in association with centrarchid fish. Trans. Am. Microsc. Soc. 95:687–693.

Etnier, D. A., and W. C. Starnes. 1993. The fishes of Tennessee. University of Tennessee Press, Knoxville.

Eure, H. 1976. Seasonal abundance of *Neoechinorhynchus cylindratus* taken from largemouth bass (*Micropterus salmoides*) in a heated reservoir. Parasitology 73:355–370.

Faber, D. J. 1963. Larval fish from the pelagial region of two Wisconsin lakes. Doctoral dissertation, University of Wisconsin, Madison.

Fahay, M. P. 1978. Biological and fisheries data on American eel, *Anguilla rostrata* (LeSueur). NMFS, NOAA, Tech. Ser. Rep. 17:82. Sandy Hook Laboratory, Northeast Fisheries Center.

Farmer, C. G., and D. C. Jackson. 1998. Air-breathing during activity in the fishes *Amia calva* and *Lepisosteus oculatus*. J. Exp. Biol. 201(7):943–948.

Fearnow, E. C. 1925. Goldfish. Their care in small aquaria and ponds. U.S. Comm. Fish. Rep. 1924 (Append. 7): 445–458.

Felley, J. D., and L. G. Hill. 1983. Multivariate assessment of environmental preferences of cyprinid fishes of the Illinois River, Oklahoma. Am. Midl. Nat. 109:209–221.

Finger, T. R., and E. M. Stewart. 1987. Response of fishes to flooding regime in lowland hardwood wetlands. Pages 86–104 *in* W. J. Matthews and D. C. Heins, editors. Community and evolutionary ecology of North American stream fishes. University of Oklahoma Press, Norman.

Fish, M. P. 1932. Contributions to the early life history of sixty-two species of fishes from Lake Erie and its tributaries. U.S. Bur. Fish. Bull. 47:293–398.

Flemer, D. A., and W. S. Woolcott. 1966. Food habits and distribution of the fishes of Tuckahoe Creek, Virginia, with special emphasis on the bluegill, *Lepomis m. macrochirus* Rafinesque. Chesapeake Sci. 7:75–89.

Fletcher, D. E. 1993. Nest association of dusky shiners (*Notropis cummingsae*) and redbreast sunfish (*Lepomis auritus*), a potentially parasitic relationship. Copeia 1993(1):159–167.

Fletcher, D. E., and B. M. Burr. 1992. Reproductive biology, larval description, and diet of the North American bluehead shiner, *Pteronotropis hubbsi* (Cypriniformes: Cyprinidae), with comments on conservation status. Ichthyol. Explor. Freshw. 3:193–218.

Fletcher, D. E., E. E. Dakin, B. A. Porter, and J. C. Avise.

2004. Spawning behavior and genetic parentage in the pirate perch (*Aphredoderus sayanus*), a fish with an enigmatic reproductive morphology. Copeia 2004:1–10.
Fletcher, D. E., and S. D. Wilkins. 1999. Glue secretion and adhesion by larvae of sailfin shiner (*Pteronotropis hypselopterus*). Copeia 1999:274–280.
Fletcher, D. E., S. D. Wilkins, J. V. McArthur, and G. K. Meffe. 2000. Influence of riparian alteration on canopy coverage and macrophyte abundance in southeastern USA blackwater streams. Ecol. Engineer. 15, Suppl. 1:S67–S78.
Flotemersch, J. E., D. C. Jackson, and J. R. Jackson. 1997. Channel catfish movements in relation to river channel floodplain connections. Proc. Annu. Conf. Southeast. Assoc. Fish Wildl. Agencies 51:106–112.
Fontenot, Q. C., and D. A. Rutherford. 1999. Observations on the reproductive ecology of pirate perch *Aphredoderus sayanus*. J. Freshw. Ecol. 14:545–550.
Forney, J. L. 1957. Bait fish production in New York ponds. N.Y. Fish Game J. 4:51–194.
Foster, N. R. 1974. Order Atheriniformes. Pages 115–151 *in* A. J. Lippson and R. L. Moran. Manual for identification of early developmental stages of fishes of the Potomac River estuary. Rep. PPSP-MP. Maryland Department of Natural Resources, Power Plant Siting Program.
Fowler, H. W. 1923. Spawning habits of sunfishes, basses, etc. Progr. Fish-Cult. 2:226–228.
Fox, M. G. 1994. Growth, density, and interspecific influences on pumpkinseed sunfish life histories. Ecology 75:1157–1171.
Fox, M. G., and A. J. Crivelli. 1998. Body size and reproductive allocation in a multiple spawning centrarchid. Can. J. Fish. Aquat. Sci. 55:737–748.
Freeman, B. J., and M. C. Freeman. 1985. Production of fishes in a subtropical blackwater ecosystem: the Okefenokee Swamp. Limnol. Oceanogr. 30:686–692.
Freeman, B. J., and M. C. Freeman. 2001. Criteria for suitable spawning habitat for the robust redhorse *Moxostoma robustum*. Report to the U.S. Fish and Wildlife Service. University of Georgia, Athens.
Freeman, H. W. 1958. Fish species of lower Three Runs creek on the Savannah River plant. Unpublished report, U.S. Atomic Energy Commission.
Freeman, M. C. 1995. Movements by two small fishes in a large stream. Copeia 1995:361–367.
Fuiman, L. A., J. V. Conner, B. F. Lathrop, G. L. Buynak, D. E. Snyder, and J. J. Loos. 1983. State of the art of identification for cyprinid fish larvae from eastern North America. Trans. Am. Fish. Soc. 112:319–332.
Gale, W. F., and W. G. Deutsch. 1985. Fecundity and spawning behavior of captive tessellated darters—fractional spawners. Trans. Am. Fish. Soc. 114:220–229.
Gale, W. F., and H. W. Mohr Jr. 1978. Larval drift in a large river with comparison of sampling methods. Trans. Am. Fish. Soc. 107:46–55.
Gatz, A. J. 1979. Ecological morphology of freshwater stream fishes. Tulane Stud. Zool. Bot. 21(2):91–124.
Gatz, A. J. Jr., and S. M. Adams. 1994. Patterns of movement of centrarchids in two warmwater streams in eastern Tennessee. Ecol. Freshw. Fish 3(1):35–48.
Gaymon, H. L. 1983. Biological assessment of Horse Creek, Langley Pond and Savannah River, Aiken County, South Carolina. Tech. Rep. 016-83. South Carolina Department of Health and Environmental Control, Columbia.
Geomatrix. 1993. Preliminary Quaternary and Neotectonic studies: Savannah River Site, South Carolina. Geomatrix Consultants, Inc., final report, Project no. 1988.
Gerald, J. W. 1971. Sound production during courtship in six species of sunfish (Centrarchidae). Evolution 25:75–87.
Germann, J. F., and C. D. Swanson. 1978. Food habits of chain pickerel in the Suwannee River, Georgia. Ga. J. Sci. 36:153–158.
Gibbons, J. W., and D. H. Bennett. 1971. Abundance and local movement of largemouth bass (*Micropterus salmoides*) in a reservoir receiving heated effluent from a reactor. Pages 524–527 *in* D. J. Nelson, editor. Proceedings of the 35th National Symposium on Radioecology, U.S. Atomic Energy Commission, Symp. Ser. CONF-710501-P1, NTIS, Springfield, Va.
Gibbons, J. W., D. H. Bennett, G. W. Esch, and T. C. Hazen. 1978. Effects of thermal effluent on body condition of largemouth bass. Nature 274:470–471.
Gibbons, J. W., J. T. Hook, and D. L. Forney. 1972. Winter responses of largemouth bass to heated effluent from a nuclear reactor. Progr. Fish-Cult. 34(2):88–90.
Giese, L. A., W. M. Aust, C. C. Trettin, and R. K. Kolka. 2000. Spatial and temporal patterns of carbon storage and species richness in three South Carolina Coastal Plain riparian forests. Ecol. Engineer. 15:157–170.
Gilbert, C. R. 1964. The American cyprinid fishes of the subgenus *Luxilus* (genus *Notropis*). Bull. Fla. State Mus. 8(2):95–194.
Gilbert, C. R. 1987. Zoogeography of the freshwater fish fauna of southern Georgia and peninsular Florida. Brimleyana 13:25–54.
Gilbert, C. R. 1989. Species profiles: life histories and environmental requirements of coastal fishes and invertebrates (Mid-Atlantic Bight). Atlantic and shortnose sturgeons. U.S. Fish. Wildl. Serv. Biol. Rep. 82 (11.122); U.S. Army Corps of Engineers TR EL-82-4.
Gilbert, C. R., editor. 1992. Rare and endangered biota of Florida. Volume 2: Fishes. University Press of Florida, Gainesville.

Gilbert, C. R., and R. M. Bailey. 1972. Systematics and zoogeography of the American cyprinid fish *Notropis (Opsopoeodus) emiliae*. Occas. Pap. Mus. Zool. Univ. Mich. 664.

Gilbert, C. R., and D. P. Kelso. 1971. Fishes of the Tortuguero area, Caribbean Costa Rica. Bull. Fla. State Mus. Biol. Sci. 16:1–54.

Gilbert, R. J., and J. E. Hightower. 1981. Assessment of tag losses and mortality of largemouth bass in an unfished reservoir. Project F-37-2. Georgia Cooperative Extension Fisheries Research Unit, School of Forestry Resources, University of Georgia, Athens.

Gilbert, R. J., S. Larson, and A. Wentworth. 1986. The relative importance of the lower Savannah River as a striped bass spawning area. School of Forestry Resources, University of Georgia, Athens.

Gladden, J. B., M. W. Lower, H. E. Mackey, W. L. Specht, and E. W. Wilde. 1985. Comprehensive cooling water study annual report. DP-1697. Savannah River Laboratory, E. I. Du Pont de Nemours and Co., Aiken, S.C.

Goff, G. P. 1984. Brood care of longnose gar (*Lepisosteus osseus*) by smallmouth bass (*Micropterus dolomieui*). Copeia 1984:149–152.

Gonzalez, R. J., and W. A. Dunson. 1989. Acclimation of sodium regulation to low pH and the role of calcium in the acid-tolerant sunfish *Enneacanthus obesus*. Physiol. Zool. 62:977–992.

Gonzalez, R. J., L. Milligan, A. Pagnotta, and D. G. McDonald. 2001. Effect of air breathing on acid-base and ion regulation after exhaustive exercise and during low pH exposure in the bowfin, *Amia calva*. Physiol. Biochem. Zool. 74(4):502–509.

Goodyear, C. P. 1967. Feeding habits of three species of gars, *Lepisosteus*, along the Mississippi Gulf Coast. Trans. Am. Fish. Soc. 96:297–300.

Gorham, S. W., and D. E. McAllister. 1974. The shortnose sturgeon, *Acipenser brevirostrum*, in the Saint John River, New Brunswick, Canada, a rare and possibly endangered species. Syllogeus 5:18.

Graham, J. H. 1993. Species diversity of fishes in naturally acidic lakes in New Jersey. Trans. Am. Fish. Soc. 122:1043–1057.

Graham, J. H., and J. D. Felley. 1985. Genomic coadaptation and developmental stability within introgressed populations of *Enneacanthus gloriosus* and *E. obesus* (Pisces, Centrarchidae). Evolution 39:104–114.

Graham, J. H., and R. W. Hastings. 1984. Distributional patterns of sunfishes on the New Jersey Coastal Plain. Environ. Biol. Fishes 10:137–148.

Graham, K. 1999. A review of the biology and management of blue catfish. Am. Fish. Soc. Symp. 24:37–49.

Grande, L. 1985. Recent and fossil clupeomorph fishes with materials for revision of the subgroups of clupoids. Bull. Am. Mus. Nat. Hist. 181:235–372.

Greeley, J. R. 1930a. A contribution to the biology of the horned dace, *Semotilus atromaculatus*. Doctoral dissertation, Cornell University, Ithaca, N.Y.

Greeley, J. R. 1930b. Fishes of the Lake Champlain watershed. Pages 44–87 *in* A biological survey of the Champlain watershed. Suppl. 19th Ann. Rep. N.Y. Conserv. Dept. 1929.

Green, D. M., and R. C. Heidinger. 1994. Longevity record for largemouth bass. N. Am. J. Fish. Manage. 14:464–465.

Green, T. D., and E. G. Maurakis. 2000. Comparison of larval myomere counts among species of *Nocomis* in Virginia (Actinopterygii: Cyprinidae). Va. J. Sci. 51:17–22.

Greenfield, D. 1973. An evaluation of the advisability of the release of grass carp, *Ctenpharyngodon idella*, into the natural waters of the United States. Trans. Ill. State Acad. Sci. 66:48–53.

Grier, H. J., D. P. Moody, and B. C. Cowell. 1990. Internal fertilization and sperm morphology in the brook silverside, *Labidestes sicculus* (Cope). Copeia 1990:221–226.

Griffith, J. S. 1978. Effects of low temperature on the survival and behavior of threadfin shad, *Dorosoma petenense*. Trans. Am. Fish. Soc. 107:63–70.

Grimm, W. W. 1937. The development of the fins of the goldfish (*Carassius auratus*). Doctoral dissertation, Ohio State University, Columbus.

Griswold, B. L. 1963. Food and growth of spottail shiners and other forage fishes of Clear Lake, Iowa. Proc. Iowa Acad. Sci. 70:215–223.

Gross, M. R. 1979. Cuckoldry in sunfishes (*Lepomis*: Centrarchidae). Can. J. Zool. 57:1507–1509.

Gross, M. R. 1982. Sneakers, satellites and parentals: polymorphic mating strategies in North American sunfishes. Z. Tierpsychol. 60:1–26.

Gross, M. R. 1984. Sunfish, salmon, and the evolution of alternative reproductive strategies and tactics in fish. Pages 55–75 *in* G. Potts and R. Wootton, editors. Fish reproduction: strategies and tactics. Academic Press, New York.

Gross, M. R., and A. M. MacMillan. 1981. Predation and the evolution of colonial nesting in bluegill sunfish (*Lepomis macrochirus*). Behav. Ecol. Sociobiol. 8:163–174.

Grossman, G. D., and R. E. Ratajczak Jr. 1998. Long-term patterns of microhabitat use by fish in a southern Appalachian stream from 1983 to 1992: effects of hydrological period, season and fish length. Ecol. Freshw. Fish 7:108–131.

Grunwald, C., J. Stabile, J. R. Waldman, R. Gross, and

I. Wirgin. 2002. Population genetics of shortnose sturgeon *Acipenser brevirostrum* based on mitochondrial DNA control region sequences. Mol. Ecol. 11:1885–1898.

Grussing, M. D., D. R. DeVries, and R. A. Wright. 1999. Stock characteristics and habitat use of catfishes in regulated sections of 4 Alabama rivers. Proc. Annu. Conf. Southeast. Assoc. Fish Wildl. Agencies 53:15–34.

Gu, B., C. L. Schelske, and M. V. Hoyer. 1996. Stable isotopes of carbon and nitrogen as indicators of diet and trophic structure of the fish community in a shallow hypereutrophic lake. J. Fish Biol. 49:1233–1243.

Guest, W. C., R. W. Drenner, S. T. Threlkeld, F. D. Martin, and J. D. Smith. 1990. Effects of gizzard shad and threadfin shad on zooplankton and young-of-year white crappie production. Trans. Am. Fish. Soc. 119:529–536.

Guillory, V. 1979a. Analysis of prey selection in chain pickerel. Proc. Annu. Conf. Southeast. Assoc. Fish Wildl. Agencies 33:507–517.

Guillory, V. 1979b. Life history of the chain pickerel in a central Florida lake. Fish Bull. 8. Florida Game and Freshwater Fish Commission, Tallahassee.

Guillory, V. 1979c. Species assemblages of fish in Lake Conway. Fla. Sci. 42(3):158–161.

Gunning, G. E., and W. M. Lewis. 1955. The fish population of a spring-fed swamp in the Mississippi bottoms of southern Illinois. Ecology 36:552–558.

Gutowski, M. J., and J. R. Stauffer. 1993. Selective predation by *Noturus insignis* (Richardson) (Teleostei, Ictaluridae) in the Delaware River. Am. Midl. Nat. 129(2):309–318.

Guy, C. S., R. D. Schultz, and C. A. Cox. 2002a. Ecology and management of white bass. N. Am. J. Fish. Manage. 22:606–608.

Guy, C. S., R. D. Schultz, and C. A. Cox. 2002b. Variation in gonad development, growth, and condition of white bass in Fall River Reservoir, Kansas. N. Am. J. Fish. Manage. 22:643–651.

Guy, C. S., D. W. Willis, and J. J. Jackson. 1994. Biotelemetry of white crappies in a South Dakota glacial lake. Trans. Am. Fish. Soc. 123:63–70.

Gwinner, H. R., H. J. Cathey, and F. J. Bulow. 1975. A study of two populations of introduced redeye bass, *Micropterus coosae* Hubbs and Bailey. J. Tenn. Acad. Sci. 50:102–105.

Haag, W. R., and M. L. Warren. 1997. Host fishes and reproductive biology of six freshwater mussel species from the Mobile Basin, USA. J. N. Am. Benthol. Soc. 16:576–585.

Hale, R. S. 1999. Growth of white crappies in response to temperature and dissolved oxygen conditions in a Kentucky reservoir. N. Am. J. Fish. Manage. 19:591–598.

Hales, L. S. Jr., and M. C. Belk. 1992. Validation of otolith annuli of bluegills in a southeastern thermal reservoir. Trans. Am. Fish. Soc. 121:823–830.

Hall, G. E., and R. M. Jenkins. 1954. Notes on the age and growth of the pirate perch, *Aphredoderus sayanus*, in Oklahoma. Copeia 1954(1):69.

Hall, W. J., T. I. J. Smith, and S. D. Lamprecht. 1991. Movements and habitats of shortnose sturgeon *Acipenser brevirostrum* in the Savannah River, Georgia. Copeia 1991:695–702.

Halpin, P. M., and K. L. M. Martin. 1999. Aerial respiration in the salt marsh fish *Fundulus heteroclitus* (Fundulidae). Copeia 1999:743–748.

Halverson, N. V., and 20 authors. 1997. SRS ecology. Environ. Inf. Doc. WSRC-TR-93-0223. Westinghouse Savannah River Co., Aiken, S.C.

Halyk, L. C., and E. K. Balon. 1983. Structure and ecological production of the fish taxocene of a small floodplain system. Can. J. Zool. 61:2446–2464.

Hammers, B. E., and L. E. Miranda. 1991. Comparison of methods for estimating age, growth, and related populations characters of white crappies. N. Am. J. Fish. Manage. 11:492–498.

Hammett, F. S., and D. W. Hammett. 1939. Proportional length growth of gar (*Lepidosteus platyrhincus* De Kay). Growth 3:197–209.

Hankinson, T. L. 1908. A biological survey of Walnut Lake, Michigan. Mich. State Board Geol. Surv. Rep. 1907:157–288.

Hansen, D. F. 1943. On nesting of the white crappie, *Pomoxis annularis*. Copeia 1943:259–260.

Hansen, J. M., and S. U. Qadri. 1984. Feeding ecology of age 0 pumpkinseed (*Lepomis gibbosus*) and black crappie (*Pomoxis nigromaculatus*) in the Ottawa River. Can. J. Zool. 62:613–621.

Harding, J. S., E. F. Benfield, P. V. Bolstad, G. S. Helfman, and E. B. D. Jones III. 1998. Stream biodiversity: the ghost of land use past. Proc. Nat. Acad. Sci. USA 95:14843–14847.

Hardy, J. D. Jr. 1978. Development of fishes of the Mid-Atlantic Bight: an atlas of egg, larval and juvenile stages. Volume 3: Aphredoderidae through Rachycentridae. U.S. Fish Wildl. Serv. Biol. Serv. Prog. FWS/OBS-78/12.

Harland, R. W. 1947. Observations on the breeding habits of yellow perch, *Perca flavescens* (Mitchill). Copeia 1947:199–200.

Haro, A., W. Richkus, K. Whalen, A. Hoar, W. D. Busch, S. Lary, T. Brush, and D. Dixon. 2000. Population decline of the American eel: implications for research and management. Fisheries 25:7–16.

Harrel, R. M., editor. 1997. Striped bass and other *Morone* culture. Elsevier, Amsterdam.

Harrington, R. W. Jr. 1956. An experiment on the effects of contrasting daily photoperiods on gametogenensis and reproduction in the centrarchid fish, *Enneacanthus obesus* (Girard). J. Exp. Zool. 131:204–223.

Harris, P. M., R. L. Mayden, H. S. Espinosa Perez, and F. Garcia de Leon. 2002. Phylogenetic relationships of *Moxostoma* and *Scartomyzon* (Catostomidae) based on mitochondrial cytochrome b sequence data. J. Fish Biol. 61:1433–1452.

Hass, J. J., L. A. Jahn, and R. Hilsabeck. 2001. Diet of introduced flathead catfish (*Pylodictus olivarus*) in an Illinois reservoir. J. Freshw. Ecol. 16(4):551–555.

Hazen, T. C., G. W. Esch, A. B. Glassman, and J. W. Gibbons. 1978. Relationship of season, thermal ecology and red-sore disease with various hematological parameters in *Micropterus salmoides*. J. Fish Biol. 12:491–498.

Heard, R. W. 1975. Feeding habits of white catfish from a Georgia estuary. Q. J. Fla. Acad. Sci. 38:20–28.

Hedrick, M. S., and D. R. Jones. 1999. Control of gill ventilation and air-breathing in the bowfin *Amia calva*. J. Exp. Biol. 202:87–94.

Heidinger, R. C. 1976. Synopsis of biological data on the largemouth bass: *Micropterus salmoides* (Lacépède, 1802). Bernan Associates.

Heins, D. C., and F. G. Rabito Jr. 1986. Spawning performance in North American minnows: direct evidence of the occurrence of multiple clutches in the genus *Notropis*. J. Fish Biol. 28:343–357.

Helfman, G. S., B. B. Collette, and D. E. Facey. 1997. Diversity of fishes. Blackwell Science, Malden, Mass.

Helfrich, L. A., K. W. Nutt, and D. L. Weigmann. 1991. Habitat selection by spawning redbreast sunfish in Virginia streams. Rivers 2:138–147.

Hellier, T. R. Jr. 1967. The fishes of the Santa Fe River system. Bull. Fla. Mus. Biol. Sci. 2(10):1–46.

Henry, R. L. 1979. Prey size utilization in a community of southeastern fishes. Bull. Ecol. Soc. Am. 60:113.

Hergenrader, G. L. 1980. Current distribution and potential for dispersal of white perch (*Morone americana*) in Nebraska and adjacent waters. Am. Midl. Nat. 103:404–406.

Hesse, L. W. 1994. The status of Nebraska fishes in the Missouri River, 4. Flathead catfish *Pylodictis olivaris*, and blue catfish *Ictalurus furcatus* (Ictaluridae). Trans. Nebr. Acad. Sci. 21:89–98.

Hettler, W. F., and A. J. Chester. 1990. Temporal distribution of ichthyoplankton near Beaufort Inlet, North Carolina. Mar. Ecol. Prog. Ser. 68:157–168.

Hida, T. S., and D. A. Thomson. 1962. Introduction of threadfin shad to Hawaii. Progr. Fish-Cult. 24:159–163.

Hildebrand, S. F. 1963. Family Clupeidae. Pages 257–454 *in* Fishes of the western North Atlantic. Sears Foundation for Marine Research, Mem. 1, pt. 3.

Hildebrand, S. F., and W. C. Schroeder. 1928. Fishes of the Chesapeake Bay. U.S. Bur. Fish. Bull. 43:1–388.

Hlohowsky, C. P., M. M. Coburn, and T. M. Cavender. 1989. Seven species of the herbivorous cyprinid genus, *Hybognathus* (Pisces: Cyprinidae). Copeia 1989:172–183.

Hoda, S. M. S., and H. Tsukahari. 1971. Studies on the development and relative growth in carp, *Cyprinus carpio* (Linne.). J. Fac. Agric. Kyushu Univ. 16:387–509.

Hodson, R. G. 1989. Hybrid striped bass, biology and life history. South. Reg. Aquacult. Ctr. Pub. 300.

Hoese, H. D., and R. H. Moore. 1977. Fishes of the Gulf of Mexico, Texas, Louisiana, and adjacent waters. Texas A&M University Press, College Station.

Hogan, D. C. 1977. Distribution and relative abundance of prey fish in a reservoir receiving a heated effluent. Master's thesis, University of Georgia, Athens.

Holland, B. F. Jr., and G. F. Yelverton. 1973. Distribution and biological studies of anadromous fishes offshore of North Carolina. N.C. Dep. Natl. Econ. Res. Spec. Sci. Rep. 24:132.

Holland, W. E., M. H. Smith, J. W. Gibbons, and D. H. Brown. 1974. Thermal tolerances of fish from a reservoir receiving heated effluent from a nuclear reactor. Physiol. Zool. 47:110–117.

Hollis, E. H. 1948. The homing tendency of shad. Science 108:332–333.

Holloway, A. D. 1954. Notes on the life history and management of the shortnose and longnose gars in Florida waters. J. Wildl. Manage. 18:438–449.

Hoover, J. J. 1980. Feeding ecology of two sympatric species of *Notropis* (Pisces: Cyprinidae) from the Hillsborough River. Fla. Sci. 43:25.

Hoover, J. J., S. G. George, and N. H. Douglas. 1998. The bluebarred pygmy sunfish (*Elassoma okatie*) in Georgia. Southeast. Fish. Counc. Proc. 36:7–9.

Hoover, J. J., and K. J. Killgore. 1999. Fish-habitat relationships in the streams of Fort Gordon, Georgia. Tech. Rep. EL-99-6. U.S. Army Corps of Engineers, Waterways Experiment Station, Vicksburg, Miss.

Hoover, J. J., and K. J. Killgore. 2002. Small floodplain pools as habitat for fishes and amphibians: methods for evaluation. Tech. Note ERDC TN-EMRRP-EM-03. U.S. Army Corps of Engineers, Ecosystem Management and Restoration Research Program, Engineering Research and Development Center.

Horwitz, R. J. 1978. Temporal variability patterns and the distributional patterns of stream fishes. Ecol. Monogr. 48:307–321.

Houser, A., and J. E. Dunn. 1967. Estimating the size of

threadfin shad population in Bull Shoals Reservoir from midwater trawl catches. Trans. Am. Fish. Soc. 96:176–184.

Howells, R. G., and G. P. Garrett. 1992. Status of some exotic sport fishes in Texas waters. Tex. J. Sci. 44:317–324.

Hubbs, C. 1951. Minimum temperature tolerances for fishes of the genera *Signalosa* and *Herichehys* in Texas. Copeia 1951:297.

Hubbs, C. L. 1921. An ecological study of the life history of the freshwater atherine fish *Labidestes sicculus*. Ecology 2:262–276.

Hubbs, C. L., and G. P. Cooper. 1936. Minnows of Michigan. Cranbrook Inst. Sci. Bull. 8:1–95.

Hubbs, C. L., and K. F. Lagler. 1958. Fishes of the Great Lakes region, with a new preface. University of Michigan Press, Ann Arbor.

Hubbs, C. L., and E. C. Raney. 1951. Status, subspecies, and variations of *Notropis cummingsae*, a cyprinid fish of the southeastern United States. Occas. Pap. Mus. Zool. Univ. Mich. 535:1–25.

Hubert, W. A., and D. T. O'Shea. 1992. Use of spatial resources by fishes in Grayrocks Reservoir, Wyoming. J. Freshw. Ecol. 7:219–225.

Huckins, C. J. F. 1997. Functional linkages among morphology, feeding performance, diet, and competitive ability in molluscivorous sunfish. Ecology 78(8):2401–2414.

Huckins, C. J. F., G. W. Osenberg, and G. G. Mittelbach. 2000. Species introductions and their ecological consequences: an example with congeneric sunfish. Ecol. Appl. 10(2):612–625.

Hudson, R. G., and F. E. Hester. 1975. Movements of the redbreast sunfish in Little River, near Raleigh, North Carolina. Proc. Annu. Conf. Southeast. Assoc. Game Fish Comm. 29:325–329.

Huff, J. A. 1975. Life history of the Gulf of Mexico sturgeon, *Acipenser oxyrhynchus desotai*, in the Suwannee River, Florida. Fla. Mar. Res. Pub. 16:32.

Hughes, G. M. 1976. Respiration of amphibious vertebrates. Academic Press, New York.

Hughes, M. J., and D. M. Carlson. 1986. White catfish growth and life history in the Hudson River Estuary, New York. J. Freshw. Ecol. 3:407–418.

Huish, M. T. 1954. Life history of the black crappie of Lake George, Florida. Trans. Am. Fish. Soc. 83:176–193.

Huish, M. T. 1957. Food habits of three centrarchids in Lake George, Florida. Proc. Annu. Conf. Southeast. Assoc. Game Fish Comm. 11:293–302.

Huizinga, H. W., G. W. Esch, and T. C. Hazen. 1979. Histopathology of red-sore disease (*Aeromonas hydrophilia*) in naturally and experimentally infected largemouth bass *Micropterus salmoides* (Lacépède). J. Fish Dis. 2:263–277.

Humphries, E. T., and K. B. Cumming. 1973. An evaluation of striped bass fingerling culture. Trans. Am. Fish. Soc. 102:13–20.

Hunt, B. P. 1953. Food relationships between Florida gar and other organisms in the Tamiami Canal, Dade County, Florida. Trans. Am. Fish. Soc. 82:13–33.

Hunter, J. B. 1963. The reproductive behavior of the green sunfish, *Lepomis cyanellus*. Zoologica 48:13–24.

Huntsman, G. R. 1967. Nuptial tubercles in carpsuckers (*Carpiodes*). Copeia 1967:457–458.

Hurst, T. P., E. T. Schultz, and D. O. Conover. 2000. Seasonal energy dynamics of young-of-the-year Hudson River striped bass. Trans. Am. Fish. Soc. 129:145–175.

Irons, K. S., T. M. O'Hara, M. A. McClelland, and M. A. Pegg. 2002. White perch occurrence, spread, and hybridization in the middle Illinois River, upper Mississippi River system. Trans. Ill. State Acad. Sci. 95:207–214.

Isaac, J. Jr., T. M. Kimmel, R. W. Bagley, V. H. Staats, and A. Barkoh. 1998. Spawning behavior of Florida largemouth bass in an indoor raceway. Progr. Fish-Cult. 60:59–62.

Jackson, D. C. 1995. Distribution and stock structure of blue catfish and channel catfish in macrohabitats along riverine sections of the Tennessee-Tombigbee Waterway. N. Am. J. Fish. Manage. 15:845–853.

Jacobson, P. M., D. A. Dixon, W. C. Leggett, B. C. Marcy Jr., and R. R. Massengill, editors. 2004. The Connecticut River Ecological Study (1965–1973) revisited: ecology of the Lower Connecticut River, 1973–2003. Am. Fish. Soc. Monogr. 9.

Jenkins, R. E., and N. M. Burkhead. 1993. Freshwater fishes of Virginia. American Fisheries Society, Bethesda, Md.

Jenkins, R. E., L. A. Revelle, and T. Zorach. 1975. Records of the blackbanded sunfish, *Enneacanthus chaetodon*, and comments on the southeastern Virginia freshwater ichthyofauna. Va. J. Sci. 26(3):128–134.

Jennings, M. J., J. E. Claussen, and D. P. Philipp. 1997. Effect of population size structure on reproductive investment of male bluegill. N. Am. J. Fish. Manage. 17:516–524.

Jessop, B. M. 1987. Migrating eels in Nova Scotia. Trans. Am. Fish. Soc. 116:161–170.

Johnson, B. L., and D. B. Noltie. 1996. Migratory dynamics of stream-spawning longnose gar (*Lepisosteus osseus*). Ecol. Freshw. Fish 5:97–107.

Johnson, B. L., and D. B. Noltie. 1997. Demography, growth, and reproductive allocation in stream-spawning longnose gar. Trans. Am. Fish. Soc. 126:438–466.

Johnson, J. H., and D. S. Dropkin. 1995. Diel feeding chronology of six fish species in the Juniata River, Pennsylvania. J. Freshw. Ecol. 10(1):11–18.

Johnson, J. H., and E. Z. Johnson. 1984. Comparative diets of subyearling redbreast sunfish and subyearling northern redbelly dace in an Adirondack Lake. J. Freshw. Ecol. 2(6):587–591.

Johnson, T. B., and D. O. Evans. 1990. Size-dependent winter mortality of young-of-the-year white perch: climate warming and invasion of the Laurentian Great Lakes. Trans. Am. Fish. Soc. 119:301–313.

Johnston, C. E. 1994. The benefits to some minnows of spawning in the nests of other species. Environ. Biol. Fishes 40:213–218.

Johnston, C. E., and W. S. Birkhead. 1988. Spawning in the bandfin shiner, *Notropis zonistius* (Pisces: Cyprinidae). J. Ala. Acad. Sci. 59:30–33.

Johnston, C. E., and K. J. Kleiner. 1994. Reproductive behavior of the rainbow shiner (*Notropis chrosomus*) and the rough shiner (*Notropis baileyi*), nest associates of the bluehead chub (*Nocomis leptocephalus*) (Pisces: Cyprinidae) in the Alabama River drainage. J. Ala. Acad. Sci. 65:230–240.

Johnston, C. E., and L. M. Page. 1992. The evolution of complex reproductive strategies in North American minnows (*Cyprinidae*). Pages 600–621 *in* R. L. Mayden, editor. Systematics, historical ecology, and North American freshwater fishes. Stanford University Press, Stanford, Calif.

Jones, E. B. D. III, G. S. Helfman, J. O. Harper, and P. V. Bolstad. 1999. Effects of riparian forest removal on fish assemblages in southern Appalachian streams. Conserv. Biol. 13:1454–1465.

Jones, P. W., F. D. Martin, and J. D. Hardy Jr. 1978. Development of fishes of the Mid-Atlantic Bight: an atlas of egg, larval and juvenile stages. Volume 1: Acipenseridae through Ictaluridae. FWS/OBS-78/12. U.S. Department of the Interior, Fish and Wildlife Service, Washington, D.C.

Jones, T. A. 2001. Seasonal and diel movement of largemouth bass in a South Carolina stream. Master's thesis, Clemson University, Clemson, S.C.

Jones, W. J., and J. M. Quattro. 1999. Phylogenetic affinities of pygmy sunfishes (*Elassoma*) inferred from mitochondrial DNA sequences. Copeia 1999:470–474.

Jordan, D. S. 1878. A catalogue of the fishes of Illinois. Ill. State Lab. Nat. Hist. Bull. 1(2):37–70.

Jordan, D. S. 1923. A classification of fishes. Stanford Univ. Pub. 3(2):79–243.

Jordan, D. S., and B. W. Evermann. 1896–1900. The fishes of North and Middle America. Bull. U.S. Natl. Mus. 47:1–3313. Pt. 1, pp. 1–1230, 1896; pt. 2, pp. 1231–2183, 1898; pt. 3, pp. 2183a–3136, 1899; pt. 4, pp. 3137–3313, 1900.

Jordan, F., and D. A. Arrington. 2001. Weak trophic interactions between large predatory fishes and herpetofauna in the channelized Kissimmee River, Florida, USA. Wetlands 21:155–159.

Karr, J. R., K. D. Fausch, P. L. Angermeier, P. R. Yant, and I. J. Schlosser. 1986. Assessing biological integrity in running waters: a method and its rationale. Ill. Nat. Hist. Surv. Spec. Pub. 5.

Katula, R. S. 1987. Spawning of the pirate perch recollected. Am. Currents (June–September):19–20.

Katula, R. S. 1992. The spawning mode of the pirate perch. Trop. Fish Hobbyist (August):156–159.

Katula, R. S., and L. M. Page. 1998. Nest association between a large predator, the bowfin (*Amia calva*), and its prey, the golden shiner (*Notemigonus crysoleucas*). Copeia 1:220–221.

Kaya, C. M. 1973. Effects of temperature and photoperiod on seasonal regression of gonads of green sunfish, *Lepomis cyanellus*. Copeia 1973(2):369–373.

Kaya, C. M., and A. D. Hasler. 1972. Photoperiod and temperature effects on the gonads of green sunfish, *Lepomis cyanellus* (Rafinesque), during the quiescent, winter phase of its annual sexual cycle. Trans. Am. Fish. Soc. 2:270–275.

Keast, A. 1978. Feeding interrelations between age-groups of pumpkinseed (*Lepomis gibbosus*) and comparisons with bluegill (*L. macrochirus*). J. Fish. Res. Board Can. 35:12–27.

Keast, A. 1985a. Implications of chemosensory feeding in catfishes: an analysis of the diets of *Ictalurus nebulosus* and *I. natalis*. Can. J. Zool. 63:590–602.

Keast, A. 1985b. The piscivore feeding guild of fishes in small freshwater ecosystems. Environ. Biol. Fishes 12:119–129.

Keast, A., J. Harker, and D. Turnbull. 1978. Nearshore fish habitat utilization and species associations in Lake Opinicon (Ontario, Canada). Environ. Biol. Fishes 3(2):173–184.

Kelly, M. S. 1989. Distribution and biomass of aquatic macrophytes in an abandoned nuclear cooling reservoir. Aquat. Bot. 35:133–152.

Kennedy, P. K., M. L. Kennedy, and M. H. Smith. 1985. Microgeographic genetic organization of populations of largemouth bass and two other species in a reservoir. Copeia 1985:118–125.

Kerby, J. H., and E. B. Joseph. 1979. Growth and survival of striped bass and striped bass × white bass hybrids: a review of methods, advances and problems. Proc. Annu. Conf. Southeast. Assoc. Fish Wildl. Agencies 32(1978):715–726.

Kerby, J. H., and R. M. Harrell. 1990. Hybridization, genetic manipulation, and gene pool conservation of striped bass. Pages 159–190 *in* R. M. Harrell, J. H. Kerby, and R. V. Minton, editors. Culture and propagation of striped bass and its hybrids. American Fisheries Society, Southern Division, Striped Bass committee, Bethesda, Md.

Khan, M. H. 1929. Early stages in the development of goldfish (*Carassius auratus*). J. Bombay Nat. Hist. Soc. 33:614–617.

Kilgo, J. C., and J. I. Blake, editors. In press. Ecology and management of a forested landscape: fifty years of natural resource stewardship on the Savannah River Site. Island Press, Covelo, Calif.

Killgore, K. J., and J. A. Baker. 1996. Patterns of larval fish abundance in a bottomland hardwood wetland. Wetlands 16:288–295.

Killgore, K. J., R. P. Morgan II, and N. B. Rybicki. 1989. Distribution and abundance of fishes associated with submersed aquatic plants in the Potomac River. N. Am. J. Fish. Manage. 9(1):101–111.

Kindler, P. M., J. M. Bahr, M. R. Gross, and D. P. Philipp. 1991a. Hormonal regulation of parental care behavior in nesting male bluegill: do the effects of bromocriptine suggest a role for prolactin? Physiol. Zool. 64:310–322.

Kindler, P. M., J. M. Bahr, and D. P. Philipp. 1991b. The effects of exogenous 11-ketotestosterone, testosterone, and cyproterone acetate on prespawning and parental care behaviors of male bluegill. Hormones Behav. 25:410–423.

King, W., and J. Parsons. 1951. Two black bass new to Tennessee. J. Tenn. Acad. Sci. 26:113–114.

Kirkman, L. K. 1992. Cyclical vegetation dynamics in Carolina bays wetlands. Doctoral dissertation, University of Georgia, Athens.

Kittl, B. 1999. Mosquitofish imperil frogs. Discover 20(8).

Klaassen, H. E., and K. L. Morgan. 1974. Age and growth of longnose gar in Tuttle Creek Reservoir, Kansas. Trans. Am. Fish. Soc. 103:402–405.

Klarberg, D. P., and A. Benson. 1973. The food habits of the brown bullhead (*Ictalurus nebulosus*) in mine acid polluted waters. Va. J. Sci. 24(3):120.

Kline, J. L., and B. M. Wood. 1996. Food habits and diet selectivity of the brown bullhead. J. Freshw. Ecol. 11:145–151.

Kohler, C. C. 2000. A white paper on the status and needs of hybrid striped bass aquaculture in the north central region. North Central Regional Aquaculture Center. (http://ag.ansc.purdue.edu/aquanic/ncrac/wpapers).

Kolka, R. K., J. H. Singer, C. R. Coppock, W. P. Casey, and C. C. Trettin. 2000. Influence of restoration and succession on bottomland hardwood hydrology. Ecol. Engineer. 15, Suppl. 1:S131–S140.

Kolka, R. K., C. C. Trettin, E. A. Nelson, C. D. Barton, and D. E. Fletcher. 2002. Application of the EPA Wetland Research Program approach to a floodplain wetland restoration assessment. J. Environ. Monitor. Restor. 1(1): 37–51.

Konradt, A. G. 1968. Methods of breeding the grass carp *Ctenopharyngodon idella* and the silver carp, *Hypophthalmichthys molitrix*. FAO Fish Rep. 4:195–204.

Koski, R. T. 1978. Age, growth, and maturity of the hogchoker, *Trinectes maculatus*, in the Hudson River, New York. Trans. Am. Fish. Soc. 107:449–454.

Kramer, R. H., and L. L. Smith Jr. 1960. Utilization of nests of largemouth bass, *Micropterus salmoides*, by golden shiners, *Notemigonus crysoleucas*. Copeia 1960:73–74.

Kruse, C. G., C. S. Guy, and D. W. Willis. 1993. Comparison of otolith and scale age characteristics for black crappies collected from South Dakota waters. N. Am. J. Fish. Manage. 13:856–858.

Kuehne, R. A., and R. W. Barbour. 1983. The American darters. University Press of Kentucky, Lexington.

Kutkuhn, J. H. 1958. Utilization of plankton by juvenile gizzard shad in a shallow prairie lake. Trans. Am. Fish. Soc. 87:80–103.

Kwak, T. J. 1988. Lateral movement and use of floodplain habitat by fishes of the Kankakee River, Illinois. Am. Midl. Nat. 120(2):241–249.

Lachner, E. A. 1952. Studies of the biology of the cyprinid fishes of the chub genus *Nocomis* of northeastern United States. Am. Midl. Nat. 48:433–466.

Lachner, E. A., and M. L. Wiley. 1971. Populations of the polytypic species *Nocomis leptocephalus* (Girard) with a description of a new subspecies. Smithson. Contrib. Zool. 92:35.

Laerm, J., and B. J. Freeman. 1986. Fishes of the Okefenokee Swamp. University of Georgia Press, Athens.

Lagler, K. F., and V. C. Applegate. 1942. Further studies of the food of the bowfin (*Amia calva*) in southern Michigan, with notes on the inadvisability of using trapped fish in food analysis. Copeia 1942:190–191.

Lagler, K. F., and F. V. Hubbs. 1940. Food of the longnosed gar (*Lepisosteus osseus oxyurus*) and the bowfin (*Amia calva*) in southern Michigan. Copeia 1940:239–241.

Lagler, K. F., and H. Van Meter. 1951. Abundance and growth of gizzard shad, *Dorosoma cepedianum* (Lesueur), in a small Illinois lake. J. Wildl. Manage. 15:357–360.

Laird, C. A., and L. M. Page. 1996. Non-native fishes inhabiting the streams and lakes of Illinois. Ill. Nat. Hist. Surv. Bull. 35 (art. 1).

Lakly, M. B., and J. V. McArthur. 2000. Macroinvertebrate

recovery of a post-thermal stream: habitat structure and biotic function. Ecol. Engineer. 15, Suppl. 1:S87–S100.

Lambou, V. W. 1959. Fish populations of backwater lakes in Louisiana. Trans. Am. Fish. Soc. 88:7–15.

Larimore, R. W. 1957. Ecological life history of the warmouth (Centrarchidae). Ill. Nat. Hist. Surv. Bull. 27(1):1–83.

Larimore, W. R., and P. W. Smith. 1963. The fishes of Champaign County, Illinois, as affected by 60 years of stream changes. Ill. Nat. Hist. Surv. Bull. 28:299–382.

Larkin, P. A., J. G. Terpenning, and P. R. Parker. 1957. Size as a determinant of growth rate in rainbow trout, *Salmo gairdneri*. Trans. Am. Fish. Soc. 83:84–96.

Larson, S. C., B. S. Saul, and S. Schleiger. 1991. Exploitation and survival of black crappies in three Georgia reservoirs. N. Am. J. Fish. Manage. 11:604–613.

Lauder, G. V. 1983. Functional and morphological bases of trophic specialization in sunfishes (Teleostei, Centrarchidae). J. Morphol. 178:1–21.

Layher, W. G., and R. J. Boles. 1980. Food habits of the flathead catfish, *Pylodictis olivaris* (Rafinesque), in relation to length and season in a large Kansas reservoir. Trans. Kans. Acad. Sci. 83:200–214.

Layman, S. R. 1993. Life history of the Savannah darter, *Etheostoma fricksium*, in the Savannah River drainage, South Carolina. Copeia 1993:959–968.

Layzer, J. B., and R. J. Reed. 1978. Food, age and growth of the tessellated darter, *Etheostoma olmstedi*, in Massachusetts. Am. Midl. Nat. 100:459–462.

Lazzaro, X., R. W. Drenner, R. A. Stein, and J. D. Smith. 1992. Planktivores and plankton dynamics: effects of fish biomass and planktivore type. Can. J. Fish. Aquat. Sci. 49:1466–1473.

Lee, D. S., C. R. Gilbert, C. H. Hocutt, R. E. Jenkins, D. E. McAllister, and J. R. Stauffer Jr. 1980. Atlas of North American freshwater fishes. N.C. Biol. Surv. Pub. 1980-12.

Leggett, W. C. 1972. Weight loss in American shad (*Alosa sapidissima*) during the freshwater migration. Trans. Am. Fish. Soc. 101:549–552.

Leggett, W. C. 1976. The American shad (*Alosa sapidissima*), with special reference to migration and population dynamics in the Connecticut River. Pages 169–225 *in* D. Merriman and L. M. Thorpe, editors. The Connecticut River Ecological Study: the impact of a nuclear power plant. Am. Fish. Soc. Monogr. 1.

Leggett, W. C. 1977. Ocean migration rates of American shad (*Alosa sapidissima*). J. Fish. Res. Board Can. 34:1422–1426.

Leggett, W. C., and J. E. Carscadden. 1978. Latitudinal variation in reproductive characteristics of American shad (*Alosa sapidissima*): evidence for population specific life history strategies in fish. J. Fish. Res. Board Can. 35:1469–1478.

Leggett, W. C., T. F. Savoy, and C. A. Tomichek. 2004. The impact of enhancement initiatives on the structure and dynamics of the Connecticut River population of American Shad (*Alosa sapidissima*). *In* Jacobson, P. M., D. A. Dixon, W. C. Leggett, B. C. Marcy Jr., and R. R. Massengill, editors. The Connecticut River Ecological Study (1965–1973) revisited: ecology of the Lower Connecticut River, 1973–2003. Am. Fish. Soc. Monogr. 9.

Leggett, W. C., and R. R. Whitney. 1972. Water temperature and the migration of American shad. Fish. Bull. 70:659–670.

Lehtinen, R. M., N. D. Mundahl, and J. C. Madejczyk. 1997. Autumn use of woody snags by fishes in backwater and channel border habitats of a large river. Environ. Biol. Fishes 49:7–19.

Leim, A. H., and W. B. Scott. 1966. Fishes of the Atlantic coast of Canada. Fish. Res. Board Can. Bull. 155:485.

Leitman, H. M., M. R. Darst, and J. J. Nordhaus. 1991. Fishes in the forested flood plain of the Ochlockonee River, Florida, during flood and drought conditions. U.S. Geol. Surv. Water-Res. Invest. Rep. 90-4202.

Lemly, A. D. 1985. Suppression of native fish populations by green sunfish in first-order streams of Piedmont North Carolina. Trans. Am. Fish. Soc. 114:705–712.

Lenat, D. R., and J. K. Crawford. 1994. Effects of land use on water quality and aquatic biota of three North Carolina Piedmont streams. Hydrobiologia 294:185–199.

Lesko, L. T., S. B. Smith, and M. A. Blouin. 1996. The effect of contaminated sediments on fecundity of the brown bullhead in three Lake Erie tributaries. J. Great Lakes Res. 22:830–837.

Leslie, J. K., and C. A. Timmins. 2002. Description of age 0 juvenile pugnose minnow *Opsopoeodus emiliae* (Hay) and pugnose shiner *Notropis anogenus* (Forbes) in Ontario. Can. Tech. Rep. Fish. Aquat. Sci. 2397:14.

Lewis, W. M. 1952. Analysis of the gizzard shad population of Crab Orchard Lake, Illinois. Ill. Acad. Sci. Trans. 46:231–235.

Lewis, W. M., and D. Elder. 1956. The fish population of the headwaters of a spotted bass stream in southern Illinois. Trans. Am. Fish. Soc. 82:193–203.

Lide, R. F. 1991. Hydrology of a Carolina bay located on the upper Coastal Plain, western South Carolina. Master's thesis, University of Georgia, Athens.

Limburg, K. E., and R. M. Ross. 1995. Growth and mortality rates of larval American shad, *Alosa sapidissima*, in different salinities. Estuaries 18:335–340.

Limburg, K. E., and R. E. Schmidt. 1990. Patterns of fish spawning in Hudson River tributaries: response to an urban gradient. Ecology 71:1238–1245.

Lindsey, C. C., T. G. Northcote, and G. F. Hartman. 1959. Homing of rainbow trout to inlet and outlet spawning streams at Loon Lake, British Columbia. J. Fish. Res. Board Can. 20:1001–1030.

Lippson, A. J., and R. L. Moran. 1974. Manual for identification of early developmental stages of fishes of the Potomac River estuary. Rep. PPSP-MP-13. Maryland Department of Natural Resources, Power Plant Siting Program.

Lobb, M. D. III, and D. J. Orth. 1991. Habitat use by an assemblage of fish in a large warmwater stream. Trans. Am. Fish. Soc. 120:65–78.

Loesch, J. G., and W. A. Lund Jr. 1977. A contribution to the life history of the blueback herring, *Alosa aestivalis*. Trans. Am. Fish. Soc. 106:583–589.

Loftus, W. F., and J. A. Kushlan. 1987. Freshwater fishes of southern Florida. Bull. Fla. State Mus. Biol. Sci. 31:147–344.

Loftus, W. F., J. A. Kushlan, and S. A. Voorhees. 1984. Status of the mountain mullet in southern Florida. Fla. Sci. 47:256–263.

Looney, G. L. 1998. Hormone induced ovulation of robust redhorse (*Moxostoma robustum*). Presented at the 1998 Southern Division of the American Fisheries Society Midyear Meeting, Lexington, Ky. (http://www.sdafs.org/meetings/98sdafs/aquac/looney.htm).

Loos, J. J., and L. A. Fuiman. 1978. Subordinate taxa of the genus *Notropis*: a preliminary comparative survey of their developmental traits. Proceedings of the Freshwater Larval Fish Workshop, I. Southeastern Electric Exchange.

Loos, J. J., L. A. Fuiman, N. R. Foster, and E. K. Jankowski. 1979. Notes on the early life histories of cyprinid fishes of the upper Potomac River. Pages 93–139 *in* R. Wallus and C. W. Voigtlander, editors. Proceedings of a Workshop on Freshwater Larval Fishes. Tennessee Valley Authority, Norris, Tenn.

Lopinot, A. 1972. White amur, *Ctenopharyngodon idella*. Memo 37:2. Illinois Department of Conservation, Division of Fisheries Management.

Lovell, R. G., and M. J. Maceina. 2002. Population assessment and minimum length limit evaluations for white bass in four Alabama reservoirs. N. Am. J. Fish. Manage. 22:609–619.

Ludwig, A., L. Debus, D. Lieckfeldt, I. Wirgin, N. Benecke, I. Jenneckens, P. Williot, J. R. Waldman, and C. Pitra. 2002. When the American sea sturgeon swam east. Nature 419:447–448.

Lukas, J. A., and D. J. Orth. 1993. Reproductive ecology of redbreast sunfish *Lepomis auritus* in a Virginia stream. J. Freshw. Ecol. 8(3):235–244.

Lydeard, C., M. C. Wooten, and M. H. Smith. 1991. Occurrence of *Gambusia affinis* in the Savannah and Chattahoochee drainages: previously undescribed geographic contacts between *G. affinis* and *G. holbrooki*. Copeia 1991:1111–1116.

Lyons, J. 1989. Changes in the abundance of small littoral-zone fishes in Lake Mendota, Wisconsin. Can. J. Zool. 67:2910–2916.

Lyons, J., and S. Navarro-Perez. 1990. Fishes of the Sierra de Manantlan, west-central Mexico. Southwest. Nat. 35:32–46.

Maciolek, J. A. 1984. Exotic fishes in Hawaii and other islands of Oceania. Pages 131–161 *in* W. R. Courtenay Jr. and J. R. Stauffer Jr., editors. Distribution, biology, and management of exotic fishes. Johns Hopkins University Press, Baltimore, Md.

Mackiewicz, M. D., D. E. Fletcher, S. D. Wilkins, J. A. DeWoody, and J. C. Avise. 2002. A genetic assessment of parentage in a natural population of dollar sunfish (*Lepomis marginatus*) based on microsatellite markers. Mol. Ecol. 11:1877–1883.

Madenjian, C. P., R. L. Knight, M. T. Bur, and J. L. Forney. 2000. Reduction in recruitment of white bass in Lake Erie after invasion of white perch. Trans. Am. Fish. Soc. 129:1340–1353.

Malloy, R., and F. D. Martin. 1982. Comparative development of redfin pickerel (*Esox americanus americanus*) and the eastern mudminnow (*Umbra pygmaea*). Pages 70–72 *in* C. F. Bryan, J. V. Conner, and F. M. Truesdale, editors. The Fifth Annual Larval Fish Conference. Louisiana Cooperative Fisheries Research Unit.

Manooch, C. S. III. 1984. Fisherman's guide—fishes of the southeastern United States. North Carolina State Museum of Natural History, Raleigh.

Mansfield, P. J. 1984. Reproduction by Lake Michigan fishes in a tributary stream. Trans. Am. Fish. Soc. 113:231–237.

Mansueti, A. J. 1963. Some changes in morphology during ontogeny in the pirateperch, *Aphredoderus s. sayanus*. Copeia 1963:546–547.

Mansueti, A. J., and J. D. Hardy. 1967. Development of fishes of the Chesapeake Bay region: an atlas of egg, larval, and juvenile stages. Natural Resources Institute, University of Maryland, Baltimore.

Mansueti, R. J. 1961. Movements, reproduction, and mortality of the white perch, *Roccus americanus*, in the Patuxent estuary. Chesapeake Sci. 2(3–4):142–205.

Mansueti, R. J. 1962. Eggs, larvae, and young of the hickory shad, *Alosa mediocris*, with comments on its ecology in the estuary. Chesapeake Sci. 3:173–205.

Mansueti, R. J. 1964. Eggs, larvae, and young of the white perch, *Roccus americanus*, with comments on its ecology in the estuary. Chesapeake Sci. 5(1–2):3–45.

Mansueti, R., and H. J. Elser. 1953. Ecology, age and growth of the mud sunfish, *Acantharchus pomotis*, in Maryland. Copeia 1953:117–118.

Mansueti, R., and R. Pauli. 1956. Age and growth of the northern hogchoker, *Trinectes maculatus maculatus*, in the Patuxent River, Maryland. Copeia 1956:60–62.

Marcy, B. C. Jr. 1969. Age determinations from scales of *Alosa pseudoharengus* (Wilson) and *Alosa aestivalis* (Mitchell) in Connecticut waters. Trans. Amer. Fish. Soc. 98(4):622–630.

Marcy, B. C. Jr. 1972. Spawning of the American shad, *Alosa sapidissima*, in the lower Connecticut River. Chesapeake Sci. 13:116–119.

Marcy, B. C. Jr. 1976. Fishes of the lower Connecticut River and the effects of the Connecticut Yankee Plant. Pages 61–113 *in* D. Merriman and L. M. Thorpe, editors. The Connecticut River Ecological Study: the impact of a nuclear power plant. Am. Fish. Soc. Monogr. 1.

Marcy, B. C. Jr., J. A. Bower, J. B. Gladden, H. M. Hickey, M. P. Jones, H. E. Mackey, and J. J. Mayer. 1994. Remediation of a large contaminated reactor cooling reservoir: resolving an environmental/regulatory paradox. Proc. Annu. Meet. Natl. Assoc. Environ. Professionals (New Orleans, La.) 19:665–676.

Marcy, B. C. Jr., and S. K. O'Brien-White. 1995. Fishes of the Edisto River basin, South Carolina. Rep. 6. South Carolina Department of Natural Resources, Columbia.

Marcy, B. C. Jr., and F. P. Richards. 1974. Age and growth of the white perch *Morone americana* in the lower Connecticut River. Trans. Am. Fish. Soc. 103:111–120.

Markham, J. L., D. L. Johnson, and R. W. Petering. 1991. White crappie summer movements and habitat use in Delaware Reservoir, Ohio. N. Am. J. Fish. Manage. 11:504–512.

Marshall, N. B. 1947. Studies on the life history and ecology of *Notropis chalybaeus* (Cope). J. Fla. Acad. Sci. 9:163–188.

Martin, C. R. 1980. Movements, growth, and numbers of largemouth bass (*Micropterus salmoides*) in an unfished reservoir receiving a heated effluent. Master's thesis, University of Georgia, Athens.

Martin, F. D. 1968. Some factors influencing penetration into rivers by fishes of the genus *Cyprinodon*. Doctoral dissertation, University of Texas, Austin.

Martin, F. D., and C. Hubbs. 1973. Observations on the development of pirate perch, *Aphredoderus sayanus* (Pisces: Aphredoderidae), with comments on yolk circulation patterns as a possible taxonomic tool. Copeia 1973:377–379.

Martin, F. D., and D. A. Wright. 1987. Nutritional state analysis and its use in predicting striped bass recruitment: laboratory calibration. Am. Fish. Soc. Symp. 2:109–114.

Martin, F. D., D. A. Wright, J. C. Means, and E. M. Setzler-Hamilton. 1985. Importance of food supply and nutritional state of larval striped bass in the Potomac River estuary. Trans. Am. Fish. Soc. 114:137–145.

Massman, W. H. 1954. Marine fishes in fresh and brackish waters of Virginia rivers. Ecology 35:75–78.

Mathur, D. 1972. Seasonal food habits of the adult white crappie, *Pomoxis annularis* (Rafinesque), in Conowingo Reservoir. Am. Midl. Nat. 87:236–241.

Mathur, D. 1973a. Food habits and feeding chronology of the blackbanded darter, *Percina nigrofasciata* (Agassiz), in Halawakee Creek, Alabama. Trans. Am. Fish. Soc. 102:48–55.

Mathur, D. 1973b. Some aspects of the life history of the blackbanded darter, *Percina nigrofasciata* (Agassiz), in Halawakee Creek, Alabama. Am. Midl. Nat. 89:381–393.

Mathur, D., P. L. McCreight, and G. A. Nardacci. 1979. Variations in fecundity of white crappie in Conowingo Pond, Pennsylvania. Trans. Am. Fish. Soc. 108:548–554.

Mathur, D., and T. W. Robbins. 1971. Food habits and feeding chronology of young white crappie, *Pomoxis annularis* (Rafinesque), in Conowingo Reservoir. Trans. Am. Fish. Soc. 100:307–311.

Matthews, D. R., and W. J. Gelwick. 2002. Resource use by large catfishes in a reservoir: is there evidence for interactive segregation and innate differences? J. Fish Biol. 60:739–750.

Maurakis, E. G., W. S. Woolcott, and J. T. Magee. 1990. Pebble-nests of four *Semotilus* species. Proc. Southeast. Fish. Counc. 22:7–13.

Maurakis, E. G., W. S. Woolcott, and E. S. Perry. 1997. Description of agonistic behaviors in two species of *Nocomis*. Va. J. Sci. 48:195–202.

Maurakis, E. G., W. S. Woolcott, and M. H. Sabaj. 1991. Reproductive-behavioral phylogenetics of *Nocomis* species-groups. Am. Midl. Nat. 126:103–110.

Maurakis, E. G., W. S. Woolcott, and M. H. Sabaj. 1992. Water currents in spawning areas of pebble nests of *Nocomis leptocephalus* (Pisces: Cyprinidae). Proc. Southeast. Fish. Counc. 25:1–2.

Mayden, R. L. 1985. Biogeography of Ouachita Highland fishes. Southwest. Nat. 30:195–211.

Mayden, R. L. 1989. Phylogenetic studies of North American minnows, with emphasis on the genus *Cyprinella* (Teleostei: Cypriniformes). Misc. Pub. Mus. Nat. Hist. Univ. Kans. 80:189.

Mayer, J. J., and L. D. Wike. 1997. SRS urban wildlife. Environ. Inf. Doc. WSRC-TR-97-0093. Westinghouse Savannah River Co., Aiken, S.C.

McAuliffe, J. R., and D. H. Bennett. 1981. Observations on the spawning habits of the yellowfin shiner, *Notropis lutipinnis*. J. Elisha Mitchell Sci. Soc. 97:200–203.

McCann, J. A. 1959. Life history studies of the spottail

shiner of Clear Lake, Iowa, with particular reference to some sampling problems. Trans. Am. Fish. Soc. 88:336–343.

McCord, J. W. 1998. Investigation of fisheries parameters for anadromous fishes in South Carolina. Completion report ASFC-53 to National Marine Fisheries Service.

McCormack, B. 1967. Aerial respiration in the Florida spotted gar. Q. J. Fla. Acad. Sci. 30:68–72.

McCort, W. D., K. K. Patterson, S. W. Oliver, and S. S. Novak. 1984. Fish populations in Pond C, a review of the literature. SWED-84-0501. Savannah River Ecology Laboratory, Aiken, S.C.

McCulley, H. H. 1962. The relationship of the Percidae and the Centrarchidae to the Serranidae as shown by the anatomy of their scales. Am. Zool. 2:247.

McFarlane, R. W., F. F. Frietsche, and R. D. Miracle. 1978. Impingement and entrainment of fishes at the Savannah River Plant: an NPDES 316b demonstration. DP-1494. E. I. du Pont de Nemours and Co., Savannah River Laboratory, Aiken, S.C.

McIlwain, T. D. 1970. Stomach contents and length-weight relationships of chain pickerel (*Esox niger*) in south Mississippi waters. Trans. Am. Fish. Soc. 99:439–440.

McInerny, M. C., and D. J. Degan. 1991. Dynamics of a black crappie population in a heterogeneous cooling reservoir. N. Am. J. Fish. Manage. 11:525–533.

McKenzie, D. J., and D. J. Randall. 1990. Does *Amia calva* estivate? Fish Physiol. Biochem. 8(2):147–158.

McLarney, W. O., D. G. Engstrom, and J. H. Todd. 1974. Effects of increasing temperature on social behavior in groups of yellow bullhead (*Ictalurus natalis*). Environ. Poll. 7:111–119.

McLean, B., P. T. Singley, D. M. Lodge, and R. A. Wallace. 1982. Synchronous spawning of threadfin shad. Copeia 1982:952–955.

McMahon, T. E. 1982. Habitat suitability index models: creek chub. FWS/OBS-82/10.4. U.S. Department of the Interior, Fish and Wildlife Service, Washington, D.C.

McNeely, D. L. 1987. Niche relationships within an Ozark stream cyprinid assemblage. Environ. Biol. Fishes 18:195–208.

McSwain, L. E., and R. M. Gennings. 1972. Spawning behavior of the spotted sucker *Minytrema melanops* (Rafinesque). Trans. Am. Fish. Soc. 101:739–740.

Meador, M. R., A. G. Eversole, and J. S. Bulak. 1984. Utilization of portions of the Santee River system by spawning blueback herring. N. Am. J. Fish. Manage. 4:155–163.

Meehan, W. E. 1910. Experiments in sturgeon culture. Trans. Am. Fish. Soc. 39:85–91.

Meffe, G. K. 1991. Failed invasion of a southeastern blackwater stream by bluegills: implications for conservation of native communities. Trans. Am. Fish. Soc. 120:333–338.

Meffe, G. K., D. L. Certain, and A. L. Sheldon. 1988. Selective mortality of post-spawning yellowfin shiners, *Notropis lutipinnis* (Pisces: Cyprinidae). Copeia 1988:853–858.

Meffe, G. K., and A. L. Sheldon. 1988. The influence of habitat structure on fish assemblage composition in southeastern blackwater streams. Am. Midl. Nat. 120(2):225–239.

Melvin, G. D., M. J. Dadswell, and J. D. Martin. 1986. Fidelity of American shad, *Alosa sapidissima* (Clupeidae), to its river of previous spawning. Can. J. Fish. Aquat. Sci. 43:640–646.

Menhinick, E. F. 1991. The freshwater fishes of North Carolina. North Carolina Wildlife Resources Commission, Raleigh.

Merriner, J. V. 1971. Development of intergeneric centrarchid hybrid embryos. Trans. Am. Fish. Soc. 100(4):611–618.

Mettee, M. F. 1974. A study on the reproductive behavior, embryology, and larval development of the pygmy sunfishes of the genus *Elassoma*. Doctoral dissertation, University of Alabama, Tuscaloosa.

Mettee, M. F., P. E. O'Neil, and J. M. Pierson. 1996. Fishes of Alabama and the Mobile Basin. Oxmoor House, Birmingham, Ala.

Meyers, C. D., and R. J. Muncy. 1962. Summer food and growth of chain pickerel, *Esox niger*, in brackish waters of the Severn River, Maryland. Chesapeake Sci. 3:125–128.

Middaugh, D. P. 1981. Reproductive ecology and spawning periodicity of the Atlantic silverside, *Menidia menidia* (Pisces: Atherinidae). Copeia 1981:766–776.

Middaugh, D. P., R. G. Domey, and G. I. Scott. 1984. Reproductive rhythmicity of the Atlantic silverside. Trans. Am. Fish. Soc. 113:472–478.

Middaugh, D. P., and M. J. Hemmer. 1987. Reproductive ecology of the tidewater silverside, *Menidia peninsulae* (Pisces: Atherinidae) from Santa Rosa Island, Florida. Copeia 1987:727–732.

Middaugh, D. P., M. J. Hemmer, and Y. Lamadrid-Rose. 1986. Laboratory spawning cues in *Menidia beryllina* and *M. peninsulae* (Pisces, Atherinidae) with notes on survival and growth of larvae at different salinities. Environ. Biol. Fishes 15:107–117.

Middaugh, D. P., G. I. Scott, and J. M. Dean. 1981. Reproductive behavior of the Atlantic silverside, *Menidia menidia* (Pisces, Atherinidae). Environ. Biol. Fishes 6:269–276.

Migdalski, E. C. 1962. Angler's guide to freshwater sportfishes of North America. Ronald Press, New York.

Millard, M. J. 1981. Comparative larval development of the pugnose minnow, *Notropis emiliae*, and the taillight shiner, *Notropis maculatus* (Pisces: Cyprinidae). Master's thesis, Louisiana State University, Baton Rouge.

Miller, H. C. 1963. The behavior of the pumpkinseed sunfish, *Lepomis gibbosus* (Linnaeus), with notes on the behavior of other species of *Lepomis* and the pygmy sunfish, *Elassoma evergladei*. Behaviour 22:88–151.

Miller, G. L., and G. H. Clemmer. 1980. Morphometric analysis of Gulf Coast populations of *Enneacanthus gloriosus* (Centrarchidae). ASB Bull. 27(2):51 (abstract).

Miller, R. J. 1964. Behavior and ecology of some North American cyprinid fishes. Am. Midl. Nat. 72:313–357.

Miller, R. R. 1963. Genus *Dorosoma* Rafinesque 1820; gizzard shads, threadfin shad. Pages 443–451 *in* Fishes of the western North Atlantic. Sears Foundation for Marine Research, Mem. 1, pt. 3.

Miller, S. J., D. D. Fox, L. A. Bull, and T. D. McCall. 1990. Population dynamics of black crappie in Lake Okeechobee, Florida, following suspension of commercial harvest. N. Am. J. Fish. Manage. 10:98–105.

Milstein, C. B. 1981. Abundance and distribution of juvenile *Alosa* species off southern New Jersey. Trans. Am. Fish. Soc. 110:306–309.

Minckley, W. E. 1963. The ecology of a spring stream, Doe Run, Meade County, Kentucky. Wildl. Monogr. 11.

Ming, A. D. 1968. Life history of the grass pickerel, *Esox americanus vermiculatus*, in Oklahoma. Bull. Okla. Fish Res. Lab. 8:66.

Mittelbach, G. G. 1984. Predation and resource partitioning in two sunfishes (Centrarchidae). Ecology 65:449–513.

Mittelbach, G. G., and C. W. Osenberg. 1993. Stage-structured interactions in bluegill: consequences of adult resource variation. Ecology 74(8):2381–2394.

Mittelbach, G. G., C. W. Osenberg, and P. C. Wainwright. 1992. Variation in resource abundance affects diet and feeding morphology in the pumpkinseed sunfish (*Lepomis gibbosus*). Oecologia 90:8–13.

Mittelbach, G. G., and L. Persson. 1998. The ontongeny of piscivory and its ecological consequences. Can. J. Fish. Aquat. Sci. 55:1454–1465.

Mohler, H. J., F. W. Whicker, and T. G. Hinton, 1997. Temporal trends of ^{137}Cs in an abandoned reactor cooling reservoir. J. Environ. Radioact. 37:251–268.

Moore, G. A., and W. E. Burris. 1956. Description of the lateral-line system of the pirate perch, *Aphredoderus sayanus*. Copeia 1956:18–22.

Monzyk, F. P., W. E. Kelso, and D. A. Rutherford. 1997. Characteristics of woody cover used by brown madtoms and pirate perch in Coastal Plain streams. Trans. Am. Fish. Soc. 126:665–675.

Morgan, G. D. 1954. The life history of the white crappie (*Pomoxis annularis*) of Buckeye Lake, Ohio. J. Sci. Lab. Denison Univ. 43(6–8):113–144.

Moroz, V. N. 1968. Description of the spawning stock, spawning and fertility of carp from the Kiliya Delta of the Danube. Probl. Ichthyol. 8(3):414–421.

Morse, J. C., J. W. Chapin, D. D. Herlong, and R. S. Harvey. 1980. Aquatic insects of Upper Three Runs Creek, Savannah River Plant, South Carolina. Part 1: Orders other than Diptera. J. Ga. Entomol. Soc. 15(1):73–101.

Morse, J. C., J. W. Chapin, D. D. Herlong, and R. S. Harvey. 1983. Aquatic insects of Upper Three Runs Creek, Savannah River Plant, South Carolina. Part 2: Diptera. J. Ga. Entomol. Soc. 18(3):303–316.

Moser, M. L., M. Bain, M. R. Collins, H. Haley, B. Kynard, J. C. O'Herron II, G. Rogers, and T. S. Squiers. 2000. A protocol for use of shortnose and Atlantic sturgeons. NOAA Tech. Mem. NMFS-OPR-18.

Moser, M. L., and S. W. Ross. 1995. Habitat use and movements of shortnose and Atlantic sturgeons in the lower Cape Fear River, North Carolina. Trans. Am. Fish. Soc. 124:2125–234.

Moshenko, R. W., and J. H. Gee. 1973. Diet, time and place of spawning, and environment occupied by creek chub (*Semotilus atromaculatus*) in the Mink River, Manitoba. J. Fish. Res. Board Can. 30:357–362.

Moyle, P. B. 2002. Inland fishes of California, revised and expanded edition. University of California Press, Berkeley.

Mullan, J. W., L. Applegate, and W. C. Rainwater. 1968. Food of logperch (*Percina caprodes*) and brook silverside (*Labidestes sicculus*) in a new and old Ozark reservoir. Trans. Am. Fish. Soc. 97:300–305.

Mulligan, T. J., F. D. Martin, R. A. Smucker, and D. A. Wright. 1987. A method for stock identification based on elemental composition of striped bass, *Morone saxatilis* (Walbaum) otoliths. J. Exp. Mar. Biol. Ecol. 114:241–248.

Muncy, R. J. 1962. Life history of the yellow perch, *Perca flavescens*, in estuarine waters of Severn River, a tributary of Chesapeake Bay, Maryland. Chesapeake Sci. 3:143–159.

Mundahl, N. D., C. Melnytschuk, D. K. Spielman, J. P. Harkins, K. Funk, and A. M. Bilicki. 1998. Effectiveness of bowfin as a predator on bluegill in a vegetated lake. N. Am. J. Fish. Manage. 18:286–294.

Munger, C. R., G. R. Wilde, and B. J. Follis. 1994. Flathead catfish age and size at maturation in Texas. N. Am. J. Fish. Manage. 14:403–408.

Muoneke, M. I., C. C. Henry, and O. E. Maughan. 1992. Population structure and food habits of white crappie

Pomoxis annularis Rafinesque in a turbid Oklahoma reservoir. J. Fish Biol. 41:647–654.

Murdy, E. O., R. S. Birdsong, and J. A. Musick. 1997. Fishes of the Chesapeake Bay. Smithsonian Institution Press, Washington, D.C.

Music, J. L. Jr. 1981. Assessment of Georgia's 1980 commercial shad season. Georgia Department of Natural Resources, Coastal Resources Division, Brunswick.

Muska, C. F., and R. A. Matthews. 1983. Biological assessment for the shortnose sturgeon, *Acipenser brevirostrum* (Lesueur, 1818), in the Savannah River Plant. DPST-83-754. E. I. du Pont de Nemours and Co., Savannah River Laboratory, Aiken, S.C.

Myers, J. J., and C. C. Kohler. 2000. Acute responses to salinity for sunshine bass and palmetto bass. N. Am. J. Aquacult. 62:195–202.

Nahamura, M. 1969. Cyprinid fishes of Japan—studies of the life history of cyprinid fishes of Japan. Res. Inst. Nat. Resour. Spec. Pub. 4:455.

National Science Center's Fort Discovery. 2002. Southeast fisheries. Educational Technology Training Center—NSC, Augusta, Ga.

NCDENR [North Carolina Department of Environment and Natural Resources]. 2001. Standard operating procedure biological monitoring: stream fish community assessment & fish tissue. Division of Water Quality, Water Quality Section, Environmental Sciences Branch, Biological Assessment Unit, Raleigh.

Neill, W. T, 1950. An estivating bowfin. Copeia 1950:240.

Nelson, E. A., R. K. Kolka, C. C. Tretin, and J. Wisniewski, editors. 2000. Restoration of the severely impacted riparian wetland system. Ecol. Engineer. 15, Suppl. 1:S67–S78.

Nelson, J. S. 1968. Life history of the brook silverside *Labidestes sicculus*, in Crooked Lake, Indiana. Trans. Am. Fish. Soc. 97:293–296.

Nelson, J. S. 1994. Fishes of the world, 3rd edition. John Wiley and Sons, New York.

Nesbit, D. H., and G. K. Meffe. 1993. Cannibalism frequencies in wild populations of the eastern mosquitofish (*Gambusia holbrooki*: Poeciliidae) in South Carolina. Copeia 1993:867–870.

Nestler, J. M., R. A. Goodwin, T. H. Cole, D. Degan, and D. Dennerline. 2002. Simulating movement patterns of blueback herring in a stratified southern impoundment. Trans. Am. Fish. Soc. 131:55–69.

Netsch, N. F. 1967. Food and feeding habits of longnose gar in central Missouri. Proc. Annu. Conf. Southeast. Assoc. Game Fish Comm. 18(1964):506–511.

Netsch, N. F., and A. Witt Jr. 1962. Contributions to the life history of the longnose gar (*Lepisosteus osseus*) in Missouri. Trans. Am. Fish. Soc. 91:251–262.

Neves, R. J., and L. Depres. 1979. The oceanic migration of American shad, *Alosa sapidissima*, along the Atlantic Coast. Fish. Bull. 17:199–212.

Newcomb, B. A. 1989. Winter abundance of channel catfish in the channelized Missouri River, Nebraska. N. Am. J. Fish. Manage. 9:195–202.

Nichols, P. R. 1959. Extreme loss in body weight on an American shad (*Alosa sapidissima*). Copeia 1959:343–344.

Nickum, J. G. 1988. Guidelines for use of fishes in field research. Fisheries 13(2):16–23.

Nielsen, L. A., and D. L. Johnson. 1983. Fisheries techniques. American Fisheries Society, Bethesda, Md.

Nikolsky, G. V. 1963. The ecology of fishes. Academic Press, New York and London.

Northcote, T. G., and R. J. Paterson. 1960. Relationship between number of pyloric caeca and length of juvenile rainbow trout. Copeia 1960:248–250.

Ogburn, M. V., D. M. Allen, W. K. Michener. 1988. Fishes, shrimps, and crabs of the North Inlet estuary, SC: a four year seine and trawl survey. Belle W. Baruch Institute for Marine Coastal Sciences, University of South Carolina.

Okada, Y. 1959–1960. Studies of the freshwater fishes of Japan. Prefectural University of Mie Tsu, Mie Prefecture, Japan.

Oliviera, K. 1999. Life history characteristics and strategies of the American eel, *Anguilla rostrata*. Can. J. Fish. Aquat. Sci. 56:795–802.

Olmsted, L. L., and D. G. Cloutman. 1979. Life history of the flat bullhead, *Ictalurus platycephalus*, in Lake Norman, North Carolina. Trans. Am. Fish. Soc. 108:38–42.

Olson, M. H. 1996. Ontogenetic niche shifts in largemouth bass: variability and consequences for first-year growth. Ecology 77(1):179–190.

Olson, M. H., G. G. Mittelbach, and C. W. Osenberg. 1995. Competition between predator and prey: resource-based mechanisms and implications for stage-structured dynamics. Ecology 76(6):1758–1771.

Osenberg, C. W., G. G. Mittelbach, and P. C. Wainwright. 1992. Two-stage life histories in fish: the interaction between juvenile competition and adult performance. Ecology 73:255–267.

Osteen, D. V., A. G. Eversole, and R. W. Christie. 1989. Spawning utilization of an abandoned ricefield by blueback herring. Pages 552–565 *in* R. R. Sharitz and J. W. Gibbons, editors. Freshwater wetlands and wildlife. Office of Science Technical Information, Oak Ridge, Tenn.

Page, L. M. 1983. Handbook of darters. T. F. H. Publications, Neptune City, N.J.

Page, L. M. 1985. Evolution of reproductive behaviors in percid fishes. Ill. Nat. Hist. Surv. Bull. 33(3):275–295.

Page, L. M., and H. L. Bart Jr. 1989. Egg mimics in darters (Pisces: Percidae). Copeia 1989:514–517.

Page, L. M., and B. M. Burr. 1991. A field guide to freshwater fishes of North America north of Mexico. Houghton Mifflin, Boston.

Page, L. M., and C. E. Johnston. 1990a. Spawning in the creek chubsucker, *Erimyzon oblongus*, with a review of spawning behavior in suckers (Catostomidae). Environ. Biol. Fishes 27:265–272.

Page, L. M., and C. E. Johnston. 1990b. The breeding behavior of *Opsopoeodus emiliae* (Cyprinidae) and its phylogenetic implications. Copeia 1990:1176–1180.

Paller, M. H. 1987. Distribution of larval fish between macrophyte beds and open channels in a southeastern floodplain swamp. J. Freshw. Ecol. 4:191–200.

Paller, M. H. 1994. Relationships between fish assemblage structure and stream order in South Carolina Coastal Plain streams. Trans. Am. Fish. Soc. 123:150–161.

Paller, M. H. 1995. Relationships among number of fish species sampled, reach length surveyed, and sampling effort in South Carolina Coastal Plain streams. N. Am. J. Fish. Manage. 15:110–120.

Paller, M. H. 1997. Recovery of a reservoir fish community from drawdown related impacts. N. Am. J. Fish. Manage. 17:726–733.

Paller, M. H. n.d. Long-term development of the fish assemblage in a southeastern reservoir. Unpublished manuscript.

Paller, M. H., and S. A. Dyer. 1999. Biotic integrity of streams in the Savannah River Site integrator operable units. WSRC-TR-00112. Savannah River Technology Center, Aiken, S.C.

Paller, M. H., J. B. Gladden, and J. H. Heuer. 1992. Development of the fish community in a new South Carolina reservoir. Am. Midl. Nat. 128:95–114.

Paller, M. H., J. H. Heuer, L. A. Kissick, and H. G. Mealing. 1988. L-Lake fish: L-Lake/Steel Creek biological monitoring program, January 1986–December 1987. ECS-SR-65. Prepared by Environmental and Chemical Sciences, Inc., for Savannah River Laboratory, E. I. du Pont de Nemours and Co., Aiken, S.C.

Paller, M. H., J. W. Littrell, and E. L. Peters. 1999. Ecological half-lives of ^{137}Cs in fishes from the Savannah River Site. Health Physics 77:392–402.

Paller, M. H., J. O'Hara, and D. V. Osteen. 1985. Annual report on the Savannah River aquatic ecology program, September 1983–August 1984. Volume 2: Ichthyoplankton. DPST-85-377. Environmental and Chemical Sciences, Inc., for Savannah River Laboratory, E. I. du Pont de Nemours and Co., Aiken, S.C.

Paller, M. H., J. O'Hara, D. V. Osteen, W. Specht, and H. Kania. 1984. Annual report on the Savannah River aquatic ecology program—September 1982 to August 1983. Volume 1: Adult fish. DPST-84-252. Environmental and Chemical Sciences, Inc., for Savannah River Laboratory, E. I. du Pont de Nemours and Co., Aiken, S.C.

Paller, M. H., M. J. M. Reichert, and J. M. Dean. 1996. The use of fish communities to assess environmental impacts in South Carolina Coastal Plain streams. Trans. Am. Fish. Soc. 125:633–644.

Paller, M. H., M. J. M. Reichert, J. M. Dean, and J. C. Seigle. 2000. Use of fish community data to evaluate restoration success in a riparian stream. Ecol. Engineer. 15, Suppl. 1:S171–S187.

Paller, M. H., and B. M. Saul. 1985. Final report on the adult fish and ichthyoplankton of Par Pond and Pond B: January 1984–June 1985. ECS-SR-22. Prepared by Environmental and Chemical Sciences, Inc., for E. I. du Pont de Nemours and Co., Aiken, S.C.

Paller, M. H., and B. M. Saul. 1986. Effects of thermal discharges on the distribution and abundance of adult and juvenile fishes in the Savannah River and selected tributaries. ECS-SR-28. Environmental and Chemical Sciences, Inc., Aiken, S.C. 257 pp.

Paller, M. H., B. H. Saul, and D. W. Hughes. 1986a. The distribution of ichthyoplankton in thermal and nonthermal creeks and swamps on the Savannah River Plant. February–July 1985. ECS-SR-25. Environmental and Chemical Sciences, Inc., for E. I. du Pont de Nemours and Co., Aiken, S.C.

Paller, M. H., B. M. Saul, and D. V. Osteen. 1986b. Distribution and abundance of ichthyoplankton in the midreaches of the Savannah River and selected tributaries. DPST-86-798. Environmental and Chemical Sciences, Inc., for Savannah River Laboratory, E. I. du Pont de Nemours and Co., Aiken, S.C.

Panek, F. M., and C. R. Cofield. 1978. Fecundity of bluegill and warmouth from a South Carolina blackwater lake. Progr. Fish-Cult. 40(2):67–68.

Pardue, G. B. 1983. Habitat suitability index models: alewife and blueback herring. FWS/OBS-82/10.58. U.S. Department of the Interior, Fish and Wildlife Service, Washington, D.C.

Pardue, G. B. 1993. Life history and ecology of the mud sunfish (*Acantharchus pomotis*). Copeia 1993:533–540.

Pardue, G. B., and F. E. Hester. 1967. Variation in growth rate of known-age largemouth bass (*Micropterus salmoides* Lacépède) under experimental conditions. Proc. Southeast. Assoc. Game Fish Comm. 20:300–310.

Parker, B. R., and W. G. Franzin. 1991. Reproductive biology of the quillback, *Carpiodes cyprinus*, in a small prairie river. Can. J. Zool. 69:2133–2139.

Parker, B. R., and W. G. Franzin. 1994. Age determination and growth of quillback from Dauphin Lake, Manitoba. Can. Tech. Rep. Fish. Aquat. Sci. 1977.

Parker, E. D., M. F. Hirshfield, and J. W. Gibbons. 1973. Ecological comparisons of thermally affected aquatic environments. J. Water Poll. Control Fed. 45:726–733.

Parker, N. C., and B. A. Simco. 1975. Activity patterns, feeding and behavior of the pirate perch, *Aphredoderus sayanus*. Copeia 1975:572–574.

Parker, W. D. 1971. Preliminary studies on sexing adult largemouth bass by means of an external characteristic. Progr. Fish-Cult. 33:54–55.

Parsons, J. W. 1954. Growth and habits of the redeye bass. Trans. Am. Fish. Soc. 83:202–211.

Partridge, D. G., and D. R. DeVries. 1999. Regulation of growth and mortality in larval bluegills: implications for juvenile recruitment. Trans. Am. Fish. Soc. 128:625–638.

Pearson, W. D., and B. J. Pearson. 1989. Fishes of the Ohio River. Ohio J. Sci. 89:181–187.

Pearson, W. D., G. A. Thomas, and A. L. Clark. 1979. Early piscivory and timing of the critical period in post-larval longnose gar at mile 571 of the Ohio River. Trans. Ky. Acad. Sci. 40:122–128.

Peer, D. L. 1966. Relationship between size and maturity in the spottail shiner, *Notropis hudsonius*. J. Fish. Res. Board Can. 23:455–457.

Pekovitch, A. W. 1979. Distribution and some life history aspects of the shortnose sturgeon (*Acipenser brevirostrum*). Hazleton Environmental Sciences Corp., Northbrook, N.Y.

Pellet, T. D., G. J. Van Dyck, and J. V. Adams. 1998. Seasonal migration and homing of channel catfish in the lower Wisconsin River, Wisconsin. N. Am. J. Fish. Manage. 18:85–95.

Perry, W. G., and D. C. Carver. 1973. Length at maturity and total length–collarbone length conversions for channel catfish, *Ictalurus punctatus*, and blue catfish, *Ictalurus furcatus*, collected from the marshes of southwest Louisiana. Proc. Ann. Conf. Southeast. Assoc. Game Fish Comm. 26(1972):541–553.

Peterson, M. S., and S. T. Ross. 1987. Morphometric and meristic characteristics of a peripheral population of *Enneacanthus*. Proc. Southeast. Fishes Counc. 17:1–4.

Peterson, M. S., and S. J. VanderKooy. 1997. Distribution, habitat characterization, and aspects of reproduction of a peripheral population of bluespotted sunfish *Enneacanthus gloriosus* (Holbrook). J. Freshw. Ecol. 12:151–161.

Peterson, T. L. 1996. Seasonal migration in the southern hogchoker, *Trinectes maculatus fasciatus* (Archiridae). Gulf Res. Rep. 9:169–176.

Peterson-Curtis, T. L. 1996. Partial life history of southern hogchokers, *Trinectes maculatus fasciatus*, in the back bay of Biloxi, Mississippi. Gulf Mex. Sci. 2:81–88.

Pezold, F. L., and R. J. Edwards. 1983. Additions to the Texas marine ichthyofauna, with notes on the Rio Grande estuary. Southwest. Nat. 28:102–105.

Pflieger, W. L. 1975. The fishes of Missouri. Missouri Department of Conservation, Jefferson City.

Phillip, D. A. T. 1993. Reproduction and feeding of the mountain mullet *Agonostomus monticola*, in Trinidad, West Indies. Environ. Biol. Fishes 37:47–55.

Phillips, C. 1958. An unusually colored garfish, *Lepisosteus platyrhincus*. Copeia 1958(4):331.

Pierce, C. L., K. A. Musgrove, J. Ritterpusch, and N. E. Carl. 1987. Littoral invertebrate abundance in bluegill spawning colonies and undisturbed areas of a small pond. Can. J. Zool. 65:2066–2071.

Pierce, R. J., T. E. Wissing, and B. A. Magrey. 1981. Aspects of feeding ecology of gizzard shad in Acton Lake, Ohio. Trans. Am. Fish. Soc. 110:391–395.

Pigg, J. 1998. Melanism in longnose gar, *Lepisosteus osseus* (Linnaeus) (Lepisosteidae). Proc. Okla. Acad. Sci. 78:123.

Pipas, J. C., and F. J. Bulow. 1998. Hybridization between redeye bass and smallmouth bass in Tennessee streams. Trans. Am. Fish. Soc. 127:141–149.

Piper, R. G., I. B. McElwain, L. E. Orme, J. P. McCraren, L. G. Fowler, and J. R. Leonard. 1982. Fish hatchery management. U.S. Department of the Interior, Fish and Wildlife Service, Washington, D.C.

Pope, K. L., and D. W. Willis. 1997. Environmental characteristics of black crappie (*Pomoxis nigromaculatus*) nesting sites in two South Dakota waters. Ecol. Freshw. Fish 1997:183–189.

Poulson, T. L. 1963. Cave adaptation in amblyopsid fishes. Am. Midl. Nat. 70:257–290.

Poulson, T. L., and W. B. White. 1969. The cave environment. Science 165:971–980.

Powles, P. M., D. Parker, and R. Reid. 1977. Growth, maturation, and apparent and absolute fecundity of creek chub, *Semotilus atromaculatus* (Mitchill), in the Kawartha Lakes region, Ontario. Can. J. Zool. 55:843–846.

Quattro, J. M., T. W. Greig, D. K. Coykendall, B. W. Bowen, and J. D. Baldwin. 2002. Genetic issues in aquatic species management: the shortnose sturgeon (*Acipenser brevirostrum*) in the southeastern United States. Conserv. Genet. 3:155–166.

Quattro, J. M., W. J. Jones, J. M. Grady, and F. C. Rohde. 2001b. Gene-gene concordance and the phylogenetic relationships among rare and widespread pygmy sunfishes (genus *Elassoma*). Mol. Phylogenet. Evol. 18:217–226.

Quattro, J. M., W. J. Jones, and F. C. Rohde. 2001a. Evolutionarily significant units of rare pygmy sunfishes (genus *Elassoma*). Copeia 2001:514–520.

Quinn, T., G. W. Esch, T. C. Hazen, and J. W. Gibbons. 1978. Long range movement and homing by largemouth bass (*Micropterus salmoides*) in a thermally altered reservoir. Copeia 1978:542–545.

Rabito, F. G. Jr., and D. C. Heins. 1985. Spawning behaviour and sexual dimorphism in the North American cyprinid fish *Notropis leedsi*, the bannerfin shiner. J. Nat. Hist. 19:155–163.

Rahel, F. J., and W. A. Hubert. 1991. Fish assemblages and habitat gradients in a Rocky Mountain–Great Plains stream: biotic zonation and additive patterns of community change. Trans. Am. Fish. Soc. 120:319–332.

Raibley, P. T., D. Blodgett, and R. E. Sparks. 1995. Evidence of grass carp (*Ctenopharyngodon idella*) reproduction in the Illinois and upper Mississippi rivers. J. Freshw. Ecol. 10:65–74.

Rakocinski, C. F., M. S. Peterson, S. J. VanderKooy, and G. J. Crego. 1997. Biodiversity patterns of littoral tidal river fishes in the Gulf Plain region of Mississippi. Gulf Mex. Sci. 1997:2–16.

Raney, E. C. 1939. The breeding habits of the silvery minnow, *Hybognathus regius* (Girard). Am. Midl. Nat. 21:674–680.

Raney, E. C. 1942. Propagation of the silvery minnow (*Hybognathus nuchalis regius* (Girard)) in ponds. Trans. Am. Fish. Soc. 71:215–218.

Raney, E. C. 1947. *Nocomis* nests used by other breeding cyprinid fishes in Virginia. Zoologica 32:125–132.

Raney, E., and E. Lachner. 1943. Age and growth of Johnny darters, *Boleosoma nigrum olmstedi* (Storer) and *Boleostoma longimanum* (Jordan). Am. Midl. Nat. 29:229–238.

Rasmussen, R. P. 1980. Egg and larval development of the brook silverside from the Peace River, Florida. Trans. Am. Fish. Soc. 109:407–416.

Reagan, R. E., and W. M. Wingo. 1985. Southern flounder — species profiles: life histories and environmental requirements of coastal fishes and invertebrates (Gulf of Mexico). Biol. Rep. 82(11:30). U.S. Department of the Interior, Fish and Wildlife Service, Washington, D.C.

Reichert, M. J. M. 1995. Index of biotic integrity: results of sampling the fish assemblage of 13 creeks at the Savannah River Site area in the fall and winter of 1994/95. Report to Westinghouse Savannah River Co., Aiken, S.C.

Reichert, M. J. M., and H. W. Van der Veer. 1991. Settlement, abundance, growth and mortality of juvenile flatfish in a subtropical tidal estuary (Georgia, U.S.A.). Neth. J. Sea. Res. 27:375–391.

Reighard, J. 1901. Some further notes on the breeding habits of *Amia*. Rep. Mich. Acad. Sci. 3:80–81.

Reighard, J. 1903. The natural history of *Amia calva* Linnaeus. Pages 59–109 *in* E. L. Mark Anniversary Volume. Henry Holt, New York.

Reighard, J. 1910. Methods of studying the habits of fishes, with an account of the breeding habits of the horned dace. Bull. U.S. Bur. Fish. 28:1111–1136.

Reighard, J. 1931. Observations on the breeding habits of fishes at Havana, Illinois 1910–1911. Bull. Ill. State Lab. Nat. Hist. 9:405–416.

Reighard, J., and J. Phelps. 1908. The development of the adhesive organ and head mesoblast of *Amia*. J. Morphol. 19(2):469–496.

Relyea, K., and B. Sutton. 1973. Cave dwelling yellow bullheads in Florida. Q. J. Fla. Acad. Sci. 36:31–34.

Rettig, J. E. 1998. Variation in species composition of the larval assemblage in four southwest Michigan lakes: using allozyme analysis to identify larval sunfish. Trans. Am. Fish. Soc. 127:661–668.

Reynolds, W. W., and M. E. Casterlin. 1978. Ontogenetic changes in preferred temperature and diel activity of the yellow bullhead, *Ictalurus natalis*. Comp. Biochem. Physiol. 59:409–411.

Rice, J. A., J. E. Breck, S. M. Bartell, and J. F. Kitchell. 1983. Evaluating the constraints of temperature, activity and consumption on growth of largemouth bass. Environ. Biol. Fishes 9:263–275.

Richards, W. J. 1966. Systematics of the percid fishes of the *Etheostoma thalassinum* species group with comments on the subgenus *Etheostoma*. Copeia 1966:823–838.

Richardson, R. E. 1913. Observations on the breeding habits of fishes at Havana, Illinois, 1910 and 1911. Bull. Ill. State Lab. Nat. Hist. 9:405–416.

Rivas, L. R. 1966. The taxonomic status of the cyprinodontid fishes *Fundulus notti* and *F. lineolatus*. Copeia 1966:353–354.

Robins, C. R., R. M. Bailey, C. E. Bond, J. R. Brooker, E. A. Lachner, R. N. Lea, and W. B. Scott. 1991. A list of common and scientific names of fishes from the United States and Canada, 5th edition. Am. Fish. Soc. Spec. Pub. 20.

Robins, C. R., and G. C. Ray. 1986. A field guide to the Atlantic Coast fishes, North America. Houghton Mifflin, New York.

Robinson, B. W., and D. S. Wilson. 1996. Genetic variation and phenotypic plasticity in a polymorphic population of pumpkinseed sunfish (*Lepomis gibbosus*). Evol. Ecol. 10:631–652.

Robinson, B. W., D. S. Wilson, and A. S. Margosian. 2000. A pluralistic analysis of character release in pumpkinseed sunfish (*Lepomis gibbosus*). Ecology 81(10):2799–2812.

Robison, H. W. 1977. Distribution, habitat notes, and status of the ironcolor shiner, *Notropis chalybaeus* (Cope), in Arkansas. Ark. Acad. Sci. Proc. 31:92–94.

Robison, H. W. 1978. Distribution and habitat of the taillight shiner, *Notropis maculatus* (Hay), in Arkansas. Ark. Acad. Sci. Proc. 32:68–70.

Robison, H. W., and T. M. Buchanan. 1984. Fishes of Arkansas. University of Arkansas Press, Fayetteville.

Rohde, F. C., and R. G. Arndt. 1987. Two new species of pygmy sunfishes (Elassomatidae, *Elassoma*) from the Carolinas. Proc. Acad. Nat. Sci. Phila. 139:65–85.

Rohde, F. C., R. G. Arndt, D. G. Lindquist, and James F. Parnell. 1994. Freshwater fishes of the Carolinas, Virginia, Maryland and Delaware. University of North Carolina Press, Chapel Hill.

Ross, M. R. 1976. Nest-entry behavior of female creek chubs (*Semotilus atromaculatus*) in different habitats. Copeia 1976:378–380.

Ross, M. R. 1977. Aggression as a social mechanism in the creek chub (*Semotilus atromaculatus*). Copeia 1977:393–397.

Ross, R. M., W. A. Lellis, R. M. Bennett, and C. S. Johnson. 2001. Landscape determinants of nonindigenous fish invasions. Biol. Invasions 3:347–361.

Ross, S. T. 2001. The inland fishes of Mississippi. University of Mississippi Press, Jackson.

Ross, S. T., and J. A. Baker. 1983. The response of fishes to periodic spring floods in a southeastern stream. Am. Midl. Nat. 109(1):1–14.

Rubenstein, D. J. 1981a. Individual variation and competition in the Everglades pygmy sunfish. J. Anim. Ecol. 50:337–350.

Rubenstein, D. J. 1981b. Population density, resource patterning, and territoriality in the Everglades pygmy sunfish. Anim. Behav. 29:155–172.

Rudacille, J. B., and C. C. Kohler. 2000. Aquaculture performance comparison of sunshine bass, palmetto bass, and white bass. N. Am. J. Aquacult. 62:114–124.

Rutherford, D. A., K. R. Gelwicks, and W. E. Kelso. 2001. Physiochemical effects of the flood pulse on fishes in the Atchafalaya River basin, Louisiana. Trans. Am. Fish. Soc. 130:276–288.

Ryder, J. A. 1883. Observations on the absorption of the yolk, the food, feeding, and development of embryo fishes, comprising some investigations conducted at the Central Hatchery, Armory Building, Washington, D.C., in 1882. U.S. Comm. Fish. Bull. 2:179–205.

Ryder, J. A. 1890. The sturgeons and sturgeon industries of the eastern coast of the United States, with an account of experiments bearing upon sturgeon culture. U.S. Comm. Fish. Bull. 8(1888):231–326.

Sabaj, M. H., E. G. Maurakis, and W. S. Woolcott. 2000. Spawning behaviors in the bluehead chub, *Nocomis leptocephalus*, river chub, *N. micropogon*, and central stoneroller, *Campostoma anomalum*. Am. Midl. Nat. 144:187–201.

Sadzikowski, M. R., and D. C. Wallace. 1976. A comparison of the food habits of size classes of three sunfishes [*Lepomis macrochirus* Rafinesque, *L. gibbosus* (Linnaeus) and *L. cyanellus* Rafinesque]. Am. Midl. Nat. 95(1):220–225.

Saecker, J. R., and W. S. Woolcott. 1988. The redbreast sunfish (*Lepomis auritus*) in a thermally influenced section of the James River, Virginia. Va. J. Sci. 39(1):1–17.

Sammons, S. M., and P. W. Bettoli. 1998. Larval sampling as a fisheries management tool: early detection of year-class strength. N. Am. J. Fish. Manage. 18:137–143.

Sammons, S. M., P. W. Bettoli, D. A. Isermann, and T. N. Churchill. 2002. Recruitment variation of crappies in response to hydrology of Tennessee reservoirs. N. Am. J. Fish. Manage. 22:1393–1398.

Savoy, T. F., V. A. Crecco, and B. C. Marcy Jr. 2004. American shad (*Alosa sapidissima*) early life history and recruitment in the Connecticut River: a 40 year summary. *In* Jacobson, P. M., D. A. Dixon, W. C. Leggett, B. C. Marcy Jr., and R. R. Massengill, editors. The Connecticut River Ecological Study (1965–1973) revisited: ecology of the Lower Connecticut River, 1973–2003. Am. Fish. Soc. Monogr. 9.

SCDNR [South Carolina Department of Natural Resources]. 1992. Langley Pond fish health assessment. South Carolina Department of Natural Resources, Columbia.

SCDNR [South Carolina Department of Natural Resources]. 1997. Langley Pond largemouth bass sampling. South Carolina Department of Natural Resources, Columbia.

Schael, D. M., L. G. Rudstam, and J. R. Post. 1991. Gape limitation and prey selection in larval yellow perch (*Perca flavescens*), freshwater drum (*Aplodinotus grunnines*), and black crappie (*Pomoxis nigromaculatus*). Can. J. Fish. Aquat. Sci. 48:1919–1925.

Schaffter, R. G. 1997. Growth of white catfish in California's Sacramento San Joaquin Delta. Calif. Fish Game 83:57–67.

Schalles, J. F., R. R. Sharitz, J. W. Gibbons, G. J. Leversee, and J. N. Knox. 1989. Carolina bays of the Savannah River Plant. SRO-NERP-18. Savannah River Ecology Laboratory, Aiken, S.C.

Schaus, M. H., M. J. Vanni, and T. E. Wissing. 2002. Biomass-dependent diet shifts in omnivorous gizzard shad: implications for growth, food web and ecosystem effects. Trans. Am. Fish. Soc. 131:40–54.

Schemske, D. W. 1974. Age, length and fecundity of the creek chub, *Semotilus atromaculatus* (Mitchill), in central Illinois. Am. Midl. Nat. 92:505–509.

Schlosser, I. J. 1987. A conceptual framework for fish communities in small warmwater streams. Pages 17–24 *in* W. J. Matthews and D. C. Heines, editors. Community and evolutionary ecology of North American stream fishes. University of Oklahoma Press, Norman.

Schmidt, R. E., and W. R. Whitworth. 1979. Distribution and habitat of the swamp darter (*Etheostoma fusiforme*) in southern New England. Am. Midl. Nat. 102:408–413.

Schneberger, E. 1937. Food of small dogfish, *Amia calva*. Copeia 1937:61.

Schneider, J. C. 1998. Fate of dead fish in a small lake. Am. Midl. Nat. 140:192–196.

Schramm, H. L. Jr. 1989. Formation of annuli in otoliths of bluegills. Trans. Am. Fish. Soc. 118:546–555.

Schubel, J. R., and B. C. Marcy Jr. 1978. Power plant entrainment: a biological assessment. Academic Press, New York.

Schultz, D. L. 1999a. Comparison of lipid levels during spawning in annual and perennial darters of the subgenus *Boleosoma*, *Etheostoma perlongum*, and *Etheostoma olmstedi*. Copeia 1999:906–916.

Schultz, D. L. 1999b. Population structure, reproduction, and lipid cycling in the dusky shiner (*Notropis cummingsae*) in contrasting streams. Copeia 1999(3):669–683.

Schwartz, F. 1963. The freshwater minnows of Maryland. Md. Conserv. 102:19–29.

Schwartz, F. J. 1964. Natural salinity tolerances of some freshwater fishes. Underw. Nat. 2:13–15.

Schwartz, F. J. 2002. Occurences [*sic*] of elopiform fishes of the genera *Elops*, *Megalops* and *Albula* in North Carolina. J. N.C. Acad. Sci. 118:86–90.

Scott, W. B. 1954. Freshwater fishes of eastern Canada. University of Toronto Press, Toronto, Canada.

Scott, W. B., and E. J. Crossman. 1973. Freshwater fishes of Canada. Fish. Res. Board Can. Bull. 184:1–966.

Secor, D. H. 2002. Estuarine dependency and life history evolution in temperate sea basses. *In* Proceedings of the 70th Anniversary International Symposium of the Japanese Scientific Fisheries Society.

Secor, D. H., T. E. Gunderson, and K. Karlsson. 2000. Effects of salinity and temperature on growth performance in anadromous (Chesapeake Bay) and non-anadromous (Santee-Cooper) strains of striped bass, *Morone saxatilis*. Copeia 2000:291–296.

Seghers, B. H. 1981. Facultative schooling behavior in the spottail shiner (*Notropis hudsonius*): possible costs and benefits. Environ. Biol. Fishes 61:21–24.

Seidensticker, E. P. 1987. Food selection of alligator gar and longnose gar in a Texas reservoir. Proc. Annu. Conf. Southeast. Assoc. Fish Wildl. Agencies 14:100–104.

Setzler, E. M., W. R. Boynton, K. V. Wood, H. H. Zion, L. Lubbers, K. K. Mountford, P. Frere, L. Tucker, and J. A. Mihursky. 1980. Synopsis of biological data on striped bass, *Morone saxatilis* (Walbaum). NOAA Tech. Rep. NMFS Circ. 433.

Setzler-Hamilton, E. M., D. A. Wright, F. D. Martin, C. V. Millsaps, and S. I. Whitlow. 1987. Nutritional state analysis and its use in predicting striped bass recruitment: field studies. Am. Fish. Soc. Symp. 2:115–128.

Shao, B. 1997. Nest association of pumpkinseed, *Lepomis gibbosus*, and golden shiner, *Notemigonus crysoleucas*. Environ. Biol. Fishes 50:41–48.

Shealy Environmental Services, Inc. 1990. Fish population assessment of Langley, Vaucluse, and Clearview ponds, Aiken County, South Carolina. Report to the Graniteville Co., Graniteville, S.C.

Sheldon, A. L., and G. K. Meffe. 1993. Multivariate analysis of feeding relationships of fishes in blackwater streams. Environ. Biol. Fishes 37:161–171.

Sheldon, A. L., and G. K. Meffe. 1995. Short-term recolonization by fishes of experimentally defaunated pools of a coastal plain stream. Copeia 1995(4):828–837.

Shelton, W. L., and B. G. Grinstead. 1973. Hybridization between *Dorosoma cepedianum* and *D. petenense* in Lake Texoma, Oklahoma. Proc. Annu. Conf. Southeast. Assoc. Fish Game Comm. 26:506–510.

Shelton, W. L., C. D. Riggs, and L. G. Hill. 1982. Comparative reproductive biology of the threadfin and gizzard shad in Lake Texoma, Oklahoma-Texas. Pages 47–51 *in* C. F. Bryan, J. V. Conner, and F. M. Truesdale, editors. The Fifth Annual Larval Fish Conference. Louisiana Cooperative Fisheries Research Unit.

Shepherd, M. E., and M. T. Huish. 1978. Age, growth, and diet of the pirate perch in a Coastal Plain stream in North Carolina. Trans. Am. Fish. Soc. 107:457–459.

Shields, J. D., N. D. Woody, A. S. Dicks, G. J. Hollod, J. Schalles, and G. J. Leversee. 1982. Locations and areas of ponds and Carolina bays at the Savannah River Plant. Savannah River Laboratory, E. I. du Pont de Nemours and Co., Aiken, S.C.

Shireman, J. V., and C. R. Smith. 1983. Synopsis of biological data on the grass carp. FAO Fish. Synop. 135. Food and Agriculture Organization of the United Nations, Rome.

Shireman, J. V., R. L. Stetler, and D. E. Colle. 1978. Possible use of the lake chubsucker as a baitfish. Progr. Fish-Cult. 40:33–34.

Siefert, R. E. 1968. Reproductive behavior, incubation and mortality of eggs, and postlarval food selection in the white crappie. Trans. Am. Fish. Soc. 97(3):252–259.

Siefert, R. E. 1969. Characteristics for separation of white and black crappie larvae. Trans. Am. Fish. Soc. 98(2):326–328.

Sigler, W. F. 1955. An ecological approach to understanding Utah's carp populations. Proc. Utah Acad. Sci. 32:95–104.

Sigler, W. F. 1958. The ecology and use of the carp in Utah. Utah Agric. Exp. Stn. Bull. 405.

Siler, J. R. 1975. The distribution of fishes in two cooling reservoirs with different heat loads. Master's thesis, University of Georgia, Athens.

Siler, J. R., and J. P. Clugston. 1975. Largemouth bass under conditions of extreme thermal stress. Pages 333–341 *in* H. Clepper, editor. Black bass biology and management. Proceedings of the National Symposium on the Biology and Management of the Centrachid Basses. Sport Fishing Institute, Washington, D.C.

Simon, T. P., and R. Wallus. 1989. Contributions to the early life histories of gar (Actinopterygii: Lepisosteidae) in the Ohio and Tennessee river basins with emphasis on larval development. Trans. Ky. Acad. Sci. 50:59–74.

Slastenenko, E. P. 1958. The freshwater fishes of Canada. Kiev Printers, Toronto.

Smallwood, W. M., and M. B. Derrickson. 1933. The development of the carp, *Cyprinus carpio*. II. The development of the liver-pancreas, the islands of Langerhans, and the spleen. J. Morphol. 55:15–28.

Smith, B. A. 1971. The fishes of four low-salinity tidal tributaries of the Delaware River estuary. Master's thesis, Cornell University.

Smith, D. G., and R. F. Stearley. 1989. The classification and scientific names of rainbow and cutthroat trout. Fisheries 14:4–10.

Smith, H. M. 1907. The fishes of North Carolina. N.C. Geol. Econ. Surv. 2. E. M. Uzzell and Co., Raleigh.

Smith, H. M. 1909. Japanese goldfish, their varieties and cultivation, a practical guide to the Japanese method of goldfish culture for amateurs and professionals. W. F. Roberts Co., Washington, D.C.

Smith, L. L., and R. H. Kramer. 1964. The spottail shiner in Lower Red Lake, Minnesota. Trans. Am. Fish. Soc. 93:35–45.

Smith, M. H., K. T. Scribner, J. D. Hernandez, and M. C. Wooten. 1989. Demographic, spatial and temporal genetic variation in *Gambusia*. Ecology and evolution of livebearing fishes (Poeciliidae). Pages 235–257 *in* G. K. Meffe and F. F. Snelson, editors. The ecology and evolution of livebearing fishes (Poeciliidae). Prentice-Hall, Englewood Cliffs, N.J.

Smith, P. W. 1979. The fishes of Illinois. University of Illinois Press, Urbana.

Smith, S. M., M. J. Maceina, V. H. Travnichek, and R. A. Dunham. 1995. Failure of quantitative phenotypic characteristics to distinguish black crappie, white crappie, and their first-generation hybrid. N. Am. J. Fish. Manage. 15:121–125.

Smith, T. I. J. 1985. The fishery, biology, and management of Atlantic sturgeon, *Acipenser oxyrinchus*, in North America. Environ. Biol. Fishes 14:61–72.

Smith, T. I. J., M. C. Collins, W. C. Post, and J. W. McCord. 2002a. Stock enhancement of shortnose sturgeon: a case study. Trans. Am. Fish. Symp. 28:31–44.

Smith, T. I. J., E. K. Dingley, and D. E. Marchette. 1980. Induced spawning and culture of Atlantic sturgeon. Progr. Fish.-Cult. 42:147–151.

Smith, T. I. J., D. E. Marchette, and G. F. Ulrich. 1984. The Atlantic sturgeon fishery in South Carolina. N. Am. J. Fish. Manage. 4:164–176.

Smith, T. I. J., J. W. McCord, M. R. Collins, and W. C. Post. 2002b. Occurrence of stocked shortnose sturgeon *Acipenser brevirostrum* in non-targeted rivers. J. Appl. Ichthyol. 18:470–474.

Smithson, E. B., and C. E. Johnston. 1999. Movement patterns of stream fishes in an Ouachita Highlands stream: an examination of the restricted movement paradigm. Trans. Am. Fish. Soc. 128:847–853.

Snodgrass, J. W., A. L. Bryan Jr., J. W. Ackerman, and J. Burger. 1998. Vertebrates collected from isolated wetlands at the Savannah River Site: 1994–1997. Final report to the Set-Aside program. Savannah River Ecology Laboratory, University of Georgia, Aiken, S.C.

Snodgrass, J. W., A. L. Bryan Jr., R. F. Lide, and G. M. Smith. 1996. Factors affecting the occurrence and structure of fish assemblages in isolated wetlands of the Upper Coastal Plain, U.S.A. Can. J. Aquat. Sci. 53:443–454.

Snodgrass, J. W., and G. K. Meffe. 1998. Influence of beavers on stream fish assemblages: effects of pond age and watershed position. Ecology 79:928–942.

Snodgrass, J. W., and G. K. Meffe. 1999. Habitat use and temporal dynamics of blackwater stream fishes in and adjacent to beaver ponds. Copeia 1999(3):628–639.

Snyder, D. E., M. B. Mulhall Snyder, and S. C. Douglas. 1977. Identification of golden shiner, *Notemigonus crysoleucas*, spotfin shiner, *Notropis spilopterus*, and fathead minnow, *Pimephales promelas*, larvae. J. Fish. Res. Board Can. 34:1397–1409.

Snyder, D. J., and M. S. Peterson. 1999. Life history of a peripheral population of bluespotted sunfish *Enneacanthus gloriosus* (Holbrook), with comments on geographic variation. Am. Midl. Nat. 141:345–357.

Specht, W. L., and M. H. Paller. 2001. Instream biological assessment of NPDES point source discharges at the

Savannah River Site, 2000. WSRC-TR-2001-00145. Savannah River Technology Center, Aiken, S.C.

SREL [Savannah River Ecology Laboratory]. 1991. Summary report of Par Pond research. Savannah River Ecology Laboratory, University of Georgia, Aiken, S.C.

Stacey, P. B., and D. Chiszar. 1978. Body color pattern and the aggressive behavior of male pumpkinseed sunfish (*Lepomis gibbosus*) during the reproductive season. Behaviour 64:271–297.

Stanley, J. G., and D. S. Danie. 1983. Species profiles: life histories and environmental requirements of coastal fishes and invertebrates (North Atlantic). White perch. FWS/OBS-82/11.7. National Coastal Ecosystems Team, Division of Biological Services, U.S. Fish and Wildlife Service, Washington, D.C.

Stanley, J. G., W. W. Miley, and D. L. Sutton. 1978. Reproductive requirements and likelihood for naturalization of escaped grass carp in the United States. Trans. Am. Fish. Soc. 107:119–128.

Starrett, W. C. 1950. Distribution of the fishes of Boone County, Iowa, with special reference to the minnows and darters. Am. Midl. Nat. 43:112–127.

Starrett, W. C. 1951. Some factors affecting the abundance of minnows in the Des Moines River, Iowa. Ecology 32:13–24.

Sterba, G. 1962. Freshwater fishes of the world. Translated from German. Visa Books, London.

Sterba, G., and D. W. Tucker. 1973. Freshwater fishes of the world. English translation and revised edition. TFH Publications, Neptune City, N.J.

Sternburg, J. G. 1986. Spawning the blackbanded sunfish. Am. Currents (January).

Stevens, R. E. 1958. The striped bass of the Santee-Cooper Reservoir. Proc. Annu. Conf. Southeast. Assoc. Game Fish Comm. 11(1957):253–264.

Stevens, R. E. 1959a. The black and white crappie of the Santee-Cooper Reservoir. Proc. Annu. Conf. Southeast. Assoc. Game Fish Comm. 12:158–168.

Stevens, R. E. 1959b. The white and channel catfishes of the Santee-Cooper Reservoir and Tailrace Sanctuary. Proc. Annu. Conf. Southeast. Assoc. Game Fish Comm. 13:203–219.

Stier, D. J., and J. H. Crance. 1985. Habitat suitability index models and instream flow suitability curves: American shad. Biol. Rep. 82:34. U.S. Fish and Wildlife Service, Washington, D.C.

Stoeckel, J. N., and R. J. Neves. 2000. Methods for hatching margined madtom eggs. N. Am. J. Aquacult. 62:42–47.

Storck, T. W., D. W. Dufford, and K. T. Clement. 1978. The distribution of limnetic fish larvae in a flood control reservoir in central Illinois. Trans. Am. Fish. Soc. 107:419–424.

Street, M. W. 1969. Fecundity of the blueback herring in Georgia. Contrib. Ser. 17:15. Georgia Game and Fish Commission, Marine Fisheries Division.

Street, M. W., and J. G. Adams. 1969. Aging of hickory shad and blueback herring in Georgia by the scale method. Contrib. Ser. 18:13. Georgia Game and Fish Commission, Marine Fisheries Division.

Streever, W. J., and T. L. Crisman. 1993. A comparison of fish populations from natural and constructed freshwater marshes in central Florida. J. Freshw. Ecol. 8:149–153.

Sublette, J. E., M. D. Hatch, and M. Sublette. 1990. The fishes of New Mexico. University of New Mexico Press, Albuquerque.

Sulak, K. J. 1975. Cleaning behavior in the centrarchid fishes, *Lepomis macrochirus* and *Micropterus salmoides*. Anim. Behav. 23:331–334.

Sutherland, A. B., J. L. Meyer, and E. P. Gardiner. 2002. Effects of land cover on sediment regime and fish assemblage structure in four southern Appalachian streams. Freshw. Biol. 47:1791–1805.

Suttkus, R. D. 1956. First record of the mountain mullet, *Agonostomus monticola* (Bancroft), in Louisiana. Proc. La. Acad. Sci. 19:43–46.

Suttkus, R. D. 1963. Order Lepisostei. Pages 61–88 *in* Fishes of the western North Atlantic. Sears Foundation for Marine Research, Yale University, Mem. 3, pt. 1.

Suttkus, R. D., and M. F. Mettee. 2001. Analysis of four species of *Notropis* included in the subgenus *Pteronotropis* Fowler, with comments on relationships, origin, and dispersion. Geol. Surv. Bull. 170:1–50.

Suttkus, R. D., B. A. Porter, and B. J. Freeman. 2003. The status and infraspecific variation of *Notropis stonei*, Fowler. Proc. Am. Philos. Soc. 147(4):354–376.

Suttkus, R. D., and J. S. Ramsey. 1967. *Percina aurolineata*, a new percid fish from the Alabama River system and a discussion of ecology, distribution, and hybridization of darters of the subgenus *Hadropterus*. Tulane Stud. Zool. Bot. 28:1–24.

Sutton, D. L. 1977. Grass carp (*Ctenopharyngodon idella* Val.) in North America. Aquat. Bot. 3:157–164.

Swee, V. B., and H. R. McCrimmon. 1966. Reproductive biology of carp, *Cyprinus carpio*. L., in Lake St. Lawrence, Ontario. Trans. Am. Fish. Soc. 95:372–380.

Swift, C. C., C. R. Gilbert, S. A. Bortone, G. H. Burgess, and R. W. Yerger. 1986. Zoogeography of the freshwater fishes of the southeastern United States: Savannah River to Lake Pontchartrain. Pages 213–266 *in* C. H. Hocutt and E. O. Wiley, editors. The zoogeography of North American freshwater fishes. John Wiley and Sons, New York.

Swift, C., R. W. Yerger, and P. R. Parrish. 1977. Distribution and natural history of the fresh and brackish water fishes of the Ochlockonee River, Florida and Georgia. Bull. Tall Timbers Res. Stn. 20.

Taber, C. A. 1969. Distribution and identification of larval fish in the Buncombe Creek Arm of Lake Texoma with observations on spawning habits and relative abundance. Doctoral dissertation, University of Oklahoma, Norman.

Taborsky, M. 1994. Sneakers, satellites, and helpers: parasitic and cooperative behavior in fish reproduction. Adv. Stud. Behav. 23:1–100.

Taubert, B. D. 1977. Early morphological development of the green sunfish, *Lepomis cyanellus*, and its separation from other larval *Lepomis* species. Trans. Am. Fish. Soc. 106(5):445–448.

Taylor, B. E., J. M. Aho, D. L. Mahoney, and R. A. Estes. 1991. Population dynamics and food habits of bluegill (*Lepomis macrochirus*) in a thermally stressed reservoir. Can. J. Fish. Aquat. Sci. 48(5):768–775.

Tharratt, R. C. 1959. Food of yellow perch, *Perca flavescens* (Mitchill), in Saginaw Bay, Lake Huron. Trans. Am. Fish. Soc. 88:330–331.

Thomas, M. E., A. G. Eversole, and D. W. Cooke. 1992. Impacts of water rediversion on the spawning utilization on a formerly impounded ricefield by blueback herring. Wetlands 12:22–27.

Thomason, C. S., D. E. Allen, and J. S. Crane. 1993. A fisheries study of the Edisto River, SC. Study completion report, Federal Aid Projects F-32 and F-30. South Carolina Wildlife and Marine Resources Department, Columbia.

Thomson, J. M. 1978. *Agonostomas monticola* (Bancroft, 1836). *In* W. Fischer, editor. FAO species identification sheets for fishery purposes, western central Atlantic (fishing area 31), volume 3.

Thorp, J. H., L. D. Goldsmith, J. A. Polgreen, and L. M. Mayer. 1989. Foraging patterns of nesting and nonnesting sunfish (Centrarchidae: *Lepomis auritus* and *L. gibbosus*). Can. J. Fish. Aquat. Sci. 46:1342–1346.

Thorpe, J. E. 1977. Morphology, physiology, behavior, and ecology of *Perca fluviatilis* L. and *Perca flavescens* (Mitchill). J. Fish. Res. Board Can. 34:1504–1514.

Tin, H. T. 1982. Family Centrarchidae, sunfishes. Pages 524–580 *in* N. A. Auer, editor. Identification of larval fishes of the Great Lakes basin with emphasis on the Lake Michigan drainage. Spec. Pub. 82-3. Great Lakes Fishery Commission, Ann Arbor, Mich.

Todd, J. H., J. Atema, and J. E. Bardach. 1967. Chemical communications in social behavior of a fish, the yellow bullhead. Science 158:672–673.

Todd, T. N. 1986. Occurrence of white bass–white perch hybrids in Lake Erie. Copeia 1986:196–199.

Toetz, D. W. 1966. The change from endogenous to exogenous sources of energy in bluegill sunfish larvae. Invest. Ind. Lakes Streams 7:115–146.

Trautman, M. B. 1957. Fishes of Ohio. Ohio State University Press, Columbus.

Trautman, M. B. 1981. The fishes of Ohio with illustrated keys, revised edition. Ohio State University Press, Columbus.

Trautman, M. B., and D. K. Gartman. 1974. Re-evaluation of the effects of man-made modifications on Gordon Creek between 1887 and 1973 and especially as regards its fish fauna. Ohio J. Sci. 74:162–173.

Travnichek, V. H., M. J. Maceina, S. M. Smith, and R. A. Dunham. 1996. Natural hybridization between black and white crappies (*Pomoxis*) in 10 Alabama reservoirs. Am. Midl. Nat. 135:310–316.

Travnichek, V. H., M. J. Maceina, M. C. Wooten, and R. A. Dunham. 1997. Symmetrical hybridization between black crappie and white crappie in an Alabama reservoir based on analysis of the cytochrome-b gene. Trans. Am. Fish. Soc. 126:127–132.

Trexler, J. C. 1995. Restoration of the Kissimmee River: a conceptual model of past and present fish communities and its consequences for evaluating restoration success. Restor. Ecol. 3:195–210.

Tsai, C. 1972. Life history of the eastern Johnny darter, *Etheostoma olmstedi* Storer, in cold tailwater and sewage-polluted water. Trans. Am. Fish. Soc. 101:80–88.

Tsai, C., and E. C. Rainey. 1974. Systematics of the banded darter, *Etheostoma zonale* (Pisces: Percidae). Copeia 1974(1):1–24.

Turner, P. R., and R. C. Summerfelt. 1970. Food habits of adult flathead catfish, *Pylodictus olivaris* (Rafinesque), in Oklahoma reservoirs. Proc. Annu. Conf. Southeast. Assoc. Game Fish Comm. 24:387–401.

Turner, T. F., J. C. Trexler, G. L. Miller, and K. E. Toyer. 1994. Temporal and spatial dynamics of larval and juvenile fish abundance in a temperate floodplain river. Copeia 1994:174–183.

Turner, W. R., and G. N. Johnson. 1973. Distribution and relative abundance of fishes in Newport River, North Carolina. NOAA Tech. Rep. NMFS-SSRF-666. National Marine Fisheries Service, Beaufort, N.C.

Tyler, J. D., J. R. Webb, T. R. Wright, J. D. Hargett, K. J. Mask, and D. R. Schucker. 1994. Food habits, sex ratios, and size of longnose gar in southwestern Oklahoma. Proc. Okla. Acad. Sci. 74:41–42.

Ulrich, G., N. Chipley, J. W. McCord, D. Cupka, J. L. Music Jr., and R. K. Mahood. 1978. Development of fishery management plans for selected anadromous fishes in South Carolina and Georgia. Spec. Pub. 14. Marine Resources Center, South Carolina Wildlife and Marine Resources Department, Columbia.

Underhill, A. H. 1949. Studies on the development, growth and maturity of the chain pickerel, *Esox niger* (Lesueur). J. Wildl. Manage. 13:377–391.

U.S. Army Corps of Engineers. 1987. Design and construction report: Steel Creek Dam, Savannah River Operations Office, Department of Energy, Aiken, South Carolina, vol. 1. U.S. Army Corps of Engineers, Savannah District, Savannah, Ga.

U.S. Fish and Wildlife Service. 2001. New Savannah Bluff Lock and Dam: benefits of restoring natural river flow. Georgia and South Carolina Ecological Services Fact Sheet.

Vadas, R. L. 1990. The importance of omnivory and predator regulation of prey in freshwater fish assemblages of North America. Environ. Biol. Fishes 27:285–302.

Van Den Avyle, M. J. 1982. Species profiles: life histories and environmental requirements of coastal fishes and invertebrates (south Atlantic)—American eel. FWS/OBS-82/11.24, U.S. Fish and Wildlife Service; TR EL-82-4, U.S. Army Corps of Engineers.

VanderKooy, K. E., C. F. Rakocinski, and R. W. Heard. 2000. Trophic relationships of three sunfishes (*Lepomis* spp.) in an estuarine bayou. Estuaries 23(5):621–632.

Van Eenennaam, J. P., S. I. Doroshov, G. P. Moberg, J. W. Watson, D. S. Moore, and J. Linares. 1996. Reproductive condition of the Atlantic sturgeon (*Acipenser oxyrinchus*) in the Hudson River. Estuaries 19:769–777.

Vanicek, D. 1961. Life history of the quillback and highfin carpsuckers in the Des Moines River. Proc. Iowa Acad. Sci. 68:238–246.

Van Olst, J. C., and J. M. Carlberg. 1990. Commercial culture of hybrid striped bass: status and potential. Aquacult. Mag. 1990(1):49–59.

Van Winkle, W., P. Anders, D. H. Secor, and D. Dixon. 2002. Biology, management, and protection of North American sturgeon. Amer. Fish. Soc. Symp. 28.

Verma, P. 1970. Normal stages in the development of *Cyprinus carpio* Var., *communis* L. Acta Biol. Acad. Sci. Hung. 21:207–218.

Vernon, E. H., and R. G. McMynn. 1957. Scale characteristics of yearling coastal cutthroat trout and steelhead trout. J. Fish. Res. Board Can. 14:203–212.

Vives, S. P. 1990. Nesting ecology of hornyhead chub *Nocomis biguttatus*, a keystone species in Allequash Creek, Wisconsin. Am. Midl. Nat. 124:46–56.

Vladykov, V. D. 1955. A comparison of the Atlantic sea sturgeon with a new subspecies from the Gulf of Mexico (*Acipenser oxyrhynchus desotai*). J. Fish. Res. Board Can. 12:754–761.

Vladykov, V. D. 1964. Quest for the true breeding area of the American eel (*Anguilla rostrata* Lesueur). J. Fish. Res. Board Can. 21:1523–1530.

Vladykov, V. D., and J. R. Greeley. 1963. Order Acipenseroidei. Pages 24–60 *in* Fishes of the western north Atlantic. Sears Foundation for Marine Research, Mem. 1, pt. 3.

Wagner, C. C., and E. L. Cooper. 1963. Population density, growth, and fecundity of the creek chubsucker, *Erimyzon oblongus*. Copeia 1963:350–357.

Walburg, C. H. 1975. Food of young-of-year channel catfish in Lewis and Clark Lake, a Missouri River reservoir. Am. Midl. Nat. 93:218–221.

Waldrip, L. 1992. Grass carp increasing in Trinity River. Texas Parks and Wildlife Department, 20 November 1992, 5–7.

Wales, J. H. 1941. Development of steelhead trout eggs. Calif. Fish Game 27:250–260.

Wallace, C. R. 1972. Spawning behavior of *Ictalurus natalis* (LeSueur). Tex. J. Sci. 24:307–310.

Wallin, J. E. 1989. Bluehead chub (*Nocomis leptocephalus*) nests used by yellowfin shiners (*Notropis lutipinnis*). Copeia 1989(4):1077–1080.

Wallin, J. E. 1992. The symbiotic nest association of yellowfin shiners, *Notropis lutipinnis*, and bluehead chubs, *Nocomis leptocephalus*. Environ. Biol. Fishes 33:287–292.

Walsh, S. J., and B. M. Burr. 1984. Life history of the banded pygmy sunfish, *Elassoma zonatum* (Jordan) (Pisces: Centrarchidae), in western Kentucky. Bull. Ala. Mus. Nat. Hist. 8:31–52.

Walters, D. M., and B. J. Freeman. 2000. Distribution of *Gambusia* (Poeciliidae) in a southeastern river system and the use of fin ray counts for species determination. Copeia 2000:555–559.

Wang, J. C. S., and R. J. Kernehan. 1979. Fishes of the Delaware estuaries: a guide to the early life histories. Ecological Analysts, Towson, Md.

Ward, H. C., and E. M. Leonard. 1954. Order of appearance of scales in the black crappie, *Poxomis nigromaculatus*. Proc. Okla. Acad. Sci. 33:138–140.

Warren, M. L. Jr. 1992. Variation of the spotted sunfish, *Lepomis punctatus* complex (Centrarchidae): meristics, morphometrics, pigmentation, and species limits. Ala. Mus. Nat. Hist. Bull. 12. University of Alabama, Tuscaloosa.

Warren, M. L. Jr., P. L. Angermeier, B. M. Burr, and W. R. Haag. 1997. Decline of a diverse fish fauna: patterns of imperilment and protection in the southeastern United States. Pages 105–164 *in* G. W. Benz and D. E. Collins, editors. Aquatic fauna in peril: the southeastern perspective. Spec. Pub. 1. Southeast Aquatic Research Institute, Decatur, Ga.

Warren, M. L. Jr., and B. M. Burr. 1994. Status of freshwater fishes of the United States: overview of an imperiled fauna. Fisheries 19:6–17.

Warren, M. L. Jr., B. M. Burr, and C. A. Taylor. 1991. Aspects of reproduction in a captive population of the state threatened ironcolor shiner (*Notropis chalybaeus*). Final report submitted to Division of Natural Heritage, Illinois Department of Natural History.

Warren, M. L. Jr., and 11 authors. 2000. Diversity, distribution, and conservation status of the native freshwater fishes of the southern United States. Fisheries 25:7–31.

Waters, T. F. 1995. Sediment in streams: sources, biological effects, and controls. Am. Fish. Soc. Monogr. 7.

Weaver, L. A., and G. C. Garman. 1994. Urbanization of a watershed and historical changes in stream fish assemblage. Trans. Am. Fish. Soc. 123:162–172.

Webb, J. F., and D. D. Moss. 1968. Spawning behavior and age and growth of white bass in Center Hill Reservoir, Tennessee. Proc. Annu. Conf. Southeast. Assoc. Game Fish Comm. 21:343–357.

Webb, M. A., H. S. Elder, and R. G. Howells. 1994. Grass carp reproduction in the lower Trinity River, Texas. Pages 29–32 *in* Proceedings of the Grass Carp Symposium, 7–9 March 1994, Gainesville, Fla. U.S. Army Corps of Engineers, Waterways Experiment Station, Vicksburg, Miss.

Webster, D. A. 1942. The life histories of some Connecticut fishes. Pages 122–227 *in* L. Thorpe, editor. A fishery survey of important Connecticut lakes. Bull. Conn. Geol. Nat. Hist. Surv. 63.

Weller, R. R., and C. Robbins. 1999. Food habits of flathead catfish in the Altamaha River system, Georgia. Proc. Annu. Conf. Southeast. Assoc. Fish Wildl. Agencies 53:35–41.

Weller, R. R., and J. D. Winter. 2001. Seasonal variation in home range size and habitat use of flathead catfish in Buffalo Springs Lake, Texas. N. Am. J. Fish. Manage. 21:792–800.

Wells, L., and R. House. 1974. Life history of the spottail shiner (*Notropis hudsonius*) in southeastern Lake Michigan, the Kalamazoo River, and western Lake Erie. Res. Rep. 78. Bureau of Sport Fisheries and Wildlife, Washington, D.C.

Werner, E. E., and D. J. Hall. 1976. Niche shifts in sunfishes: experimental evidence and significance. Science 191:404–406.

Werner, E. E., and D. J. Hall. 1977. Competition and habitat shift in two sunfishes (Centrarchidae). Ecology 58:869–876.

Werner, R. G. 1972. Bluespotted sunfish, *Enneacanthus gloriosus*, in Lake Ontario drainage, New York. Copeia 1972:878–879.

Werner, R. G. 1980. Freshwater fishes of New York State. Syracuse University Press, Syracuse, N.Y.

Whicker, F. W., T. G. Hinton, D. J. Wiquette, and J. Seel. 1993b. Health risks to hypothetical residents of a radioactively contaminated lake bed. Presented at Meeting the Challenge: Environmental Remediation Conference, U.S. Department of Energy, Augusta, Ga., 24–28 October 1993. Savannah River Ecology Laboratory, Aiken, S.C.

Whicker, F. W., D. J. Niquette, and T. G. Hinton. 1993a. To remediate or not: a case history. Pages 473–485 *in* Proceedings of the Health Physics Society of America meeting, Coeur d'Alene, Ida.

Whicker, F. W., J. E. Pinder III, J. W. Bowling, J. J. Alberts, and I. L. Brisbin Jr. 1990. Distribution of long-lived radionuclides in an abandoned reactor cooling reservoir. Ecol. Monogr. 60:471–496.

White, D. S., and K. H. Hagg. 1977. Foods and feeding habits of the spotted sucker, *Minytrema melanops* (Rafinesque). Am. Midl. Nat. 98:137–146.

Whitehurst, D. K. 1981. Seasonal movements of fishes in an eastern North Carolina swamp stream. Pages 182–190 *in* L. A. Krumholz, editor. The Warm Water Streams Symposium. South. Div. Am. Fish. Soc. Spec. Pub. 1.

Whiteside, B. G. 1964. Biology of the white crappie, *Pomoxis annularis*, in Lake Texoma, Oklahoma. Master's thesis, Oklahoma State University, Stillwater.

Whiteside, B. G., and C. Berkhouse. 1992. Some new collection locations for six fish species. Tex. J. Sci. 44:494.

Whiteside, L. A., and B. M. Burr. 1986. Aspects of the life history of the tadpole madtom, *Noturus gyrinus* (Siluriformes: Ictaluridae), in southern Illinois. Ohio J. Sci. 86:153–160.

Whitworth, W. R. 1969. List of fishes from the Savannah River Plant. Unpublished manuscript, University of Connecticut, Storrs.

Whitworth, W. R., P. L. Berrien, and W. T. Keller. 1968. Freshwater fishes of Connecticut. Bull. Conn. Geol. Nat. Hist. Surv. 101.

Wigginton, J. D., B. G. Lockaby, and C. C. Trettin. 2000. Soil organic matter formation and sequestration across a forested floodplain chronosequence. Ecol. Engineer. 15: 141–156.

Wike, L. D. 1987. An examination of the basis for poor growth performance in the hybrid grass carp. Doctoral dissertation, University of Ilinois, Urbana-Champaign.

Wike, L. D., and 14 authors. 1994. SRS ecology. Environ. Inf. Doc. WSRC-TR-93-496. Westinghouse Savannah River Co., Aiken, S.C.

Wike, L. D., W. L. Specht, H. E. Mackey, M. H. Paller, E. W. Wilde, and A. S. Dicks. 1989. SRS ecology, reactor operations. Environ. Inf. Doc. WSRC-RP-89-816, vol. 2. Westinghouse Savannah River Co., Aiken, S.C.

Wilbur, R. L. 1969. The redear sunfish in Florida. Fish. Bull. 5:64. Florida Game and Freshwater Fish Commission, Tallahassee.

Wilcox, J. F. 1960. Experimental stockings of Rio Grande blue catfish, a subspecies of *Ictalurus furcatus*, in Lake J. B. Thomas, Colorado City Lake, Nasworthy Lake, Lake Abiline, and Lake Trammel. Dingell-Johnson Project F-5-R-7, Job E-2, job completion report. Texas Game and Fish Commission, Austin.

Wilde, E. W. 1985. Compliance of the Savannah River Plant P-Reactor cooling system with environmental regulations: demonstrations in accordance with sections 316(a) and (b) of the Federal Water Pollution Control Act of 1972. DP-1708. Savannah River Laboratory, E. I. du Pont de Nemours and Co., Aiken, S.C.

Wilde, E. W., editor and compiler. 1987. Comprehensive cooling water study final report. Volume 7: Ecology of Par Pond. DP-1739-7. E. I. du Pont de Nemours and Co., Aiken, S.C.

Wilde, E. W., and L. J. Tilly. 1985. Influence of P-Reactor operation on the aquatic ecology of Par Pond: a literature review. DP-1698. Savannah River Laboratory, E. I. du Pont de Nemours and Co., Aiken, S.C.

Wiley, E. O. 1977. The phylogeny and systematics of the *Fundulus nottii* species group (Teleostei: Cyprinodontidae). Occas. Pap. Mus. Nat. Hist. Univ. Kans. 66:1–31.

Wiley, M. J., and L. D. Wike. 1986. Energy balances of diploid, triploid, and hybrid grass carp. Trans. Am. Fish. Soc. 115:853–863.

Williamson, K. L., and P. C. Nelson. 1985. Habitat suitability index models and instream flow suitability curves: gizzard shad. Biol. Rep. 82(10.112). U.S. Fish and Wildlife Service, Washington, D.C.

Willis D. W., C. P. Paukert, and B. G. Blackwell. 2002. Biology of white bass in eastern South Dakota glacial lakes. N. Am. J. Fish. Manage. 22:627–636.

Wiltz, J. W. 1981. Savannah River fish population study and impingement prediction for Plant Vogtle, Burke County, Georgia. Report to Georgia Power Co., Atlanta.

Wiltz, J. W. 1993. A survey of the fishes of Beaverdam Creek and the adjacent Savannah River, Burke County, Georgia. Georgia Power Co., Atlanta.

Winkelman, D. L. 1993. Growth, lipid use, and behavioral choice in two prey fishes, dollar sunfish (*Lepomis marginatus*) and mosquitofish (*Gambusia holbrooki*), under varying predation pressure. Doctoral dissertation, University of Georgia, Athens.

Winkelman, D. L. 1996. Reproduction under predatory threat: trade-offs between nest guarding and predator avoidance in male dollar sunfish (*Lepomis marginatus*). Copeia 1996:845–851.

Winn, H. E. 1958. Observations on the reproductive habits of darters (Pisces-Percidae). Am. Midl. Nat. 59:190–212.

Woods, L. P., and R. F. Inger. 1957. The cave, spring and swamp fishes of the family Amblyopsidae of central and eastern United States. Am. Midl. Nat. 58:232–256.

Woodward, R. L., and T. E. Wissing. 1976. Age, growth, and fecundity of the quillback (*Carpiodes cyprinus*) and highfin (*C. velifer*) carpsuckers in an Ohio stream. Trans. Am. Fish. Soc. 195:411–415.

Woolcott, W. S. 1962. Intraspecific variation in the white perch, *Roccus americanus* (Gmelin). Chesapeake Sci. 3:94–113.

Wooten, M. C., K. T. Scribner, and M. H. Smith. 1988. Genetic variability and the systematics of *Gambusia* in the southeastern United States. Copeia 1988:283–289.

Wright, A. H., and A. A. Allen. 1913. The fauna of Ithaca, New York: fishes. Zool. Field Noteb. (Ithaca, N.Y.):4–6.

WSRC [Westinghouse Savannah River Company]. 1995. SRS geology and hydrology. Environ. Inf. Doc. WSRC-TR-95-0046. Westinghouse Savannah River Co., Aiken, S.C.

Yeager, B. L., and R. T. Bryant. 1983. Larvae of the longnose gar, *Lepisosteus osseus*, from the Little River in Tennessee. J. Tenn. Acad. Sci. 58:20–22.

Yerger, R. W., and K. Relyea. 1968. The flat-headed bullhead (Pisces: Ictaluridae) of the southeastern United States, and a new species of *Ictalurus* from the Gulf Coast. Copeia 1968:361–384.

Zimmerman, L. C., E. A. Standora, and J. R. Spotila. 1989. Behavioral thermoregulation of largemouth bass (*Micropterus salmoides*): response of naïve fish to the thermal gradient in a nuclear reactor cooling reservoir. J. Therm. Biol. 14(3):123–132.

Zimpfer, S. P., C. F. Bryan, and C. H. Pennington. 1987. Factors associated with the dynamics of grass carp larvae in the lower Mississippi River Valley. Pages 102–108 *in* R. D. Hoyt, editor. Proceedings of the Tenth Annual Larval Fish Conference. Am. Fish. Soc. Symp. 2.

Index

Note: Page numbers in *italics* refer to illustrations; **boldface** page numbers refer to the main discussion of a species or family.

About the Authors

BARTON C. MARCY JR.

Barton C. Marcy Jr. has 40 years of experience in the environmental impact assessment and fisheries research field. He was an environmental project ecologist for 10 years studying the impacts of the Connecticut Yankee Atomic Power Plant (the first U.S. nuclear plant) on the ecology of 193 km of the Connecticut River. He was an office manager and director of environmental programs for 20 years with NUS Corporation, an international environmental consulting firm. He was a research assistant at the University of Connecticut Marine Laboratory and a regional biologist with the Connecticut State Board of Fish and Game. Bart is the author of or a contributor to seven books and more than 50 scientific journal/publications on environmental impact assessment, fisheries, and mitigation. He holds an undergraduate degree from Wake Forest University and a graduate degree in marine ecology/ichthyology from the University of Connecticut. He is a Certified Environmental Professional (CEP) and Certified Fisheries Scientist. He currently serves as Senior Fellow Scientist and lead for regulatory training and National Environmental Policy Act programs in Environmental Services at the Westinghouse Savannah River Co., Aiken, S.C.

DEAN E. FLETCHER

Dean Fletcher's interest in fishes began while he was growing up along the Illinois River as the grandson of a commercial fisherman. He received his formal education at Southern Illinois University, where he earned B.S. and M.S. degrees in zoology under the guidance of Brooks M. Burr. He was further influenced early in his career while working with Gary K. Meffe at the University of Georgia's Savannah River Ecology Laboratory. Dean is currently research coordinator of the Fish Ecology Program at SREL, where he has conducted extensive field studies of fishes in the middle Savannah River basin for the past 15 years. He has studied the reproductive biology of many local fish species and published papers in refereed professional journals on the ecology of fishes and stream habitat disturbance. His current work in the Savannah River basin and southern Spain includes the study of fish reproductive strategies and interactions between fish and their habitat.

F. DOUGLAS MARTIN

F. Douglas Martin received a B.S. in zoology from Louisiana State University and a Ph.D. in zoology from the University of Texas at Austin. He has been on the faculty or an adjunct faculty member of Indiana University, the University of Puerto Rico—Mayaguez, the University of Maryland, St. Mary's College of Maryland, and the University of North Texas. He has also worked for the Maryland Department of Natural Resources Tidewater Fisheries Administration and Texas Parks and Wildlife. He has published more than 30 refereed papers in professional journals, is a coauthor of three volumes, and has written four invited chapters for books. He currently serves as principal scientist in the Environmental Analysis Section, Environmental Sciences and Technology Department of the Savannah River National Laboratory.

MICHAEL H. PALLER

Michael Paller received his Ph.D. in zoology from Southern Illinois University, where he studied fisheries biology and aquaculture. Since 1983 he has worked in South Carolina conducting extensive studies of the fishes in the Savannah River and numerous streams and reservoirs. He has published more than 90 refereed papers and reports concerning fish ecology, environmental assessment, and aquaculture. He is currently a consulting scientist in the Environmental Analysis Section of the Savannah River National Laboratory and serves on the faculty of Augusta State University.

MARCEL J. M. REICHERT

Marcel Reichert grew up in the Netherlands. He graduated from the University of Amsterdam with a degree in biology, and received his Ph.D. from the University of Groningen. After working as a research associate at the Royal Netherlands Institute for Sea Research he moved to the United States in 1992 to become a member of the research faculty at the University of South Carolina. His research has focused on various aspects of fish ecology in the freshwater and marine environments, including age and growth, bioenergetics, and recruitment processes. Since 1994 he has been involved in a variety of fish-related research projects at the Savannah River Site. His research has been published in international journals and in reports to various government and private organizations.

DAVID E. SCOTT

David Scott conducts research at SREL on the population ecology of amphibians that use seasonal wetlands. He received a B.S. in biology from Wofford College and an M.S. in environmental sciences from the University of Virginia. During the past 15 years his photos of a wide variety of plant and animal species have been used to illustrate scientific articles in numerous journals, magazines, and books, including cover photos for *Science*, *Bioscience*, *Herpetologica*, *Amphibian Conservation*, the *Handbook of Turtles*, and *Fundamentals of Ecotoxicology*. Additional photos have appeared in *Natural History*, *National Geographic*, and *Discover* magazines, as well as the Encyclopedia Britannica. He notes that no cooperative fish were eaten in the making of this book.